Sixteenth Conference of the Euro-American Consortium for Promoting the Application of Mathematics in Technical and Natural Sciences 2024

AA007199

Journal of Physics: Conference Series Volume 2910

Albena, Bulgaria
21-26 June 2024

ISBN: 979-8-3313-1175-9
ISSN: 1742-6588

Printed from e-media with permission by:

Curran Associates, Inc.
57 Morehouse Lane
Red Hook, NY 12571

Some format issues inherent in the e-media version may also appear in this print version.

This work is licensed under a Creative Commons Attribution 3.0 International Licence.
Licence details: http://creativecommons.org/licenses/by/3.0/.

No changes have been made to the content of these proceedings. There may be changes to pagination and minor adjustments for aesthetics.

Printed with permission by Curran Associates, Inc. (2025)

For permission requests, please contact the Institute of Physics
at the address below.

Institute of Physics
Dirac House, Temple Back
Bristol BS1 6BE UK

Phone: 44 1 17 929 7481
Fax: 44 1 17 920 0979

techtracking@iop.org

Additional copies of this publication are available from:

Curran Associates, Inc.
57 Morehouse Lane
Red Hook, NY 12571 USA
Phone: 845-758-0400
Fax: 845-758-2633
Email: curran@proceedings.com
Web: www.proceedings.com

TABLE OF CONTENTS

Journal of Physics: Conference Series **2910** (2024) 011001 doi:10.1088/1742-6596/2910/1/011001

Preface: Application of Mathematics in Technical and Natural Sciences 16th International Conference – AMiTaNS'24

The Euro-American Consortium for Promoting the Application of Mathematics in Technical and Natural Sciences organized its 16th annual conference of the series AMiTaNS (Application of Mathematics in Technical and Natural Sciences). The event took place during June 21-26, 2024 in the Congress Centre of Black-Sea resort of Albena, Bulgaria. Its aim was to foster existing and new personal contacts, scientific discussion as well as to provide a forum for the exchange of ideas and experience between specialists in the fields of Mathematics, Physics, Biology, Ecology, Economics, Computer Science and Technology. The event became a world conference covering 15 time zones between Chile and Nebraska through Texas, Alabama, Florida, the UK, Europe, Cyprus, Russia, and China. The conference was managed from the Congress Centre in Albena where 40 participants and delegates were present in person and the other 40 attended the event via Zoom video platform.

Interest in this conference is growing year by year. This year we registered around 100 talks and scientific reports, more than 80 participants – contributors and delegates from 12 countries around the world, representing the continents of Europe, North and South America, and Asia. About 30% of them attended the conference for the first time. The accepted talks were delivered and attended both on-location at the big hall and Corolla hall in Flamingo Grand hotel and on-line. Every day the morning sessions started by plenary lectures followed by contributed sessions on the following topics: Applied Analysis, Mathematical Physics, Numerical Methods, Scientific Computing, E- and M-learning. As part of the program, leaders of their respective fields organized four special sessions: *Integrable Systems and Their Application, Interdisciplinary Frontiers of Nonlinear Science, Math Modeling and Applications in Bioinformatics, Math Modeling in Finance.* A number of very interesting presentations were delivered, which sparked stimulating scientific discussions both on- and off-line. The reader can find comprehensive and detailed information about the talks and reports including their records on the conference website at http://2024.eac4amitans.eu.

All contributed manuscripts were reviewed by anonymous independent referees via Morressier platform, and the approved revised manuscripts were accepted for publication in the present volume. The continued success of AMiTaNS enhances its reputation as an established annual meeting with an eminent and continuously growing core of contributors. Yet, the hybrid format turned out very flexible both for the organizers and contributors. It makes the conference accessible to all those who wish to attend either in person or remotely. We would like to thank the plenary speakers, who accepted our invitation and contributed to the conference through their elucidating lectures. We would also like to thank the special sessions organizers, hosts and co-hosts for the high scientific quality of the talks they selected, and for ensuring the smooth running of the conference in hybrid format. We are deeply indebted to the scores of anonymous referees all over the world, who approached their duties very professionally and thoroughly, and helped us to maintain high scientific standards. We highly appreciate the involvement of the members of the Program Committee and Steering Committee, who supported AMiTaNS from its inception, and worked tirelessly throughout the intensive labors of advertising, arranging, organizing, and reviewing. Special thanks go to our colleagues from the University of Essex at Colchester, UK, who worked with dedication so that parallel Zoom conference sessions could be held completely on-line.

Content from this work may be used under the terms of the Creative Commons Attribution 4.0 licence. Any further distribution of this work must maintain attribution to the author(s) and the title of the work, journal citation and DOI.
Published under licence by IOP Publishing Ltd

Journal of Physics: Conference Series **2910** (2024) 011001 doi:10.1088/1742-6596/2910/1/011001

We are very grateful to our publisher Institute of Physics, Bristol, UK for the helpful cooperation during the preparation of this volume.

Coordinator and *Zoom host* of AMiTaNS'24 and
Editor of the proceedings volume,
M.D. Todorov
Sofia, September 2024

Journal of Physics: Conference Series **2910** (2024) 011001 doi:10.1088/1742-6596/2910/1/011001

Program Committee

AZMY ACKLEH
University of Louisiana at Lafayette, LA, USA

CHRISTO CHRISTOV †
University of Louisiana at Lafayette, LA, USA

BOYAN DIMITROV
Kettering University, Flint, MI, USA

ARDESHIR GURAN
Institute of Structronics, Ottawa, and McGill University, Montreal, Canada

PEDRO M. JORDAN
Naval Research Laboratory, Stennis, MS, USA

HRISTO KOJOUHAROV
University of Texas at Arlington, TX, USA

ALEXANDER KURGANOV
Southern University of Science and Technology, Shenzhen, P.R. China

RAYTCHO LAZAROV
Texas A&M University, College Station, TX, USA

ANDREI LUDU
Embrie-Riddle Aeronautical University, Daytona Beach, FL, USA

ALEXANDER MIKHAILOV
University of Leeds, the UK

PETER MINEV
University of Alberta, Canada

NECTARIOS PAPANICOLAOU
University of Nicosia, Cyprus

ANDREI POLYANIN
Institute for Problems in Mechanics, Moscow, Russia

BARBARA PRINARI
University of Buffalo, New York, USA

ZHIJUN (GEORGE) QIAO
University of Texas Rio Grande Valley, Edinburg, TX, USA

MIGUEL A.F. SANJUAN
Universidad Rey Juan Carlos, Spain

Journal of Physics: Conference Series **2910** (2024) 011001 doi:10.1088/1742-6596/2910/1/011001

VLADIMIR SHAYDUROV
Institute of Mathematical Modeling RAS, Krasnoyarsk, Russia
and Tianjin University of Finance and Economics, P. R. China

BRIAN STRAUGHAN
University of Durham, UK

MICHAIL TODOROV
Institute of Mathematics and Informatics, BAS, Sofia, Bulgaria

MANUEL G. VELARDE
Universidad Complutense, Madrid, Spain

RUNZHANG XU
Harbin Engineering University, Harbin, P.R. China

SANICHIRO YOSHIDA
Southeastern Louisiana University, Hammond, LA, USA

Organizer

Euro-American Consortium for Promoting the Application of Mathematics in the Technical and Natural Sciences Plc

Sponsor

– Ultra Quantum Inc., Huntsville, Alabama, USA

Journal of Physics: Conference Series **2910** (2024) 011001 doi:10.1088/1742-6596/2910/1/011001

List of Participants

K. ANGELOW
Institute of Mathematics and Informatics, BAS
Sofia
BULGARIA

S. APOSTOLOV
St. Kliment Ohridski University of Sofia
BULGARIA

S. ASSENOVA
Institute of Robotics, BAS
Sofia
BULGARIA

P. BOKES
Comenius University
Bratislava
SLOVAKIA

G. BOYADZHIEV
Institute of Mathematics and Informatics, BAS
Sofia
BULGARIA

Yu. BOZHKOV
Universidade Estadual de Campinas
Campinas - SP
BRAZIL

JINCAI CHANG
North China University of Science and Technolgy
Tangshan, Hebei
P R CHINA

XIAOMIN CHEN
Beijing University of Technolog
P R CHINA

YUYING CHEN
Henan Agricultural University
Zhengzhou
P R CHINA

A. CHOWDHURY
Savannah State University, GA
USA

Journal of Physics: Conference Series **2910** (2024) 011001 doi:10.1088/1742-6596/2910/1/011001

M. CHRISTOU
University of Nicosia
CYPRUS

JIFENG CHU
Shanghai Normal University
P R CHINA

YUJIAO CUI
Beijing University of Technology
P R CHINA

D. DANKOV
Institute of Mechanics, BAS
Sofia
BULGARIA

D. DANTCHEV
Institute of Mechanics, BAS
Sofia
BULGARIA
and Max-Planck Institute fuer Intelligente Systems
Stuttgart
GERMANY

B. DIMITROV
Institute of Nuclear Research and Nuclear Energetics, BAS
& Institute for Advanced Physical Studies
Sofia
BULGARIA

A. FEDOSEYEV
Ultra Quantum Inc.
Huntsville, AL
USA

G. GEORGIEV
Chernorizec Hrabar Free University of Varna
BULGARIA

I.R. GEORGIEV
Angel Kanchev University of Rousse
and Institute of Mathematics and Informatics, BAS, Sofia
BULGARIA

S. GEORGIEV
Angel Kanchev University of Rousse
and Institute of Mathematics and Informatics, BAS, Sofia
BULGARIA

 doi:10.1088/1742-6596/2910/1/011001

V. GERDJIKOV
Institute for Nuclear Research and Nuclear Energy
and Institute of Mathematics and Informatics, BAS, Sofia
BULGARIA

V. GLADKOVA
Angel Kanchev University of Rousse
BULGARIA

G. GRAHOVSKI
University of Essex
Colchester
UK

Ya. GURIMSKAYA
Solestial Inc.
Tempe, AR
USA

V. GUSHCHIN
Institute for Computer Aided Design of RAS
Moscow
RUSSIA

V. GUEORGUIEV
Institute for Advanced Physical Studies
Sofia
BULGARIA

G. HAYNATZKI
University of Nebraska Medical Center
Omaha, NE
USA

R. HAYNATZKI
University of Nebraska Medical Center
Omaha, NE
USA

I. HRISTOV
St. Kliment Ohridksi University of Sofia
BULGARIA

R. HRISTOVA
St. Kliment Ohridksi University of Sofia
BULGARIA

LIYAN JIA
North China University of Science and Technology
Tangshan
P R CHINA

Journal of Physics: Conference Series **2910** (2024) 011001 doi:10.1088/1742-6596/2910/1/011001

B. IDIRIZOV
Angel Kanchev University of Ruse
BULGARIA

N. ILIEVA
Institute of Information and Communication Technologies
& Institute of Mathematics and Informatics, BAS
Sofia
BULGARIA

A. IVANOV
Paisii Hilendarski University of Plovdiv
BULGARIA

R. IVANOV
Technological University Dublin
IRELAND

M. KOLEV
University of Architecture, Civil Engineering and Geodesy
Sofia
BULGARIA

O. KOUNCHEV
Institute of Mathematics and Informatics, BAS
Sofia
BULGARIA

N. KOVACHEV
Angel Kanchev University of Ruse
BULGARIA

Z. KOVACHEVA
Institute of Mathematics and Informatics, BAS
Sofia
BULGARIA

M. LAZAROVA
Technical University of Sofia
BULGARIA

SHUXIA LI
University of Texas Rio Grande Valley
Edinburg, TX
USA

DONGMEI LIU
North China University of Science and Technolgy
Tangshan, Hebei
P R CHINA

NURI MA
Beijing University of Technology
P.R. CHINA

V. MIHOVA
Angel Kanchev University of Ruse
BULGARIA

A. MIKHAILOV
University of Leeds
UK

I. NASKINOVA
University of Architecture, Civil Engineering and Geodesy
Sofia
BULGARIA

PENGCHENG NIU
Northwestern Polytechnical University
Xi'an Shaanxi
P R CHINA

O. OLEVSKYI
Oles Honchar National University
Dnipro
UKRAINE

V. OLEVSKYI
Ukrainian State University of Chemical Technology
Dnipro
UKRAINE

Tz. OSTROMSKY
Institute of Information and Communication Technologies, BAS
Sofia
BULGARIA

N. PAPANICOLAOU
University of Nicosia
CYPRUS

T. PLACHETKA
Comenius University
Bratislava
SLOVAKIA

ZHIJUN (GEORGE) QIAO
University of Texas Rio Grande Valley
Edinburg, TX
USA

E. REYES
Universidad de Santiago de Chile
Santiago
CHILE

YUNHE SHENG
Jilin University
P R CHINA

G. SHYSHKANOVA
Zaporizhzhia Polytechnic National University
Zaporizhzhia
UKRAINE
and Humboldt-Universitaet zu Berlin
GERMANY

G. SIMEONOV
Institute of Mathematics and Informatics, BAS
Sofia
BULGARIA

D. SLAVCHEV
Institute of Information and Communication Technologies, BAS
Sofia
BULGARIA

M. STOENCHEV
Technical University of Sofia
BULGARIA

RAN SUN
Beijing University of Technology
P R CHINA

XIANGYU TIAN
Beijing University of Technology
P R CHINA

M. TODOROV
Institute of Mathematics and Informatics, BAS
Sofia
BULGARIA

V. TODOROV
Institute of Information and Communication Technologies, BAS&
Institute of Mathematics and Informatics, BAS
Sofia
BULGARIA

S. TRANEV
Prof. Asen Zlatarov University
Bourgas
BULGARIA

V. TRANEVA
Prof. Asen Zlatarov University
Bourgas
BULGARIA

T. VALCHEV
Institute of Mathematics and Informatics, BAS
Sofia
BULGARIA

V. VATCHEV
University of Texas, Rio Grande Valley
Rancho Viejo, TX
USA

CHENXI WANG
North China University of Science and Technology
Tangshan
P R CHINA

DENGSHAN WANG
Beijing Normal University
P R CHINA

NA WANG
Henan University
Kaifeng, P R CHINA

QIAN WANG
Beijing University of Technology
P R CHINA

XIYIN WANG
North China University of Science and Technology
Tangshan
P R CHINA

WANG YAO
North China University of Science and Technology
Tangshan
P R CHINA

HONGWANG YUAN
North China University of Science and Technology
Tangshan
P R CHINA

SHOUMING ZHOU
Chongqing Normal University
P R CHINA

W. ZUNIGA_GALINDO
University of Texas Rio Grande Valley
Brownsville, TX
USA

Peer Review Statement

All papers published in this volume have been reviewed through processes administered by the Editors. Reviews were conducted by expert referees to the professional and scientific standards expected of a proceedings journal published by IOP Publishing.

- **Type of peer review:** Single Anonymous
- **Conference submission management system:** Morressier
- **Number of submissions received:** 60
- **Number of submissions sent for review:** 52
- **Number of submissions accepted:** 50
- **Acceptance Rate (Submissions Accepted / Submissions Received × 100):** 83.3
- **Average number of reviews per paper:** 1.0612244897959184
- **Total number of reviewers involved:** 35
- **Contact person for queries:**
  Name: Michail Todorov
  Email: mditod@gmail.com
  Affiliation: EAC4AMiTaNS - Mathematical Modelling and Numerical Methods

Content from this work may be used under the terms of the Creative Commons Attribution 4.0 licence. Any further distribution of this work must maintain attribution to the author(s) and the title of the work, journal citation and DOI.
Published under licence by IOP Publishing Ltd

Journal of Physics: Conference Series **2910** (2024) 012001 doi:10.1088/1742-6596/2910/1/012001

Topological, Differential Geometry Methods and Modified Variational Approach for Calculation of the Propagation Time of a Signal, Emitted by a GPS-Satellite and Depending on the Full Set of 6 Kepler Parameters

Bogdan G. Dimitrov[1,2]

[1]Institute of Nuclear Research and Nuclear Energetics, Bulgarian Academy of Sciences, 72 Tzarigradsko shaussee, 1784 Sofia, Bulgaria
[2]Institute for Advanced Physical Studies , Sofia Tech Park, 111 Tzarigradsko shaussee, 1784 Sofia, Bulgaria

E-mail: [1]dimitrov.bogdan.bogdan@gmail.com

E-mail: [2]bogdan.dimitrov@iaps.institute

Abstract. In preceding publications a mathematical approach has been developed for calculation of the propagation time of a signal, emitted by a moving along an elliptical orbit satellite and accounting also for the General Relativity Theory (GRT) Effects. So far, the formalism has been restricted to one dynamical parameter (the true anomaly or the eccentric anomaly angle). In this paper the aim is to extend the formalism to the case, when also the other five Kepler parameters will be changing and thus, the following important problem can be stated: if two satellites move on two space-distributed orbits and they exchange signals, how can the propagation time be calculated? This paper requires the implementation of differential geometry and topological methods. In this approach, the action functional for the propagation time is represented in the form of a quadratic functional in the differentials of the Kepler elements, consequently the problem is related to the first and second quadratic forms from differential geometry. Such an approach has a clear advantage, because if the functional is written in terms of Cartesian coordinates X, Y, Z, the extremum value after the application of the variational principle is shown to be the straight line - a result, known from differential geometry, but not applicable to the current problem of signal exchange between satellites on different orbits. So the known mapping from celestial mechanics is used, when by means of a transformation the 6 Kepler parameters are mapped into the cartesian coordinates X, Y, Z. This is in fact a submersion of a manifold of 6 parameters into a manifold of 3 parameters. If a variational approach is applied with respect to a differential form in terms of the differentials of the Kepler parameters, the second variation will be different from zero and the Stokes theorem can be applied, provided that the second partial derivatives of the Cartesian coordinates with respect to the Kepler parameters are assumed to be different from zero. ' From topology this requirement is equivalent to the existence of the s.c. Morse functions (non-degenerate at the critical points). In the given case it has been shown that Morse function cannot exist with respect to each one of the Kepler parameters- Morse function cannot be defined with respect to the omega angle.

Content from this work may be used under the terms of the Creative Commons Attribution 4.0 licence. Any further distribution of this work must maintain attribution to the author(s) and the title of the work, journal citation and DOI.
Published under licence by IOP Publishing Ltd

1 Introduction

1.1 Basic facts about the standard Shapiro delay formulae and its implementation in contemporary space experiments

Propagation of signals (radio, light or laser) in the gravitational field of the near-Earth space is an important theory in applied gravitational theory, since it is related to many astrophysical experiments in the near-Earth space and also in the deep space (such as LISA, for example). [1]. The main ingredient of this theory is the Shapiro delay formulae

$$T_{AB} = \frac{R_{AB}}{c} + \frac{2GM_E}{c^3} \ln\left(\frac{r_A + r_B + R_{AB}}{r_A + r_B - R_{AB}}\right) \quad , \tag{1}$$

which expresses the signal propagation time $T_{AB} = T_B - T_A$ between two space-time points and is calculated from the null cone equation

$$ds^2 = -c^2\left(1 + \frac{2V}{c^2}\right)(dT)^2 + \left(1 - \frac{2V}{c^2}\right)\left((dx)^2 + (dy)^2 + (dz)^2\right) = 0 \quad , \tag{2}$$

obtained after setting up the infinitesimal metric element equal to zero. In formulae (1) $G_{\oplus}M_{\oplus}$ is the geocentric gravitational constant, $M_{\oplus}$ is the Earth mass, $G_{\oplus}$ is the gravitational constant, the standard potential of the Earths gravitational field $V = \frac{G_{\oplus}M_{\oplus}}{r}$ is considered without taking into account the harmonics. The coordinates of the emitting and of the receiving satellite are correspondingly $\mid x_A(t_A) \mid = r_A$ and $\mid x_B(t_B) \mid = r_B$, $R_{AB} = \mid x_A(t_A) - x_B(t_B) \mid$ is the Euclidean distance between the signal - emitting satellite and the signal - receiving satellite. We shall denote also by $t = TCG$ the Geocentric Coordinate Time (TCG). The second term in formulae Eq. (1) is the Shapiro time delay term, accounting for the signal delay due to the curved space-time.The Shapiro delay term has an important physical meaning - it means that in a curved space-time due to the General Relativity effects, a signal travels a longer time because of the curved signal trajectory.

Since the discovery of the Shapiro delay formulae it has found numerous applications in VLBI interferometry (see also the review article [2] by Sovers, Fanselow and Jacobs) and also in creation of models for relativistic reference frames for the GRAIL mission (Gravity Recovery and Interior Laboratory) [3] accurate to 1 $\mu m/\sec$ (1 *micrometer* per sec*ond*; 1 micrometer= $10^{-6}\ m$) and also for the GRACE-Follow-on mission (Gravity Recovery and Climate Experiment) [4]. For the second model, the relativistic change of the phase of the signal has been computed due to the changing (geodesic!) distance between the emitting satellite and the receiving satellite. In fact, in these papers [3], [4] and [5] the difference between the emitting time and the receiving time turns out to be a relativistic observable, dependent on the geodesic distance, travelled by light - in such a case it contains also the relativistic logarithmic correction, typical for the Shapiro delay formulae. In the third paper [5] (for the case of the ACES experiment - Atomic Clock Ensemble in Space) of the series of these papers the difference between the two times is shown to be

$$T_2 - T_1 = \frac{\mid r_2 - r_1 \mid}{c} + (1+\gamma)\sum_b \frac{GM_b}{c^3} \ln \frac{r_1^b + r_2^b + r_{12}^b}{r_1^b + r_2^b - r_{12}^b} + O(c^{-5}) \quad , \tag{3}$$

where the summation is over the "deflecting" bodies (in the case-the Moon, the Earth and the Sun), causing the Shapiro time delay.

1.2 Uncertainties in the Shapiro formulae due to some approximations in the integration

Now it is important to stress that eq. (3), as well as (1) are obtained from the null cone equation after performing an integration over a variable r, related to the distance between the transmitter and the receiver. However, in the known review article by Ashby from 2003 [7] it was proved that at the level of several millimeters, spatial curvature effects have to be considered, which is evident from the simple equality

$$\int dr\left[1 + \frac{GM_E}{c^2 r}\right] \approx r_2 - r_1 + \frac{GM_E}{c^2} \ln\frac{r_2}{r_1} \quad . \tag{4}$$

In [7] the second term was estimated to be $4.43 \times \ln(4.2) \approx 6.3\ mm$. Also in the thesis [8] the approximation $cdt = \left(1 + \frac{GM_E}{c^2 r}\right) \parallel dr \parallel$ has been used, where the infinitesimal vector dr is not related to the signal trajectory, represented by the vector $\overrightarrow{R} = \overrightarrow{r} - \overrightarrow{r}_A$, where $\overrightarrow{r}_A$ is the vector, associated with the position of emission and $\overrightarrow{r}$ - with a position of a point in a Geocentric Reference Frame. In other words, the integration is not along $\parallel dR \parallel$, because the approximation $\parallel dR \parallel \approx \parallel dr \parallel$ has been assumed and thus the integration is along dr. A similar approximation has been used in the PhD thesis [9].

All these approximations are incompatible with the uncertainty $1\ \mu m/\sec$ of determination of the Relativistic Reference Frame in [3], but also in contemporary investigations of the gravitational field by means of resolving the gravitational redshift [10] across a millimeter scale atomic sample. This means that a clock will tick differently at the top and bottom of a sample of 100 000 ultra-cold strontium atoms [11], i.e. the clock ticks slowly at lower elevations.

1.3 Brief review of the approach for calculation of the propagation time by using the more general transformation in celestial mechanics and the true anomaly angle f as the only dynamical parameter

The accuracy of determination of the propagation time evidently will depend also on the choice of the integration variable. For example, in the monograph [12] and in many other books on celestial mechanics (see, for example [13]) a transformation from the six Kepler parameters $(f, \Omega, \omega, i, a, e)$ to the cartesian coordinates (x, y, z) has been used

$$x = \frac{a(1-e^2)}{1+e\cos f}\left[\cos\Omega\cos(\omega+f) - \sin\Omega\sin(\omega+f)\cos i\right] \quad , \tag{5}$$

$$y = \frac{a(1-e^2)}{1+e\cos f}\left[\sin\Omega\cos(\omega+f) + \cos\Omega\sin(\omega+f)\cos i\right] \quad , \tag{6}$$

$$z = \frac{a(1-e^2)}{1+e\cos f}\sin(\omega+f)\sin i \quad , \tag{7}$$

where (a, e) are the large semi-major axis of the satellites elliptical orbit and eccentricity respectively, $r = \frac{a(1-e^2)}{1+e\cos f}$ is the radius-vector in the orbital plane, the angle Ω of the longitude of the right ascension of the ascending node is the angle between the line of nodes and the direction to the vernal equinox, the argument of perigee (periapsis) ω is the angle within the orbital plane from the ascending node to perigee in the direction of the satellite motion $\left(0 \leq \omega \leq 360^0\right)$. The angle i is the inclination of the orbit with respect to the equatorial plane and the true anomaly angle f geometrically represents the angle between the line of nodes and the position vector $\vec{r}$ on the orbital plane. The angle f is related to the motion of the satellite and is assumed to be the only changing dynamical parameter meaning that the space-distributed orbit does not change during the motion of the satellite. Under such assumptions and taking into account that $dr = \alpha(f)df$ (where $\alpha(f) \equiv -\frac{a(1-e^2)e\sin f\ df}{(1+e\cos f)^2}$), in the paper [14] the propagation time $\widetilde{T}$ in the null cone equation (2) has been expressed by a complicated expression

$$\widetilde{T} = G(f) = -\frac{2G_{\oplus}M_{\oplus}}{c^3}.\frac{n}{(1-e^2)^{\frac{3}{2}}}\widetilde{T}_1 + T_2^{(2)} \quad , \tag{8}$$

where $\widetilde{T}_1$ and $T_2^{(2)}$ is a combination of elliptic integrals of the second and of the fourth order. The exact expressions can be found in the review article [15], but also will be given in Appendix B of this paper. Further in this paper they will participate in the expressions in the proposed variational formalism, based on the Gauss-Ostrogradsky and the Stokes theorems, but their concrete analytical forms will not be taken into account. Formulae (8) is the propagation time for the signal, while the true anomaly angle f changes from some value f_0 to some final value f_1. In other words, this formulae does not presume that there is any transmission of signals between a signal-emitting and a signal-receiving satellite because the second point of signal perception is not related to any satellite, it is determined only by the distance, travelled by the signal for the calculated propagation time $\widetilde{T}$. Formulae (8) expresses the propagation time $\widetilde{T}$ as a function only of the dynamical parameter, which is the true anomaly angle f [14].

This means also that all the other (five) remaining parameters of the orbit $(\Omega, \omega, i, a, e)$ do not change during the motion of the satellite. For such a case, it can easily be found that

$$\sqrt{(dx)^2 + (dy)^2 + (dz)^2} = \sqrt{\left(v_f^x\right)^2 + \left(v_f^y\right)^2 + \left(v_f^z\right)^2}\,df = v_f df \quad , \tag{9}$$

where the velocity v_f, associated to the true anomaly angle f is given by

$$v_f = v = \frac{na}{\sqrt{1-e^2}}\sqrt{1+e^2+2e\cos f} \quad . \tag{10}$$

Making use of the null cone equation (2) and also of the approximation $\beta = \frac{2V}{c^2} = \frac{2G_{\oplus}M_{\oplus}}{c^2 a} \ll 1$, one can obtain the general formulae for the propagation time in the form (8).

1.4 The other not yet investigated case of transmission of signals (with account of the General Relativity effects) between satellites on different space-distributed orbits

The more interesting case will be when a signal is emitted by a satellite on a space-distributed orbit and is received (percepted) by a satellite on another space-distributed orbit. Then there will be two dynamical parameters f_1 and f_2, related to the motion of both satellites. The integration in the expression for the propagation time will be over the infinitesimal distance element dR_{AB} between the two satellites, since it involves the differentials of both sets of cartesian coordinates (x_1, y_1, z_1) and (x_2, y_2, z_2) or the two sets of Kepler parameters $(f_1, \Omega_1, \omega_1, i_1, a_1, e_1)$ and $(f_2, \Omega_2, \omega_2, i_2, a_2, e_2)$ for the transformations (5) - (7), written respectively for the indices "1" and "2".

1.5 Autonomous navigation as an impetus for developing intersatellite communications

In the book [18] autonomous navigation is defined as referring to "processes in which the spacecraft without the support from the ground-based TT&C (Telemetry & Tracking Control) system for a long time, relying on its onboard devices, obtains all kinds of measurement data; determines the navigation parameters like orbit, time and attitude". In this monograph DORIS and PRARE navigation systems are determined as non-autonomous, because "DORIS system can determine the spacecraft's orbit with high accuracy, but both need to exchange information with the ground stations". Thus, the main aim of autonomous navigation is to reduce the dependence of spacecrafts on the ground TT&C network and in this way to enhance the capability of the systems anti-jam and autonomous survivability. It is evident also that the determination of the satellite's exact location through the reception, processing, and transmitting of ranging signals between the different satellites.

Autonomous GPS navigation system is necessary in view of the proposal of researchers from the University of Texas, Aerospace Corporation, National Bureau of Standards, International Business Machines Corporation (IBM) and Rockwell Automation Inc. to monitor nuclear explosions, based on GPS inter-satellite communication link.

The development of inter-satellite laser communication systems in space - orbit technology that enable super-high-speed data transfers at rates greater than 1 *Gbps* is widely applied also in.cube-/nanosatellite platforms such as CubeLCT, AeroCube-7B/C, CLICK, LINCS-A/B, SOCRATES and LaserCube [19]. However, in order to establish laser communications, high performance of arc-second level pointing system is required, and this is a difficulty for the nanosatellite platforms. It can be supposed why there is such a difficulty - since the data-transmission rate of 1*Gbps* between the two nanosatellites is at an intersatellite range of 1000 km, the effects of curving the trajectory of the laser signal may be considerable. In publications, related to small satellite optical links [20] it has been admitted that the s.c. "pointing error" arizes not only due to tracking sensors and mechanical vibrations, but also due to the base motion of the satellite. This fact, together with the ranging of signals and the pointing error illustrate the assumption that not only the motion of two satellites is important, but the effect of curving of the signal trajectory. The paper [20] also asserts that for a GPS constellation with 24 satellites, there are a total of $8 - 16$ links - they can be forward and backward links in the same orbit, but as well as lateral links between adjacent orbits. Moreover, the distance of GPS inter-satellite crosslink can reach 49 465km.

2 Overview of the aims and approaches of this paper

The main goal of this paper is to present a theoretical model for inter-satellite communications between two satellites on different space-distributed orbits, characterized by two different sets and Kepler parameters, accounting for the General Relativity effect of curving the signal trajectory. In future, the aim will be also to find the optimal position of the satellites for transmission of signals with a minimal propagation time. From an experimental point of view the importance of developing such approaches is described in the introduction in view of the rapidly developing laser broadband and large capacity communication links in free space, especially image transmissions.

The peculiarity of the proposed approach is that while the first publications on the implementation of General Relativity in GPS communications focused on space-ground links for the most simple case of plane Kepler orbits (characterized only by the eccentric anomaly angle E and constant ellipticity e , here the problem is more complicated and is related to the communication between satellites on two different orbits. Each orbit is characterized by six dynamically changing Kepler parameters.

The main proposal in this paper is to implement ideas from differential geometry and topology for the solution of this problem. Thus it may be asserted that the problem is also non-trivial from a mathematical point of view. Namely, if the motion of the first satellite and the emission of the signal is characterized by the parameters $(f_1, \Omega_1, \omega_1, i_1, a_1, e_1)$, defining a manifold M_1, and the motion of the second satellite and the processes of perception of the signal - by the parameters $(f_2, \Omega_2, \omega_2, i_2, a_2, e_2)$, defining a second manifold M_2, then can the combined motion of the two satellites and the processes of signal-exchanges be characterized by one manifold? It will be shown that it is possible to introduce such a manifold, characterized by the two sets of (total) 12 Kepler parameters, but an additional parameter s should be introduced. This turned out to be possible due to the fulfillment of the Whitney theorem in differential geometry. This parameter parametrizes the two sets of 12 Kepler parameters, which enter the expressions for the two sets of cartesian coordinates (x_1, y_1, z_1) and (x_2, y_2, z_2).

This theoretical setting clearly demonstrates the necessity to implement differential geometry and topology approaches in this problem about intersatellite communications. That is why, particular attention in the paper is paid to notions such as topological and metric spaces and manifolds, homeomorphisms (topological equivalence), manifolds with boundaries. It has been shown also that the standard Shapiro delay formulae can be modified, and thus the modified formulaes (26) and (27) have been obtained. The boundaries of the manifold M_1 were found by applying the concrete definitions. Since the manifold is a three-dimensional one, the boundaries were found to be three two-dimensional planes of the variables (x, y), (x, z) and (y, z).

Further the aim was to develop a variational formalism, based on the Stokes and the Gauss - Ostrogradsky theorem and accounting for higher-order variations. The corresponding variational action was found with derivatives of the first, second and third order. The idea about higher-order variational calculus is not new and it comes from older books such as [46] and many others. In particular, second-order variations allow to take into account non-fixed, "moving boundaries" in the corresponding integrals. This is convenient in view of the fact that these boundaries will be related to the motion of the two satellites and further to the necessity of "optimizing" the process of transmission of signals. The variational model is constructed by means of variation of the parameters u_i and v_j, denoting two different subsets of Kepler parameters of one satellite

$$u_i = (f_., a_., e_.) \quad i = 1, 2, 3 \text{ and } v_j = (\Omega_., i_., \omega_.) \quad j = 1, 2, 3) \tag{11}$$

or the two different sets of Kepler parameters, corresponding to the first and to the second satellite respectively

$$u_i = (f_{.1}, a_{1.}, e_{1}, \Omega_{1.}, i_{1.}, \omega_1) \quad i = 1, 2, ...4, 5, 6 \tag{12}$$

and

$$v_j = (f_{2.}, a_{2.}, e_2, \Omega_{2.}, i_{2.}, \omega_{2.}) \quad j = 1, 2, ..., 4, 5, 6 \quad . \tag{13}$$

3 Some basic notions from topology, needed for the development of the theory in this paper

This study makes use of some standard notions from differential geometry, but also from topology These two branches from geometry are of course related, but it is important to keep in mind that there is an important difference. For example, a topological manifold is a set of points of the Euclidean space, for which the notion of proximity is defined [42]. It should be noted, however, that proximity is not related to properties of metric spaces, which shall also be defined further. To be more precise, a topological manifold is the set of points of the Euclidean space, defined by the equations

$$F_i(x_1, x_2, ...x_{\overline{n}}) = 0 \quad , i = 1, 2, ...m \quad , \tag{14a}$$

$$f_j(x_1, x_2, ...x_{\overline{n}}) \geq 0 \quad , j = 1, 2, ...l \quad . \tag{15a}$$

This (topological) manifold, defined by the equations (14a) is a sub-manifold of a dimension $n = \overline{n} - m$. The notion of a sub-manifold is also defined as a topological property, and not related to any metric properties. Another typical notions for topological manifolds are boundedness, connectedness,compactness, continuity. The notions of a sub-manifold and a boundary,as further will be explained, are also not related to any metric properties.

It is evident that topological spaces are generalizations of metric spaces and as such they are defined as a class of open sets of E (the Euclidean space), satisfying the following conditions [38]:

1. The intersection of a finite number of open sets is an open set X.
2. The union of any number of open sets is an open set. The class of open sets is called a topology.

These two conditions are supplemented by a third one in [31]: The empty set and the whole set X must be open.

The definition about topological space by itself is sufficient to define the important notions of continuity and homeomorphicity.

A map $f : X \longrightarrow Y$ of one topological space to another is continuous if the complete inverse image $f^{-1}(U) \subset X$ of every open set $U \subset Y$ is open also in X [31]. Two topological spaces E and F are homeomorphic (or topologically equivalent) if there is a one-to-one mapping $\varphi : E \longrightarrow F$ of E onto F such that both φ and the inverse one φ^{-1} are continuous (see also [38]). A simple application of homeomorphicity, given in [38] is the following one: Since the hemisphere is homeomorphic to a disc, two pairs of opposite points on the boundary can be glued together and thus the $n-$dimensional projective plane RP^n can be obtained. This can be illustrated by the following simple formulae

$$R^{n+1}\backslash\{0\} \to S^n \to RP^n \quad . \tag{16}$$

This definition makes use of the topological notion about "manifolds with boundary", which shall be clarified further. Instead of the term "gluing", often the term "identifying" is used. For example, in order to visualize RP^2, one may take the two-dimensional sphere S^2 and identify pairs of opposite points below the equator and above the equator.

Now an interesting property of a topological space M can be defined,which enables to define a differentiable manifold [43]. If

A. The topological space M is provided with a family of pairs $\{(M_\alpha, \Phi_\alpha)\}$, where Φ_α are homeomorphisms from M_α to an open subset of O_α of R^n i.e. $\Phi_\alpha : M_\alpha \mapsto O_\alpha$, and M_α are a family of open sets, which cover $M : \cup M_\alpha = M$.

B. Given M_α, M_β such that $M_\alpha \cap M_\beta \neq \emptyset$, the map $\Phi_\beta \circ \Phi_\alpha^{-1}$ from the subset $\Phi_\alpha(M_\alpha \cap M_\beta)$ of R^n to the subset $\Phi_\beta(M_\alpha \cap M_\beta)$ of R^n is infinitely differentiable (written as C^∞),

then the family $\{(M_\alpha, \Phi_\alpha)\}$ satisfying A. and B. is called an atlas.

An important feature of the topological space M is the Hausdorff property, which however should be additionally imposed and is independent from the metric space. The Hausdorff property turns out to be important, if one would like to determine the dimension of M, in view of the following

Definition 1 *[45]: A topological manifold is a Hausdorff, second countable (meaning that it has a countable basis) and locally Euclidean space.*

Here the two important notions of Hausdorff space and locally Euclidean space have to be clarified. The Hausdorff property [43] means that if there are two distinct points x, y belonging to the topological space M, then there exists a pair of open sets O_x and O_y, such that $O_x \cap O_y = \emptyset$. This means also that there exist small enough disjoint open sets O_x and O_y, which contain x and y respectively. The respective definition, given in [45] is analogous.

Further, a topological space M is locally Euclidean of dimension n if every point p in M has a neighborhood U such that there is a homeomorphism Φ from U onto an open subset of R^n [45]. The pair $(U, \Phi : U \to R^n)$ is called a chart, Φ-a coordinate map or a coordinate system on U and U- a coordinate neighborhood or a coordinate open set. This definition is the essence of the s.c. invariance of dimension, according to which for $n \neq m$ an open subset of R^n is not homeomorphic to an open set of R^m. This topic will be essential for the topological notions of embedding, submersion and immersion, which shall further be explained with reference to the different mappings $f : (f, \Omega, \omega, i, a, e) \to (x, y, z)$ which can be defined, concerning the transformations (5) - (7). In particular, these mappings will depend on the number of the Kepler elements, which will change during the process of transmission of a signal between two satellites on two space-dependent orbits.

Unlike the topological notion of a homeomorphism, which generates equivalence classes between topological spaces, the notion of a metric space is a particular kind of a topological space, where the open sets are provided by a distance function $d(x, y)$, which corresponds to our intuitive notion for the distance between points. Of course, $d(x, y)$ can be generalized also to the $3D-$case. The distance function $d(x, y)$ has the following properties [43], [45] and is expressed by the simple formulae $d(x, y) = \sqrt{(x_1 - x_2)^2 + (y_1 - y_2)^2}$, for the $n-$dimensional space $d(p, q) = \sqrt{\sum\limits_{i=1}^{n} (p^i - q^i)^2}$:

1A. $d(x, y) \geq 0$.
2A. $d(x, y) = 0 \leftrightarrow x = y$.

3A.If the point $z \in M$, then the triangle inequality $d(x,z) \leq d(x,y)+d(y,z)$ is fulfilled. A topological space, satisfying these three conditions is a metric space. In addition, if M is also a Riemannian manifold, then the distance is $ds^2 = g_{ij}dx^i dx^j$ $(i,j=1,2....4)$ and the points a,b are connected by a piecewise differentiable curve C, parametrized by the functions $x^i(s)$, then the distance L_{ab} between the points will be defined as

$$L_{ab} = \int_a^b g_{ij}\frac{dx^i}{ds}\frac{dx^j}{ds}ds \quad . \tag{17}$$

Then the distance function in the above definition $d(x,y)$ will be defined as the infimum or least upper bound of L_{ab} as it varies over all such curves $x^i(s)$ i.e. $d(a,b) = \inf L_{ab}(C)$.

An important clarification will be given below, which is related to the present research.

3.1 Application of metric and topological spaces in the present formalism

The analogy of the distance function $d(x,y)$ is the introduced in Appendix A formulae (181) and formulae (182) for the differential dR^2_{AB} for the case, when f_1 and $_{f_2}$ are the only dynamical parameters. Since formulae (181) can be written also as

$$R^2_{AB} = r_1^2 + r_2^2 - 2(\overrightarrow{\mathbf{r}}_1.\overrightarrow{\mathbf{r}}_2) \quad , \tag{18}$$

where the vectors $\overrightarrow{\mathbf{r}}_1$ and $\overrightarrow{\mathbf{r}}_2$ are defined in the natural way , $\overrightarrow{\mathbf{r}}_1 = (x_1,y_1,z_1)$ and $\overrightarrow{\mathbf{r}}_2 = (x_2,y_2,z_2)$, the following problem arizes: suppose that for the two topological manifolds one can define the two open balls $B_1(p_1,r_1)$ and $B_2(p_2,r_2)$ with centers at the points $p_1 \in R_1^3$ and $p_2 \in R_2^3$ (the lower indices "1" and "2" mean that these are two different Euclidean manifolds) such that

$$B_1(p_1,r_1) = \{x_1,y_1,z_1 \in R_1^3 \mid d(x_1,y_1,z_1,p_1) < r_1\} \quad , \tag{19}$$

$$B_2(p_2,r_2) = \{x_2,y_2,z_2 \in R_1^3 \mid d(x_2,y_2,z_2,p_2) < r_2\}. \tag{20}$$

Then, in order to apply the definition for an open set in the meaning of an open ball (see [45]), can one define the open ball $B(p,r) = \{x,y,z \in R_1^3 \times R_2^3 \mid d(x,y,z,p) < R_{AB}\}$? This problem is outside the scope of the present paper, but should be investigated also for the general case when the two manifolds R_1^3 and R_2^3 are expressed in terms of the two sets of Kepler parameters $(f_1,\Omega_1,\omega_1,i_1,a_1,e_1)$ and $(f_2,\Omega_2,\omega_2,i_2,a_2,e_2)$.

3.2 A new proposition for a generalized Shapiro delay formulae, based on the differential geometry formalism

3.2.1 The additional condition for equality of the two differentials Using the definition about the metric space, let us derive the general formulae for the differential of the propagation time T for the signal between two satellites on two different space-distributed orbits. Remembering the null cone equation (2) and formulae (26) from the review paper [15], one can obtain

$$dT \approx \frac{1}{c}\int_{path}\sqrt{\delta_{ij}dx^i dx^j} + \frac{1}{c^3}\int_{path} 2U\sqrt{\delta_{ij}dx^i dx^j} \quad , \tag{21}$$

where the integration is performed over the s.c. "slant range" $d\rho = \sqrt{\delta_{ij}dx^i dx^j}$, called also the "geometric path". The Earths gravitational potential U will be taken in the standard form $U = \frac{G_{\oplus}M_{\oplus}}{R}$. But for this case, we may define the as the infinitesimal geometric path dR_{AB} on the line, joining the signal-emitting and the signal-receiving satellite. Further argumentation will be provided in favour of the statement that imposing the equality $d\rho = dR_{AB}$ is very convenient.

If the coordinates x,y,z are parametrized by the parameter s, i.e. $x=x(s)$, $y=y(s)$ and $z=z(s)$ and remembering the previously introduced notations $x=x_2-x_1$, $y=y_2-y_1$ and $z=z_2-z_1$, one can represent $d\rho$ as

$$d\rho = \sqrt{(\dot{x})^2 + (\dot{y})^2 + (\dot{z})^2}ds \tag{22}$$

and

$$dR^2_{AB} = 2(x_2-x_1)(dx_2-dx_1) + 2(y_2-y_1)(dy_2-dy_1) + 2(z_2-z_1)(dz_2-dz_1) =$$

$$= \frac{1}{R_{AB}}\left[xdx + ydy + zdz\right] = \frac{1}{2R_{AB}}\frac{d}{ds}\left(x^2+y^2+z^2\right) = \frac{1}{2R_{AB}}\frac{dR^2_{AB}}{ds} = \frac{dR_{AB}}{ds} \quad . \tag{23}$$

The last simple formulae $dR^2_{AB} = \frac{dR_{AB}}{ds}$ is related to the fact that the infinitesimal Euclidean distance dR^2_{AB} is equal to the infinitesimal distance $\frac{dR_{AB}}{ds}$ along the curve $R_{AB}(s)$, parametrized by the parameter s. In fact, formulae (23) can be written in the same way as in the monograph by Postnikov [21] (see also [30])

$$l(s) = \int_{s_0}^{s} \mid \dot{\overrightarrow{\mathbf{R}}}_{AB}(s) \mid ds \quad , \tag{24}$$

where the coordinates of the vector $\dot{\overrightarrow{\mathbf{R}}}_{AB}(s)$ are $\dot{\overrightarrow{\mathbf{R}}}_{AB}(s) = (\dot{x}(s), \dot{y}(s), \dot{z}(s))$ and thus $\mid \dot{\overrightarrow{\mathbf{R}}}_{AB}(s) \mid = \sqrt{\dot{x}^2(s) + \dot{y}^2(s) + \dot{z}^2(s)}$. It is known that the parameter s in the general case is not a natural parameter, meaning that it is not equal to the length of the curve. It would have been a natural parameter only if $\mid \dot{\overrightarrow{\mathbf{R}}}_{AB}(s) \mid = 1$ and thus $l(s) = s - s_0$. This basic fact from the differential geometry of curves and surfaces will be of primary importance, when further some theorems about the geodesic and the null cone equations will be given.

Now one can see the subtle difference - in the infinitesimal sense the distance between two points along the curve is the same as the distance between the two points on the straight line, which connects the two points. In particular, this is the simple intuitive reasoning for requiring the fulfillment of the equality

$$d\rho = \sqrt{\delta_{ij} dx^i dx^j} = dR_{AB} \quad , \tag{25}$$

which can be written also as

$$\sqrt{(\dot{x})^2 + (\dot{y})^2 + (\dot{z})^2} ds = \frac{1}{2R_{AB}} d(x^2+y^2+z^2) = \tag{26}$$

$$= dR^2_{AB} = 2r_1 dr_1 + 2r_2 dr_2 - 2(\overrightarrow{\mathbf{dr}}_1.\overrightarrow{\mathbf{r}}_2) - 2(\overrightarrow{\mathbf{r}}_1.\overrightarrow{\mathbf{dr}}_2) \quad , \tag{27}$$

where expression (18) for dR^2_{AB} has been used. Note also that formulae (21) is not necessarily to be written on the base of the assumption (25) $d\rho = dR_{AB}$ - at the end of this section such a possibility will also be discussed.

3.2.2 Modified Shapiro delay formulae with the additional condition It should be kept in mind that $2r_1 dr_1$ and $2r_2 dr_2$ are scalar multiplication of functions, while $(\overrightarrow{\mathbf{dr}}_1.\overrightarrow{\mathbf{r}}_2)$ and $(\overrightarrow{\mathbf{r}}_1.\overrightarrow{\mathbf{dr}}_2)$ are vector multiplications of two vectors, for example

$$(\overrightarrow{\mathbf{dr}}_1.\overrightarrow{\mathbf{r}}_2) = x_2 dx_1 + y_2 dy_1 + z_2 dz_1 \quad , (\overrightarrow{\mathbf{r}}_1.\overrightarrow{\mathbf{dr}}_2) = x_1 dx_2 + y_1 dy_2 + z_1 dz_2 \tag{28}$$

and the two set of functions (x_1, y_1, z_1) and (x_2, y_2, z_2) are given by formulaes (5) - (7), supplemented by the corresponding indices "1" and "2". A known property from celestial mechanics (see for example [12] and [13]) is that the scalar functions r_1 and r_2 represent the corresponding Euclidean distances

$$x_1^2 + y_1^2 + z_1^2 = r_1^2 \quad , \quad x_2^2 + y_2^2 + z_2^2 = r_2^2 \quad , \tag{29}$$

which depend on the corresponding true anomaly angles f_1 and f_2 and also on (e_1, a_1) and (e_2, a_2)

$$r_1 = \frac{a_1(1-e_1^2)}{1+e_1\cos f_1} \quad , \quad r_2 = \frac{a_2(1-e_2^2)}{1+e_2\cos f_2} \quad . \tag{30}$$

If all the orbital elements (i.e. the two sets of orbital parameters) $(f_1, \Omega_1, \omega_1, i_1, a_1, e_1)$ and $(f_2, \Omega_2, \omega_2, i_2, a_2, e_2)$ are varying during the process of emission and perception of the signal between the two satellites (i.e.all these two sets of Kepler elements depend on the parameter s and thus represent functions), this has also to be taken in account when calculating the differentials dr_1 and dr_2 in formulae (27) and also $(\overrightarrow{\mathbf{r}}_1.\overrightarrow{\mathbf{dr}}_2)$ and $(\overrightarrow{\mathbf{dr}}_1.\overrightarrow{\mathbf{r}}_2)$). We shall not present in details these lengthy calculations, these formulaes and the further development of the new approach will be given in another publication. .

Journal of Physics: Conference Series **2910** (2024) 012001 doi:10.1088/1742-6596/2910/1/012001

However, when an integration along the parameter s is performed along a curve, joining two non-infinitesimal points, corresponding to the parameter values s_1 and s_2, the integration will not give an "infinitesimal result". This would mean that the first term in expression (21) for the differential of the propagation time dT

$$dT \approx \frac{1}{c}\int\limits_{path}\sqrt{(\dot{x})^2+(\dot{y})^2+(\dot{z})^2}ds+\frac{2G_{\oplus}M_{\oplus}}{c^3}\int\limits_{R_{0A}}^{R_{0B}}\frac{dR}{R} \tag{31}$$

will no longer give a constant Euclidean distance, divided by c, as is in the standard Shapiro formulae. In (31) we have denoted the under-integral variable in the second integral as $d\overline{\rho} = \sqrt{\delta_{ij}dx^i dx^j ds} = dR$ and the path integral will be from R_{01} to R_{02} (R_{01} and R_{02} are the distances from the first point and the second point to the zero-point, the origin of the coordinate system and thus $R_{AB} = R_{0B} - R_{0A}$)

$$\frac{2G_{\oplus}M_{\oplus}}{c^3}\int\limits_{R_{0A}}^{R_{0B}}\frac{dR}{R} = \frac{2G_{\oplus}M_{\oplus}}{c^3}\ln(R_{0B}-R_{0A}) = \frac{2G_{\oplus}M_{\oplus}}{c^3}\ln(R_{AB}) \quad . \tag{32}$$

This second term might be considered to be a generalization of the second term in the standard Shapiro formulae $T_{AB} = \frac{R_{AB}}{c} + \frac{2GM_E}{c^3}\ln\left(\frac{r_A+r_B+R_{AB}}{r_A+r_B-R_{AB}}\right)$ (1), but this should be understood in the most general sense. The reason is that formulae (21) is for the differential dT, which depends on the two sets of Euclidean coordinates (in total 6 coordinates-or in the case, on the coordinates $x = x_2 - x_1$, $y = y_2 - y_1$) or one might also consider a case, when dT will depend on the two sets of Kepler parameters. Moreover, the formulae (31) is valid when the condition (25) is also considered, and it represents a complicated quadratic expression in the differentials dx^i and dx^j.

3.2.3 Modified Shapiro delay formulae without any additional conditions Equation (21) can be written also without applying the condition $d\rho = \sqrt{\delta_{ij}dx^i dx^j} = dR_{AB}$ (25).Then expression (31) will be rewritten as

$$dT \approx \frac{1}{c}\int\limits_{path}\sqrt{(\dot{x})^2+(\dot{y})^2+(\dot{z})^2}ds+\frac{2G_{\oplus}M_{\oplus}}{c^3}\int\frac{\sqrt{(\dot{x})^2+(\dot{y})^2+(\dot{z})^2}}{R_{AB}}dR_{AB} \quad . \tag{33}$$

Now making use of the formulaes (27) and (28), it can easily be proved that

$$dR_{AB} = \frac{1}{R_{AB}}\left[xdx + ydy + zdz\right] \tag{34}$$

Thus, the whole expression (33) can be written as

$$dT \approx \frac{1}{c}\int\limits_{path}\sqrt{(\dot{x})^2+(\dot{y})^2+(\dot{z})^2}ds$$

$$+\frac{2G_{\oplus}M_{\oplus}}{c^3}\int\frac{\sqrt{(\dot{x})^2+(\dot{y})^2+(\dot{z})^2}}{[x^2+y^2+z^2]}\left[xdx+ydy+zdz\right] \quad . \tag{35}$$

This is the expression in the general case. If the parametrization $x = x(s)$, $y = y(s)$ and $z = z(s)$ is taken into account (also $dT = \int dTds$), then the propagation time $T = T_2 - T_1$ will be given by an integral of the type $T \approx \int F(x, y, z, \dot{x}, \dot{y}, \dot{z})ds$, where the under-integral function acquires the form

$$F(x,y,z,\dot{x},\dot{y},\dot{z}) = \frac{1}{c}\sqrt{(\dot{x})^2+(\dot{y})^2+(\dot{z})^2}$$

$$+\frac{2G_{\oplus}M_{\oplus}}{c^3}\frac{\sqrt{(\dot{x})^2+(\dot{y})^2+(\dot{z})^2}}{[x^2+y^2+z^2]}\left[x\dot{x}+y\dot{y}+z\dot{z}\right] \quad . \tag{36}$$

The first integral in (35) is known from differential geometry [30], [31], [36], [32] (and many others) and its relation to the physical essence of the propagation of signals will be investigated in the next section, related also to some basic theorems from differential geometry.

The second term in (36) is more complicated and it will be investigated in another paper, since it will be interesting to include this second term in the usual variational formalism, when the action functional derived from the null cone equation will be minimized and the Euler Lagrange equations with account of this second term have to be solved. Then it will be interesting to compare the results from the usual variational formalism with the results from the higher-order variational formalism, which will be proposed in the following sections of this paper.

4 Minimization of the first term in the action functional for the propagation time

4.1 Mathematical derivation of the extremal function - the straight line

Now we shall make the first step towards solving the following physical problem: suppose that a satellite on one space-distributed orbit sends a signal towards another satellite (also on a space-distributed orbit). Then what is the optimal configuration (i.e. disposition of the satellites), so that the transmission will be for the least time.

This problem requires the minimization of the functional (35), in particular we shall investigate the minimization only of the first term. Since (35) originally has been obtained by the null cone equation (2), this minimization will require the implementation of a variational principle, based on the Euler-Lagrange equations. So in this section we shall minimize the action functional

$$T = \int \left[\sqrt{(\dot{x})^2 + (\dot{y})^2 + (\dot{z})^2} + \lambda(x,y,z) R_{AB}(x,y,z) \right] ds \quad , \tag{37}$$

where we have denoted $x = x_1 - x_2$, $y = y_1 - y_2$, $z = z_1 - z_2$, supplemented by the Lagrange multiplied $\lambda(x,y,z)$ and the distance between the space points (x_1, y_1, z_1) and (x_2, y_2, z_2) is $R_{AB} = \sqrt{(x_1 - x_2)^2 + (y_1 - y_2)^2 + (z_1 - z_2)^2}$ as a constraint function. Additionally we introduce the notation for the length function L_1

$$L_1 = \sqrt{(\dot{x})^2 + (\dot{y})^2 + (\dot{z})^2} \quad , \tag{38}$$

where, as usual, the dot "." denotes a differentiation along the parameter s. Note that if in accord with the standard differential geometry formalism these coordinates This length function L_1 measures the distance between the two points along the arc of two points on a given curve, not necessarily assuming that s is a "natural parameter".

The action functional will have a minimum if and only if the under-integral function satisfies the Euler-Lagrange equations

$$\frac{\delta L}{\delta x} = 0 = \lambda_x R_{AB} + \lambda \frac{x}{R_{AB}} + \left(\frac{\ddot{x}}{L_1} - \frac{1}{L_1^3} \frac{d}{ds} \frac{\dot{x}^3}{3} \right) \quad , \tag{39}$$

$$\frac{\delta L}{\delta y} = 0 = \lambda_y R_{AB} + \lambda \frac{y}{R_{AB}} + \left(\frac{\ddot{y}}{L_1} - \frac{1}{L_1^3} \frac{d}{ds} \frac{\dot{y}^3}{3} \right) \quad , \tag{40}$$

$$\frac{\delta L}{\delta z} = 0 = \lambda_z R_{AB} + \lambda \frac{z}{R_{AB}} + \left(\frac{\ddot{z}}{L_1} - \frac{1}{L_1^3} \frac{d}{ds} \frac{\dot{z}^3}{3} \right) \quad . \tag{41}$$

Multiplying the first equation by $\dot{x}$, the second equation by $\dot{y}$ and the third equation by $\dot{z}$ and summing up, one can obtain

$$(\lambda . R_{AB})_s + \frac{dL_1}{ds} - \frac{1}{L_1^3} \left[\dot{x}^3 \ddot{x} + \dot{y}^3 \ddot{y} + \dot{z}^3 \ddot{z} \right] = 0 \quad . \tag{42}$$

From the expressions for the functions R_{AB} and L_1 after differentiating one can obtain

$$L_1 \frac{dL_1}{ds} = \dot{x}\ddot{x} + \dot{y}\ddot{y} + \dot{z}\ddot{z} \ , \quad R_{AB} \frac{dR_{AB}}{ds} = x\dot{x} + y\dot{y} + \ z\dot{z} \quad . \tag{43}$$

If the first differentiated equation in (43) is multiplied by x , the second differentiated equation - by $\ddot{x}$ and the resulting equations are equated, then

$$x L_1 \frac{dL_1}{ds} - \ddot{x} R_{AB} \frac{dR_{AB}}{ds} = x\dot{y}\ddot{y} + x\dot{z}\ddot{z} - \ddot{x}y\dot{y} - \ddot{x}z\dot{z} \ , \tag{44}$$

doi:10.1088/1742-6596/2910/1/012001

which can be rewritten as

$$xL_1\frac{dL_1}{ds} - \ddot{x}R_{AB}\frac{dR_{AB}}{ds} = x(L_1\frac{dL_1}{ds} - \dot{x}\ddot{x}) - \ddot{x}(R_{AB}\frac{dR_{AB}}{ds} - x\dot{x}) \quad \Longrightarrow x\dot{x}\ddot{x} = 0 \ . \tag{45}$$

Similarly one can obtain the equations $y\dot{y}\ddot{y} = 0$ and $z\dot{z}\ddot{z} = 0$, from the three equations it follows

$$xyz\frac{d}{ds}\left[(\dot{x})^2 + (\dot{y})^2 + (\dot{z})^2\right] = 0 \quad \Rightarrow L_1^2 = const = C_1 \tag{46}$$

Now we rewrite the three dynamical equations (39), (40) and (41):

$$(\lambda R_{AB})_x + \frac{1}{C_1}\ddot{x} = 0 \ , \ (\lambda R_{AB})_y + \frac{1}{C_1}\ddot{y} = 0 \ , \quad (\lambda R_{AB})_z + \frac{1}{C_1}\ddot{z} = 0 \tag{47}$$

Multiplying the first equation by $\dot{x}$, the second equation by $\dot{y}$ and the third equation by $\dot{z}$ and summing up, one can obtain

$$(\lambda R_{AB})_s + \frac{1}{C_1}\left(\dot{x}\ddot{x} + \dot{y}\ddot{y} + \dot{z}\ddot{z}\right) = 0 \quad . \tag{48}$$

Since we have proved that $L_1\frac{dL_1}{ds} = \dot{x}\ddot{x} + \dot{y}\ddot{y} + \dot{z}\ddot{z} = const$, it follows that $\lambda R_{AB} = const_1 s + const_2$, i.e. $R_{AB}(x, y, z) = \sqrt{x^2 + y^2 + z^2}$ is the straight line! In the case, the straight line will be between the signal-emitting satellite and the signal-receiving satellite and

$$R_{AB} = \sqrt{(x_{1.} - x_2)^2 + (y_{1.} - y_2)^2 + (z_{1.} - z_2)^2} = const_1 s + const_2 \quad , \tag{49}$$

where $x_{1.}, x_2, y_{1.}, y_2, z_{1.}z_2$ are expressed from (5), (6) and (7) in terms of the two sets of Kepler parameters $(f_{.1}, a_{1.}, e_{1.}, \Omega_{1.}, i_{1.}, \omega_{1.})$ and $(f_{2.}, a_{2.}, e_{2.}, \Omega_{2.}, i_{2.}, \omega_{2.})$. So in terms of all these variables, this will be a complicated 12−dimensional surface.

4.2 Geometrical justification of the obtained result

The obtained formulae (49) may seem elementary and trivial, but it has an important physical meaning, which will justify the application of a more specific variational formalism.

Let us first understand what does this formulae implies: it means that the manifold defined in terms of the space variables $x = x_1 - x_2$, $y = y_1 - y_2$, $z = z_1 - z_2$ contains a straight line. But then one can notice that there is a special theorem in differential geometry (see the contemporary textbook [32]):

Theorem 2 *If a surface contains a straight line then this line (endowed with the natural parameter s) is a geodesic.*

In fact, we started from the action functional (37) $T = \int\left[\frac{1}{c}\sqrt{(\dot{x})^2 + (\dot{y})^2 + (\dot{z})^2} + \lambda(x, y, z)R_{AB}(x, y, z)\right]ds$ (supplemented with the constraint function, which however did not influence the variational formalism), then we applied a variational formalism and we found as extremals the straight lines! Since the action functional $T = \frac{1}{c}\int\sqrt{(\dot{x})^2 + (\dot{y})^2 + (\dot{z})^2}ds$ is a consequence of the original null cone equation (2), there is an analogy with a theorem, cited in the review paper [15] and proved in the monograph by Fock [22].The theorem is the following:

Let $F = \frac{1}{2}g_{\alpha\beta}\frac{dx^\alpha}{ds}\frac{dx^\beta}{ds}$, $F = 0$ is the null cone equation and the extremum of the integral $s = \int_{s_1}^{s_2} Lds$ is searched ($L = \sqrt{2F}$), then the Lagrange equation of motion will give

$$\frac{dF}{ds} = 0 \quad \Longrightarrow F = const \Longrightarrow \frac{d}{ds}\frac{\partial F}{\partial \dot{x}_\alpha} - \frac{\partial F}{\partial x_\alpha} = 0 \quad . \tag{50}$$

From the last equation the geodesic equations $(\alpha, \beta, \nu = 0, 1, 2, 3)$

$$\frac{d^2x_\nu}{ds^2} + \Gamma^\nu_{\alpha\beta}\frac{dx_\alpha}{ds}\frac{dx_\beta}{ds} = 0 \tag{51}$$

can be obtained. Since the constant in (50) can be set up to zero, this would mean that the geodesic equation (51) is compatible with the null-cone equation $F = \frac{1}{2} g_{\alpha\beta} \frac{dx^{\alpha}}{ds} \frac{dx^{\beta}}{ds} = 0$. It is important to acquire the proper understanding of this statement - it means that every solution of the null equation (including in the proper parametrization of the differentials, related to the trajectory of motion of the satellite) is also a solution of the light-like geodesic equation. In the "backward direction", however the statement is not true, meaning that there might be another solutions of the geodesic equation, which are not solutions of the null-cone equation. This problem needs further investigation.

The meaning of the geodesic as the shortest path, joining two given points (but only in the local sense, i.e a sufficiently short part of the geodesics) is confirmed also by the following theorem in the monograph by Prasolov [32]. Let us present the exact formulation of this theorem:

Theorem 3 *Any point p on a surface S has a neighborhood U such that the length of a geodesic γ going from the point p to some point q and lying entirely in the neighborhood U does not exceed the length of any curve α on S, joining the points p and q. Moreover, if the lengths of the curves γ and α are equal, then γ and α coincide as non-parametrizable curves.*

In fact, what does it mean that "the length of the geodesics from point p to point q does not exceed the length of any other curve" (but only locally, i.e. in the vicinity of the given neighborhood)? This means that this geodesic line is minimal. So this confirms the theorem in another, third well-known book on differential geometry by Mishtenko and Fomenko [36]:

Theorem 4 *The geodesic line $\gamma : [a, b] \rightarrow M^n$ is minimal if it is not longer than any smooth path, joining the endpoints $\gamma(a)$ and $\gamma(b)$.*

So if we have for example the functional $E(\gamma(s)) = \int \mid \dot{\gamma}^2(s) \mid ds = \int g_{ij}(x)\dot{x}^i(s)\dot{x}^j(s)ds$ for the null cone equation, and we are searching for the extremals, these extremals will be derived from the geodesics- the multitude of the solutions for the extremals from the geodesics turns out to be larger in comparison with the one for the other local curves in the vicinity, in particular for the functional $E(\gamma(s))$ for the null-cone equation.

On the base of combining the results in the previously mentioned monographs by Fock [22] and [32], one easily establishes the validity of the following theorem in [36]:

Theorem 5 *The extremals of the length functional $L(\gamma(s)) = \int \sqrt{g_{ij}(x)\dot{x}^i(s)\dot{x}^j(s)}ds$ are smooth functions, derived from the geodesics (by means of smooth change of the parameter along the curve). Then each extremal of the functional $E(\gamma(s)) = \int \mid \dot{\gamma}^2(s) \mid ds = \int g_{ij}(x)\dot{x}^i(s)\dot{x}^j(s)ds$ (called also action functional of the trajectory) is also an extremal of the functional $L(\gamma(s)) = \int \sqrt{g_{ij}(x)\dot{x}^i(s)\dot{x}^j(s)}ds$. However, each extremal of the functional $L(\gamma(s))$ is not necessarily an extremal of the functional $E(\gamma(s))$.*

There is a simple inequality $L^2(\gamma(s)) \leq E(\gamma(s))$ [36] in confirmation of the above theorem, which can be immediately proved by means of the Schwartz inequality

$$\left(\int_0^1 f(s)g(s)ds \right) \leq \left(\int_0^1 f^2(s)ds \right) \left(\int_0^1 g^2(s)ds \right) \tag{52a}$$

for the following functions [36]

$$f(s) \equiv 1 \quad g(s) \equiv \mid \dot{\gamma}(s) \mid \quad . \tag{53}$$

It should be noted that the parameter s in the book [22] is defined simply as a "space parameter along a curved line". In the above formulaes, s is also a space parameter. But as noted in [36], when $g(s) = const$, i.e. $\gamma(s) = const$ and the parameter s is proportional to the length of the arc of the curved line, an equality can be derived $L^2(\gamma(s)) = E(\gamma(s))$.

4.3 Physical interpretation of the obtained constancy of the Euclidean distance and the length function as a result of the applied variational formalism

The constancy of the length function $L_1 = \sqrt{(\dot{x})^2 + (\dot{y})^2 + (\dot{z})^2}$ and of $R_{AB}(x,y,z)$ is non-acceptible from a physical point of view, because the signal trajectory should not be a straight line due to the General Relativity effects of curving the trajectory. Of course, for the moment we have minimized only the first term $\frac{1}{c}\int\limits_{path} \sqrt{(\dot{x})^2 + (\dot{y})^2 + (\dot{z})^2}ds$ in the action functional (35) for the generalized Shapiro delay formulae. Due to this reason it is necessary to investigate also the second term in (33) and (35).

However, investigation of the second term will not give an answer to some fundamental questions, which naturally appear, concerning the Shapiro delay formulae (1). On purely physical considerations it is a very crude approximation to assume that there will exist a (even not so small) space-time area, where the light (or radio or laser) ray will move along a straight line (the first term $\frac{R_{AB}}{c}$ in (1) and so, it will not experience any influence from the gravitational field. This statement follows of course from the Equivalence Principle in General Relativity Theory. However, the Equivalence Principle is local in its essence, while in the original derivation of the Shapiro delay formulae (1) the distance R_{AB} is not small. The approximation, from a fundamental point of view is that the gravitational field causes only a small deviation from the straight line trajectory of the signal. In the papers [16] and [17] a formulae was obtained for the square of the geodesic distance $\widetilde{R}^2_{AB}$ as compared to the square of the Euclidean distance R^2_{AB} (see also formulae (163) in the review [15])

$$\widetilde{R}^2_{AB} - R^2_{AB} = -\frac{1}{2}(a_1^2 + a_2^2) + e_1 e_2 a_1 a_2$$
$$-\frac{1}{4}(a_1^2 e_1^2 + a_2^2 e_2^2) + 2a_1 a_2\sqrt{(1-e_1^2)(1-e_2^2)} \quad . \tag{54}$$

Although only for the partial case of two satellites on two plane elliptical orbits, in [16] and [17] it was proved that the right - hand side of (54) is positive, consequently $\widetilde{R}^2_{AB} > R^2_{AB}$ and $\widetilde{R}^2_{AB}$ really represents the geodesic distance, travelled by the signal. What is also interesting is that the last 4 numerical terms depend only on constant parameters e_1, e_2, a_1, a_2 and not on the dynamical parameters - the eccentric anomaly angles E_1 and E_2. This means that these terms describe a deviation from the straight line and thus are not related to the process of "curving" the trajectory, which will be described by the dynamical parameters E_1 and E_2 (if this was the case)

If not in the infinitesimal sense and according to formulae (??) $l(s) = \int\limits_{s_0}^{s} \mid \overrightarrow{\mathbf{R}}_{AB}(s) \mid ds$, the distance along the curve $\overrightarrow{\mathbf{R}}_{AB}(s) = (\dot{x}(s), \dot{y}(s), \dot{z}(s))$ will be greater than the distance R_{AB} between the points on the straight line. So the "non-local" character of the action of the gravitational field will be more correctly "approximated" by the term $\frac{1}{c}\int\limits_{path} \sqrt{(\dot{x})^2 + (\dot{y})^2 + (\dot{z})^2}ds$ rather than the simple term $\frac{R_{AB}}{c}$. In support of this assertion, the following citation from the monograph by Fock [22] can be reminded: "The essence of the Equivalence Principle is that by means of introducing a locally-geodesic (freely-falling) reference system, the Galilean space-time is restored in an infinitely small space. However, in no way this enables one to make any conclusions how to discern acceleration and gravitation in a finite space-time area."

It has been also asserted by Fock [22] that: "In any non-local interpretation the approximate equivalence of the gravitational field and the acceleration is restricted. This "equivalence" can take place only for weak and homogeneous gravitational fields and also for slow motions." The reason is that the formulae (21) $dT \approx \frac{1}{c}\int\limits_{path} \sqrt{\delta_{ij}dx^i dx^j} + \frac{1}{c^3}\int\limits_{path} 2U\sqrt{\delta_{ij}dx^i dx^j}$, from which we started for the derivation of the modified Shapiro formulae, in its original form has been derived under the assumption $\frac{U}{c^2} \ll 1$. It is equivalent to the following two inequalities [22]

$$U \ll c^2 \quad , \tag{55}$$

$$\left(\frac{dx}{dt}\right)^2 + \left(\frac{dy}{dt}\right)^2 + \left(\frac{dz}{dt}\right)^2 = v^2 \ll c^2 \quad . \tag{56}$$

Under the fulfillment of these inequalities, from the original metric (in fact, this is the metric, used in this paper, from which the null cone equation (2) was derived)

$$ds^2 = (c^2 - 2U)\,(dt)^2 - (1 + \frac{2U}{c^2})((dx)^2 + (dy)^2 + (dz)^2) \quad , \tag{57}$$

only the approximate metric

$$ds^2 = (c^2 - 2U)\,(dt)^2 - (1 + \frac{2U}{c^2})\left((dx)^2 + (dy)^2 + (dz)^2\right) \tag{58}$$

can be derived, and this is the mathematical expression of the approximation, pointed out by Fock [22]. In our notations, instead of "t" we have written the propagation time T. Now let us see whether it will be correct to keep the physical meaning of (56) as the Newtonian velocity. If this is so, (56) can be written as $v^2 = \left[(\dot{x})^2 + (\dot{y})^2 + (\dot{z})^2\right](\frac{ds}{dT})^2 \ll c^2$, or

$$\left[(\dot{x})^2 + (\dot{y})^2 + (\dot{z})^2\right] \ll c^2 \left(\frac{dT}{ds}\right)^2 \quad . \tag{59}$$

But $\frac{dT}{ds}$ is the function $F(x, y, z, \dot{x}, \dot{y}, \dot{z}) = \frac{1}{c}\sqrt{(\dot{x})^2 + (\dot{y})^2 + (\dot{z})^2} + \frac{2G_{\oplus}M_{\oplus}}{c^3}\frac{\sqrt{(\dot{x})^2+(\dot{y})^2+(\dot{z})^2}}{[x^2+y^2+z^2]}\left[x\dot{x} + y\dot{y} + z\dot{z}\right]$ in (36) and neglecting the second term in the expansion for c^2F^2, which is of order $O(c^{-4})$, we obtain

$$1 \ll 1 + \frac{G_{\oplus}M_{\oplus}}{c^2}\frac{d}{ds}\ln(x^2 + y^2 + z^2) \quad . \tag{60}$$

This is however not realistic, because the second term should be large number. Since this cannot be the case, since numerous calculations have proved that only because of the coefficient ,$\frac{G_{\oplus}M_{\oplus}}{c^3} \approx 10^{-9}$ (see the paper [14] and the cited papers) it turns out that the definition (56) is incorrect. The reason is that the velocity should be defined only as a tangent vector to the curve, and then the locality condition of Fock should be defined as

$$(\dot{x})^2 + (\dot{y})^2 + (\dot{z})^2 \ll c \quad . \tag{61}$$

Then if the second term in (36) is much less than the first one and inequality (61) is fulfilled, then from

$$\frac{2G_{\oplus}M_{\oplus}}{c^3}\frac{\sqrt{(\dot{x})^2 + (\dot{y})^2 + (\dot{z})^2}}{[x^2 + y^2 + z^2]}\left[x\dot{x} + y\dot{y} + z\dot{z}\right] \ll \frac{1}{c}\sqrt{(\dot{x})^2 + (\dot{y})^2 + (\dot{z})^2} \tag{62}$$

it will follow that $\frac{2G_{\oplus}M_{\oplus}}{c^3}\frac{\sqrt{(\dot{x})^2+(\dot{y})^2+(\dot{z})^2}}{[x^2+y^2+z^2]}\left[x\dot{x} + y\dot{y} + z\dot{z}\right] \ll 1$ or

$$\frac{G_{\oplus}M_{\oplus}}{c^3}\frac{d}{ds}\ln(x^2 + y^2 + z^2) \ll \frac{1}{c} \quad . \tag{63}$$

This is much more realistic and compatible with the fact that the second Shapiro delay term should be a very small one.

5 Differential geometry notions about manifolds with boundary - basic definitions and related notions and new results about the boundaries of the manifold (5)-(7) of the celestial mechanics transformations

The application of the Shapiro formulae in this paper is based on the transformations (5) - (7) for the space coordinates x, y, z, depending on the 6 Kepler parameters. These transformations have not been investigated from a differential geometry point of view, and they have interesting geometrical properties, depending on the fact which variables will be considered dynamical ones and which will be kept constant in the process of signal propagation from one satellite to another. For example, if only the two true anomaly angles f_1 and f_2 are the dynamical parameters, we will have the mapping $F_1 : (f_1, f_2) \rightarrow (x, y, z)$ (again, here we denote $x = x_2 - x_1, y = y_2 - y_1, z = z_2 - z_1$), which is a mapping from a two-dimensional manifold to a three-dimensional manifold. We can define, however, also the mapping $F_2 : (f_1, \Omega_1, \omega_1, i_1, a_1, e_1; f_2, \Omega_2, \omega_2, i_2, a_2, e_2) \rightarrow (x, y, z)$ or $F_2 : (f_1, \Omega_1, \omega_1, i_1, a_1, e_1; f_2, \Omega_2, \omega_2, i_2, a_2, e_2) \rightarrow$

$(x_1, y_1, z_1; x_2, y_2, z_2)$-these are mappings from a 12− dimensional manifold to a three-dimensional manifold (x, y, z) or to a six-dimensional manifold. So in the first case the mapping F_1 is from a lower-dimensional (2−dimensional manifold) to a higher-dimensional manifold, while in the second case of the mapping F_2 the mapping is from a higher-dimensional to a lower-dimensional manifold. These different kinds of mappings are related to the notions of immersions, submersions, embeddings, which will be clarified in the next sections and are related to the Stokes and the Gauss - Ostrogradsky theorems.

5.1 Basic definitions about the boundary of a manifold and some relations to other definitions
One of the basic definitions is given in the textbook [36]:

Definition 6 *A manifold M with a boundary is characterized by the property that there exists an atlas $\{U_\alpha\}$ with local coordinates $(x^1_\alpha, x^2_\alpha, ..., x^n_\alpha)$ so that for each map the strict inequality $x^n_\alpha > 0$ is fulfilled for the inner points and the equality $x^n_\alpha = 0$ is fulfilled at the boundary points. The set of boundary points ∂M is a smooth manifold with a dimension $(n-1)$ (i.e. one dimension less than the dimension of the original manifold).*

In the more strict mathematical sense, a manifold with a boundary is a metric space M, for which an atlas of charts $\{U_\alpha\}$ and coordinate homeomorphisms $\varphi_\alpha : U_\alpha \to V_\alpha \subset \mathbb{R}^n_+$ (V_α is an open set in $\mathbb{R}^n_+$ and $\mathbb{R}^n_+$ is the semi-space, denoted as $x^n \geq 0$, $(x^1, x^2, ...x^n) \in \mathbb{R}^n$) exist such that the functions for coordinate changes

$$\varphi_\beta \varphi_\alpha^{-1} : V_{\alpha\beta} = \varphi_\alpha(U_\alpha \cap U_\beta) \to V_{\beta\alpha} = \varphi_\beta(U_\alpha \cap U_\beta) \tag{64}$$

are smooth functions of class C^r, $r = 1, 2, ...\infty$.

This definition however does not reveal the relation of the boundary of the manifold with the inverse function theorem as a basic requirement for the existence of the boundary. This means that if a vector function $f : R^n \to R^k$ is defined, then it should satisfy the inverse function theorem as a necessary requirement (condition) for the existence of a boundary. [36] Namely, if at all points $P \in M = \{f(P) = 0\}$, the differential df has a maximal rank.

This means also that if the system of $n-$equations is investigated

$$f^1(x^1, x^2...x^n) = 0 \;.. \;, \quad f^{k-1}(x^1, x^2...x^n) = 0 \;, \; .... f^k(x^1, x^2...x^n) = 0 \tag{65}$$

and $\overline{M}$ is the set of solutions of this system, and if also at each point $P \subset. \ \overline{M}$ the rank of the determinant $\| \frac{\partial f^i}{\partial x^s} \|$, $1 < i \leq k-1$, $1 \leq s \leq n$ is equal to $(k-1)$, then $\overline{M}$ is a manifold with a boundary.

Now let us give a simple example about a manifold with a boundary - this is the closed ball $D^m \subset R^m$, defined by the inequality $(x^1)^2 + (x^2)^2 + (x^m)^2 \leq 1$. The $m-$cube $[0, 1]^m$ is homeomorphic to D^m and is thus also a manifold with a boundary.

5.1.1 Definition for a boundary of a manifold in terms of semi-spaces The definition in another monographs, for example in [37] implies a slightly different formulation.

Definition 7 *A manifold with a boundary is a Hausdorff space M with a countable basis of open sets and the subspace of $\mathbb{R}^n$*

$$H^n = \{x = (x_1, x_2, ...x_n) \in R^n) \mid x^n \geq 0\} \tag{66}$$

with the relative topology of R^n and

$$\partial H^n = .\{x \in H^n \mid x^n = 0\} \tag{67}$$

denotes the boundary of H^n. The boundary ∂H^n carry the metric topology derived from the metric of R^n and ∂H^n is homeomorphic to R^{n-1} by the map

$$(x_1, x_2, ...x_{n-1}) \mapsto (x_1, x_2, ..., x_{n-1}, 0) \tag{68}$$

.The diffeomorphisms of open sets of H^n take boundary points to boundary points and interior points to interior points. This means also that by means of the homeomorphisms φ_n a point $p \subset M$ is contained either 1. In an open set U of $H^n - \partial H^n$ or 2. to an open set U of H^n with $\varphi(p) \subset .\partial H^n$ (i.e. a boundary point of H^n) The set of points in $H^n - \partial H^n$ is in fact $\{x = (x_1, x_2, ...x_n) \in R^n) \mid x^n \geq 0\}$ and in [38] it is called "a closed half-space". This is the set of vectors $x \in R^n$, satisfying the positive vector multiplication $x.y \geq 0$, where $y \in R^m$ is a fixed non-zero vector. Now one can see the close similarity of the definition for $H^n - \partial H^n$ or H^n with $x^n \geq 0$ with the definition of convex spaces, which exist in optimization problems [24]. In [24] sets of the kind $H_{p\beta} = \{x \in R^n \mid < p.x >= \beta\}$ are called hyperplanes and they cause the semi-spaces

$$H^{+}_{p\beta} = \{x \in R^n \mid < p.x > \ \geq \beta\} \quad , \quad H^{--}_{p\beta} = \{x \in R^n \mid < p.x > \ \leq \beta\} \quad . \tag{69}$$

Optimization problems are investigated in mathematics and apart from gravity theory, but in the previous sections it was demonstrated that they may appear in problems, related to the optimal configuration for the propagation of a signal between two moving satellites on different space-distributed orbits, with account also of the General Relativity effect of curving the trajectory of the signal.

5.1.2 The notion of convexity in differential geometry and its relation to the notion of a boundary It is very interesting that the notion of convexity appears not only in optimization theory, but also in differential geometry and it turns out that it is related also to the notion of a boundary. In the course of Thorpe [25] on differential geometry the following definition is given for a convex surface

An oriented $n-$dimensional hyper-surface S in $R^{n+1}$is called convex (or a convex one as a whole), if at each point $p \in S$ the hyper-surface is contained in one of the following semi-spaces

$$H^{+}_{p} = \{q \in R^{n+1} \quad : (q-p).N(p) \geq 0\} \quad , \tag{70}$$

or

$$H^{-}_{p} = \{q \in R^{n+1} \quad : (q-p).N(p) \leq 0\} \quad . \tag{71}$$

On the base of this definition in [25], another definition about a strictly convex surface is given, which evidently has a relation to the notion of a boundary, given for example by (67) $\partial H^n = . \{x \in H^n \mid x^n = 0\}$

Definition 8 *If the surface S is convex at each point p and $S \cap H_p = \{p\}$ at each point $p \in S$, where*

$$H_{p} = \{q \in R^{n+1} \quad : (q-p).N(p) = 0\} \quad , \tag{72}$$

then the hyper-surface S is called strictly convex.

The vector $N(p)$ is an important characteristic of the surface and is called the Gaussian mapping [25]. This is a mapping $N : S \rightarrow S^n$, which juxtaposes to each point $p \in S$ a point in R^{n+1}, derived after the translation of the unit normal vector p to the origin of the coordinate system. The image in the Gaussian mapping $N(s) = \{q \in S^n : q = N(p)\}$ is called also the spherical image of the oriented surface.

By means of $N(p)$, two other characteristics are defined [25].

Definition 9 *The Weingarten mapping $L_p : S_p \rightarrow S_p$ is determined as $L_p(v) := -\nabla_v N$, $\overrightarrow{v} \in S_p$. It measures the turning angle of the normal to S at the point p, moving at different speeds v and thus L_p is a measure of the curvature of the space R^{n+1} at the point p.*

The second characteristics is the s.c. normal curvature of S at the point p in the direction of the vector $\overrightarrow{v}$.

Definition 10 *In the case of $\| \overrightarrow{v} \| = 1$, the normal curvature of S at the point p in the direction of the vector $\overrightarrow{v}$ is defined as the number $k(p) := \overrightarrow{L_p}(v).\overrightarrow{v}$ and it represents the normal component of the acceleration at the point p, which is caused by the curvature of S in R^{n+1}. This means that if $k(\overrightarrow{v}) > 0$, then the surface S is "curving" in the direction of $\overrightarrow{N}$ and in the direction of $\overrightarrow{v}$, if $k(\overrightarrow{v}) < 0$ the surface S is curving in the backward direction of $\overrightarrow{N}$ and in the direction of $\overrightarrow{v}$.*

5.2 New result - calculation of the boundary manifold for the 3D manifold of the celestial mechanics transformations (5)-(7)

Further the Stokes theorem will be used, which for the concretely investigated case will relate an integral on a $3D$ space M to an integral on the two-dimensional boundary ∂M. Since the boundary manifold is one dimension less than the original $3D$ manifold, it will be in terms of the variables (x, y) (the first boundary, when $z = 0$), or in terms of the variables (x, z) (the second boundary, when $y = 0$) or in terms of the variables (y, z) (the third boundary, when $x = 0$).

5.2.1 The first boundary in terms of (x, y) and $z = 0$ In order to find the boundary, one should combine the two equations (5) for x and (6) for y so that to obtain a relation between the coordinates. Let us multiply both sides of (5) with $\cos\Omega$ and of equation (6) with $\sin\Omega$ and sum up the two equations. The obtained equation will be

$$x\cos\Omega + y\sin\Omega = r\cos(\omega+f) \quad . \tag{73}$$

Similar equation may be obtained if (5) for x is multiplied by $\sin\Omega$ and (6) for y by $(-\cos\Omega)$ and the two equations are summed up. The resulting equation will be

$$x\sin\Omega - y\cos\Omega = -r\sin(\omega+f)\cos i \quad . \tag{74}$$

Next, according to the definition $\partial H^n = .\{x \in H^n \mid x^n = 0\}$ in (67), we have to find when the condition

$$z = r\sin(\omega+f)\sin i = 0 \tag{75}$$

is fulfilled. The first case is when

$$\sin i = 0 \quad \Rightarrow i = \pi k = \pi(m+1) \tag{76}$$

and $\sin(\omega+f) \neq 0$ is also fulfilled. Then when

$$k = 0, 2, 4....... \Rightarrow k = m+1 \;\Rightarrow\; m = 1, 3..........k \text{ - even; } m\text{-odd} \tag{77}$$

we have for

$$\cos i = \cos(\pi(m+1)) = (-1)^{m+1} \quad . \tag{78}$$

Indeed, for odd numbers $m = 1, 3.....(2p+1)$, $\cos(\pi(m+1)) = (-1)^{m+1} = (-1)^{2p+2} = 1$, for even numbers $m = 2, 4, 6....2p$ one has $\cos(\pi(m+1)) = (-1)^{m+1} = (-1)^{2p+1} = -1$, which can be checked. Note for $k = 0$ one cannot define the corresponding value for m, then $\cos(\pi.0) = \cos 0 = 1$. However, instead of $i = \pi k = \pi(m+1)$ in (76) one can also define $i = \pi k = \pi(m-1)$ - then $k = 0$ will correspond to the value $m = 1$.

The other condition $\sin(\omega+f) \neq 0$ will be fulfilled when similarly to (76) one has

$$\sin(\omega+f) \neq 0 \quad \Rightarrow \omega + f \neq \pi k = \pi(m \pm 1) \quad . \tag{79}$$

Next, dividing the two equations (73) and (74) and making use of (78), one can derive the following simple equation

$$\frac{x\cos\Omega + y\sin\Omega}{x\sin\Omega - y\cos\Omega} = -\frac{tg(\omega+f)}{(-1)^{m+1}} \quad . \tag{80}$$

For odd numbers $m = 1, 3.....(2p+1)$ after some transformations and also using the simple trigonometric equalities

$$\cos(\omega+f+\Omega) = \cos(\omega+f)\cos\Omega - \sin(\omega+f)\sin\Omega \quad , \tag{81}$$

$$\sin(\omega+f+\Omega) = \sin(\omega+f)\cos\Omega + \cos(\omega+f)\sin\Omega \quad , \tag{82}$$

the equation (80) can be rewritten as

$$x\frac{\sin(\omega+f+\Omega)}{\cos(\omega+f)\cos\Omega} - y\frac{\cos(\omega+f+\Omega)}{\cos(\omega+f)\cos\Omega} = 0 \quad . \tag{83}$$

Provided that the inequalities $\cos(\omega+f) \neq 0$ and also $\cos\Omega \neq 0$ the last expression represents the equation of the straight line

$$x\sin(\omega+f+\Omega) - y\cos(\omega+f+\Omega) = 0 \quad . \tag{84}$$

The two inequalities can be fulfilled when $\omega + f \neq (2q-1)\frac{\pi}{2}$, $\Omega \neq (2p-1)\frac{\pi}{2}$. The numbers p and q can be odd or even, i.e. $p, q = 0, 1, 2, ...$ It is easy to check that $(2q-1)\frac{\pi}{2} = 0, -\frac{\pi}{2}, \frac{\pi}{2}, \frac{3\pi}{2}, \frac{5\pi}{2}$.......for $q = 0, 1, 2, 3.......$ and also $(2p-1)\frac{\pi}{2} = 0, -\frac{\pi}{2}, \frac{\pi}{2}, \frac{3\pi}{2}, \frac{5\pi}{2}$.......for $p = 0, 1, 2, 3$, correspondingly when $\omega + f \neq (2q-1)\frac{\pi}{2}$ and $\Omega \neq (2p-1)\frac{\pi}{2}$, $\cos(\omega+f) \neq 0$ and $\cos\Omega \neq 0$ will be fulfilled.

Similarly, for even numbers $m = 2, 4, 6....2p$......it can easily be proved that the analogous to (84) equation will be

$$x\sin(\omega+f-\Omega) + y\cos(\omega+f-\Omega) = 0 \quad , \tag{85}$$

Journal of Physics: Conference Series **2910** (2024) 012001 doi:10.1088/1742-6596/2910/1/012001

where the analogous to (81) trigonometric equality

$$\cos(\omega + f - \Omega) = \cos(\omega + f)\cos\Omega + \sin(\omega + f)\sin\Omega \quad , \tag{86}$$

has been used. The second case when $z = r\sin(\omega + f)\sin i = 0$, i.e. $\sin(\omega + f) = 0$ should be considered in the same way. The calculations shall not be performed here. Note also that the number m can be either odd or even, consequently either equation (84) or equation (85) is fulfilled, but not both, which means that the corresponding lines as boundaries do not intersect.

5.2.2 The second boundary in terms of (x, z) and $y = 0$ It might be expected that the calculation for the second boundary (x, z) is of the same type as for the previous case. Surprisingly, the calculation turns out to have some peculiarities and the derived expressions will be more complicated.

Let us multiply equation (5) for x by $(\sin i)$ and (7) for z by $(\cos i \sin\Omega)$ and then sum up the two equations. The obtained equation will be

$$x\sin i + \cos i\,(\sin\Omega)\;\; z = r\cos\Omega\cos(\omega + f)\sin i \quad . \tag{87}$$

Another equation can be obtained if (5) is multiplied by $(\sin(\omega + f)\sin i)$ and (7) is multiplied by $(-\cos\Omega\cos(\omega + f))$ and the resulting equations are summed up

$$\sin(\omega + f)\sin i\; x - \cos\Omega\cos(\omega + f) z$$

$$= r\sin\Omega\sin^2(\omega + f)\cos I\sin i \quad . \tag{88}$$

After dividing the two equations the following equation is obtained

$$\frac{x\sin i + \cos i\sin\Omega\; z}{\sin(\omega + f)\sin i\; x - \cos(\omega + f)\cos\Omega\; z} = -\frac{\cot g\Omega\cot g(\omega + f)}{\cos i\sin(\omega + f)} \quad . \tag{89}$$

Now we set up $y = 0$ in expression (6) for y, express $\cot g\Omega$ as

$$\cot g\Omega = -\frac{1}{tg(\omega + f)\cos i} \tag{90}$$

and substitute it in the preceding equation (89)

$$x\;.\sin i\frac{\left[\sin^2(\omega + f)\left(1 + \cos^2 i\right) - 1\right]}{\sin^2(\omega + f)\cos^2 i}$$

$$+z\;\frac{\cos\Omega\left[1 - \cos^4 i\; tg^4(\omega + f)\right]}{\cos^2 i\; tg^3(\omega + f)} = 0 \quad . \tag{91}$$

This is again an equation for a straight line, but in terms of the variables x and z and with complicated coefficient functions in terms of trigonometric functions. It can be proved that they contain functions with

$\cos 2(\omega + f)$, $\cos 2i$, $\cos 4i$.

5.2.3 The third boundary in terms of (y, z) and $x = 0$ The equation (6) for y is multiplied by $[\sin(\omega + f)\sin i]$, equation (7) for z is multiplied by $[-\sin\Omega\cos(\omega + f)]$ and after the obtained equations are summed up, the following equation is obtained

$$y\sin(\omega + f)\sin i - z\;\cos(\omega + f)\sin\Omega$$

$$= r\cos\Omega\sin^2(\omega + f)\sin i\cos i \quad . \tag{92}$$

The second equation is obtained after (6) for y is multiplied by $[\sin i]$, (7) for z is multiplied by $[-\cos i\cos\Omega]$ and after the obtained equations are summed up, the obtained equation is

$$y\sin i - z\;\cos i\cos\Omega = r\sin\Omega\sin i\cos(\omega + f) \quad . \tag{93}$$

Now setting up equation (5) for x to zero, i.e. $x = 0$, one can obtain

$$\cos i = \cot g\Omega \cot g(\omega + f) \quad . \tag{94}$$

One can calculate also $\sin i = \sqrt{1 - \cos^2 i}$ by using the trigonometric identities (82) for $\cos(\omega + f + \Omega)$ and (86) for $\cos(\omega + f - \Omega)$. The result is

$$\sin i = \frac{\sqrt{-\cos(\omega + f + \Omega)\cos(\omega + f - \Omega)}}{\sin\Omega \sin(\omega + f)} \quad . \tag{95}$$

After dividing the equations (92) and (93) and substituting the obtained expressions (94) for $\cos i$ and (95) for $\sin i$, after some transformations the following equation is obtained

$$\frac{\sin(\omega + f)\left[\sin^2\Omega - \cos^2\Omega\right]}{\sin^2\Omega}\left[Ny - z\cos(\omega + f)\right] = 0 \quad , \tag{96}$$

where we have denoted

$$N(\omega, f, \Omega) \equiv \sqrt{-\cos(\omega + f + \Omega)\cos(\omega + f - \Omega)} \quad . \tag{97}$$

In order to obtain the equation for the straight line

$$Ny - z\cos(\omega + f) = 0 \tag{98}$$

in (y, z) coordinates, one should have also the fulfillment of the three conditions, so that the straight line equation (98), obtained from (96) is correct

$$\sin(\omega + f) \neq 0 \quad , \quad \sin^2\Omega \neq 0, \;\; \sin^2\Omega - \cos^2\Omega \neq 0 \quad . \tag{99}$$

The fulfillment of the first condition was established in (79), the second condition is investigated analogously, the third condition means that

$$\sin^2\Omega - \cos^2\Omega = 2\sin^2\Omega - 1 \neq 0 \quad \Rightarrow \sin\Omega \neq \pm\frac{\sqrt{2}}{2}, \text{ i.e. } \Omega \neq 45^\circ \quad . \tag{100}$$

Much more nontrivial is the problem whether the expression $N(\omega, f, \Omega)$ in (97) should be real-valued or complex. On physical grounds, for the moment one cannot assert anything about the real-valuedness or imaginary value of $N(\omega, f, \Omega)$. However, on purely geometrical grounds, it is possible for $N(\omega, f, \Omega)$ to be imaginary - this can happen when both $\cos(\omega + f + \Omega)$ and $\cos(\omega + f - \Omega)$ are positive or when they are both negative. Then there will be the imaginary straight line $i\overline{N}y - z\cos(\omega + f) = 0$ - it will be one of the imaginary intersecting lines (the other is $i\overline{N}y + z\cos(\omega + f) = 0$) on the cone $\overline{N}^2 y^2 + z^2\cos^2(\omega + f) = 0$. A theorem, proved in the monograph [26] states that there exist exactly 5 classes of projective equivalence among the lines of second order, namely:

1. a real-valued oval $x_1^2 + x_2^2 - x_3^2 = 0$;
2. a complex oval $x_1^2 + x_2^2 + x_3^2 = 0$;
3. intersecting straight lines $x_1^2 - x_2^2 = 0$;
4. imaginary intersecting straight lines $x_1^2 + x_2^2 = 0$;
5. coinciding straight lines $x_1^2 = 0$.

6 Differential geometry notions about immersions, embeddings, submersions, critical and regular points and Morse functions - a review of the different definitions and some new applications in the current problem

We have already defined the notion of a boundary of a manifold, which represents in fact a submanifold with dimension $(n - 1)$ of the original $n-$dimensional manifold. As clarified in the preceding chapters, a new variational formalism will be applied, based on the Stokes and Gauss-Ostrogradsky theorems. These theorems relate the integration on a $n-$manifold to an integration on the boundary. But in fact, the notion of the boundary manifold, concrete examples about which were given, are closely related with some other notions in differential geometry about immersions, submersions, embeddings, also critical and regular points and Morse functions.

6.1 Immersions and the sub-manifold property

Before giving the strict definitions, let us first define what is the rank (denoted by $rank\varphi$) of the mapping $\varphi : N \rightarrow M$ between two manifolds-this is simply the dimension of the manifold N at all its points. The essence of this assertion is given in the following

Definition 11 *[27] Let the mapping $\varphi : M_1 \rightarrow M_2$ is a differentiable mapping with (U, h)-coordinate charts in the neighborhood of the point $\varphi(p)$, where $p \in M_1$. The differentiable functions on M_1 are denoted by $y^i \circ \varphi$. Let h is a local chart in M_2, g is a local chart in M_1 and in terms of the coordinate chart $(V, g = (x^1, x^2, ...x^{n_1}))$ the mapping $\varphi \equiv h \circ \varphi \circ g^{-1}$ is given by the transformation*

$$y^i = \varphi^i(x^1, x^2, ...x^{n_1}) \qquad (i = 1, 2, ...n_2) \quad . \tag{101}$$

Then the rank of the mapping φ is equal to n_1, if the rank of the Jacobian $(n_1 \times n_2)$ matrix $\left(\frac{\partial \varphi^i}{\partial x^j}\right)_{(x^1, x^2, ...x^{n_1})}$ is equal to n_1.Such a case, when at all points of the manifold M_1 the $rank\varphi = n_1$ (the dimension of the manifold M_1) and when the dimension of the manifold M_2 is $n_2 \geq n_1$ is called an immersion. (see also [37]) The subset $\varphi(M_1) = \overline{M}_1 \subset M_2$ is called an immersed sub-manifold, but the peculiar fact is that it depends on both manifolds M_1 and $\overline{M}_1$- $\overline{M}_1$ is not even a subspace of M_2 (but, can be called a subset). Thus it is not possible for the mapping φ and its inverse one φ^{-1} to be one-to-one, even in the local sense.

Naturally the following question appears: how can the image subset $\overline{M}_1$ be endowed with the property to be a sub-manifold?

Definition 12 *[37] The subset $\overline{M}_1$ can be a sub-manifold if it is the image in M_2 of a one-to-one immersion*

and $\varphi : M_1 \mapsto \overline{M}_1$ establishes an one-to-one correspondence between M_1 and the subset $\varphi(M_1) = \overline{M}_1$, i.e. the mapping $\varphi : M_1 \rightarrow \overline{M}_1$ is a diffeomorphism. Now $\overline{M}_1$ is called an immersed sub-manifold $\overline{M}_1 \in M_2$ and is endowed with the topology and C^∞ (infinitely differentiable) structure of the manifold M_2.

Let us remind also that a mapping φ is diffeomorphism if φ and its inverse one φ^{-1} are smooth. [28].

The sub-manifold property of $\overline{M}_1$ also is related to the existence of preferred coordinates relative to $\overline{M}_1$ [37]. This means that each point $p \in \overline{M}_1$ has a coordinate neighborhood U, φ on M_2 such that 1. $\varphi(p) = (0, 0......0)$; 2. $\varphi(U) = C^m_\epsilon(0)$ and 3. $\varphi(U \cap \overline{M}_1) = \{x \in C^m_\epsilon(0) \mid x^{n_1+1} = ...x^{n_2} = 0\}$. The last equality means that the immersion $\varphi : M_1 \rightarrow M_2$ on $V = U \cap \overline{M}_1$ in terms of the preferred local coordinates will be given by

$$(x^1, x^2, ...x^{n_1}) \rightarrow (x^1, x^2, ...x^{n_1}, 0, 0, ...0) \tag{102}$$

and the zeroes correspond to $x^{n_1+1} = ...x^{n_2} = 0$.

Evidently, immersions and embeddings are related to the different notions of diffeomorphism and homeomorphism. But in some sense they can also be related - if $\varphi : M_1 \mapsto M_2$ is an immersion, then each point $p \in M_1$ has a neighborhood U such that $\varphi \mid U$ (the restriction of the mapping φ on U) is an embedding of U in M_2 [37].

An example about an immersion is the mapping $F : R \rightarrow R^3 : \ F(t) = (\cos 2\pi t, \sin 2\pi t, t)$, representing the helix, lying on the unit cylinder, whose axis is the x_3-axis in R^3 [37]. The equation of the helix is $\cos^2 2\pi t + \sin^2 2\pi t = 1 \ \ , \ x_3 = t$. It can be checked that the rank of the Jacobian matrix $\frac{dF(t)}{dt} \mid_{t=1}$is equal to one. In some sense, this equation is a simplified model of the equation $x^2 + y^2 + z^2 = r^2 = \left(\frac{a(1-e^2)}{1+e\cos f}\right)^2$, which can be derived from the equations (5)-(7) in this paper.

6.2 Embeddings,regular points and sub-manifolds, critical points and zero measure manifolds, Sards theorem

6.2.1 Embedding as a partial case of the immersion Next the embedding shall be defined as a partial case of the immersion, the only difference will be that now the notion of "diffeomorphism" shall be replaced by "homeomorphism". Homeomorphicity is related to the notion of continuity, according to the definition in [38].

Definition 13 *[37] An embedding is a one-to-one immersion $\varphi : M_1 \to M_2$, which is a homeomorphism M_1 onto its image $\varphi(M_1) = \overline{M}_1$ and the image has a topology as a subspace of M_2. The image $\overline{M}_1$ is called an embedded manifold.*

Now, after the definition of the notion of sub-manifold, it turns out to be possible to define the notion about a regular manifold, which later on will be defined in a different way, by means of its relation to the critical points.

6.2.2 Regular sub-manifolds

Definition 14 *[37] A regular sub-manifold of a C^∞ manifold M_2 is any subspace of M_1 with the sub-manifold property and with the C^∞ structure that the corresponding preferred coordinate neighborhoods determine on it.*

Thus, if the C^∞ mapping $\varphi : M_1 \to M_2$ maps a sub-manifold of M_1 with the rank of the mapping $k < n_1$, and q is a point $q \in \varphi(M_1)$, then $\varphi^{-1}(q)$ is a regular sub-manifold of M_1 of dimension $n_1 - k$.

In other words, the inverse mapping $\varphi^{-1}(q)$ maps the sub-manifold of M_2 not into the initial sub-manifold of M_1 with a dimension k, but to a regular sub-manifold of M_1 of dimension $(n_1 - k)$. But then, how the initial sub-manifold of M_1 with dimension k can be characterized, if the inequality between the ranks $k < n_1$ is fulfilled?

We shall consider now the general case of the immersion $\varphi : M_1 \to M_2$ with the dimensions of the corresponding manifolds n_1 and n_2, so that $n_1 < n_2$. Note that here the sign "=" is purposefully omitted, compared to the definition of immersion (see (101)).

6.2.3 Zero measure sub-manifolds and manifolds

Definition 15 *[40] Let $\varphi : M_1 \to M_2$ is a differentiable mapping of class C^1. Then the set $\varphi(M_1)$ has a zero measure in M_2. This also means that the image $\varphi(M_1)$ does not fill out the whole manifold M_2 [31].*

The notion of "zero measure" is closely related also with the Lebesque measure theory, necessary also for introduction of the Sard theorem. For the purpose the definition of an $n-$dimensional cube is necessary.

Definition 16 *[40] Let $a = (a_1, a_2, ..., a_n)$ and $b = (b_1, b_2, ...b_n)$ are points in the $n-$dimensional space R^n, such that $a_i < b_i$ $(1 \leq i \leq n)$ and let $C(a, b)$ denotes the open cube*

$$\{(t_1, t_2, ...t_n) \in R^n \mid a_i < t_i < b_i \quad , \quad 1 \leq i \leq n\} \quad . \tag{103}$$

Let the volume of the cube $C(a, b)$ is determined as

$$vol\,[C(a, b)] = (b_1 - a_1)(b_2 - a_2).....(b_n - a_n) \quad . \tag{104}$$

Definition 17 *[40] (about a set with zero measure) Let S is a subset of R^n. The subset S is said to have a zero measure if the following two conditions are fulfilled:*

A. For every $\epsilon > 0$ a covering of the subset S exists with a countable number of open cubes C_1, C_2.......such that $\sum_{i=1}^{\infty} vol[C_i] < \epsilon$.

B. Let X is a differentiable $n-$dimensional manifold and S is a subset of X. Let also a countable number of charts $\varphi_i : U_i \longrightarrow R^n$ exists such that U_1, U_2....cover S and for every value of i the set $\varphi_i(U_i \cap S)$ has a zero measure. Then an unification of sets with zero measure will again give a set with zero measure.

This definition is used in order to define the notion about the critical points and also the well-known Sards theorem in differential geometry. For the definition of zero measure manifolds, see also the monograph by Sternberg [27].

6.2.4 Critical points, Sards theorem in differential geometry and Morse functions

Theorem 18 *[31] (Sards theorem) Let a smooth differentiable mapping $\varphi : M_1 \rightarrow M_2$ is defined with $\dim M_1 = n_1 < \dim M_2 = n_2$ and let $C(\varphi)$ denotes the set of those points $x \in M_1$, for which the differential $d\varphi_x : T_x \rightarrow T_{\varphi(x)}$ has a rank, less than $n_2 = \dim M_2$. The rank of the differential $d\varphi_x$ is in fact the rank of the Jacobian matrix $\left(\frac{\partial\varphi^\alpha}{\partial x^\beta}\right)$ (see also (101), T_x and $T_{\varphi(x)}$ denote the vectors in the tangent spaces TM_1 and TM_2 of the manifolds M_1 and M_2. The set $C(\varphi)$ will be called the set of the critical points for the mapping φ and the set $C(\varphi)$ has a zero measure in M_2.*

Also, let in the neighborhood of the point $p \in M_1$ a local coordinate system $x^1, x^2, ...x^n$ is chosen, and f is a real smooth function on M_1. Then the point p is said to be a critical point for the function f if the following condition is fulfilled [33]

$$\frac{\partial f}{\partial x^1}(p) = \frac{\partial f}{\partial x^2}(p) = ... = \frac{\partial f}{\partial x^n}(p) = 0 \quad . \tag{105}$$

Further this definition shall be applied with respect to the mapping $F : (f_1, f_2) \rightarrow (x, y, z) = (x_2 - x_1, y_2 - y_1, z_2 - z_1)$ for the celestial transformation (5) - (7), when a two-dimensional manifold M_1 of the variables f_1, f_2 (i.e.$\dim M_1 = 2$) will be mapped to the three-dimensional manifold M_2 with dimension $\dim M_2 = 3$, defined by the cartesian coordinates (x, y, z).

Critical points can be degenerate and non-degenerate ones.

Definition 19 *[31] A critical point $x \in M$ of the smooth function $f(x)$, defined on the manifold M is called a non-degenerate one, if the matrix $a_{ij} = \frac{\partial^2 f(x)}{\partial x_i \partial x_j}$ is a non-degenerate one, i.e. $\det \parallel a_{ij} \parallel \neq 0$. The quadratic form is invariant with respect to the choice of the point x and is called a second differential at the given point. In local coordinates x^i with a centre at a critical point the function f can be written as [34]*

$$f(x) = f(0) + \sum_{i \neq j} a_{ij} x_i x_j + O(\mid x \mid^3) \quad . \tag{106}$$

A critical point is called a Morse one, if the quadratic form a_{ij} is a non-degenerate one. If all the critical points are Morse ones, the function f is called a Morse function.

By means of a suitable choice of the local coordinates $(x_1, x_2, ...x_n)$ the quadratic form can be transformed to a canonical form

$$x_1^2 + x_2^2 + ... + x_k^2 - x_{k+1}^2 - .... - x_n^2 \tag{107}$$

in the neighborhood of the Morse critical point. The index of the critical point is equal to the number of negative terms (i.e. the negative inertial index))

This is the exact mathematical definition for a critical point and a Morse function. However, the problem is whether this definition is sufficient in view of the determination of Morse functions.

The definition above is given because critical points (degenerate and non-degenerate) will be studied further with respect to the transformations (5) - (7) from Kepler parameters $(f, a, e, \Omega, i, \omega)$ to the cartesian coordinates (x, y, z). Here we shall point out only some aspects of the problem, the detailed study will be presented in subsequent publications.

The literature on critical points and Morse functions is rather extensive -only a part of the references on the subject are [31], [34],[39], [38], [43], [42], [40] and especially the monograph [35].

6.2.5 Critical points in the transformation (5) - (7). Are there Morse functions with respect to some of the Kepler parameters? All the critical points and Morse functions will not be found in this paragraph, but we shall pay a special attention to the case, concerning the derivatives of x, y and z with respect to the $\Omega-$ angle. We shall restrict ourselves only to the case of one manifold M, represented in cartesian coordinates as (x, y, z) and also in terms of the 6-Kepler parameters. From expressions (5) - (7) the derivatives $\frac{\partial x}{\partial \Omega}, \frac{\partial y}{\partial \Omega}$ and $\frac{\partial z}{\partial \Omega}$ can be calculated as follows

$$\frac{\partial x}{\partial \Omega} = r\left[-\sin\Omega\cos(\omega + f) - \cos\Omega\sin(\omega + f)\cos i\right] = -y \quad , \tag{108}$$

$$\frac{\partial y}{\partial \Omega} = r\left[\cos\Omega\cos(\omega+f) - \sin\Omega\sin(\omega+f)\cos i\right] = x \quad , \tag{109}$$

$$z = \frac{a(1-e^2)}{1+e\cos f}\sin(\omega+f)\sin i \quad \Rightarrow \frac{\partial z}{\partial \Omega} = 0 \quad . \tag{110}$$

Consequently, one will have $\frac{\partial x}{\partial \Omega} = \frac{\partial y}{\partial \Omega} = \frac{\partial z}{\partial \Omega} = 0$ at the critical point (in terms of the cartesian coordinates) $(x, y, z)_{crit} = (0, 0, z)$. Let us see which are the critical values in terms of the Kepler parameters. From the first equation (108) one can express

$$\cos i = -\frac{\sin(\omega+f)}{tg\Omega\cos(\omega+f)} \tag{111}$$

and substitute in the second equation $\frac{\partial y}{\partial \Omega} = 0$. The result will be

$$\cos\Omega\cos(\omega+f) + \sin\Omega\sin(\omega+f)\frac{\sin(\omega+f)}{tg\Omega\cos(\omega+f)} \tag{112}$$

$$= \cos\Omega\left[\frac{\cos^2(\omega+f)+\sin^2(\omega+f)}{\cos(\omega+f)}\right] = \frac{\cos\Omega}{\cos(\omega+f)} = 0 \quad . \tag{113}$$

So if $\cos(\omega+f) \neq 0$, to the critical point $(0, 0, z)$ will correspond the critical point $(\cos\Omega = 0, \cos i = 0)$ This might seem to be the critical point, but in fact it is not if besides the condition $\cos(\omega+f) \neq 0$, the condition $\sin(\omega+f) \neq 0$ should also be imposed. So the critical points are $\Omega = k\frac{\pi}{2}$, but additionally $\omega + f \neq m\frac{\pi}{2}$ and also $\omega + f \neq p\pi$ $(m = 1, 2, , , , ; p = 0, 1, 2)$.

Note that in the definition for critical points it was not required that the zero point $(x, y, z) = (0, 0, 0)$ should not be a critical point. If it is required (i.e. postulated) that $(x, y, z) \neq (0, 0, 0)$ (in the sense of all the three coordinates to be different from zero), then $(0, 0, z)$ will be the critical point for $\Omega = k\frac{\pi}{2}$.

However, because of the equalities $\frac{\partial^2 x}{\partial \Omega^2} = -\frac{\partial y}{\partial \Omega}$ and $\frac{\partial^2 y}{\partial \Omega^2} = \frac{\partial x}{\partial \Omega}$ it is seen that the critical points are degenerate ones and consequently the function $f = (x, y, z)$ is not a Morse function. This is admissable in view of the Sards theorem, since if in a manifold the set of critical points can have a zero measure, then every smooth mapping $f : (f, a, e, \Omega, i, .\omega.) \rightarrow (x, y, z)$ should admit non-critical values.

The last assertion about the function f being a non-Morse function is consistent with a theorem in [31] that each Morse function on a compact manifold has only a finite number of critical points. But in the given case, if one assumes that $f = (x, y, z)$ is a Morse function, this will contradict with the infinite number of values for $\Omega = k\frac{\pi}{2}$, since the number k can be arbitrary large.

6.2.6 Regular points In contrast to the set $C(\varphi)$ of critical points, a point $p \in M_1$ will be called a regular point for the smooth mapping φ, if it is not a critical one, i.e. the rank of the mapping between the vectors in the corresponding tangent spaces $d\varphi : TM_1 \rightarrow TM_2$ is equal to $n_2 = \dim M_2$. Also, the point $q \in M_2$ will be a regular point if all its inverse images are regular points in the manifold M_1 (for the case when $\varphi^{-1}(q) = \varnothing$ the point q will also be considered to be a regular point) [31].

For the definition of critical and regular points, see also [27].

6.2.7 Submersions Let again a differentiable mapping $\varphi : M_1 \rightarrow M_2$ is given, but contrary to the case of immersion, when the ranks of the manifolds were defined by the inequality $rankM_1 = n_1 \leq rankM_2 = n_2$, for the case of submersion the inequality for the ranks is defined as $rankM_1 = n_1 \geq rankM_2 = n_2$ [37], [31]. As noted in [37], the case $rankM_1 = n_1 = rankM_2 = n_2 = rank\varphi$ is a necessary condition for establishing a diffeomorphism between the two manifolds, for the case of submersion it is impossible.

6.2.8 Critical points defined for both cases of immersions and submersions

Definition 20 *[41] Let M and N are manifolds. The point $x \in M$ is called a critical point for the smooth mapping $f : M \rightarrow N$ if the rank of the mapping f_{*x} between the tangent spaces of M and N*

$$f_{*x} : T_x M \rightarrow T_{f(x)} N \tag{114}$$

at this point is smaller than the maximum possible, i.e. small than the smaller between the dimensions of the manifolds M and N

$$rank\ f_{*x} < \min(\dim M, \dim N) \quad . \tag{115}$$

Evidently, when $\dim M > \dim N$ (the case of submersion) this definition can be applied for the definition of critical points.

6.2.9 Submersions, the Whitney theorem and some applications to the current research The case of submersion is important for the presently investigated case of the mapping

$$F : M_1(f_., a_., e_., \Omega_., i_., \omega_.) \rightarrow N(x, y, z) \quad , \tag{116}$$

when the celestial transformations (5) - (7) map the manifold M_1, related to the six Kepler parameters $(f, a, e_., \Omega_., i_{.,}\ \omega_.)$ to the manifold N, related to the three Cartesian coordinates (x, y, z).

In the present investigation we would like to consider two sets of Kepler parameters $(f_1, a_1, e_{1.}, \Omega_{1.}, i_{1.,.}\ \omega_{1.})$ and $(f_2, a_2, e_{2.}, \Omega_{2.}, i_{2.,.}\ \omega_{2.})$, characterizing the motion respectively of the signal-emitting and signal-receiving satellite. These two sets will define two manifolds $M_1(f_1, a_1, e_{1.}, \Omega_{1.}, i_{1.,.}\ \omega_{1.})$ and $M_2(f_2, a_2, e_{2.}, \Omega_{2.}, i_{2.,.}\ \omega_{2.})$ and since the aim will be to imply a variational formalism, based on all the variables, the question is whether it will be possible to define a mapping $\overline{F}$ of a manifold M, related to all the 12 Kepler parameters to the manifold N, related to the two sets of cartesian coordinates (x_1, y_1, z_1) and (x_2, y_2, z_2)? These two mappings will be the submersions

$$\overline{F}_1 : \overline{M}_1(f_1, a_1, e_{1.}, \Omega_{1.}, i_{1.,.}\ \omega_1, f_2, a_2, e_{2.}, \Omega_{2.}, i_{2.,.}\ \omega_{2.}) \rightarrow \overline{N}_1(x_1, y_1, z_1, x_2, y_2, z_2) \tag{117}$$

or

$$\overline{F}_1 : \overline{M}_1(f_1, a_1, e_{1.}, \Omega_{1.}, i_{1.,.}\ \omega_1, f_2, a_2, e_{2.}, \Omega_{2.}, i_{2.,.}\ \omega_{2.}) \rightarrow \overline{N}_2(x = x_1 - x_2, y = y_1 - y_2, z = z_1 - z_2) \quad . \tag{118}$$

In the first case this is a submersion of a 12−dimensional manifold M_1 into a 6−dimensional manifold N_1 and in the second case- a submersion of the manifold M_1 into a 3−dimensional manifold. Naturally, the 12−dimensional manifold $\overline{M}_1$ will exist if it is possible to embed or immerse both manifolds $M_1(f_1, a_1, e_{1.}, \Omega_{1.}, i_{1.,.}\ \omega_{1.})$ and $M_2(f_2, a_2, e_{2.}, \Omega_{2.}, i_{2.,.}\ \omega_{2.})$ into $\overline{M}_1$. Such a possibility is guaranteed by the known Whitney theorem, which can be proved by using the Sards theorem .

Theorem 21 *(Whitney) Each connected smooth and closed n-dimensional manifold can be smoothly embedded in an R^{2n+1}−dimensional Euclidean space or it can be immersed in an R^{2n}−dimensional Euclidean space. .*

The proof of this theorem in [31] is rather complicated and is based on the subsequent projection of a smooth embedded manifold $M \subset R^n$ on hyperplanes, so that at each projection the dimension of the enclosing space is diminished by one. As a result of an orthogonal projection $\pi_{l_0} : M \rightarrow R_{t_0}^{N-1}$ a smooth immersion is obtained so that a point and a forbidden direction (straight line) with a total of $n + (n - 1)$ parameters should have a zero measure. According to Sards theorem this is possible when $2n - 1 < N - 1$ or $2n < N$ - this is the second case of immersion. Analogous considerations are presented in [31] for the other case of an imbedding in a R^{2n+1}− dimensional space.

In the concrete case, we have $n = 6$ (the dimension of the manifold M_1 or M_2), so M_1 (or M_2) is possible to be smoothly embedded to a 2.6+1 = 13 dimensional manifold or to be smoothly immersed in an 2.6 = 12−dimensional manifold - this will be exactly the manifold $\overline{M}_1(f_1, a_1, e_{1.}, \Omega_{1.}, i_{1.,.}\ \omega_1, f_2, a_2, e_{2.}, \Omega_{2.}, i_{2.,.}\ \omega_{2.})$. For the first case of 2.6 + 1 = 13−dimensions, the manifold is supplemented by one more function - this can be, for example, the distance R_{AB} between two points $A \subset M_1$ and $B \subset M_2$. Also, this distance will be related to the parameter s, because each of the Kepler parameters in the two sets will depend on it, respectively the two sets of coordinates $(x_1, y_1, z_1, x_2, y_2, z_2)$ will also depend on it. So the mapping (117) will be

$$\overline{F}_1 : \overline{M}_1(\Gamma^{(1)}(s), \Gamma^{(2)}(s), R_{AB}(s)) \rightarrow \overline{N}(x_1, y_1, z_1, x_2, y_2, z_2) \tag{119}$$

and $\Gamma^{(1)}(s), \Gamma^{(2)}(s)$ denote the two sets of Kepler parameters

$$\Gamma^{(1)}(s) \equiv (f_1(s), a_1(s), e_{1.}(s), \Omega_{1.}(s), i_{1.}(s),.\ \omega_1(s)) \ , \ . \tag{120}$$

$$\Gamma^{(2)}(s) \equiv (\ f_2(s), a_2(s), e_{2.}(s), \Omega_{2.}(s), i_{2.}(s),.\ \omega_{2.}(s)) \quad . \tag{121}$$

Since the coordinates $x_1, y_1, z_1, x_2, y_2, z_2$ do not introduce any new parameters or functions, the dimension of the manifold $\overline{N}$ remains 6.

6.2.10 Homotopy theory and combining the approaches of algebraic topology and variational calculus
The mapping (119) is in fact an attempt to combine the formalism of variational calculus with the topological ideas. Let us remind the standard definition about homotopy [37], [31] (see also the modern textbook [44] on the application of topology in physics) in algebraic topology, which on the one hand represents a mapping between topological spaces X and Y, but one the other hand represents also an immersion $X \times I \to Y$, where I is an interval.

Definition 22 *([37]) Let F, G are continuous mappings from the topological space X to the topological space Y and $I = [0,1]$ is an unit interval. Then F is homotopic to G if there is a continuous mapping (homotopy)*

$$H : X \times I \mapsto Y \quad , \tag{122}$$

which satisfies the condition $F(x) = H(x,0)$ and $G(x) = H(x,1)$ for all $x \in X$.

A slightly different definition is given in [44], where $a, b : I \to X$ are defined to be loops (i.e. the interval I is a two-parametric one - the parameters are s and t), $F : I \times I \to X$ is a continuous mapping and it is required that

$$F(s,0) = a(s), \qquad F(s,1) = b(s) \quad \forall s \in I \quad , \tag{123}$$

$$F(0,t) = F(1,t) = x \qquad \forall t \in I \quad . \tag{124}$$

Then the connecting map F is called a homotopy between a and b (denoted as $a \sim b$).

The differences in the presently applied approach are the following ones:

1. The interval I will not be the unit interval, but the interval $I = [s_a, s_b]$, where the values $s = s_a$ and $s = s_b$ will correspond to the values of the parameter s for the Kepler parameters

$$\Gamma^{(1)}(s) \equiv (f_1(s_a), a_1(s_a), e_{1.}(s_a), \Omega_{1.}(s_a), i_{1.}(s_a),. \omega_1(s_a)) \tag{125}$$

of the signal-emitting satellite and

$$\Gamma^{(2)}(s) \equiv (f_2(s_b), a_2(s_b), e_{2.}(s_b), \Omega_{2.}(s_b), i_{2.}(s_b),. \omega_{2.}(s_b)) \tag{126}$$

of the signal-receiving satellite (both are moving along different elliptic, space - distributed orbits).

2. Further it will also be supposed that both ends of the signal trajectory are dynamically moving and will be required to find the optimal trajectory of the signal (i.e.the optimal positions of both satellites) for transmission between the satellites. If finding the signal trajectory takes into account the General Relativity effects and the Fermats principle for minimal time of propagation of the signal, finding the optimal position of the satellites on both orbits will require the application of a multi-parameter optimization theory. In other words, both ends of the signal trajectory will be dynamical, not static and fixed ones. This formalism will be developed in subsequent papers, but it may only be mentioned that for a long time ago a formalism in variational calculus has been developed - the s.c. "variational calculus with moving boundaries" [46] and many other monographs.

3. It will be impossible to use the direct product $X \times I$, because if $I = [s_a, s_b]$, the manifold X will depend on the parameter s. A new problem here is formulated when an immersion or embedding is realized between two manifolds $f : X \to Y$, if they are considered at the same time to be topological spaces and each of their elements depends also on the parameter s.

7 Higher order variational formalism, based on the Stokes theorem and differential forms

The Stokes theorem is the basic theoretical tool, which shall be used further. It is based on three important notions - differential forms of degree n (or simply $n-$form), orientation of an $n-$manifold and the previously defined notion about the boundary ∂M of a manifold M.

7.1 Definition of the Stokes theorem

Definition 23 *[37] An exterior differential form of degree n (n−form) is an alternating covariant tensor field of order n.*

Definition 24 *[37] The manifold M is orientable if a C^{∞} $n-$form $\omega = \sum_i f_i \varphi_i^{*}(dx^1 \Lambda dx^2 \Lambda .....dx^n)$ can be defined, which is not zero at any point. Then it is said that M is oriented by the choice of ω. This means that the orientation of M determines the orientation of the boundary ∂M. This holds for regular domains (sub-manifolds) of M.*

Then the Stokes theorem, relating an integral on the manifold M to an integral on the boundary is defined as follows

Theorem 25 *[37] Let M is an $n-$dimensional oriented manifold with a boundary ∂M, ω is a differential form of the $(n-1)-$degree with a compact support. Then the generalized Stokes formulae can be expressed as*

$$\int_M d\omega = \int_{\widetilde{\partial}M} i^*\omega \quad . \tag{127}$$

In the last equality $\widetilde{\partial}M$ denotes the boundary ∂M with its different possible orientations. This means that when n is even, the orientation, induced by M is positive, i.e. ∂M, but when n is odd, then the orientation will be the opposite one, i.e. $-\partial M$. Both cases can be united as $\widetilde{\partial}M = (-1)^n\partial M$. In (127) $i^*\omega$ denotes the restriction of ω to ∂M since $i : \partial M \rightarrow M$ is an inclusion mapping. In a slightly different manner, the above equality is written as $(-1)^n \int_M d\omega = \int_{\partial M}\omega$ (see also the definitions in [31], [36], [43], [38] and many others). In the second term, instead of $i^*\omega$, many authors write only ω.

7.2 Application of a variational formalism, based on higher order differential forms and the Stokes theorem

Now let us describe the variational formalism, developed in this paper. The idea is to start from the one form

$$\omega_1 = Adx + Bdy + Cdz \quad , \tag{128}$$

where A, B, C are the coefficient functions in the variated action (one can take the action (36). In other words, A, B, C will be the Euler-Lagrange equations as a result of the variation in terms of the x, y and z variables

$$A \equiv \frac{\delta L}{\delta x} + \lambda_1(\widetilde{T} - G(f)) \ ; \ B \equiv \frac{\delta L}{\delta y} + \lambda_2(\widetilde{T} - G(f)) \ ; \ C \equiv \frac{\delta L}{\delta z} + \lambda_3(\widetilde{T} - G(f)) \quad , \tag{129}$$

supplemented by the Lagrange function multipliers $\lambda_1(x, y, z), \lambda_2(x, y, z), \lambda_3(x, y, z)$. In (129) $\frac{\delta L}{\delta x}, \frac{\delta L}{\delta y}, \frac{\delta L}{\delta z}$ are the generalized variational derivatives

$$\frac{\delta L}{\delta x} \equiv \frac{\partial L}{\partial x} - \frac{d}{ds}\frac{\partial L}{\partial \dot{x}} \quad , \quad \frac{\delta L}{\delta y} \equiv \frac{\partial L}{\partial y} - \frac{d}{ds}\frac{\partial L}{\partial \dot{y}} \quad ,$$

$$\frac{\delta L}{\delta z} \equiv \frac{\partial L}{\partial z} - \frac{d}{ds}\frac{\partial L}{\partial \dot{z}} \quad . \tag{130}$$

The function $\widetilde{T}$ is equal to the function $G(f)$ (8), calculated in the paper [14] by means of higher-order elliptic integrals - the corresponding expressions are given in Appendix B. The motivation to include the function $(\widetilde{T} - G(f))$ as constraints is that if in this higher order variational formalism the propagation time is calculated, then in the partial case of only f being the dynamical parameter, the result should coincide with $\widetilde{T}$. This would mean also that the derivatives of R_{AB} with respect to the other 5 Kepler parameters should be zero. The notation for $\dot{x}$, $\dot{y}$ and $\dot{z}$ means, as usual, derivatives with respect to the parameter s.

In the usual variational formalism, A, B and C are the usual Euler-Lagrange equations (plus the constraint function). In this case A, B, C will not be equal to zero.(see also [39]) and they will be used for the derivation of the two-form by means of the Stokes theorem for the three-dimensional case

$$\int_{\partial M} [Adx + Bdy + Cdz] = \int_M [Pdy\Lambda dz + Qdz\Lambda dx + Rdx\Lambda dy] \tag{131}$$

where $P = P(x, y, z)$, $Q = Q(x, y, z)$, $R =, R(x, y, z)$ are the expressions

$$P(x, y, z) \equiv \frac{\partial C}{\partial y} - \frac{\partial B}{\partial z} \quad , \ Q(x, y, z) \equiv \frac{\partial A}{\partial z} - \frac{\partial C}{\partial x} \quad ,$$

$$R(x, y, z) \equiv \frac{\partial B}{\partial x} - \frac{\partial A}{\partial y} \quad . \tag{132}$$

It should be mentioned that these expressions are a partial case in the definition of the absolute integral invariant of the second order [39]

$$J = \int\limits_{\tilde{\Omega}} \sum_{i,j} \left(\frac{\partial F_i}{\partial x_j} - \frac{\partial F_j}{\partial x_i} \right) \delta x_i \delta x_j \quad , \tag{133}$$

where

$$J = \int\limits_{\partial\tilde{\Omega}} (F_1 \delta x_1 + ... + F_n \delta x_n) \tag{134}$$

is called the relative integral invariant. Let us also remind that $x = x_2 - x_1$, $y = y_2 - y_1, z = z_2 - z_1$.

Further, let us introduce the notations

$$(q_x, q_y, q_z) = (\frac{\partial L}{\partial x}, \frac{\partial L}{\partial y}, \frac{\partial L}{\partial z}) \quad , \quad \overrightarrow{p} = (p_x, p_y, p_z) = (\frac{\partial L}{\partial \dot{x}}, \frac{\partial L}{\partial \dot{y}}, \frac{\partial L}{\partial \dot{z}}) \quad , \tag{135}$$

$$\dot{\overrightarrow{p}} = (\dot{p_x}, \dot{p_y}, \dot{p_z}) = (\frac{d}{ds}\frac{\partial L}{\partial \dot{x}}, \frac{d}{ds}\frac{\partial L}{\partial \dot{y}}, \frac{d}{ds}\frac{\partial L}{\partial \dot{z}}) \quad . \tag{136}$$

Then it can easily be found that

$$P_1 \equiv \frac{\partial}{\partial z}\left(\frac{d}{ds}\frac{\partial L}{\partial \dot{y}}\right) - \frac{\partial}{\partial y}\frac{d}{ds}\frac{\partial L}{\partial \dot{z}} = -\frac{\partial \dot{p_z}}{\partial y} + \frac{\partial \dot{p_y}}{\partial z} = (rot \dot{\overrightarrow{p}})_x \quad . \tag{137}$$

Analogously

$$Q_1 \equiv \{z \rightarrow x; y \rightarrow z\} = (rot \dot{\overrightarrow{p}})_y \quad , \quad R_1 \equiv \{x \rightarrow y; z \rightarrow x\} = (rot \dot{\overrightarrow{p}})_z \quad . \tag{138}$$

Then, after taking into account the expressions (129) - (131) and also (135) - (138), each of the functions $P(x, y, z), Q(x, y, z), R(x, y, z)$ in (132) can be represented as a sum of four components

$$P \equiv \widetilde{P}_0 + P_1 + T^P + \widetilde{G}_1 \quad ,.Q \equiv \widetilde{Q}_0 + Q_1 + T^Q + \widetilde{G}_2 \quad , \tag{139}$$

$$R \equiv \widetilde{R}_0 + R_1 + T^R + \widetilde{G}_3 \quad . \tag{140}$$

Each one of the components is calculated to be

$$\widetilde{P}_0 \equiv \frac{\partial^2 L}{\partial y \partial z} - \frac{\partial^2 L}{\partial z \partial y}, \quad P_1 = (rot \dot{\overrightarrow{p}})_x \quad , \tag{141}$$

$$T^P \equiv \lambda_3 \widetilde{T}_y - \lambda_2 \widetilde{T}_z \quad , \quad \widetilde{G}_1 \equiv \lambda_2 \frac{\partial G}{\partial f}\frac{1}{z_f} - \lambda_3 \frac{\partial G}{\partial f}\frac{1}{y_f} \quad . \tag{142}$$

Similar are the decompositions for $\widetilde{Q}_0, Q_1, T^Q, \widetilde{G}_2$ and $\widetilde{R}_0, R_1, T^R, \widetilde{G}_3$, taking into consideration the symmetry properties $(y, z) \rightarrow (z, x) \rightarrow (x, y)$, $x \rightarrow y \rightarrow z$ and $(\lambda_3, \lambda_2) \rightarrow (\lambda_1, \lambda_3) \rightarrow (\lambda_2, \lambda_1)$ (circular symmetry $1 \rightarrow 2 \rightarrow 3 \rightarrow 1$). The functions $\widetilde{T}_x, \widetilde{T}_y, \widetilde{T}_z$ are the partial derivatives of the propagation time $\widetilde{T}$ (8), according to the formulaes in Appendix B. The functions $\widetilde{P}_0$, $\widetilde{Q}_0 \equiv \frac{\partial^2 L}{\partial z \partial x} - \frac{\partial^2 L}{\partial x \partial z}$, $\widetilde{R}_0 \equiv \frac{\partial^2 L}{\partial x \partial y} - \frac{\partial^2 L}{\partial y \partial x}$ in (139), (140) and (141) usually are put equal to zero, since the partial derivatives $\frac{\partial}{\partial x}, \frac{\partial}{\partial y}, \frac{\partial}{\partial z}$ commute. However, if they are expressed through the Kepler parameters $(f., a., e., \Omega., i., \omega)$, the corresponding commutators $\left[\frac{\partial}{\partial x}, \frac{\partial}{\partial y}\right]$, $\left[\frac{\partial}{\partial y}, \frac{\partial}{\partial z}\right]$ and $\left[\frac{\partial}{\partial x}, \frac{\partial}{\partial z}\right]$ will turn out to be nonzero. Let us remind also that the coordinates of the vector fields $\overrightarrow{u}$ and can be chosen to be

$$u_i = (f., a., e.) \quad i = 1, 2, 3) \text{ and } v_k = (\Omega., i., \omega.) \qquad k = 1, 2, 3) \quad , \tag{143}$$

but also can be chosen to be the two sets of Kepler parameters for the first (signal-emitting) and the second (signal-receiving) satellite

$$u_i = (f._1, a_{1.}, e_{1,}. \Omega_{1.}, i_{1.}, \omega_1) \quad i = 1, 2, 3, 4, 5, 6 \tag{144}$$

and

$$v_j = (f._2, a_{2.}, e_{2,}. \Omega_{2.}, i_{2.}, \omega_{2.}) \qquad j = 1, 2, 3, 4, 5, 6 \quad . \tag{145}$$

We shall give first the proof about the non-commutativity of the partial derivatives and afterwards we shall remind why the result is natural from the point of view of the Frobenius theorem, which is presented in almost every monograph on differential geometry.

Journal of Physics: Conference Series **2910** (2024) 012001 doi:10.1088/1742-6596/2910/1/012001

7.3 Non-commutativity of the partial derivatives, when expressed by derivatives, related to the Kepler parameters

Let us investigate the case of non-commutativity of the partial derivatives $\frac{\partial}{\partial y}$ and $\frac{\partial}{\partial z}$. This means that the commutator
$\left[\frac{\partial}{\partial y}, \frac{\partial}{\partial z}\right] \neq 0$, provided that the partial derivatives are expressed as

$$\frac{\partial}{\partial z} = \frac{1}{3}\sum_{k=1}^{3}\frac{1}{z_{u_k}}\frac{\partial}{\partial u_k} + \frac{1}{3}\sum_{l=1}^{3}\frac{1}{z_{v_l}}\frac{\partial}{\partial v_l} \quad , \tag{146}$$

Theorem 26 ***Proof.***

$$\frac{\partial}{\partial y} = \frac{1}{3}\sum_{i=1}^{3}\frac{1}{y_{u_i}}\frac{\partial}{\partial u_i} + \frac{1}{3}\sum_{j=1}^{3}\frac{1}{y_{v_j}}\frac{\partial}{\partial v_j} \quad . \tag{147}$$

The second derivative can easily be calculated to be

$$\frac{\partial^2}{\partial y \partial z} = \frac{1}{3}\sum_{k=1}^{3}\frac{\partial}{\partial y}\left(\frac{1}{z_{u_k}}\frac{\partial}{\partial u_k}\right) + \frac{1}{3}\sum_{l=1}^{3}\frac{\partial}{\partial y}\left(\frac{1}{z_{v_l}}\frac{\partial}{\partial v_l}\right) \quad . \tag{148}$$

Let us denote by

$$M_{ik}(u_i, u_k, y \Longleftrightarrow z) \equiv M_{ik}(u_i, u_k, y) - M_{ik}(u_i, u_k, z) \tag{149}$$

the expression

$$M_{ik}(u_i, u_k, y \Longleftrightarrow z) \equiv \frac{1}{y_{u_i}}\frac{\partial}{\partial u_i}\left(\frac{1}{z_{u_k}}\right) - \frac{1}{z_{u_i}}\frac{\partial}{\partial u_i}\left(\frac{1}{y_{u_k}}\right) = \tag{150}$$

$$= \frac{1}{y_{u_i}}\frac{1}{z_{u_k}}\left[\frac{1}{z_{u_k}}\frac{\partial^2 z}{\partial u_i \partial u_k} + \frac{1}{y_{u_i}}\frac{\partial^2 y}{\partial u_i \partial u_k}\right] = \tag{151}$$

$$= \frac{1}{y_{u_i} z_{u_k}}\left[(\ln z_{u_k})_{,u_i} + (\ln y_{u_i})_{,u_k}\right] \quad . \tag{152}$$

This is different from zero! Now the commutator $\left[\frac{\partial L}{\partial y}, \frac{\partial L}{\partial z}\right]$ *for* $\widetilde{P}_0$ *in (??) between the two partial derivatives can be calculated as*

$$\widetilde{P}_0 \equiv \frac{\partial^2 L}{\partial y \partial z} - \frac{\partial^2 L}{\partial z \partial y} = \left[\frac{\partial L}{\partial y}, \frac{\partial L}{\partial z}\right]$$

$$= \frac{1}{9}\sum_{i,k=1}^{3}\left[M_{ik}(u_i, u_k, y \Longleftrightarrow z) + M_{ik}(v_i, u_k, y \Longleftrightarrow z)\right]\frac{\partial L}{\partial u_k}$$

$$+ \frac{1}{9}\sum_{i,k=1}^{3}\left[M_{ik}(u_i, v_k, y \Longleftrightarrow z) + M_{ik}(v_i, v_k, y \Longleftrightarrow z)\right]\frac{\partial L}{\partial v_k} \quad . \tag{153}$$

In the same way, the other commutators $\left[\frac{\partial L}{\partial z}, \frac{\partial L}{\partial x}\right]$ *and* $\left[\frac{\partial L}{\partial x}, \frac{\partial L}{\partial y}\right]$ *can be calculated. Note that when we have the case of the defining equalities (144) and (145) for* u_i *and* v_j, $i, j = 1, 2, ...6$, *the summation in (153) will be from1 to 6. In any case, all the four constituents in the two sums will be different one from another, so no cancellation of terms with opposite signs will be possible.* ■

7.3.1 A partial case of the commutativity of the partial derivatives Suppose now that the inverse function theorem can be applied. This means that if for example the derivative $\frac{\partial y_i}{\partial u_k}$ or $\frac{\partial y_i}{\partial v_k}$ can be calculated,

then the derivatives $\frac{\partial u_k}{\partial y_i}$ and $\frac{\partial v_k}{\partial y_i}$ can also be calculated. Now it is important to remind briefly the formulation of the inverse function theorem, so that it is evident that in the concrete case this theorem can be applied because the dimension of the initial manifold M_1 is 6 (equal to the number of the Kepler parameters in the right-hand side of the equations (5) - (7)) and the dimension of the second manifold M_2 is equal to 3−the number of the equations.

Theorem 27 *[36] Let $f : R^n \to R^k$ is a mapping of smoothness class C^∞ of the system of equations and $M_{\overrightarrow{C}}$ is the set of solutions of the system of equations $f(\overrightarrow{x}) = \overrightarrow{C}$, $(x^1, x^2, ...x^n)$, $\overrightarrow{C} = (C^1, C^2......C^k)$ and the mapping $f : R^n \to R^k$ is determined by the functions $(f^1, f^2, ...f^k)$, where*

$$f^1(x^1, x^2, ...x^n) = C^1, ..., f^k(x^1, x^2, ...x^n) = C^k \quad . \tag{154}$$

If the rank of the Jacobi matrix for the mapping f is maximal at each point $P_0 \in M_{\overrightarrow{C}}$ (i.e.$rkdf(P_0) = k$), then

If the inverse function theorem can be applied, then $M_{\overrightarrow{C}}$ is a $(n-k)$ dimensional smooth manifold of class C^∞. Then in the neighborhood of each point $P_0 \in M_{\overrightarrow{C}}$ one can choose as local coordinates some $(n-k)$ cartesian coordinates of the embedding Euclidean space.

In other words, in the neighborhood of the non-degenerate points such local coordinates can be introduced and the "inverse derivatives" such as $\frac{\partial u_k}{\partial y_i}$ and $\frac{\partial v_k}{\partial y_i}$ can be found

From the definition of the function $M_{ik}(u_i, u_k, y \Longleftrightarrow z)$ in (149) and (150) it will follow that

$$M_{ik}(u_i, v_k, y \Longleftrightarrow z) = \frac{\partial u_i}{\partial y}.\frac{\partial}{\partial u_i}\left(\frac{\partial u_k}{\partial z}\right) - \frac{\partial u_i}{\partial z}.\frac{\partial}{\partial u_i}\left(\frac{\partial u_k}{\partial y}\right) = \tag{155}$$

$$= \frac{\partial}{\partial y}\left(\frac{\partial u_k}{\partial z}\right) - \frac{\partial}{\partial z}\left(\frac{\partial u_k}{\partial y}\right) = 0 \quad . \tag{156}$$

However, the conditions of the inverse function theorem, when the "inverse partial derivatives" can be found are not always fulfilled, and these are the critical (singular) points.

Definitions and proofs of the inverse function theorem can be found also in [30], [27], [37], [23], a particular clear formulation is presented in [39] and other books.

Consequently, for the cases when the inverse function theorem cannot be applied (in particular for critical points), such a non-commutativity may appear. In order to avoid such situations, some authors assert [37] that "the Stokes theorem holds only for smooth manifolds with a smooth boundary". But at the same time, it has been noted also that [37] "the search for reasonable domains of integration to validate the Stokes theorem leads to the concept of simplicial or polyhedral complex, i.e. a space made up of fastening together along their faces a number of simplicies (line segments, triangles, tetrahedra and their generalizations (polyhedra, cubes). Then every C^∞ manifold can be "triangulated", which means that even with considerable smoothness it can be homeomorphic to such a complex and thus, the integral becomes a sum of integrals over the pieces, which are images of simplicies, cubes and etc."

The reason for introducing this citation is that we have already found three boundaries of the manifold - the first one is the (x, y) boundary, represented by the plane equation (84) or (85), then the (x, z) plane, represented by the plane equation (91) and the third (y, z) boundary, which is represented by the plane (98). So in a further research it will be interesting to see whether the three planes will intersect and thus the intersecting points will determine the boundaries of integration in the Stokes theorem.

In fact, the "commutative case" is related to the s.c. "integrable" or "involutive" case, which is the essence of the Frobenius theorem.

7.3.2 Different formulations of the Frobenius theorem The Frobenius theorem has several definitions, which differ in their formulation.

Let us give first the most concise one [29]. Let M is a smooth manifold with a set of tangent hyperplanes. In the neighborhood of a point with local coordinates $(x_1, x_2, ...x_n)$ the set of hyperplanes is given by the differential $1-$form $\alpha = du - pdx$ (where $u = f(x)$), different from zero and where $pdx = p_1 dx_1 + ... + p_n dx_n$, $p_i = \frac{\partial f}{\partial x^i}(x)$.The function $f(x)$ is a $1-$jet function or is called also a standard contact $1-$form on the manifold $J^1(V, \mathbb{R})$, determined by the set of $(2n+1)$ numbers $(x_1, x_2, ...x_n; u; p_1, p_2, ...p_n)$.

Then the Frobenius theorem states the condition when a set of hyperplanes in $\mathbb{R}^3$ will have integral hypersurfaces. In other words, this set of hypersurfaces is fully integrable when the form $d\alpha$ in the corresponding hyperplane is zero.

Theorem 28 ***Definition 29*** *[29] The hyperplane $\alpha = 0$ is fully integrable if and only if the equality $\alpha \Lambda d\alpha = 0$ is fulfilled.*

Journal of Physics: Conference Series **2910** (2024) 012001 doi:10.1088/1742-6596/2910/1/012001

However, the notion of integrability in [23] is defined in a slightly different and more general context.

Theorem 30 ***Definition 31*** *[23] The differential system D of class C^r $(r \geq 2)$, defined on a manifold V is completely integrable if and only if it is involutive, which means that at each point $a \in V$ there exists a neighborhood U and C^r vector fields $X_1, X_2.....X_p$ (determining D on U) such that $[X_\mu, X_\nu](b) \in D(b)$.*

In case the differential form system D is specified to be

$$D(a) = \{X \in T_a V \quad : \omega_\nu(a)(X) = 0 \quad , p < \nu \leq n\} \tag{157}$$

and $\omega_{p+1}, ..., \omega_n$ are linearly independent at each point 1−forms on V. Then again in [23] the following theorem has been proved:

Theorem 32 *If in suitable coordinates the one-forms ω_ν are expressed as*

$$\omega_\nu = dx_\nu + \sum_{\mu \leq p} a_{\nu\mu} dx_\mu \quad , \quad \nu > p \quad , a_{\nu\mu} \in C(u) \quad , \tag{158}$$

then D is a differential form system, formed by the vector fields

$$X_\mu = \frac{\partial}{\partial x_\mu} - \sum_{\nu > p} a_{\nu\mu} \frac{\partial}{\partial x_\nu} \quad , \tag{159}$$

so that $\omega_\nu(X_\mu) = 0$.

If X_μ and $\frac{\partial}{\partial x_\mu}$ are the corresponding vectors in the two tangent space, we can write also the relations

$$X_\mu = \sum_{\nu=1}^{p} b_{\mu\nu} Y_\nu \quad , \qquad Y_\nu(x) = \sum_{\alpha=1}^{n} a_{\nu\alpha}(x) \left(\frac{\partial}{\partial x_\alpha}\right)_x \tag{160}$$

Then it can easily be calculated that

$$[X_\mu, X_\nu] = \sum_{m=1}^{p} \lambda_m X_m = \sum_{m=1}^{n} \xi_{m\beta} \frac{\partial}{\partial x_\beta} \quad . \tag{161}$$

Concretely for our case, instead of the local coordinates x_μ we have u_μ and v_ν, the analog of (161) is (153), instead of X_μ and X_ν we have $\frac{\partial}{\partial x_\mu}$ and $\frac{\partial}{\partial x_\nu}$ and $\xi_{ik} = M_{ik}(u_i, u_k, y \Longleftrightarrow z) + M_{ik}(v_i, u_k, y \Longleftrightarrow z)$. In fact, the commutator $\left[\frac{\partial}{\partial x_\mu}, \frac{\partial}{\partial x_\nu}\right]$ acts on the Lagrangian L and since .

$$\left[\frac{\partial}{\partial x_\mu}, \frac{\partial}{\partial x_\nu}\right] L = \frac{1}{9} \sum_{i,k=1}^{3} \left[\xi^{(1)}_{ik} \frac{\partial}{\partial u_k} + \xi^{(2)}_{ik} \frac{\partial}{\partial v_k}\right] L \neq 0 \quad , \tag{162}$$

the system is not involutive.

The involutive meaning of the Frobenius theorem in [37] is formulated with regard to finding the condition for the integrability of the system of nonlinear equations

$$\frac{\partial z}{\partial x} = g(x, y, z) \quad , \quad \frac{\partial z}{\partial y} = h(x, y, z) \quad . \tag{163}$$

Then the solution will exist if the following pair of vector fields X and Y can be constructed so that

$$X = \frac{\partial}{\partial x} + f_x(x, y) \frac{\partial}{\partial z} \quad , \quad Y = \frac{\partial}{\partial y} + f_y(x, y) \frac{\partial}{\partial z} \quad , \tag{164}$$

where $z = f(x, y)$. Then the necessary condition for the existence of an independent pair of vector fields X and Y, defining a tangent plane to the surface $z = f(x, y)$, can be expressed as

$$[X, Y] = \left(\frac{\partial^2 f}{\partial x \partial y} - \frac{\partial^2 f}{\partial x \partial y}\right) = 0 \quad . \tag{165}$$

However, it should be kept in mind that this formulation can be applied if it is possible to express the $z-$coordinate as a graph,i.c. as a function $z = f(x, y)$ of x and y . In the case of the transformations (5) - (7) this is impossible to the following reason: from the already used relation $x^2 + y^2 + z^2 = r^2 = \left(\frac{a(1-e^2)}{1+e\cos f}\right)^2$ (see also (29) and (30)) $\cos f$ can be expressed and substituted in the equations (5)- (7) for x, y and z. Then a much more complicated algebraic equation will be obtained, which even does not guarantee that it will be possible to find any dependence $z = f(x, y)$.

8 Second stage of the variational formalism, based on the Gauss-Ostrogradsky theorem

8.1 Formulation of the Gauss-Ostrogradsky theorem

Previously we started from the Stokes theorem (131) $\int\limits_{\partial M} [Adx + Bdy + Cdz] = \int\limits_{M} [Pdy\Lambda dz + Qdz\Lambda dx + Rdx\Lambda dy]$, where in the right-hand side the integration is over the manifold M and is over the $2-$form ω_2

$$\omega_2 = Pdy\Lambda dz + Qdz\Lambda dx + Rdx\Lambda dy \quad . \tag{166}$$

The idea, related to the Gauss-Ostrogradsky theorem is whether the integration in the right-hand side of (131) may be considered as an integration on some boundary ∂M of the manifold, which after applying a new differentiation $d\omega_2$ will be integral over a $3-$form

The Gauss-Ostrogradsky theorem is based on the general equation

$$\int\limits_{--\partial M} \omega_2 = - \int\limits_{--\partial M} i^*\omega_2 = \int\limits_{--M} d\omega_2 \tag{167}$$

$$= \int\limits_{M} \left[\frac{\partial P}{\partial x} + \frac{\partial Q}{\partial y} + \frac{\partial R}{\partial z}\right] dx\Lambda dy\Lambda dz \quad . \tag{168}$$

We shall not give the detailed calculation how the three-form is derived - this can be found in almost all textbooks on differential geometry such as [38], [36], [37], [27], [30] and others. Let us remind the functions P, Q and R, calculated in (139) and (140)

$$P \equiv \widetilde{P}_0 + P_1 + T^P + \widetilde{G}_1 \;,.Q \equiv \widetilde{Q}_0 + Q_1 + T^Q + \widetilde{G}_2 \;,\; R \equiv \widetilde{R}_0 + R_1 + T^R + \widetilde{G}_3 \quad . \tag{169}$$

The function T^P previously was defined as $T^P \equiv \lambda_3 T_y - \lambda_2 T_z$, analogously T^Q, T^R are defined.

8.2 Calculation of the functions in the right-hand side of the Gauss-Ostrogradsky equation

Proposition 33 *The following two equalities are fulfilled*

$$\frac{\partial P_1}{\partial x} + \frac{\partial Q_1}{\partial y} + \frac{\partial R_1}{\partial z} = 0 \quad , \tag{170}$$

$$\frac{\partial T^P}{\partial x} + \frac{\partial T^Q}{\partial y} + \frac{\partial T^R}{\partial z} = 0 \tag{171}$$

but only provided the partial derivatives $\frac{\partial}{\partial x}, \frac{\partial}{\partial y}, \frac{\partial}{\partial z}$ *commute. These two equalities will not be equal to zero, if expressed in terms of the Kepler parameters. In case the partial derivatives do not commute, expression (153) for* $\widetilde{P}_0$ *should be used (analogously,* $\widetilde{Q}_0$ *and* $\widetilde{R}_0$ *are constructed) for calculation of the expressions for* P, Q, R *in (139) and (140).*

The expression for the function P (169) in the Stokes and Gauss-Ostrogradsky theorem can be calculated by means of the partial derivatives of the cartesian coordinates (x, y, z) with respect to the two sets u_i and v_j of Kepler parameters

$$u_i = (f., a., e.) \quad i = 1, 2, 3 \text{ and } v_j = (\Omega., i., \omega.) \quad j = 1, 2, 3)$$

to be

$$P = \frac{\partial C}{\partial y} - \frac{\partial B}{\partial z} = \frac{1}{3}\sum_{i=1}^{3}(\frac{\lambda_3}{y_{u_i}} - \frac{\lambda_2}{z_{u_i}})\frac{\partial \widetilde{T}}{\partial u_i} + \frac{1}{3}\sum_{j=1}^{3}(\frac{\lambda_3}{y_{v_j}} - \frac{\lambda_2}{z_{v_i}})\frac{\partial \widetilde{T}}{\partial v_j}$$

$$+\frac{1}{3}(\frac{\lambda_2}{z_f}-\frac{\lambda_3}{y_f})\frac{\partial G}{\partial f}+[K_1(y,z)-K_1(z,y)] \quad . \tag{172}$$

In case the calculation for P is performed for the two sets of Kepler parameters $(f_{.1}, a_{1.}, e_1, \Omega_{1.}, i_{1.}, \omega_1)$ and $(f_{2.}, a_{2.}, e_2, \Omega_{2.}, i_{2.}, \omega_2)$, the summation in the two sums will be from 1 to 6 (respectively there will be 6 derivatives $\frac{\partial}{\partial u_i}$ and 6 derivatives $\frac{\partial}{\partial v_j}$) and the coefficients in front of the sums will be $\frac{1}{6}$.

Proposition 34 *The antisymmetric function $K_1(y,z)$ with respect to the variables y and z can be represented as a sum of ten variational derivatives of the first, second and the third order*

$$K_1(y,z) \equiv P_{ij}^{(1)}\frac{\partial^2 L}{\partial u_i \partial s}+P_{ij}^{(2)}\frac{\partial^2 L}{\partial v_i \partial s}+P_{ij}^{(3)}\frac{\partial^3 L}{\partial u_i \partial u_j \partial s}$$

$$+P_{ij}^{(4)}\frac{\partial^3 L}{\partial u_i \partial v_j \partial s}+P_{ij}^{(5)}\frac{\partial^3 L}{\partial v_i \partial v_j \partial s}+P_{ij}^{(6)}\frac{\partial^2 L}{\partial u_i \partial u_j}$$

$$+P_{ij}^{(7)}\frac{\partial^2 L}{\partial u_i \partial v_j}+P_{ij}^{(8)}\frac{\partial^2 L}{\partial v_i \partial v_j}+P_{ij}^{(9)}\frac{\partial L}{\partial u_i}+P_{ij}^{(10)}\frac{\partial L}{\partial v_i} \quad , \tag{173}$$

where the coefficient functions $P_{ij}^{(1)}, P_{ij}^{(2)}.......P_{ij}^{(10)}$ are given in Appendix C.

Proposition 35 *The exterior product $dy\Lambda dz$ in terms of the variables, related to the Kepler parameters can be expressed as*

$$dy\Lambda dz = \sum_{\substack{i,j=1 \\ i \pm j}}^{3}\left[\frac{D(y,z)}{D(u_i,u_j)}du_i du_j+\frac{D(y,z)}{D(v_i,v_j)}dv_i dv_j\right]$$

$$+\sum_{\substack{i,j=1 \\ i \pm j, also\ i=j}}^{3}\frac{D(y,z)}{D(u_i,v_j)}du_i dv_j \quad , \tag{174}$$

where $\frac{D(y,z)}{D(u_i,u_j)}$ is the Jacobian determinant

$$\frac{D(y,z)}{D(u_i,u_j)} \equiv \begin{vmatrix} \frac{\partial y}{\partial u_i} & \frac{\partial y}{\partial u_j} \\ \frac{\partial z}{\partial u_i} & \frac{\partial z}{\partial u_j} \end{vmatrix} = y_{u_i}z_{u_j}-y_{u_j}z_{u_i} \quad . \tag{175}$$

Taking into account the similar expressions for $\frac{D(y,z)}{D(v_i,v_j)}$ and $\frac{D(y,z)}{D(u_i,v_j)}$ in (174), it is easily proved that

$$dy\Lambda dz = N_1(y,z)-N_1(z,y) \quad , \tag{176}$$

$$N_1(y,z) \equiv \sum_{i,j=1}^{3}\left[y_{u_i}z_{u_j}du_i du_j+y_{v_i}z_{v_j}dv_i dv_j+y_{u_i}z_{v_j}du_i dv_j\right] \tag{177}$$

and $N_1(y,z)$ is an antisymmetric function with respect to the variables y and z.. Respectively,

$$\frac{D(y,z)}{D(v_i,v_j)}, \frac{D(y,z)}{D(u_i,v_j)} \quad , \quad N_1(z,x) \ , \ N_1(x,y) \tag{178}$$

are determined similarly to formulae (175).

It is important to note that with account of (177), expression (176) can be written as

$$dy\Lambda dz = \sum_{i,j=1}^{3}\left[(y_{u_i}z_{u_j}-z_{u_i}y_{u_j})du_i du_j+(y_{v_i}z_{v_j}-z_{v_i}y_{v_j})dv_i dv_j\right]$$

$$+(y_{u_i}z_{v_j}-z_{u_i}y_{v_j})du_i dv_j] \quad . \tag{179}$$

Each of the first and the second terms in the square brackets consists of an antisymmetric term in the round brackets, multiplied by a symmetric term ($du_i du_j$ or $dv_i dv_j$). It is known that such a multiplication will give zero because an antisymmetric function is multiplied by a symmetric one. Consequently, only the last (third) term will remain

$$dy\Lambda dz = \sum_{i,j=1}^{3} (y_{u_i} z_{v_j} - z_{u_i} y_{v_j}) du_i dv_j \quad . \tag{180}$$

This is a second order differential form, based on the different differentials du_i and dv_j. For the case of two satellites, the summation will be from 1 to 6. Similarly one can write the expressions for $dz\Lambda dx$ and $dx\Lambda dy$. Thus, there will be $6 \times 6 = 36$ terms in the right-hand side of the two-form $\omega_2 = Pdy\Lambda dz + Qdz\Lambda dx + Rdx\Lambda dy$ (131), where formulae (172) for P will be used (Q and R are obtained after cyclic permutation $x \to y \to z \to x$) and also (173) for the antisymmetric function $K_1(y,z)$ (respectively after cyclic permutation, also for $K_1(z,x)$ and $K_1(x,y)$). The 36 terms in the sum $\sum\limits_{i,j=1}^{6}$ will be composed of all the combinations of multiplication of differentials from the first set of Kepler parameters $(df_{.1}, da_{1.}, de_1, d\Omega_{1.}, di_{1.}, d\omega_1)$ with the differentials $(df_{2.}, da_{2.}, de_2, d\Omega_{2.}, di_{2.}, d\omega_2)$ from the second set. The detailed calculations will be performed in another paper.

9 Conclusion

In this paper a new model is presented for transmitting signals between satellites on two different space-distributed orbits, characterized by the two sets of Kepler parameters $(f_{.1}, a_{1.}, e_1, \Omega_{1.}, i_{1.}, \omega_1)$ and $(f_{2.}, a_{2.}, e_2, \Omega_{2.}, i_{2.}, \omega_2)$, with account also of the the General Relativity effect of curving the signal trajectory and the Shapiro delay formulae. First, it has been proved on the base of the Whitney theorem from differential geometry and the topological notions of immersions, embeddings and submersions that it is possible to construct a manifold, which will depend on all the 12 Kepler parameters plus the parameter s, which parametrizes cartesian coordinates x, y, z in the transformations (5) - (7). In fact, two sets of transformations (5) - (7) should be written for the variables x_1, y_1, z_1 and x_2, y_2, z_2 , but since it turned out to be possible to introduce the variables $x = x_2 - x_1$, $y = y_2 - y_1$, $z = z_2 - z_1$, the variational formalism was performed only by variation of the x, y, z variables. Secondly, it was proved that the first term in the Shapiro delay formulae $\frac{R_{AB}}{c}$ (R_{AB}–the Euclidean distance between the satellites) should be replaced by the term $\frac{1}{c}\int\limits_{path}\sqrt{(\dot{x})^2 + (\dot{y})^2 + (\dot{z})^2}ds$, which shows that the distance travelled by the signal is greater than the distance between points on the straight line. Here an important clarification should be made. In this paper a variational formalism was performed, taking into account only the term $\frac{1}{c}\int\limits_{path}\sqrt{(\dot{x})^2 + (\dot{y})^2 + (\dot{z})^2}ds$, and the result was that the straight line is the optimal trajectory for propagation of the signal. However, this does not mean that the signal propagates along a straight line. In previous papers [15], [16] and [17] it has been proved that geodesic distance, travelled by the signal is greater than the Euclidean distance. Here the straight line appears in the consideration because of the condition for "optimality" of the distance and the positions of the two satellites and is fully consistent with theorems from differential geometry. So the result by itself does not contradict to any physical laws about propagation of signals, it only suggests that in order to find any optimal positions or trajectories, one should take into account the second term $\frac{2G_{\oplus}M_{\oplus}}{c^3}\frac{\sqrt{(\dot{x})^2+(\dot{y})^2+(\dot{z})^2}}{[x^2+y^2+z^2]}\left[x\dot{x} + y\dot{y} + z\dot{z}\right]$ (see (36)) in the proposed modified Shapiro delay formulae. It should be stressed also that particular attention in the paper is paid to the approximations in deriving the standard Shapiro formulae, since the differential in the null cone equation can be determined in several ways. As a result of a not so restrictive condition, the modified Shapiro delay formulae was obtained in the form (35).

Taking into account of this second term (36) has not been performed in this paper (in the sense of solving the corresponding Euler-Lagrange equations), but the variational formalism is proposed in its most general form .

The main goal of this paper was to propose a variational formalism, based on higher order differential forms and the Stokes and Gauss Ostrogradsky formalism. Since the Stokes theorem relates an integral on a manifold M to an integral on its boundary ∂M, it was necessary to clarify the notion of a boundary, which in its classical formulation is related to the inverse function theorem [36] and sub-manifolds, convexity and semi-spaces. The last two notions are frequently used also in differential geometry of surfaces [25] and optimization theory [24]. The definition for a boundary turned out to be useful for finding the boundary of the manifold, given by the transformations (5) - (7). For this case, there are three boundaries - the $(x,y)-$boundary, given by the plane (84) $x\sin(\omega+f+\Omega)-y\cos(\omega+f+\Omega)=0$ or the plane (85) $x\sin(\omega+f-\Omega)+y\cos(\omega+f-\Omega)=0$ (and the condition for a boundary $z=0$); the (x,z)-boundary (91) $x .\sin i\frac{\left[\sin^2(\omega+f)\left(1+\cos^2 i\right)-1\right]}{\sin^2(\omega+f)\cos^2 i}+z\frac{\cos\Omega\left[1-\cos^4 i\; tg^4(\omega+f)\right]}{\cos^2 i\; tg^3(\omega+f)}=0$ (and the condition $y=0$) and the (y,z) boundary (98) $Ny-z\cos(\omega+f)=0$ with $N(\omega,f,\Omega)\equiv\sqrt{-\cos(\omega+f+\Omega)\cos(\omega+f-\Omega)}$ (97) (and the condition $x=0$). These boundaries naturally are $2-$dimensional hyperplanes, because the initial manifold is a $3-$dimensional one, and consequently the boundaries are 1 dimension less. The boundaries were found only for the case of one $3D$ manifold M_1. The boundary of the manifold $M=M_1\times M_2$ can be found if each of these equations is written for the corresponding manifolds M_1 and M_2 with indices "1" and "2" respectively, and the system, determining the (x,y,z)-coordinates of the $M-$manifold $x=x_2-x_1$, $y=y_2-y_1$, $z=z_2-z_1$ is added to them. Thus there will be a system of $9-$equations for the $9-$variables $(x_1,y_1,z_1,x_2,y_2,z_2,x,y,z)$ and the x,y,z coordinates will be expressed through the two sets of Kepler parameters.

Another peculiarity in the variational formalism, based on second- and third-order differential forms and the Stokes theorem is the non-commutativity of the partial derivatives $\frac{\partial}{\partial x},\frac{\partial}{\partial y},\frac{\partial}{\partial x}$, which means that the commutator (162) $\left[\frac{\partial}{\partial x_\mu},\frac{\partial}{\partial x_\nu}\right]L=\frac{1}{9}\sum_{i,k=1}^{3}\left[\xi_{ik}^{(1)}\frac{\partial}{\partial u_k}+\xi_{ik}^{(2)}\frac{\partial}{\partial v_k}\right]L\neq 0$ is nonzero.This is natural from the viewpoint of differential geometry since a zero commutator means that the condition for integrability and the Frobenius theorem are satisfied [37]. In the case, however, the additional terms in the variational action due to the non-commutativity have to be accounted.

Further, the function $P(x,y,z)=\frac{\partial C}{\partial y}-\frac{\partial B}{\partial z}$ has been calculated in (172), which is in the right-hand side of the Stokes theorem (131) $\int\limits_{\partial M}[Adx+Bdy+Cdz]=\int\limits_{M}[Pdy\Lambda dz+Qdz\Lambda dx+Rdx\Lambda dy]$ and also in the two-form (166) $\omega_2=Pdy\Lambda dz+Qdz\Lambda dx+Rdx\Lambda dy$. The function P is expressed in terms of the functions (x,y,z) and their derivatives with respect to the Kepler parameters. Again it is useful to remind that when the summation is changed from 1 to 6 in both sums, which include the derivatives with respect to u_i and v_j, the expression for P will contain derivatives with respect to both sets of Kepler parameters. Particularly interesting and rather complicated is the antisymmetric function $K_1(y,z)$ in (173), which contains derivatives (of first, second and third order) of the original Lagrangian with respect to u_i,v_j and the parameter s.

There is one more interesting problem, concerning the right-hand side in the Stokes theorem (131)(the integral $\int\limits_{M}[Pdy\Lambda dz+Qdz\Lambda dx+Rdx\Lambda dy]$. The functions P in (172), Q and R contain the function $K_1(y,z)$ (173) (respectively $K_1(z,x)$ and $K_1(x,y)$), and these functions contain first-order derivatives with respect to u_i and v_j and second - order derivatives with respect to u_i,u_j or v_i,v_j or u_i,v_j. In the derived expression (180) for $dy\Lambda dz$, a sum over the different combinations $du_k dv_l$ was obtained, so each derivative with respect to u_i or second-order derivative with respect to u_i and u_j is multiplied by du_k. In the variational formalism, we have not imposed any restrictions which are the concrete Kepler parameters corresponding to each u_i or to each v_j. So the problem is: in how many ways one can choose two variables u_i,u_k from among the 6 parameters for the first set, respectively for the case of second derivatives in $K_1(y,z)$ (respectively $K_1(z,x)$ and $K_1(x,y)$) - in how many ways three variables u_i,u_j,u_k can be chosen from among the total 6? Note that the variable u_k can be chosen independently from u_i and u_j (but can be also equal to one of them). Consequently, after obtaining the final formulaes in the variational formalism, one should in addition sum up over all the possible choices (permutations) of u_i,u_j,u_k (or u_i,u_k, depending on the terms in $K_1(y,z)$). The same should be performed with respect to the possible permutations of v_i,v_j,v_k (or v_i,v_k. This would require the implementation of group theory and shall be developed in further publications.

10 Appendix A: Expression for the square of the Euclidean distance between two satellites on two different space-distributed orbits

The differential dR_{AB}^2 of the square of the Euclidean distance, where

$$R_{AB}^2=(x_1-x_2)^2+(y_1-y_2)^2+(z_1-z_2)^2 \tag{181}$$

and (x_1, y_1, z_1) and (x_2, y_2, z_2) are given by the transformations (5), (6) and (7) with indices "1" and "2" correspondingly:

$$dR_{AB}^2 = \{\frac{2e_1 \sin f_1 a_1^2(1-e_1^2)}{(1+e_1\cos f_1)^3} + \frac{2a_1 a_2(1-e_1^2)(1-e_2^2)}{(1+e_1\cos f_1)^2(1+e_2\cos f_2)}[K_1(1,2)+$$

$$+K_2(1,2)+K_3(1,2)+K_4(1,2)]\}df_1 + \{1 \longleftrightarrow 2\}df_2 \ , \tag{182}$$

where the expression $\{1 \longleftrightarrow 2\}$ is the same as in the first bracket $\{\}$ on the first line of (182), but with interchanged indices 1 and 2, i.e.$1 \longleftrightarrow 2$. Expression (182) in fact is the expression for the differential dR_{AB}^2 in terms of the differentials df_1 and df_2 (provided that only f_1 and f_2 change, the remaining five Kepler parameters in the two sets $(\Omega_1, \omega_1, i_1, a_1, e_1)$ and $(\Omega_2, \omega_2, i_2, a_2, e_2)$ remain constant,but remain in the formulae), i.e.

$$dR_{AB}^2 = \frac{\partial\left(R_{AB}^2\right)}{\partial f_1}df_1 + \frac{\partial\left(R_{AB}^2\right)}{\partial f_2}df_2 = \widetilde{K}_1(\Gamma^{(1)}, \Gamma^{(2)})df_1 + \widetilde{K}_2(\Gamma^{(2)}, \Gamma^{(1)})df_2 \quad , \tag{183}$$

where again fo convenience the notations $\Gamma^{(1)} = (f_1, \Omega_1, \omega_1, i_1, a_1, e_1)$ and $\Gamma^{(2)} = (f_2, \Omega_2, \omega_2, i_2, a_2, e_2)$ have been introduced. The expressions for $K_1(1,2), K_2(1,2), K_3(1,2), K_4(1,2)$ in (182) are presented below. The corresponding expressions for $K_1(2,1), K_2(2,1), K_3(2,1), K_4(2,1)$ in $\{1 \longleftrightarrow 2\}df_2$ are the same, but with interchanged indices $1 \longleftrightarrow 2$

The terms $K_1(1,2)$ do not contain the inclination angles i_1 and i_2

$$K_1(1,2) \equiv -e_1 \sin f_1 \cos(\omega_1 + f_1)\cos(\omega_2 + f_2)\cos(\omega_1 + f_1 + \Omega_1 - \Omega_2)+$$

$$+e_1 \cos f_1 \sin(\omega_1 + f_1)\cos(\omega_2 + f_2)\cos(\Omega_1 - \Omega_2)+$$

$$+\sin(\omega_1 + f_1)\cos(\omega_2 + f_2)\cos(\Omega_1 - \Omega_2) \quad . \tag{184}$$

The terms $K_2(1,2)$ contain the inclination angles i_1 and i_2 in terms with the multipliers $\cos i_1 \cos i_2$ and $\sin i_1 \sin i_2$

$$K_2(1,2) \equiv -e_1 \sin(\omega_2 + f_2)\cos i_1 \cos i_2[\sin f_1 \cos(\Omega_1 - \Omega_2)\sin(\omega_1 + f_1)+$$

$$+\cos(\omega_1 + f_1)\cos f_1 \cos(\Omega_1 - \Omega_2) + \cos(\omega_1 + f_1)\sin f_1 \sin(\Omega_1 - \Omega_2)] \quad . \tag{185}$$

Terms $K_3(1,2)$ contain $\cos i_1$ and not containing $\cos i_2$

$$K_3(1,2) \equiv [e_1 \cos f_1 + 1]\cos(\omega_1 + f_1)\cos(\omega_2 + f_2)\sin(\Omega_1 - \Omega_2)\cos i_1 \quad . \tag{186}$$

The terms $K_4(1,2)$ contain the multiplier $\cos i_2$ and do not contain $\cos i_1$

$$K_4(1,2) \equiv [-e_1 \cos f_1 + 1]\sin(\omega_1 + f_1)\sin(\omega_2 + f_2)\sin(\Omega_1 - \Omega_2)\cos i_2 \quad . \tag{187}$$

11 Appendix B: The propagation time for a signal, emitted by a satellite on a space-distributed orbit, expressed in terms of higher-rank elliptic integrals

Formulae (8) $\widetilde{T} = G(f) = -\frac{2G_{\oplus}M_{\oplus}}{c^3}.\frac{n}{(1-e^2)^{\frac{3}{2}}}\widetilde{T}_1 + T_2^{(2)}$ for the propagation time has been derived in [14], but in the review paper [15] the different parts of the expression have been collected in one formulae. The expression for $\widetilde{T}_1$ in (8) is

$$\widetilde{T}_1 = . - 2i\frac{na}{c}q^{\frac{3}{2}}\int \frac{y^2 dy}{\sqrt{(1-y^2)(1-q^2y^2)}} \quad , \tag{188}$$

where $y = \sqrt{\frac{1}{q}.\frac{(1+e\cos E)}{(1-e\cos E)}}$, $q = \frac{1-e}{1+e}$ and $\tan\frac{f}{2} = \sqrt{\frac{1-e\cos f}{1+\cos f}} = \sqrt{\frac{1+e}{1-e}}\tan\frac{E}{2}$. The second integral T_2 is

$$T_2 = -\frac{2G_{\oplus}M_{\oplus}}{c^3}.\frac{n}{(1-e^2)^{\frac{3}{2}}}\widetilde{T}_1 + T_2^{(2)} \quad , \tag{189}$$

where $T_2^{(2)}$ is a combination of the elliptic integral $\widetilde{J}_2^{(4)}(\widetilde{y},q)$ of the second order and the elliptic integral of the fourth order $\widetilde{J}_4^{(4)}(\widetilde{y},q)$

$$T_2^{(2)} = . - inq^{\frac{5}{2}}\frac{2G_{\oplus}M_{\oplus}}{c^3}(1+e^2)\widetilde{J}_2^{(4)}(\widetilde{y},q) + inq^{\frac{3}{2}}\frac{2G_{\oplus}M_{\oplus}}{c^3}\frac{(1+e^2)}{(1-e^2)}\widehat{J}_4^{(4)}(\widehat{y},q) \ . \tag{190}$$

The lower indice denotes the order of the integral (determined by the degree of $\widetilde{y}$ or $\widehat{y}$ in the nominator) and the upper indice in the bracket () is related to the degree of the polynomial under the square root in the denominator - in the case it is $\left(1-\widetilde{y}^2\right)\left(1-q^2\widetilde{y}^2\right)$ or $\left(\widehat{y}^2-1\right)\left(1-q^2\widehat{y}^2\right)$. The corresponding elliptic integrals of the second and of the fourth order are

$$\widetilde{J}_2^{(4)}(\widetilde{y},q) = \int \frac{\widetilde{y}^2 d\widetilde{y}}{\sqrt{(1-\widetilde{y}^2)(1-q^2\widetilde{y}^2)}} \quad , \quad \widehat{J}_4^{(4)}(\widehat{y},q) = \frac{q^5}{i}\int \frac{\widehat{y}^4 d\widehat{y}}{\sqrt{(\widehat{y}^2-1)(1-q^2\widehat{y}^2)}} \quad , \tag{191}$$

In [15] it was proved that the integrals $\widetilde{T}_1$ and $T_2^{(2)}$ are imaginary quantities, so that the propagation time $T=G(f)$ in expression (8) will be a real quantity, as it should be.

12 Appendix C: Coefficient functions $P_{ij}^{(1)}, P_{ij}^{(2)} ....P_{ij}^{(10)}$ in the expression for the function $K_1(y,z)$ in (172)

$$P_{ij}^{(1)}(y,z) \equiv \frac{1}{9}\sum_{i,j=1}^{3}\left[\frac{1}{z_{u_j}}\frac{\partial}{\partial u_j}\left(\frac{\dot{y}}{\ddot{y}}.\frac{1}{y_{u_i}}\right) + \frac{1}{z_{v_j}}\frac{\partial}{\partial v_j}\left(\frac{\dot{y}}{\ddot{y}}.\frac{1}{y_{u_i}}\right)\right] \Longrightarrow \text{variat.deriv.}\frac{\partial^2 L}{\partial u_i \partial s} \ , \tag{192}$$

$$P_{ij}^{(2)}(y,z) \equiv \frac{1}{9}\sum_{i,j=1}^{3}\left[\frac{1}{z_{u_j}}\frac{\partial}{\partial u_j}\left(\frac{\dot{y}}{\ddot{y}}.\frac{1}{y_{v_i}}\right) + \frac{1}{z_{v_j}}\frac{\partial}{\partial v_j}\left(\frac{\dot{y}}{\ddot{y}}.\frac{1}{y_{v_i}}\right)\right] \Longrightarrow \text{variat.deriv.}\frac{\partial^2 L}{\partial v_i \partial s} \ , \tag{193}$$

$$P_{ij}^{(3)}(y,z) \equiv \frac{1}{9}\sum_{i,j=1}^{3}\frac{1}{z_{u_i}}\frac{\dot{y}}{\ddot{y}}.\frac{1}{y_{u_j}} \Longrightarrow \text{variat.deriv.}\frac{\partial^3 L}{\partial u_i \partial u_j \partial s} \ , \tag{194}$$

$$P_{ij}^{(4)}(y,z) \equiv \frac{1}{9}\sum_{i,j=1}^{3}\left[\frac{1}{z_{v_j}}\frac{\dot{y}}{\ddot{y}}.\frac{1}{y_{u_i}} + \frac{1}{z_{u_i}}\frac{\dot{y}}{\ddot{y}}.\frac{1}{y_{v_j}}\right] \Longrightarrow \text{variat.deriv.}\frac{\partial^3 L}{\partial u_i \partial v_j \partial s} \ , \tag{195}$$

$$P_{ij}^{(5)}(y,z) \equiv \frac{1}{9}\sum_{i,j=1}^{3}\frac{1}{z_{v_i}}\frac{\dot{y}}{\ddot{y}}.\frac{1}{y_{v_j}} \Longrightarrow \text{variat.deriv.}\frac{\partial^3 L}{\partial v_i \partial v_j \partial s} \ , \tag{196}$$

$$P_{ij}^{(6)}(y,z) \equiv \frac{1}{9}\sum_{i,j=1}^{3}\frac{1}{y_{u_i}}T^{(j)}(\dot{z},\ddot{z},\dddot{z},z_{v_j}) \Longrightarrow \text{variat.deriv.}\frac{\partial^2 L}{\partial u_i \partial u_j} \ , \tag{197}$$

$$P_{ij}^{(7)}(y,z) \equiv \frac{1}{9}\sum_{i,j=1}^{3}\left[\frac{1}{y_{v_j}}T^{(i)}(\dot{z},\ddot{z},\dddot{z},z_{u_i}) + \frac{1}{y_{u_i}}T^{(j)}(\dot{z},\ddot{z},\dddot{z},z_{v_j})\right] \Longrightarrow \text{variat.deriv.}\frac{\partial^2 L}{\partial u_i \partial v_j} \ , \tag{198}$$

$$P_{ij}^{(8)}(y,z) \equiv \frac{1}{9}\sum_{i,j=1}^{3}\frac{1}{y_{v_j}}T^{(i)}(\dot{z},\ddot{z},z_{u_i}) \Longrightarrow \text{variat.deriv.}\frac{\partial^2 L}{\partial v_i \partial v_j} \ , \tag{199}$$

$$P_{ij}^{(9)}(y,z) \equiv \frac{1}{3}\sum_{i,j=1}^{3}\left[\frac{1}{y_{v_j}}\frac{\partial T^{(i)}(\dot{z},\ddot{z},\dddot{z},z_{u_i})}{\partial v_j} + \frac{1}{y_{u_i}}\frac{\partial T^{(j)}(\dot{z},\ddot{z},\dddot{z},z_{u_i})}{\partial u_j}\right] \Longrightarrow \text{variat.deriv.}\frac{\partial^{\cdot} L}{\partial u_i} \ , \tag{200}$$

$$P_{ij}^{(10)}(y,z) \equiv \frac{1}{3}\sum_{i,j=1}^{3}\left[\frac{1}{y_{u_j}}\frac{\partial T^{(i)}(\dot{z},\ddot{z},\dddot{z},z_{u_i})}{\partial u_j} + \frac{1}{y_{v_j}}\frac{\partial T^{(j)}(\dot{z},\ddot{z},\dddot{z},z_{v_i})}{\partial v_j}\right] \Longrightarrow \text{variat.deriv.}\frac{\partial^{\cdot} L}{\partial v_i} \ , \tag{201}$$

everywhere by $T^{(i)}(\dot{z},\ddot{z},\dddot{z},z_{u_i})$ we have denoted the expression

$$T^{(i)}(\dot{z},\ddot{z},\dddot{z},z_{u_i}) \equiv \frac{1}{3}\frac{\dot{z}}{\ddot{z}}\frac{1}{(z_{u_i})^2}\dot{z}_{u_i} + \frac{1}{3}\frac{\dot{z}\dddot{z}}{(\ddot{z})}\frac{1}{z_{u_i}} \ . \tag{202}$$

14.3. Acknowledgements
The author is grateful to the organizers of the Sixteenth Conference of the Euro-American Consortium for Promoting the Application of Mathematics in Technical and Natural Sciences (AMITaNS24) June 21-26, 2024, Albena resourt, Bulgaria and especially to Prof. Michail Todorov for the opportunity to participate in this conference.

References

[1] Shapiro I 1964 Fourth test of General Relativity Phys. Rev. Lett. 13789

[2] Sovers O, Fanselow J and Jacobs C 1998 Rev. Mod. Phys. **70** 1393

[3] Turyshev S, Toth V, and Sazhin M 2013 Phys. Rev. **D87** 024020 (Preprint arXiv: 1212.0232v4 [gr-qc])

[4] Turyshev S, Sazhin M, and Toth V 2014 Phys. Rev. **D89** 105029 (Preprint arXiv: 1402.7111v1 [qr-qc])

[5] Turyshev S, Yu N, and Toth V 2016 Phys. Rev. **D93** 045027 (Preprint arXiv: 1512.09019v2 [gr-qc])

[6] A. D. Ludlow, M. M. Boyd, and J. Ye 2015 Rev. Mod. Phys. **87** 637

[7] Ashby N 2003 *Relativity in the Global Positioning System* Living Reviews in Relativity **61** (Preprint https://link.springer.com/content/pdf/10.12942

[8] Duchayne L 2*008 Transfert de temps de haute performance : le Lien Micro-Onde de la mission ACES* Physique mathematique PhD Thesis Observatoire de Paris Preprint https://tel.archives-ouvertes.fr/tel00349882/document; also https://theses.hal.science/tel-00349882

[9] Gulklett M 2003 *Relativistic effects in GPS and LEO* PhD Thesis, University of Copenhagen, Denmark, available at https://www.yumpu.com/en/document/view/4706552/

[10] Bothwell T 2022 *Resolving the gravitational redshift in a millimetre-scale atomic sample*, Nature DOI: 10.1038/s41586-021-04349-7. www.nature.com/articles/s41586-021-04349-7

[11] Aeppli A et al 2024 *A clock with* 8.10^{-19} *systematic uncertainty* DOI: 10.48550/arxiv.2403.10664; also https://phys.org/news/2024-07-world-accurate-precise-atomic-clock.html

[12] Kopeikin S, Efroimsky M, and Kaplan G 2011 *Relativistic Celestial Mechanics of the Solar System,* (Wiley-VCH, New York: Wiley-VCH)

[13] Gurfil P and Seidelman P 2016 *Celestial Mechanics and Astrodynamics: Theory and Practice.* Astrophysics and Space Science Library 436 (Berlin Heidelberg: Springer-Verlag)

[14] Dimitrov B 2022 Elliptic Integrals for Calculation of the Propagation Time of a Signal, Emitted and Received by Satellites on One Orbit AIP Conf. Proc. **2522** 978-0-7354-4361-7 (Preprint gen-ph/2201.13208v2)

[15] Dimitrov B 2023 J. Phys.: Conf. Ser. **2675** 012026 DOI 10.1088/1742-6596/2675/1/012026 review article, open access at https://https://iopscience.iop.org/article/10.1088/1742-6596/2675/1/012026

[16] Dimitrov B 2017 *Two null gravitational cones in the theory of GPS-intersatellite communications between two moving satellites. I. Physical and mathematical theory of the space-time interval and the geodesic distance on intersecting null cones* Preprint https://arxiv.org/abs/1712.01101v3

[17] Dimitrov B 2019 *New Mathematical Models of GPS Intersatellite Communications in the Gravitational Field of the Near-Earth Space* AIP Conf. Proc. **2075** 040007 (Preprint https://arxiv.org/abs/1909.12753v1)

[18] Shuai P 2021 *Understanding Pulsars and Space Navigations Navigation Science and Technology* vol.**11** (Singapore:National Defence Industry Press + Springer Nature Pte Ltd)

[19] Kim G, Park S, Seong S, Lee J, Choi S, Kim Y, Ryu H, Lee S, Choi J, Han S 2023 *Concept of Laser Crosslink Systems Using Nanosatellites in Formation Flying* Acta Astronautica **211** 877

[20] Zaman I, Boyraz O 2020 *Impact of receiver architecture on small satellite optical link in the presence of pointing jitter* Applied Optics **59** 10177

[21] Postnikov M 1987 *Smooth Manifolds Semestr III. Lectures on Geometry* (Moscow: Editorial Publishing House for Physical and Mathematical Literature)

[22] Fock V 1959 *Theory of Space, Time and Gravitation* (New York:Pergamonn Press)

[23] Narasimhan R 1968 *Analysis on Real and Complex Manifolds* (Amsterdam:North-Holland Publishing Company)

[24] Suharev A, Timohov A and Fedorov V 2005 *A Course on Optimization Methods* 2nd ed. (Moscow:Fizmatlit)

[25] Thorpe J 1979 *Elementary Topics in Differential Geometry* (New York-Heidelberg Berlin: Springer)

[26] Fedorchuk V 1990 *A Course of Analytical Geometry and Linear Algebra* (Moscow:Publishing House of the Moscow University)

[27] Sternberg S 1964 *Lectures on Differential Geometry* (New Jersey:Prentice Hall, Inc. Englewood Cliffs)

[28] Arnold V 1984 *Ordinary Differential Equations* Third Edition (Moscow:Nauka Publishing House)

[29] Arnold V 1988 *Geometrical Methods in the Theory of Ordinary Differential Equations Grundlehren der Mathematischen Wissenschaften* GL vol 250 (New York:Springer Science+Business Media)

[30] Dubrovin B, Fomenko A and Novikov S 1992 *Modern Geometry-Methods and Applications. Part I. Geometry of Surfaces, Transformation Groups and Fields. Graduate Texts in Mathematics 93* (New York:Springer Science+Business Media LLC)

[31] Dubrovin B, Fomenko A and Novikov S 1985 *Modern Geometry-Methods and Applications. Part II. The Geometry and Topology of Manifolds. Graduate Texts in Mathematics 93* (New York:Springer Science+Business Media LLC)

[32] Prasolov V 2022 *Differential Geometry* (Springer Nature Switzerland AG)

[33] Milnor J 1963 *Morse Theory Based on lecture notes by M. Spivak and R. Wells* (Princeton New Jersey:Princeton University Press)

[34] Vasiliev V A 2001 *Introduction to Topology Student Mathematical Library vol 14* (Rhode Island:American Mathematical Society)

[35] Novikov S P, Taimanov I A 2005 *Contemporary Geometrical Structures and Fields* (Moscow:Moscow Centre for Continuous Mathematical Education)

[36] Mishtenko A, Fomenko A. 2000 *A Course of Differential Geometry and Topology* (Moscow:Factorial Press)

[37] Boothby W 1975 *An Introduction to Differentiable Manifolds and Riemannian Geometry* (New York San Francisco London:Academic Press)

[38] Schwarz A 1994 *Topology for Physicists Grundlehren der Mathematischen Wissenschaften* 308 (Berlin Heidelberg:Springer-Verlag))

[39] Trofimov V, Fomenko A 1995 *Algebra and Geometry of Integrable Hamiltonian Equations* (Moscow:Factorial Publishing House)

[40] Golubitsky M , Guillemin V 1973 *Stable Mappings and Their Singularities* (New York Berlin:Springer-Verlag)

[41] Arnold V, Varchenko A and Gusein-Zade S 2004 *Singularities of Differentiable Maps* (Moscow:Moscow Centre for Continuous Mathematical Education)

[42] Shapiro M, Olshanetsky M 1999 *Lectures on Topology for Physicists* (Publishing House "Udmurts University")

[43] Nash C Sen S 1983 *Topology and Geometry for Physicists* (London New York:Academic Press)

[44] Nakahara M 2003 *Geometry, Topology and Physics Graduate Study Series in Physics* (Bristol and Philadelphia:Institute of Publishing Ltd)

[45] TuL 2011 *An Introduction to Manifolds* Sec. Ed. (New York:Springer Science+Business Media, lLC)

[46] Gelfand I Fomin S 1963 *Calculus of Variations* (New Jersey Englewood Cliffs:Prentice-Hall Inc.)

Journal of Physics: Conference Series **2910** (2024) 012002 doi:10.1088/1742-6596/2910/1/012002

Air pollution modelling of accidents involving hazardous substances

Tzvetan Ostromsky [1*], Kiril Alexiev [1] and Stefan Parvanov [2]

[1] Department of Parallel Algorithms and Machine Learning with Neurotechnology Laboratory, Institute of Information and Communication Technologies, Bulgarian Academy of Sciences, Sofia, Bulgaria
[2] Faculty of Fire Safety and Civil Protection, Academy of the Ministry of Interior, Sofia, Bulgaria

*E-mail: tz.ostromsky@gmail.com

Abstract. Air pollution modeling for accidents involving hazardous substances is a critical aspect of emergency response and environmental protection. This modeling aims to predict the dispersion and impact of toxic substances released into the atmosphere during accidents, such as industrial spills, transport accidents, chemical plant incidents or acts of terrorism. Such dangers are among the challenges which the modern society is facing and has to overcome. In this study we propose a systematic approach for organizing monitoring, collecting data from mobile sensors, creating situation development scenarios, modelling the spread of potential toxic-element pollution in dependence with the current meteorological conditions. The UNI-DEM model for simulation of the pollutants' dispersion and transport in the air, as well as models for spread of leaking gas and liquid vapors are involved in this study. Finally, depending on the comprehensive analysis of the modeling results, an adequate response to such severe situations can be proposed. Assessment of safety, healthcare and environmental impacts should be the final result and the most socially important application of our work. The calculated pollutant concentrations can be compared with regulatory standards or health-based guidelines to assess potential health risks to exposed populations and environmental impacts on ecosystems. The importance of this information for emergency response planning, public health interventions, and environmental management strategies is beyond dispute.

1. Aims of the work

The paper aims to propose a systematic approach to the problem of early identifying, evaluating the danger and proper response to the peak pollution of the atmosphere caused by some disaster. Some of the major challenges the modern society is facing are related to the spread of harmful gasses, dust particles, aerosols and other dangerous substances in the air. Such threads can appear as a result of industrial accidents, terrorist attacks or natural disasters.

The fact that the listed events are undesirable does not mean that they are impossible and our society must be prepared to deal with them. The actions how to prevent these events are not considered in the article. The main activities of monitoring organization, data collection and processing and the mathematical models used for air pollutant transport are reviewed. The aim is to outline the main mathematical tools and models that are used in the processing of signals, data and information. Special attention is paid to the difficulties that must be overcome to make informed decisions in real time in the harsh conditions of such an incident. A real-time

Content from this work may be used under the terms of the Creative Commons Attribution 4.0 licence. Any further distribution of this work must maintain attribution to the author(s) and the title of the work, journal citation and DOI.
Published under licence by IOP Publishing Ltd

computer system has to be developed for integration/fusion and visualization of heterogeneous data from hundreds or even thousands of sensors, filtering outliers and minimizing noise, estimating basic parameters, adapting pre-prepared models to specific conditions, feeding the models with true data and obtaining information about the dynamics of pollutant transfer development. Without good preliminary preparation it can not be implemented in the case of a specific incident. This article focuses on the main components of such systems. The goal is to outline the possible range of problem tasks that have to be solved in incidents related to the air transfer of pollutants and for the solution of which the relevant mathematical and software tools should be prepared in advance. The article also lists the main fundamental achievements in these areas, with the help of which such a system should be created.

2. Monitoring system

2.1 Technology of the modern monitoring systems

The quick and efficient solution of the problems arising as a result of industrial accidents, terrorist attacks or natural disasters requires the maintenance of a Disaster management system with sufficient material and human resources and rights granted by the law. The tasks of this system are numerous, but in this publication we will focus only on some of them, regarding the creation of a monitoring system for collecting data and information about the accident that have occurred, the area/volume of the affected territory, the degree of danger for local residents, the dynamics of observed process and many more. For these purposes, the Disaster management system deploys a monitoring system. An example of such system is illustrated in Figure 2.1.

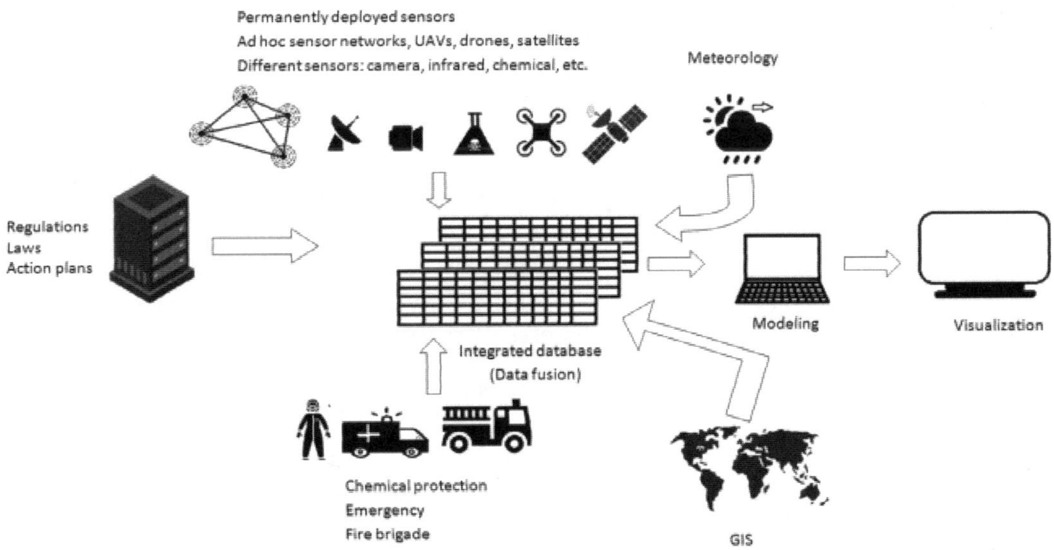

Figure 2.1. Various monitoring tools and their integration into a monitoring system.

In the case of particularly dangerous contamination and radioactivity, this system consists entirely of deployed ad hoc sensor networks. Sensor networks can include different types of sensors such as simple visual surveillance cameras, thermal cameras to evaluate ongoing processes with heat release/absorption, detection of crowded people, chemical sensors to measure the concentration of certain pollutants, inertial sensors to measure vibrations, earthquake/shock and accelerations, radar sensors to determine the movement of earth layers, remote monitoring of pulse and breathing and many others. The sensors can be directly located in the area of the accident, along its perimeter or on mobile objects - mobile robots, drones, airplanes, satellites and others. In more gentle conditions, suitably equipped people can be used to collect information and data. Additional information aiding the process of forecasting pollution dynamics is provided by local meteorological offices.

Below we will focus on the concept of an intelligent sensor network.

2.2 Sensors and sensor networks

The sensor network architecture can be designed to perform a specific task (Figure 2.2). A more flexible approach implies dynamically changing sensor networks, with different numbers of active sensors at any moment in time, and different tasks to solve. These networks enable the study of an area or process in an ad hoc deployment. This provides great flexibility to the sensor network, but at the expense of this, the demands on developers increase significantly. The main problems that arise are the ability to work with inaccurate, incomplete and often contradictory information from sensors of the same or different types, reliable estimation of the parameters of the observed process or phenomenon based on this information, creation of specialized protocols for establishing sustainable behavior of the network of sensors and optimal exchange of information between them. In the case of networks with wireless technologies for data exchange, a significant problem is the optimization of the energy costs of the sensor nodes in order to preserve the system's operability as long as possible.

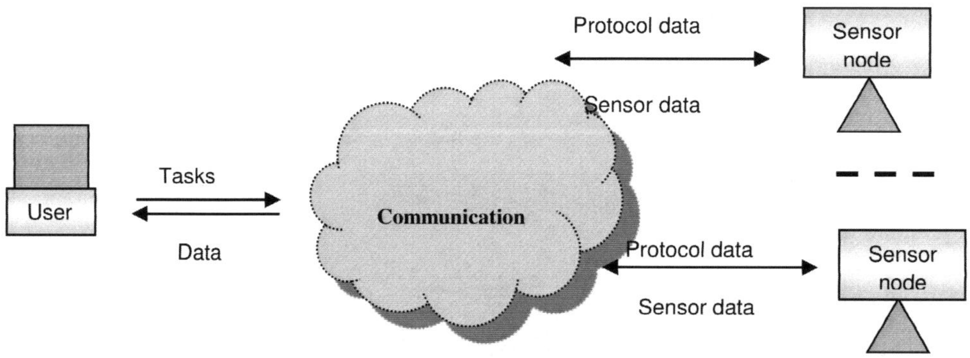

Figure2.2. Generalized representation of a sensor network

Sensor networks can be generally represented as an interaction of four main components – sensors, the studied environment, algorithms for processing the received signals and data, and communication equipment/protocols.

According to the infrastructure they use, sensor networks can be conditionally divided into several groups. The first group uses the established network infrastructure. It can be deployed in very short time. The second group belongs to the server-client type, in which the sensors gain access to the network environment through access points. The sensors use radio communication to communicate with access points and have freedom of positioning in the space around the access point (usually no more than 80-100m). In the presence of full coverage of the monitored space with a system of access points, the sensors have complete freedom of positioning. Typically, access points use a copper or fiber connection with the rest of the nodes in the network. Sensor networks that do not use any pre-deployed infrastructures are called ad hoc networks. These networks have the greatest flexibility and possibility of use in the most diverse conditions. With them, the exchange of information takes place entirely in the radio range. Each sensor node also has the functionality of a router.

In terms of stability, wireless sensor networks successfully meet the requirements for durability and reliability in harsh conditions thanks to the new advances in materials. Sensors embedded in the modern materials are able to withstand repeated exposure to corrosive gases and high temperatures. However, in case of disaster conditions the sensor components have to be designed to withstand particularly extreme conditions. In addition, a very high reliability is required in order to prevent loss of observation of the monitored space. Currently there are many wireless technologies with preventive maintenance and recovery capabilities to meet these challenges.

Sensor networks offer several advantages with respect to single sensors. Some of the more important ones are:

- High reliability of the information about the monitored event/process. Two or more sensors provide information about the studied process. In this way, the information uncertainty in our knowledge of the situation is reduced;
- As a result of the integration of measurements from multiple sensors (the so-called data fusion), the signal-to-noise ratio increases, which in turn increases the probability of detection;
- The reliability of the entire system is improved. In the case of unavailable sensors, sensors in a non-working state, or ineffective ones, there is a high probability that at least one sensor will provide the necessary information;
- It is easier to cover completely the monitored space;
- A continuous mode of monitoring can be implemented - 24/7;
- A network of cheaper sensors can achieve the same or a higher level of monitoring quality than a single very expensive sensor.

In the intelligent sensor networks there are significantly higher requirements for individual sensor nodes and involvement of nodes in the processing of information. A possible architecture of a smart sensor node is shown in Figure 2.3.

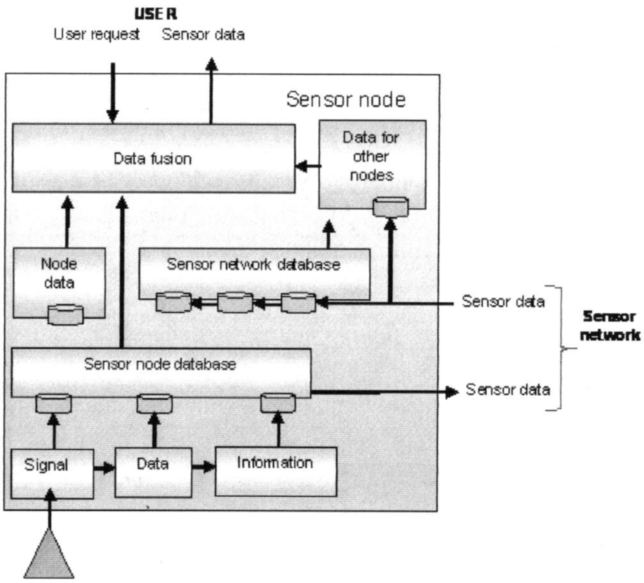

Figure 2.3. Generalized representation of a sensor node

2.3 Processing and validation of sensor data

The environment includes useful signals (from the studied object) and signals generated by various types of noise received by the sensors. Due to the limited accuracy of the sensors, additional noise from the sensor receiver (most often) is added to these signals. In some cases, the noise is multiplicative, which significantly complicates signal processing. From this input information, distinguished by its inaccuracy, incompleteness and contradiction (two sensors can claim two different things), all useful information must be extracted and the parameters of the observed object, process or phenomenon must be estimated with maximum accuracy. This is the main task of the processing algorithms.

Processing algorithms are distinguished as signal, data and information processing algorithms. Signal processing deals with the detection, analysis, and manipulation of audio, video, images, and many other types of analog and digital signals. Signal processing algorithms are used to remove noise and detect, locate or recognize the objects under study. Unnecessary information should be deleted. The choice of threshold and detection algorithm largely determines the performance of the entire system. For example, choosing a low detection threshold minimizes the probability of not detecting useful information, but at the expense of this, a huge amount of unnecessary information enters the system, which overloads the processor. Other tasks that are performed in addition to noise removal are signal amplification, modulation, demodulation, error correction, enhancement of the quality of the signals, features extraction, etc.

Data processing algorithms estimate the parameters of the observed objects. When evaluating a single object, statistical methods are usually used. A serious problem to solve is the removal of those from the data that implausibly reflect the behavior of the object. Much more difficult is the case of simultaneous observation of several objects at the same time [9,14]. If the objects have not clear identificators/labels it is not know to which object the measurements belong. This problem is known as the data association problem. The computational complexity of this task arises exponentially on the number of measurements being obtained. Data processing algorithms are often considered pooling algorithms because they use a sample of measurements to estimate individual parameters. The most popular of the model based data processing algorithms are Kalman and non-linear Kalman filters, AR and ARMA estimators, various transformations, correlation algorithms, coherent filtration methods, contour detection algorithms (Sobel, Prewitt, Canny, Roberts, Laplacian, Hough, Radon), data association algorithms, hybrid scoring algorithms, multi-model approaches, and much more. [9,10,12,13].

Information processing algorithms serve as a source of identification of the object of surveillance. They are based on our prior knowledge of objects stored in databases and various types of decision rules such as Bayes, Demster-Shaffer, etc. [11].

3. Modelling the spread of potential toxic-element pollution in the air

There are many possible models for the dispersion of gases in the atmosphere. They can be classified based on various characteristics, such as the way the pollutant is generated (instantaneous, continuous source), or the spatial type of the source (point, line, area, volume source), terrain morphology and atmospheric conditions, composition of pollutants (chemical, radioactive, etc.), their state (solid, liquid, gaseous), or their scale (local, medium-scale, or large-scale). From a scientific standpoint, the most systematic classification is based on the mathematical approach applied in these models.

3.1 Main categories of air pollution models

In general, most of the air pollution models used nowadays can be classified in the following three categories:

- ***Empirical models*** – These models are based entirely on experimental observations and data (possibly interpolated and/or extrapolated), without rigorous mathematical analysis of the atmospheric processes.
- ***Lagrangian models*** – In these models the pollutants' clouds are tracked individually, treating them as individual particles that move along with the air currents.
- ***Eulerian models*** – These models consider and evaluate the pollution within a fixed control volume unit, calculating changes in concentration and distribution of various dangerous substances over time and space

3.2 The Danish Eulerian Model (DEM)

The Danish Eulerian Model (DEM) is a large-scale air pollution model, used to calculate the concentrations of various dangerous pollutants and other species over a large geographical region (4800 × 4800 km), see [1]. It takes into account the main physical, chemical and photochemical processes between the studied species, the emissions, the quickly changing meteorological conditions. The Danish Eulerian Model is mathematically represented by a large system of partial differential equations /PDE/ (1), in which the unknown concentrations of a large number of chemical species (pollutants and other chemically active components) take part. The main atmospheric processes (advection, diffusion, chemical reactions, emissions and deposition) are represented as separate additive terms in the right-hand-side of this system.

$$(1)\ \frac{\partial c_s}{\partial t} = -\frac{\partial(uc_s)}{\partial x} - \frac{\partial(vc_s)}{\partial y} - \frac{\partial(wc_s)}{\partial z} + \frac{\partial}{\partial x}(K_x\frac{\partial c_s}{\partial x}) + \frac{\partial}{\partial y}(K_y\frac{\partial c_s}{\partial y}) + \frac{\partial}{\partial z}(K_z\frac{\partial c_s}{\partial z}) + \\ + E_s + Q_s(c_1, c_2, \ldots, c_q) - (k_{1s} + k_{2s})c_s\ ; \quad s=1, 2, \ldots, q$$

The parameters' meaning in the above system (1) is as follows:

q - number of equations (equal to the number of chemical species),
c_s - concentrations of the chemical species,
u, v, w - components of the wind along the coordinate axes,
K_x, K_y, K_z - diffusion coefficients,
E_s - emissions in the space domain,
k_{1s}, k_{2s} - coefficients of dry and wet deposition respectively ($s = 1,\ldots, q$),
$Q_s(c_1, c_2,\ldots, c_q)$ - non-linear functions that describe the chemical reactions between the species.

The PDE system (1) is too complicated to be solved directly. For the purpose of its effective numerical solution, it is split into submodels (stages in the subsequent computer code) in accordance with the main physical/chemical processes as well as with the specific numerical methods used to solve them. In this way three separate submodels are obtained, namely (i) horizontal advection and diffusion (2); (ii) chemistry, emissions, and deposition (3); and (iii) vertical exchange (4) submodels, described rigorously by the next formulae:

$$(2)\ \frac{\partial c_s^{[1]}}{\partial t} = -\frac{\partial(uc_s^{[1]})}{\partial x} - \frac{\partial(vc_s^{[1]})}{\partial y} + \frac{\partial}{\partial x}(K_x\frac{\partial c_s^{[1]}}{\partial x}) + \frac{\partial}{\partial y}(K_y\frac{\partial c_s^{[1]}}{\partial y})$$

$$(3)\ \frac{\partial c_s^{[2]}}{\partial t} = E_s + Q_s\left(c_1^{[2]}, c_2^{[2]}, \ldots, c_q^{[2]}\right) - (k_{1s} + k_{2s})c_s^{[2]}$$

$$(4)\ \frac{\partial c_s^{[3]}}{\partial t} = -\frac{\partial(wc_s^{[3]})}{\partial z} + \frac{\partial}{\partial z}(K_z\frac{\partial c_s^{[3]}}{\partial z})$$

High stability requirements are imposed to the algorithms, used in the numerical solution of the submodels. More detailed description on the numerical methods used as well as on their parallel implementation can be found in [2-8].

3.3 Modeling the spread of gas flowing from a point source

A point source is understood as the outflow of gas or vapor through an orifice with a constant diameter over time.

The main equation in the methodology has the form:

$$(5)\ c(x,y,z,t) = \frac{Q}{(2\pi)^{3/2}\sigma_x\sigma_y\sigma_z} e^{-\frac{(x-u_a t)^2}{2\sigma_x^2}} e^{-\frac{y^2}{2\sigma_y^2}} e^{-\frac{(h-z)^2}{2\sigma_z^2}}$$

where the following notation and dimensions are used:

$C(x,y,z)$ – concentration at a point with coordinates x, y, and z [g/m³];
Q – mass flow rate from the source [g/s];
u – wind speed [m];
σx, σy, σz – coefficients of horizontal and vertical dispersion [m];
h – discharge height [m].

The dispersion parameters σ_x, σ_y, and σ_z can be evaluated by either following the rules of the statistical theory, or by using the empirical data fit approach [15].

3.4 Modeling the spread of toxic vapors evaporating from liquid spills

The methodology applies to emergency spills of liquids, where the area of the spill remains constant during the event. The rate of liquid evaporation m_g from an exposed surface is determined by the following equation:

$$(6)\ m_g = M\,P_s(0.352 + 0.786\,w_a)10^3 \qquad [\mathrm{kg/m^2h}]$$

where the following notation and dimensions are used:

M – mass of the liquid [kg],
P_s – partial pressure of the liquid [mm Hg];
ω_a – wind speed [m/s].

4. Validation and calibration

As usual for any modeling approach, its accuracy and reliability needs validation against observational data from past disasters (if available) and calibration to account for uncertainties in model parameters and input data. These procedures ensure that the model's predictions match real-world conditions and can be trusted for decision-making, such as emergency responses or long-term risk assessments. Validation is the process of confirming that the examined model accurately predicts air pollution levels under different conditions. While calibration focuses on fitting the model to past data, validation checks its performance in new, independent scenarios. Here there are some of the main characteristics of the validation process:

Independent dataset analysis: The model is validated using an independent dataset that was not used in the calibration process. This ensures the model's predictions are robust and can generalize to different conditions. Such data can come from a different time period, location, or under different weather conditions.

Comparison of Predicted vs. Measured Concentrations: Validation includes comparing the model's predicted pollutant concentrations to actual measurements in the independent dataset. Statistical metrics, such as the root mean square error (RMSE), mean absolute error (MAE), or correlation coefficients, are used to quantify the accuracy of the model.

Model Refinement: If the model performs poorly during validation, further refinement may be needed. This could involve recalibrating parameters, improving the representation of physical or chemical processes, or incorporating additional data sources.

Sensitivity and Uncertainty Analysis: Sensitivity analysis is often used to determine how sensitive the model's outputs are to changes in different parameters. This helps to identify which parameters need further refinement. Uncertainty analysis assesses how uncertainties in input data or model assumptions affect the final predictions. This is crucial for understanding the model's limitations and confidence in its predictions.

Calibration means adjusting the model parameters to ensure that the simulation results closely match observed data. It is an iterative process that fine-tunes the model by comparing its outputs with actual measurements of pollution levels in the atmosphere. Key issues in the process of calibration are:

Data Collection (of meteorological data, as well as historical air quality data). Information about wind speed, temperature, humidity, and other weather conditions is also crucial for providing initial and boundary conditions in the numerical solution of some air pollution models(see Section 3.2).

Parameter Adjustment (including Source Emission Rates, Dispersion Coefficients, Chemical Transformation Rates, etc.).

5. Assessment of safety, healthcare and environmental impacts

The assessment of consequences for individuals exposed to surrounding spaces contaminated with hazardous gases or vapours is based on the concentration of the dangerous substance and the duration of exposure. Two systems widely used globally for assessing the consequences of emergency releases of gases or vapours are known by the abbreviations AEGL [16,17] and IDLH [18,19].

The AEGL (Acute Exposure Guideline Levels) is a 3-level system of indicative values of exposure to certain toxic substance, each level representing increasing severity of health effects based on exposure concentration and time duration. These levels are summarized as follows:

- AEGL 1 - concentration of the substance above which it is predicted that the general population, including some sensitive individuals, is unlikely to experience more than mild, reversible discomfort or irritation. However, the symptoms are transient and reversible after discontinuation of exposure.
- AEGL 2 - concentration of the substance above which it is predicted that the population, including some susceptible individuals, may develop irreversible serious and long-term adverse health effects or an impaired ability to escape. This level accounts for more significant but non-lethal health impacts, such as respiratory distress, neurological effects, or more severe irritation.
- AEGL 3 - concentration of the substance above which it is predicted that the population, including some sensitive individuals, may experience life-threatening effects and death. This is the most severe level, indicating critical health risks for the general population upon exposure.

The IDLH (Immediately Dangerous to Life or Health) system was defined by the US National Institute for Occupational Safety and Health (NIOSH) [20] as "exposure to airborne contaminants that is likely to cause death or immediate or delayed permanent adverse health effects or prevent escape from such an environment". It is a system based on several factors, including acute toxicity data, human and animal exposure data, the ability to detect or escape from exposure, and the physical characteristics of the substance (e.g., flammability, explosiveness). An IDLH value represents the highest concentration of a chemical in the air from which a person could escape within 30 minutes without experiencing any life-threatening or irreversible health effects. Key factors in IDLH determination are:

- ***Toxicity:*** Chemicals are evaluated based on their acute toxicity, including lethal doses (LD50) and lethal concentrations (LC50) from animal studies, and any available human exposure data.
- ***Physical Hazards:*** Combustibility, reactivity, or explosiveness of a substance may influence its IDLH rating.
- ***Human Exposure Limits:*** The system incorporates occupational exposure limits like NIOSH Recommended Exposure Limits (RELs) or OSHA Permissible Exposure Limits (PELs) to compare safe long-term exposure levels with IDLH thresholds.

6. Conclusions and plans for future work

In this study we examined the main problems related to air pollution engineering for accidents with hazardous substances. A computerized system for signal, data and information processing and modeling should be developed in advance. The main components of this system are described. Some fundamental models for air pollution transport are presented. The critical points in implementation are outlined - the unification / fusion of disparate information from heterogeneous sensors and modeling of the pollutant distribution process.

The real model, its validation and numerical experiments are yet to be done.

Acknowledgments

This work is supported by the Bulgarian NSF project "Efficient methods for modelling, optimization and decision making" (Grant No. KP-06-N52/5 , 2021) and by the NSP SD program, which has received funding from the Ministry of Education and Science of the Republic of Bulgaria under the grant agreement no. D01-74/19.05.2022.

References

[1] Christensen J., The Danish Eulerian hemispheric model — a three-dimensional air pollution model used for the arctic. Atmospheric Environment, 31(24), 1997, 4169-4191.
[2] Dimov I., Monte Carlo Methods For Applied Scientists. World Scientific, 2007.
[3] Dimov I. et al., A Super-Convergent Stochastic Method Based on the Sobol Sequence for Multidimensional Sensitivity Analysis in Environmental Protection. Axioms, 2023, 12(2):146.
[4] Dimov I. et al., Studying the sensitivity of pollutants' concentrations caused by variations of chemical rates. J. Comput. Appl. Math. 235, 2010, 391–402.
[5] Hvidtfeldt U. A. et al., Evaluation of the Danish AirGIS air pollution modeling system against measured concentrations of PM2.5, PM10, and black carbon. Environmental Epidemiology 2(2), 2018, pe014.
[6] Ostromsky Tz. et al., Air pollution modeling, sensitivity analysis, and parallel implementation. Int. Journal of Environment and Pollution 46 (1-2), 2011, 83–96.
[7] Ostromsky Tz. and Zlatev Z., Parallel Implementation of a Large-scale 3-D Air Pollution Model. Large Scale Scientific Computing (S. Margenov, J. Wasniewski, P. Yalamov, Eds.) Spinger 2179, 2001, 309–316.
[8] Ostromsky Tz. and Zlatev Z., Flexible Two-level Parallel Implementations of a Large Air Pollution Model. Numerical Methods and Applications (I.Dimov, I.Lirkov, S. Margenov, Z. Zlatev eds.) LNCS 2542, 2002, 545–554.
[9] Bar-Shalom Y., Multitarget multisensor tracking: Advanced applications, Artech House, 1990.
[10] Alexiev, K., Simulation of Distributed Sensor Networks, International Conference Automatic and Informatics'04, 6-8 October 2004, Sofia, Bulgaria, pp.255-258.
[11] Brooks R., Iyengar, S., Multi-Sensor Fusion: Fundamentals and Applications with Software, Prentice Hall, New Jersey, 1998.
[12] Murilo da S. Dantasa, Reinaldo R. Rosa, Nilson Sant'Anna, Moacyr G. Cereja Jr, Thalita B. Veronese, Silvia Bianchi, Julia C. Rosa, Kiril M. Alexiev and Jos´e D. S. da Silva, "The VLADA White Paper: Building an Active Virtual Lab for Advanced Data Analysis", Journal of Computational Interdisciplinary Sciences, ISSN 1983-8409, Brazil, pp. 47-56, doi: 10.6062/jcis.2011.02.01.0031
[13] McMichael, Jarrad, G., Williams, Kennett, M., Modelling, simulation and estimation of situation histories, Int. Conf. Fusion'04, Sweden, pp. 928-935.
[14] Alexiev K., Nikolova I., "Methods for Data and Information Fusion", Scientific Support for Decision Making in Security Sector, ed. O.Kounchev, R.Willems, V.Shalamanov, T.Tsachev, IOS Press, ISSN 1874-6268, pp.101-117.
[15] Methods for the calculation of physical effects – "Yellow Book", CPR 14E, 1997, 4.33.
[16] W.F. ten Berge, A. Zwart, L.M. Appelman, Concentration-time mortality response relationship of irritant and systemically acting vapours and gases, Journal of Hazardous Materials, Vol. 13, Issue 3, 1986, pp. 301-309, ISSN 0304-3894, doi: 10.1016/0304-3894(86)85003-8.
[17] Webcite of the US Environmental Protection Agency (EPA): https://www.epa.gov/aegl

Journal of Physics: Conference Series **2910** (2024) 012002 doi:10.1088/1742-6596/2910/1/012002

[18] Ludwig, HR; Cairelli, SG; Whalen, JJ, Documentation for immediately dangerous to life or health concentrations (IDLHs), PB94-195047 National Institute for Occupational Safety and Health, Cincinnati, OH, 1994, 530 p.. https://www.cdc.gov/niosh/idlh/idlhintr.html

[19] Dotson GS, Maier A, Parker A, Haber L, Immediately dangerous to life or health (IDLH) value profile: methacrylo-nitrile., PB2017-204 Department of Health and Human Services, Centers for Disease Control and Prevention, National Institute for Occupational Safety and Health, Cincinnati, OH, 2017.

[20] Webcite of the US National Institute for Occupational Safety and Health (NIOSH): https://www.cdc.gov/niosh/index.html

Study on Impact Load Mitigation for High-Speed Water Entry Projectiles Using Aluminum Foam Buffering Structure

Xiangying Guo[1*],Xiangyu Tian[2] and Tian Nan[3]

[1]Beijing Key Laboratory of Nonlinear Vibrations and Strength of Mechanical Structures, School of Mathematics, Statistics and Mechanics, Beijing University of Technology,Beijing,100124,China
[2]Beijing Key Laboratory of Nonlinear Vibrations and Strength of Mechanical Structures,College of Mechanical & Energy Engineering,Beijing University of Technology, Beijing,100124, China
[3]Beijing Key Laboratory of Nonlinear Vibrations and Strength of Mechanical Structures, School of Mathematics, Statistics and Mechanics, Beijing University of Technology,Beijing,100124,China

*E-mail: guoxy@bjut.edu.cn

Abstract. This paper investigated the fluid-solid coupling nonlinear problem of water entry by a high-speed projectile, considering cavitation, air cushion effect, and water vapor generated by the high-speed relative motion between the projectile and the fluid. To this end, a sandwich structure containing aluminum foam attached piston was studied and designed. The numerical simulation of the entry process was analyzed by LS-DYNA. The results show that the sandwich structure reduces the peak impact load. 87.47% reduction in peak force during water entry and aluminum foam has good cushioning effect.

1. Introduction

The study of water entry phenomena can be traced back to the late 19th and early 20th centuries. Early scholars such as Worthington, Cole[1], and Mallock[2] conducted experiments to preliminarily explain the phenomena of low-speed impact on various liquids, observing the formation of cavities, splashing, and exploring the relationship between the drag force and density of spheres during water entry. In 1932, Wagner, building upon von Karman's work, proposed the small deadrise angle flat plate theory and employed the generalized Bernoulli equation to obtain the pressure distribution on the wetted surface of wedge-shaped bodies, laying the theoretical foundation for subsequent research on water entry impact loads[3].

In the mid-to-late 20th century, systematic experimental studies were carried out by researchers like Anderson, May, and Woodhull. They analyzed the effects of various factors, such as projectile size, water entry velocity, and air pressure, on the water entry process[4]. They also investigated the formation and evolution stages of cavities, their influence on the free surface, and proposed the ideal cavity model[5].

In the past twenty years, with the rapid development of domestic computer technology and the increasing investment in scientific research, the research on the problem of the projectile into the water has also made significant progress.In 2000, Zhang et al. used numerical methods to study the hydroelastic effect of a certain shape of elastic body in the process of water impact.The method of linearizing the discrete Bernoulli equation is used to couple the finite element equation and the boundary element equation with each other, and the coupled equations of motion for solving the flow field and the elastic force are obtained[6].In 2001, Shi[7,8] used high-speed photography to experimentally obtain the changes in the water entry attitude of the projectile under different water entry conditions. They further changed the shape of the projectile and studied the generation and

Content from this work may be used under the terms of the Creative Commons Attribution 4.0 licence. Any further distribution of this work must maintain attribution to the author(s) and the title of the work, journal citation and DOI.
Published under licence by IOP Publishing Ltd

development of air bubbles during water entry under various influencing factors.In his published work, he summarizes and compares in detail the results of his own research and that of his predecessors on the problem of water entry[9].In 2008, Yi used FLUENT as a platform to calculate the development process of the vacuole of a projectile when it enters the water at high speed[10,11], and found that the change of the angle of attack of the projectile will have different effects on the symmetry of its vacuole.In 2014, Ma[12] utilized the numerical value method to analyze the regularity and quantitative analysis of the change process of the load during the high-speed entry of the projectile into the water.In 2016, Zhang Wei[13] measured the water entry process of two kinematic body models, flat head and spherical head, by using pressure sensors through the method of experimental research,and obtained the attenuation characteristics of the impact pressure of the two kinematic bodies entering water.In 2017, Guo[14] preset the projectile launching velocity at 100m/s~150m/s to experimentally study the water entry process, and summarized the theoretical model of the expansion of the vacuole using the experimental data.In 2020, Wang Zhen[15] used LS-DYNA software as a platform to carry out numerical simulation of the water entry process of the projectile with a velocity close to 200m/s, and obtained the peak load and the change rule of the load of the projectile under different water entry angles and angles of attack.

2. Stress-Strain Response of Foam Aluminum in LS-DYNA

Foam aluminum can be categorized into open-cell and closed-cell varieties. Open-cell foam aluminum finds applications in noise reduction and filtration processes, among others. In contrast, closed-cell foam aluminum exhibits good energy absorption capabilities, particularly for small pore diameters, where its energy absorption capacity is significantly enhanced. The advent of closed-cell foam aluminum with smaller pore sizes has led to notable improvements in its energy dissipation performance.Due to the high energy absorption efficiency of this material, it was chosen as the damping device for this thesis.

The energy absorption efficiency of aluminum foam is contingent upon the porosity of the foam, which is in turn influenced by the cell size of the foam. The contribution of Chen resulted in a reduction of the pore size of the foam from 3~8mm to 1~2mm. Furthermore, stress-strain curves of the foam materials with different pore sizes were obtained following a quasi-static uniaxial compression test. This resulted in an increase in the plateau stress from less than 5MPa to 10~20MPa, which significantly enhanced the energy-absorbing capacity of the aluminum foam[16]. The stress-strain curve of aluminum foam #8, chosen as the buffer device for this paper and obtained from the quasi-static compression test conducted on a microcomputer controlled electronic universal testing machine (model WDW-200E), is presented in Figure 1.Upon compression, the foam material exhibits a stress-strain relationship that differs from that of conventional materials. It can be roughly divided into three regions: the linear elastic region, the platform region, and the dense region. The energy-absorbing effect of the closed-cell aluminum foam is primarily observed in the platform region.

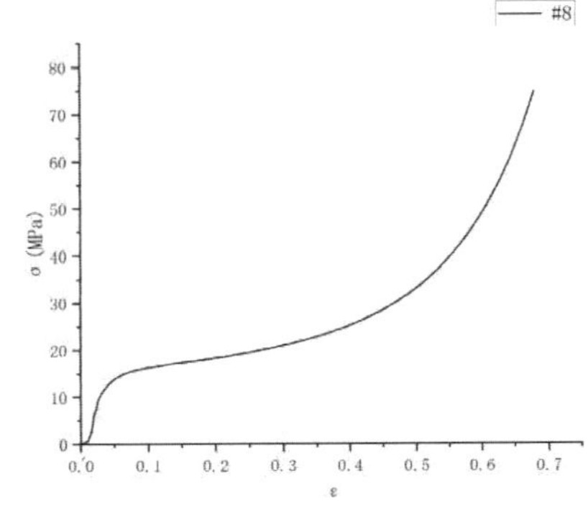

Figure 1. The stress-strain curve of the No. 8 foam aluminum

In this study, material 163-MODIFIED_CRUSHABLE_FOAM was selected as the basis, which was corrected by inputting the stress-strain curve. Two rigid planes were set up at the top and bottom to compress the foam aluminum at a speed of 1 m/s. The performance of the obtained #8 foam aluminum in the LS-DYNA simulation environment is shown in Figure 2. The comparison of the simulated stress-strain curve with the actual curve, as shown in Figure 3.The comparison of the stress-strain curve obtained from the simulation software and the experimental data for aluminum foam #8 reveals that while the former was slightly lower than the latter, the overall trend and scale are closely aligned, thus satisfying the requirements for the subsequent simulations.

Figure 2. The quasi-static compression test in the LS-DYNA simulation environment.

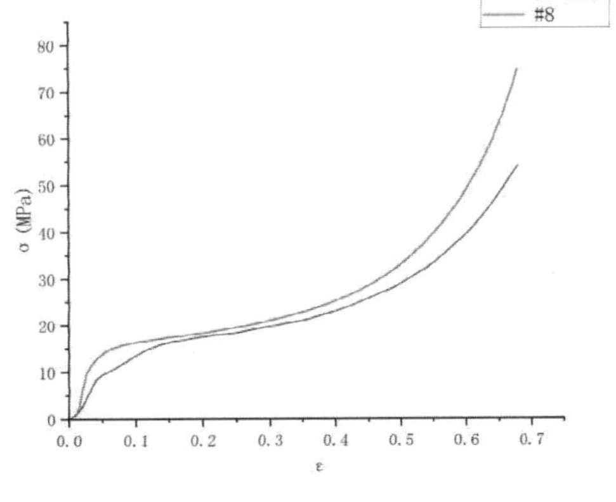

Figure 3. The comparison of the stress-strain curve of No. 8 material.

3. Numerical Model Setup

The LS-DYNA general-purpose finite element program with explicit algorithms is capable of simulating strong nonlinear problems such as collisions in the real world, which are widely used in various industries. Considering investigating the supersonic water entry problem and obtaining the response of a flat-bottomed projectile under high-intensity load in a short period.To tackle this nonlinear problem, an approach involving explicit integral formulations was adopted.This study examined how the aluminum foam buffer influenced the variation of peak forces on the protected projectile during water impact.

3.1 ALE algorithm for the coupling of fluid and solid phases.

There are three methods for achieving fluid-solid coupling in LS-DYNA simulation software: the co-nodal algorithm, the contact algorithm, and the Eulerian-Lagrangian coupling algorithm (ALE), which refers to the simultaneous simulation of the interactions and motions of multiple materials.The ALE framework employed allowed for simulating solids, liquids, gases, and other materials by combining the advantages of the traditional Lagrangian description, where the mesh deforms with the fluid, and the Eulerian description, where the mesh is fixed and the fluid passes through. This numerical method is widely used in computational fluid dynamics (CFD) [17,18]. The ALE algorithm provides an effective means to deal with the problems of moving boundaries and large deformations.It plays an influential role in the simulation of problems such as high-speed water entry.

In this study, all computational models are meshed using eight-node hexahedral cells. During the simulation, the air and water domains were modeled using the Multi-Media Arbitrary Lagrangian-Eulerian (ALE) algorithm, which permits the incorporation of multiple media within the model grid. Furthermore, a Lagrangian grid is utilized for the construction of the structural mesh. It is possible to overlay the structural body mesh with the air domain mesh when constructing the model.

In order to the transfer of mechanical variables between the solid and the fluid, a penalty function constraint is operated in this study. This function permits a minimal degree of mesh penetration. The extent of penetration between the Lagrangian and Eulerian meshes is monitored by tracing the penetration between the two, with the fluid force being proportional to the amount of penetration,which is given by

$$F = k_i d = P\frac{KA}{V}d \tag{1}$$

where k_i is the stiffness coefficient, it is based on the fluid and structural body node mass model characteristics.The bulk modulus K and volume V of the penetrated unit are represented by K and V, respectively. The average area of the structural unit connecting the nodes, A, is also included in the equation. P is the penalty function factor, which is the value of the PFAC in LS-DYNA for calculating the force distributed between the Lagrangian and ALE parts. This parameter is crucial for the overall outcome of the calculation, as it affects the negative volume of the calculation grid and the unexpected termination of the calculation when it is too large. Conversely, an excessively small value would weaken the coupling force between the master and slave segments, resulting in significant leakage across the fluid-structure interface during the simulation.Following a series of attempts, it was ultimately determined that the PFAC value in this example is 0.1.

The governing equations underlying the ALE algorithm encompass mass, momentum, and energy conservation equations.

$$\frac{\partial \rho}{\partial t} =- \rho\frac{\partial v_i}{\partial x_i} - \omega_i\frac{\partial \rho}{\partial x_i} \tag{2}$$

$$\rho\frac{\partial v_i}{\partial t} = \sigma_{ij,j} + \rho b_i - \rho\omega_i\frac{\partial v_i}{\partial x_j} \tag{3}$$

$$\rho\frac{\partial E}{\partial t}=\sigma_{ij}v_{i,j}+\rho b_i v_i-\rho\omega_i\frac{\partial e}{\partial x_j} \quad (4)$$

where ρ is the fluid density; t is the time; v_i is the fluid velocity; ω_i is the convective velocity; x_i is the Eulerian coordinates; σ_{ij} is the stress tensor; b_i is the volumetric force acting on the fluid, which is small and can be neglected in this algorithmic example; and e is the specific total energy.

3.2 Materials and Equations of State

In the simulation environment setup, the flat-bottomed rotary projectile, air, and water are given materials, respectively. The simulation model of the projectile buffer device is comprised of three distinct components: the piston in contact with the water, the foam aluminum, and the cylinder. The cylinder serves as the primary structural element, with a mass of 650 kg. The foam aluminum is composed of a novel composite blowing agent derived from aluminum foam with a small cell size. The material models for water and air are defined using the MAT_NULL designation, and the unit system employed for simulation is kg-mm-ms, as illustrated in the table below.

Table 1. Definition of water and air in the *MAT_NULL.

*MAT_NULL	Mass density	Pressure cutoff	Viscosity coefficient
air	$1.300e^{-09}$	$-1.000e^{-19}$	$2.000e^{-11}$
water	$1.000e^{-06}$	$-1.000e^{-08}$	$8.684e^{-10}$

The equation of state for water is defined using *EOS_GRUNEISEN, which defines the pressure of a compressed material,which can be written in the form

$$p=\frac{\rho_0 C^2\mu\left[1+\left(1-\frac{\gamma_0}{2}\right)\mu-\frac{a}{2}\mu^2\right]}{\left[1-(S_1-1)\mu-S_2\frac{\mu^2}{\mu+1}-S_3\frac{\mu^3}{(\mu+1)^2}\right]^2}+(\gamma_0+a\mu)E \quad (5)$$

Where γ_0 represents the dimensionless Gruneisen constant,while the dimensionless parameter a denotes the first-order volume correction term for γ_0. C represents the velocity intercept of the $v_s(v_p)$ curve, with the units of velocity. S_1, S_2, and S_3 are the dimensionless coefficients associated with the slope of the $v_s(v_p)$ curve;and

$$\mu=\frac{\rho}{\rho_0}-1 \quad (6)$$

Where E denotes the internal energy;The following table describes the parameter values:

Table 2. Equation of state parameters for water.

*EOS_GRUNEISEN	C	S_1	S_2	S_3	GAMMA0	A	E_0	V_0
water	1647	1.921	-0.097	0	0.35	0	$2.895e^{-04}$	1.0

The equation of state for air is defined using the EOS_LINEAR_POLYNOMIAL. The selected equation of state is linear in internal energy. The pressure p is given by

$$p=C_0+C_1\mu+C_2\mu^2+C_3\mu^3+\left(C_4+C_5\mu+C_6\mu^2\right)E \quad (7)$$

The linear polynomial equation of state can be employed to model gases using the gamma law equation of state which can be written in the form

$$C_0=C_1=C_2=C_3=C_6 \quad (8)$$

$$C_4 = C_5 = \gamma - 1 \tag{9}$$

$$\gamma = \frac{C_p}{C_v} \tag{10}$$

Where γ is the ratio of specific heats.For a perfect gas, the pressure can be given by

$$p = (\gamma - 1)\frac{\rho}{\rho_0}E = (\gamma - 1)\frac{e_{v0}}{v_r} \tag{11}$$

E carries the units of pressure, where E is expressed as the product of the specific internal energy e and the reference volume V_0, while v_r denotes the ratio of the reference density ρ_0 to the current density ρ.The specific parameter values are presented in the table below.

Table 3. Equation of state parameters for air.

*EOS_LINEAR_ POLYNOMIAL	C_0,C_1 C_2, C_3,C_6	C_4	C_5	E_0	V_0
water	0	0.4	0.4	$2.5e^{-04}$	1.0

4.Model validation and setup

4.1 Verification of LS-DYNA Modeling

To ensure that the environment in which LS-DYNA is built is reasonable and that the forces of air and water on the rigid body are correctly transmitted throughout the impact process, it is necessary to experimentally corroborate the plausibility of the environment.

The equation of state and material parameters constructed above were utilized to construct the 7600mm×4500mm×2600mm water model, which was developed by Sheng-Lun Chuang's experiment [19]. For the simulations, a rigid rectangular body was considered, with one of its faces (650 mm × 500 mm) impacting the water at a velocity of 1.524m/s. The mass of the rigid body with the guide rail is 234.7kg. Based on the experimental data presented in Figure 4 for different impact velocities, it can be observed that when the rigid rectangular body impacts the water at a velocity of 1.524 m/s, the resulting pressure on its bottom face reaches 117.2 KPa., and the maximal impact acceleration obtained in the LS-DYNA simulation environment is $0.174\,\mathrm{mm/ms^2}$, as shown in Figure 5. After calculation, it was determined that the peak force was 122 KPa. After data comparison, the experimental data basically match the simulated values in the same order of magnitude.

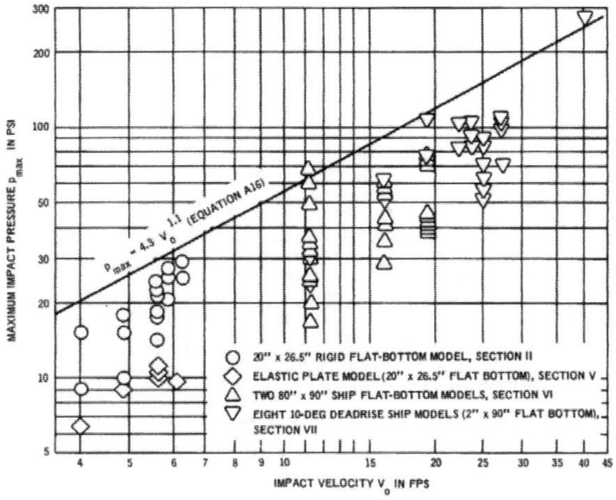

Figure 4. Theoretical and experimental data obtained[19].

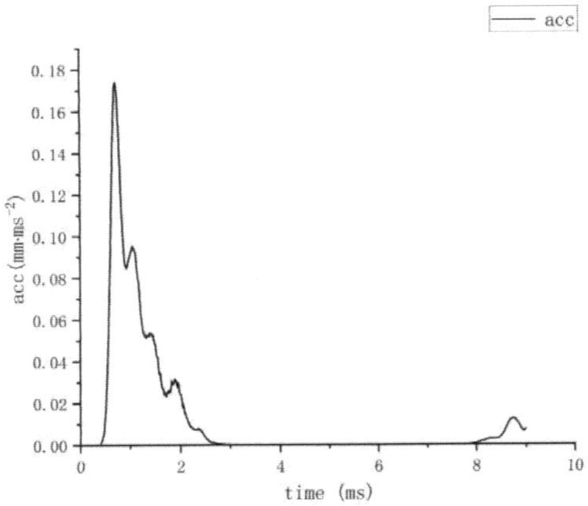

Figure 5. Results of rigid body acceleration simulation.

4.2. Model Setup and Time-Step Verification

The model should be constructed by the specifications depicted in Figure 6, which illustrates the Complete model and the Cross-sectional view of the model, respectively.

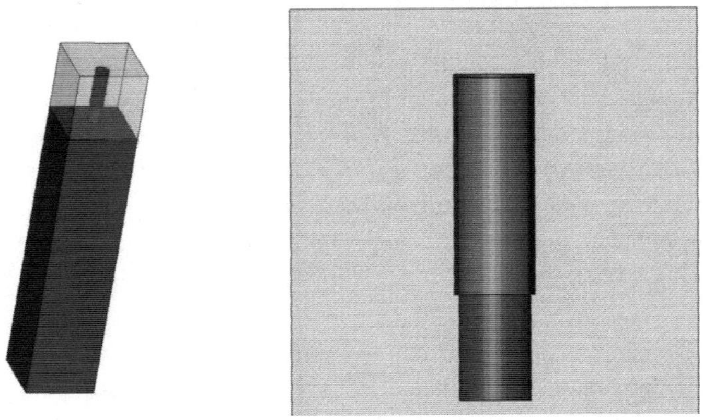

(a)Complete model (b)Cross-sectional view of the model

Figure 6. Model established.

 doi:10.1088/1742-6596/2910/1/012003

The dimensions of the air domain are 500mm x 500mm x 500mm, while those of the water domain are 500mm x 500mm x 2000mm. The grid size is 5mm, and the projectile cushioning device begins to impact downward at Mach I speed at a distance of 20mm from the water surface. It is necessary to reserve sufficient air in the domain because the air cushion effect must be taken into account, which is more in line with the actual working conditions. The piston is a non-deformable rigid body that extends out of the cylinder, ensuring sufficient travel to fully compress the foam aluminum within the cylinder. The dimensions of the piston are as follows: a piston with a radius of 44mm, which is the same as the radius of the aluminum foam in the cylinder; and a length of 210mm. The outer diameter of the cylinder is 50mm, while the inner diameter is 46.475mm. This indicates that the inner diameter of the cylinder is slightly larger than the radius of the aluminum foam, which is a consideration for the possibility of transverse deformation after compression collapse of the aluminum foam. The complete model comprises 5,154,864 nodes, which necessitates significant computational time and cost. To reduce the computational burden, the symmetry of the model was exploited by utilizing a quarter of the original model, as illustrated in Fig. 7. This approach enabled the reflection-free boundary to be set to infinity, while the lower boundary was fixed, thus facilitating the smooth execution of the entire simulation.

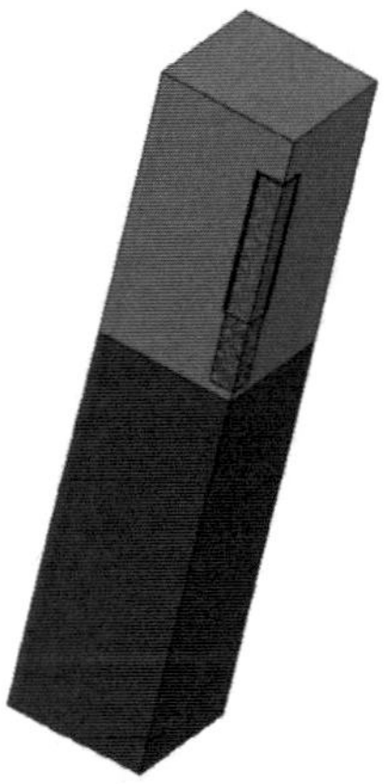

Figure 7. Simplified 1/4 model established.

Given that the problem under investigation is a high-speed collision problem involving the coupling of multiple fluids, it is imperative to select an appropriate time step.

In the context of high-speed problems, LS-DYNA's TIMESTEP is typically set to 0.67. However, in the presented calculation example, due to the speed of 340 m/s, a significant volume of leakage occurs when 0.67 is employed, as illustrated in Fig. 8. One of the criteria for assessing the degree of realism of the coupling is the magnitude of the slip energy. In the 'glstat' post-processing, the sliding energy is identified, and the sliding energies corresponding to different time steps are illustrated in Fig. 9.(the comparison curves of sliding energies corresponding to 0.1, 0.2, 0.3, and 0.67). It can be observed that the sliding energy is stabilized near 0 when the time step is taken to be 0.1, and thus, the time step is ultimately selected to be 0.1.

 IOP Publishing
doi:10.1088/1742-6596/2910/1/012003

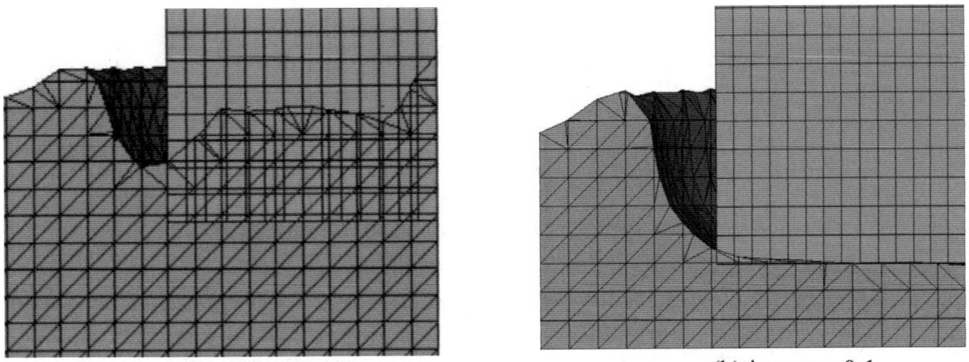

(a)time step 0.67 (b)time step 0.1

Figure 8. Volume leak comparison for different time step.

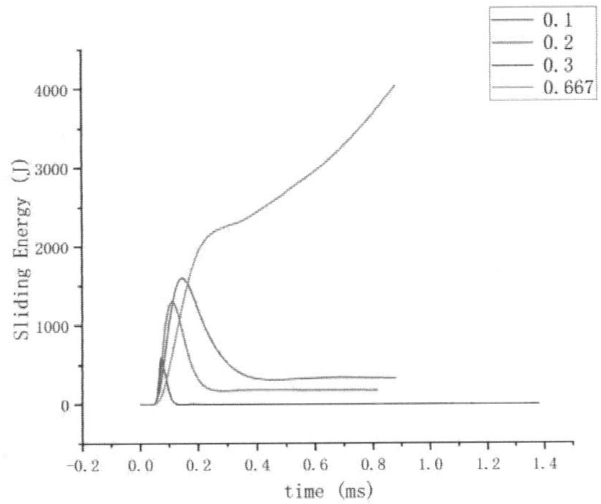

Figure 9. Sliding energy for varying time-step increments.

5.Simulation results analysis

Considering the simulation environment described above, the acceleration of the piston in contact with the water and the acceleration of the cylinder were measured for this buffer device at a speed of 340 m/s, and the acceleration-time curve are shown below.

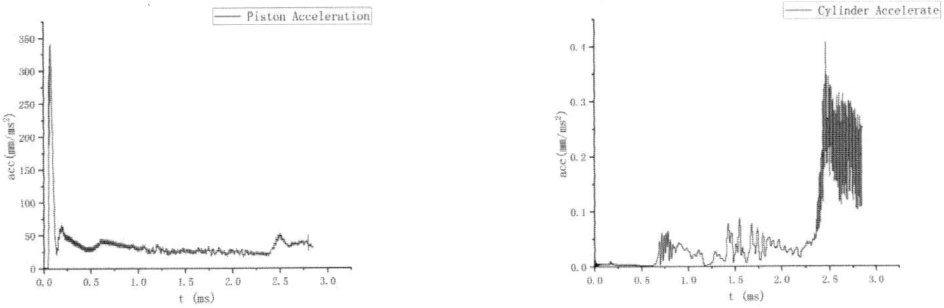

(a)Piston (b)Cylinder body

Figure 10. Acceleration profiles comparison

It can be seen that the acceleration of the piston in direct contact with the water increases sharply at 13μs after contact with the water, at which time the piston acceleration reaches a peak of 221 mm/ms^2, at the same time the piston is subjected to the force of the water on it reaches a peak. Due to the presence of aluminum foam, the cylinder is subjected to acceleration with hysteresis, and the peak acceleration of the cylinder is 0.407 mm/ms^2, which is obtained at 2.46ms. The F-t diagrams of the piston and the cylinder parts are obtained by Newton's second law, respectively, as shown in Fig11. It can be seen that the force on the piston increases suddenly and then decreases sharply. There is a smooth period before the cylinder hits the water, which is taken in the interval of 1.25ms~2.15ms, and the average value of the force on the piston during this period is calculated to be 38.57 KN. This result matches the findings of Lee et al. [20].Force analysis shows that after impacting the water surface, the projectile transitions into a stable navigational phase.During this phase, the pressure exerted on the structure's entry interface into the water is comparatively low, and the magnitude of the resistance force encountered remains consistent as it enters the water.At about 2.5ms, the piston is completely pressed into the cylinder, the force fluctuation occurs, and then the piston force tends to stabilize. In the period of 2.61ms~2.85ms, the average value of 54.73KN is taken, which corresponds to the force of the cylinder body.In the interval of 1.25ms~2.15ms, the average value of the force is 5.29KN, which can be seen that the foam aluminum is in play in this period, and the average force of the cylinder body is reduced by 86.28% compared with that of the piston, so the force of the cylinder body is much smaller than that of the piston.In the period of about 2.5ms, there is a fluctuation of the force, and then the piston force tends to stabilize. Around 2.5ms, a peak force of 66.14KN is observed on the cylinder body. This peak occurs when the aluminum foam buffer has exhausted its effectiveness, allowing the cylinder body to directly impact the water. Compared to the peak force of 528.03KN experienced by the piston during direct water impact, the peak force on the cylinder body is reduced by 87.47% due to the buffer's energy absorption.During the period of 2.61ms-2.85ms, the force on the cylinder body stabilizes, with an average value of 39KN. Ultimately, the forces on both the piston and the cylinder body converge to approximately 40KN. The force values during these stages are summarized in Table 4.

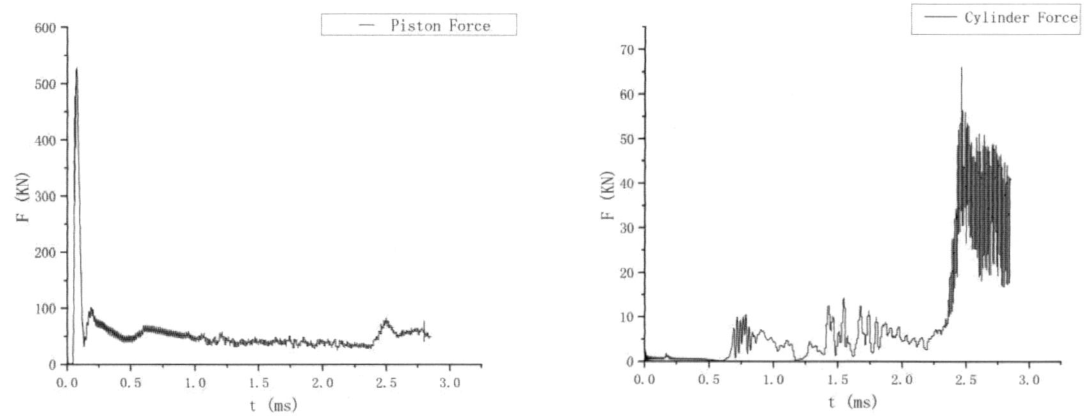

(a)Piston (b)Cylinder body

Figure 11. Force profiles comparison

Table 4. Compare forces at each stage.

	Peak force	Average force at stabilization(1.25ms~2.15ms)	Average final stabilizing force(2.61ms~2.85)
Piston	528.03KN	38.57KN	54.73KN
Cylinder	66.14KN	5.29KN	39KN

6.Conclusion

The finite element analysis software LS-DYNA was applied to study the peak force in the process of high-speed water entry of the cylinder, and a structure of foam aluminum in the cylinder was designed. The applied foam aluminum material has good energy absorption characteristics, and the platform stress is higher than that of the general foam aluminum[16].Through analyzing the forces experienced by the piston and cylinder during the water entry process of this structure, in conjunction with the force variations and final trends over the entire water entry event, it is evident that the aluminum foam exhibits a commendable energy absorption capability, thereby effectively mitigating the peak impact load. It is noteworthy that this cushioning effect is also contingent upon the impact area; in scenarios where the protected projectile possesses a larger impact area, the dimensions of this buffering device can be adjusted accordingly to ensure an adequate protective effect for different projectiles.

References

[1] Worthington A M, Cole R S. IV. Impact with a liquid surface studied by the aid of instantaneous photography. Paper II[J]. Philosophical Transactions of the Royal Society of London. Series A, Containing Papers of a Mathematical or Physical Character, 1900, 194(252-261): 175-199.

[2] Mallock H R A. Sounds produced by drops falling on water[J]. Proceedings of the Royal Society of London. Series A, Containing Papers of a Mathematical and Physical Character, 1918, 95(667): 138-143.

[3] Wagner H. Phenomena associated with impacts and sliding on liquid surfaces[J]. Z. Angew. Math. Mech, 1932, 12(4): 193-215.

[4] Gilbarg D, Anderson R A. Influence of atmospheric pressure on the phenomena accompanying the entry of spheres into water[J]. Journal of Applied Physics, 1948, 19(2): 127-139.

[5] May A. Vertical entry of missiles into water[J]. Journal of applied physics, 1952, 23(12): 1362-1372.

[6] Zhihong Zhang, Jiannong Gu, Wujie Fan, & Li Jialian. (2000). The hydroballistic modeling method for high-speed water entry of a rotary projectile, (6), 1-5.

[7] Shi. H. H, Itoh. M, Takami. T. Optical observation of the supercavitation induced by high-speed water entry[J]. Journal of Fluids Engineering-Transactions of the ASME, 2000, 122(4): 806-810.

[8] Shi. H. H, Kume. M. Underwater acoustics and cavitating flow of water entry[J]. Acta Mechanica Sinica, 2004, 20(4): 374-382.

[9] Honghui Shi. Compressibility and High-Speed Multiphase Flow[M]. Zhong guo ke xue ji shu da xue chu ban she,2014.

[10] Wenjun Yi, et al. Characteristics of Cavity Shape for Underwater High-speed Projectile at Small Attack Angle[J].Journal of Nanjing University of Science And Technology,32.4 (2008): 464-467.

[11] Tianhong Xiong,et al. Effect of Cavitator Parameters for Underwater Projectiles on Natural Supercavity[J]. Journal of Nanjing University of Science And Technology, 2008, 32(5): 545-548.

[12] Qingpeng Ma. Study on the Multiphase Flow Field Characteristics of the High-Speed Projectile Water Entry Process[D],Harbin Institute of Technology, 2014.

[13] ZHANG Wei,HUANG Wei,REN Peng,YE Nan,LI Dacheng.The underwater shock wave characteristics caused by high speed horizontal water entry projectiles[J].Journal of Harbin Institute of Technology,2016,48(4):37.DOI:10.11918/j.issn.0367-6234.2016.04.006.

Journal of Physics: Conference Series **2910** (2024) 012003 doi:10.1088/1742-6596/2910/1/012003

[14] Guo Zitao,et al.Characteristics of velocity attenuation and cavity expansion induced by horizontal water-entry of truncated-ogive nosed projectiles[J]. Explosion and Shock Waves, 2017, 37(4): 727-733.
[15] WANG Zhen,WU Maolin,DAI Wenliu.Study on Load Characteristics of High-speed Water-entry of Large Caliber Projectile[J].Journal of Ballistics,2020,32(1):15-22.
[16] Cheng Y, Li Y, Chen X, et al. Compressive properties and energy absorption of aluminum foams with a wide range of relative densities[J]. Journal of Materials Engineering and Performance, 2018, 27: 4016-4024.
[17] Dangyong SHi,et al. Explicit Dynamic Analysis Based on ANSYS/LS-DYNA 8.1 [M]. Tsinghua University publishing house co., ltd,2005.
[18] Shaoqing Shi. Engineering Applications of ANSYS/LS-DYNA in Explosion and Impact [M]. China Architecture & Building Press, 2011.
[19] Chuang S L. Investigation of impact of rigid and elastic bodies with water[M]. The Catholic University of America, 1969.
[20] Lee M, Longoria R G, Wilson D E. Cavity dynamics in high-speed water entry[J]. Physics of Fluids, 1997, 9(3): 540-550.

Periodic orbits and bifurcation of a two-degree-of freedom dry friction coupled vibration system

Jing Li[1*], Yujiao Cui[1] and Shaotao Zhu[2*]

[1]Interdisciplinary Research Institute, School of Mathematics, Statistics and Mechanics, Beijing University of Technology, Beijing, China
[2]College of Computer Science, Beijing University of Technology, Beijing, China

E-mail: leejing@bjut.edu.cn

Abstract. This article explores the dynamic behavior of a two-degree-of freedom dry friction coupled vibration system. By constructing a generalized Poincaré map and obtaining the necessary and sufficient conditions for the existence of invariant cone in the system, the linearization principle of Poincaré map is equivalently replaced by the study of a global monodromy matrix. Based on the Floquet theory, the stability of the invariant cone consists of periodic orbits is proved, and the bifurcation phenomenon and persistence of the invariant cone under parameter changes are explored. The distribution configuration of the trajectory is given through numerical simulation, verifying the accuracy of the theory. This provides a theoretical basis for parameter optimization and vibration control of high-dimensional dry friction vibration systems in practical engineering applications.

keywords. Dry friction, Periodic orbit, Invariant cone, Bifurcation

1 Introduction

In practical engineering fields such as mechanical vibration, neural networks, and control systems, there are many nonsmooth factors that cannot be ignored, such as collisions, dry friction, and impacts, which lead to many complex phenomena in the system. At the same time, many classic dynamic theories of smooth system are no longer fully applicable, and there is an urgent need to explore and analyze the dynamic characteristics of piecewise smooth system from both theoretical and practical perspectives, so as to make the study of nonsmooth system closer to real-world engineering problems [1, 2]. The vibration system with dry friction, as a highly representative Filippov type piecewise smooth discontinuous system, reveals its dynamic evolution mechanism, which is of great significance for the safety and durability of parts in mechanical systems [3].

Dry friction systems exhibit bifurcation phenomena different from general smooth systems, which are usually related to the position of the trajectory and switching surface. [4, 5, 6] studied some complex bifurcation in nonsmooth dynamical systems, such as discontinuous induced bifurcation, border-collision bifurcation, and sliding bifurcation unique to discontinuous system, etc. Most papers focus on planar or three-dimensional piecewise smooth system [7, 8, 9], but the theoretical research and numerical calculations of higher dimensional dry friction application models are still challenging. To further investigate the dynamic behavior of piecewise smooth system, it is necessary to find effective methods, but the presence of nonsmooth factors can violate previous rules. Poincaré map method is an important method, [10, 11] established the Poincaré map to study the theory of periodic solutions in piecewise smooth system. Therefore, we construct a generalized Poincaré map and conduct a dimensionality reduction study on high-dimensional dry friction model by studying the existence of invariant cone

Content from this work may be used under the terms of the Creative Commons Attribution 4.0 licence. Any further distribution of this work must maintain attribution to the author(s) and the title of the work, journal citation and DOI.
Published under licence by IOP Publishing Ltd

similar to the traditional central manifold concept. Based on the approximation of piecewise smooth linear systems to the original nonlinear problem, the study of dynamic characteristics is transformed into the study of the existence and stability of invariant cone [12].

Based on this, the dynamic characteristics of a high-dimensional dry friction coupled vibration system is investigated. A generalized Poincaré map based on the motion type of trajectory is constructed, the existence conditions of an invariant cone consists of periodic orbits is obtained, in addition, the linearization of Poincaré map with a global monodromy matrix is replaced, and we obtain the stability properties of invariant cone based on Floquet theory, analyze the bifurcation phenomenon under parameter changes, discuss the persistence problem of invariant cone under parameter disturbances, and provide the geometric configuration to verify the correctness of the theory.

2 Periodic solutions and bifurcation of a two-degree-of freedom dry friction coupled vibration system

2.1 *Establishment of the dry friction model*

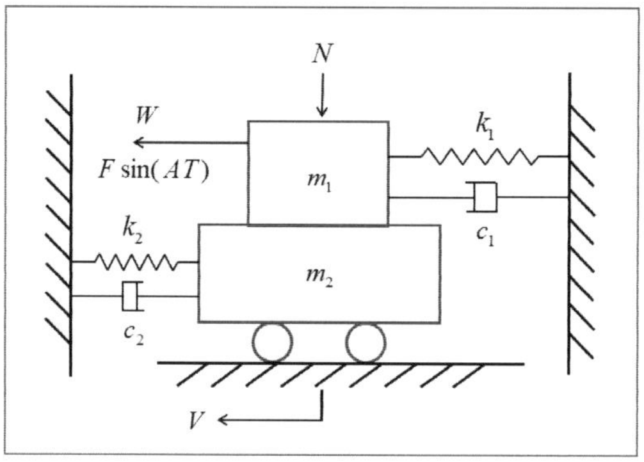

Figure 1: Illustration of a 2-DOF dry friction coupled vibration model.

Figure 1 is a model diagram of a two-degree-of freedom dry friction coupled vibration system, where the mass m_1 and m_2 are connected to the right and left rigid surface through an elastic coefficient k_1, k_2 and a damping coefficient c_1, c_2. In addition, m_1 is subjected to a positive pressure N and a harmonic excitation force $F\sin(AT)$. W and V are displacements generated by the mass of m_1 and m_2. Considering that the relative velocities of m_1 and m_2 are not 0, the two masses can slide relative to each other, resulting in sliding friction f_0. According to Newton's Second Law, the motion behavior is written as a mathematical model as follows

$$\begin{aligned}&m_1\ddot{W}+c_1\dot{W}+k_1W+f_0\,\mathrm{sgn}(\dot{W}-\dot{V})-F\sin(AT)=0,\\&m_2\ddot{V}+c_2\dot{V}+k_2V+f_0\,\mathrm{sgn}(\dot{V}-\dot{W})=0.\end{aligned}\tag{1}$$

Further perform dimensionless transformation on the parameters of system, so that

$$\beta_m=\frac{m_2}{m_1},\ \beta_k=\frac{k_2}{k_1},\ \beta_c=\frac{c_2}{c_1},\ \eta_1=\frac{f_0}{F},\ \eta_2=\frac{f_0m_1}{Fm_2},\ a=A\sqrt{\frac{m_1}{k_1}},\ t=T\sqrt{\frac{k_1}{m_1}}$$
$$\beta_1=\frac{k_2m_1}{k_1m_2},\ \beta_2=\frac{c_2m_1}{c_1m_2},\ \gamma=\frac{c_1}{2\sqrt{m_1k_1}},\ w=\frac{Wk_1}{F},\ v=\frac{Vk_1}{F}.$$

At this point, the differential equation for relative motion of the system is transformed into

$$\begin{aligned}&\ddot{w}+2\gamma\dot{w}+w+\eta_1\,\mathrm{sgn}(\dot{w}-\dot{v})-\sin(at)=0,\\&\ddot{v}+2\beta_2\gamma\dot{v}+\beta_1v+\eta_2\,\mathrm{sgn}(\dot{v}-\dot{w})=0.\end{aligned}\tag{2}$$

Let $w = \xi_1, \dot{w} = \xi_2, v = \xi_3, \dot{v} = \xi_4$, the system can be transformed into

$$\begin{aligned}
\dot{\xi}_1 &= \xi_2, \\
\dot{\xi}_2 &= -\xi_1 - 2\gamma\xi_2 - \eta_1 \operatorname{sgn}(\xi_2 - \xi_4) + \sin(at), \\
\dot{\xi}_3 &= \xi_4, \\
\dot{\xi}_4 &= -\beta_1\xi_3 - 2\beta_2\gamma\xi_4 - \eta_2 \operatorname{sgn}(\xi_4 - \xi_2).
\end{aligned} \tag{3}$$

A hyperplane function $h(\xi) = n^T(\xi)\cdot\xi = \xi_2 - \xi_4$ can be defined, and $n(\xi) = (0, 1, 0, -1)^T$, at this point, switch manifold $H_0 = \{y \in R^4 \,|\xi_2 - \xi_4 = 0\}$, $H^+ = \{\xi \in R^4_+ |\xi_2 - \xi_4 > 0\}$ and $H^- = \{\xi \in R^4_- |\xi_2 - \xi_4 < 0\}$. From this, the system can be summarized as

$$\dot{\xi}(t) = f(t, \xi(t)) = \begin{cases} f^+(t, \xi(t)) = A^+\xi + \psi^+(t), & \xi(t) \in H^+, \\ \overline{co}\{f^-(t, \xi(t)), f^+(t, \xi(t))\} = A_s\xi + \psi_s(t), & \xi(t) \in H_0, \\ f^-(t, \xi(t)) = A^-\xi + \psi^-(t), & \xi(t) \in H^-. \end{cases} \tag{4}$$

Based on the above steps, we transform the two-degree-of freedom dry friction coupled vibration model into a four-dimensional piecewise smooth system.

2.2 *The existence, stability, and bifurcation of periodic orbits and invariant cone*

Based on the motion mode of the trajectory, consider the situation where the trajectory undergoes direct transition, the region that the trajectory passes through is recorded as H_c, as shown in Figure 2. We obtain the eigenvalues of $A^\pm$, which are $-\gamma \pm \sqrt{\gamma^2 - 1}$ and $-\beta_2\gamma \pm \sqrt{\beta_2^2\gamma^2 - \beta_1}$. Without losing generality, assuming that $\rho^\pm(t^\pm(\xi), \xi)$ is the solution curve, the flow starts from $\xi_0 = (\xi_{01}, \xi_{02}, \xi_{03}, \xi_{04})^T \in \partial H_c^-$, after time t^-, it reaches a new region H_c^+, i.e. $P(\xi) : H_c^- \to H_c^-$, $T = t_+ + t_-$, the corresponding Poincaré maps in H_c^+ and H_c^- have been obtained.
Here

$$\begin{aligned}
P^-(\xi, \psi(\xi)) &= e^{A^-t^-}\xi + \int_0^{t^-} e^{A^-(t^- - s)}\psi(\rho^-(s, \xi), t^-)\mathrm{d}s, \\
P^+(\xi, \psi(\xi)) &= e^{A^+t^+}\xi + \int_0^{t^+} e^{A^+(t^+ - s)}\psi(\rho^+(s, \xi), t^+)\mathrm{d}s,
\end{aligned} \tag{5}$$

where

$$P^\pm(\xi, \psi(\xi)) = \begin{pmatrix} P_1^\pm(\xi, \psi(\xi)) \\ P_2^\pm(\xi, \psi(\xi)) \\ P_3^\pm(\xi, \psi(\xi)) \\ P_4^\pm(\xi, \psi(\xi)) \end{pmatrix}.$$

For the convenience of analysis, system (6) is treated as an undisturbed system (i.e. $\psi^\pm(t) = 0$), and the system expressions in two subspaces is given

$$\dot{\xi}(t) = f^\pm(t, \xi(t)) = A^\pm\xi, \quad \pm h(\xi) > 0, \tag{6}$$

further

$$\begin{aligned}
H_c^+ &= \{\xi \in H_0 | -\xi_1 - 2\gamma\xi_2 + \beta_1\xi_3 + 2\beta_2\gamma\xi_4 > 0\}, \\
H_c^- &= \{\xi \in H_0 | -\xi_1 - 2\gamma\xi_2 + \beta_1\xi_3 + 2\beta_2\gamma\xi_4 < 0\}.
\end{aligned}$$

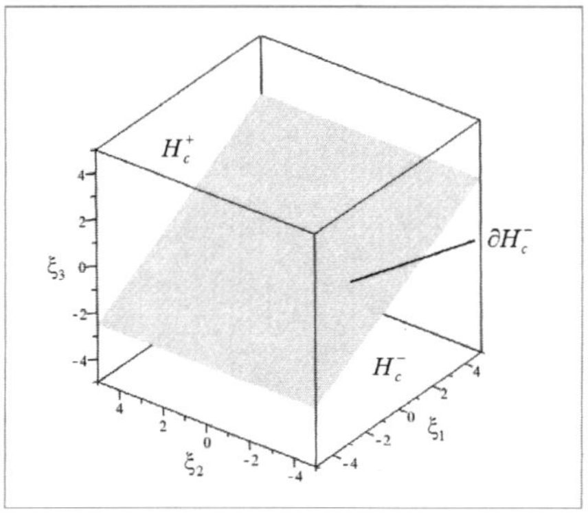

Figure 2: Regional division of the switching manifold contained in the system.

The time $t^-(\xi_0)$ for the trajectory starting from $\xi_0 \in \partial H_c^-$ to return to manifold H_0 is determined by the positive solution of the following equation

$$P_2^-(\xi_0) - P_4^-(\xi_0) = 0. \tag{7}$$

Theorem 2.1. *Assume that the flow starts from $\xi_0 \in \partial H_c^-$ and enters the intersection region H_c^-, $n^T(\xi)f^-(\xi_0) = 0$, reaches the region H_c^+ after time t^- (i.e. $\xi_1 = P^-(\xi_0) \in H_c^+$), t^- satisfies equation (7), then the system (6) has a two-zonal invariant cone consists of periodic orbits if and only if the following conditions are true*

(*a*) $P_1^+(P_1^-(\xi_0), P_2^-(\xi_0), P_3^-(\xi_0), P_4^-(\xi_0)) = \xi_{01}$,

(*b*) $P_2^+(P_1^-(\xi_0), P_2^-(\xi_0), P_3^-(\xi_0), P_4^-(\xi_0)) = \xi_{02}$,

(*c*) $P_3^+(P_1^-(\xi_0), P_2^-(\xi_0), P_3^-(\xi_0), P_4^-(\xi_0)) = \xi_{03}$,

(*d*) $P_4^+(P_1^-(\xi_0), P_2^-(\xi_0), P_3^-(\xi_0), P_4^-(\xi_0)) = \xi_{04}$.

Proof. According to the assumed conditions, the trajectory passes through the negative subspace at time t^- satisfying equation (7) from the initial point $\xi_0 \in \partial H_c^-$, the complete definition of $P^{\pm}(\xi)$ in H^+ and H^- is obtained. then the generalized map $P(\xi) = P^+(P^-(\xi))$ can be constructed. The study of periodic orbit can be transformed into the study of the fixed point of the Poincaré map. Through research and calculation, it has been found that the existence problem of an invariant cone can be solved only when the system satisfies conditions (*a*)-(*d*). the generalized map $P(\xi)$ contains a unique fixed point $\xi_0 \in \partial H_c^-$, thus generates a two-zonal invariant cone containing periodic orbits at ξ_0. At this point, $P(\xi)$ generates an invariant cone consists of periodic orbits at the fixed point ξ_0. □

Next, to further verify the correctness of the conclusion, we select a special set of parameters for numerical simulation, i.e. $\beta_1 = 4$, $\beta_2 = 1$, $\gamma = 0.0015$, it can be observed that the system has an invariant cone consists of periodic orbits. Figure 3(a) and Figure 3(b) show the geometric configuration of the trajectory when it undergoes direct transition.

Journal of Physics: Conference Series **2910** (2024) 012004 doi:10.1088/1742-6596/2910/1/012004

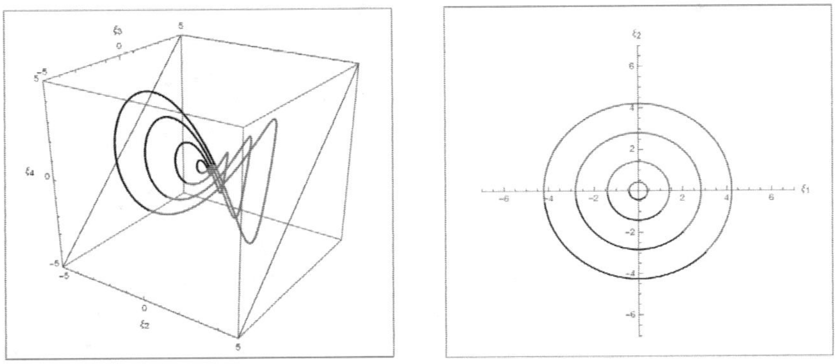

(a) Three-dimensional geometric explanation of the existence conditions of an invariant cone consists of periodic orbit.

(b) Two-dimensional geometric explanation of the existence condition of an invariant cone consists of periodic orbit.

Figure 3: The existence of an invariant cone consists of periodic orbits.

Theorem 2.2. *Assume that the system* (6) *has an invariant cone consists of periodic orbits (i.e. Theorem* 3.1 *holds), then the invariant cone is unstable under the selected parameters.*

Proof. We construct a global monodromy matrix to equivalently replace the research on the linearization $\frac{\partial P(\xi_0)}{\partial \xi}$ of $P(\xi)$. Through calculation, the eigenvalues of $\frac{\partial P(\xi_0)}{\partial \xi}$ are given by $eig(\frac{\partial P(\xi)}{\partial \xi}(\xi \in \partial H_c^-)) = (0.995, 0.995, 0.340 + 0.005i, 4.10 - 0.035i)^T$. Therefore, the stability of invariant cone can be judged by its properties, except for 1 and μ, the eigenvalues of $\frac{\partial P(\xi)}{\partial \xi}$ are defined as μ_i. Assuming $|\mu_i| \leq \min\{1, \mu\}, i = 1, 2$, the invariant cone is stable, otherwise it is unstable. It can be concluded that the invariant cone mentioned above is unstable. □

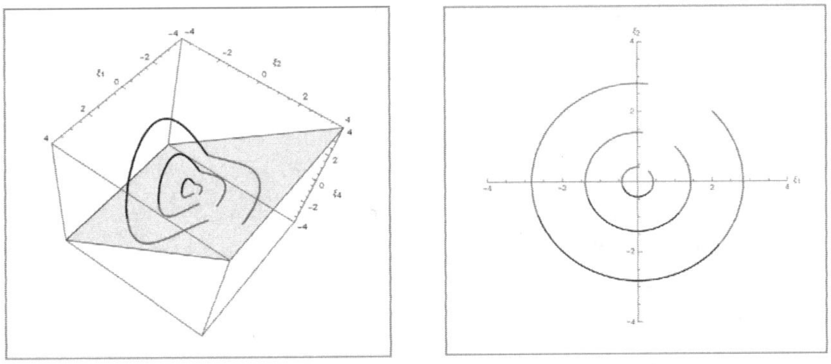

(a) A three-dimensional geometric explanation of bifurcation caused by parameter changes.

(b) A two-dimensional geometric explanation of bifurcation caused by parameter changes.

Figure 4: The existence of invariant cones is disrupted.

Finally, we discuss the persistence issue of invariant cone with parameter changes. We take parameter β_1 as the bifurcation parameter, i.e. $\beta_1 = 5$, $\beta_2 = 1$, $\gamma = 0.0015$, as shown in Figure 4(a) and Figure 4(b), as the parameter changes, the existence of invariant cone can be disrupted, and the periodic orbits can disappear.

Journal of Physics: Conference Series **2910** (2024) 012004 doi:10.1088/1742-6596/2910/1/012004

3 Conclusions

The study of dynamic characteristics such as periodic solutions and bifurcation of nonsmooth dynamical system has always been a topic of great concern in many engineering fields. This article investigates the periodic solutions and bifurcation theory of a two-degree-of freedom dry friction coupled vibration system. Firstly, a global Poincaré map is established to describe the trajectory behavior of the system during direct transition. Secondly, sufficient and necessary conditions for the existence of an invariant cone is obtained, further the stability properties and bifurcation situations are explored. When considering the stability of invariant cone, we equivalently replace the study of linearization of Poincaré map with the method of constructing a global monodromy matrix. Finally, numerical simulations are conducted to verify the accuracy of the theory.

Acknowledgements. This research was supported by the National Key R&D Program of China (Grant No.2022YFB3806000). This research was supported by the National Natural Science Foundation of China (Grant Nos.12272011 and 12332001).

References

[1] Gouveia M R A, Llibre J, Novaes D D, et al. Piecewise smooth dynamical systems: Persistence of periodic solutions and normal forms. Journal of Differential Equations, 2016, 260(7): 6108-6129.

[2] Grzesikiewicz W, Wakulicz A. Dynamics of non-smooth mechanical system. Machine Dynamics Problems, 1999, 23: 25-37.

[3] Chen H, Duan S, Tang Y, et al. Global dynamics of a mechanical system with dry friction. Journal of Differential Equations, 2018, 265(11): 5490-5519.

[4] Pi D, Yu J, Zhang X. On the sliding bifurcation of a class of planar Filippov systems. International Journal of Bifurcation and Chaos, 2013, 23(03): 1350040.

[5] Simpson D J W. Border-collision bifurcations in R^n. Siam Review, 2016, 58(2): 177-226.

[6] Li J, Guo Z, Zhu S, et al. Bifurcation of periodic orbits and its application for high-dimensional piecewise smooth near integrable systems with two switching manifolds. Communications in Nonlinear Science and Numerical Simulation, 2023, 116: 106840.

[7] Li L, Huang L. Concurrent homoclinic bifurcation and Hopf bifurcation for a class of planar Filippov systems. Journal of Mathematical Analysis and Applications, 2014, 411(1): 83-94.

[8] Kalmár-Nagy T, Csikja R, Elgohary T A. Nonlinear analysis of a 2-DOF piecewise linear aeroelastic system. Nonlinear Dynamics, 2016, 85: 739-750.

[9] Li T, Chen X. Degenerate grazing-sliding bifurcations in planar Filippov systems. Journal of Differential Equations, 2020, 269(12): 11396-11434.

[10] Carmona V, Freire E, Fernandez-Garcia S. Periodic orbits and invariant cones in three-dimensional piecewise linear systems. Discrete and Continuous Dynamical Systems, 2015, 35(01): 59-72.

[11] Küpper T. Invariant cones for non-smooth dynamical systems. Mathematics and Computers in Simulation, 2008, 79(4): 1396-1408.

[12] Hosham, H A. Bifurcation of periodic orbits in discontinuous systems. Nonlinear Dynamics, 2017, 87(1): 135-148.

Melnikov analysis of Type I subharmonic orbits in quadruple-well coupled SD oscillators

Ran Sun[1], **Jing Li**[1,*], **Shaotao Zhu**[2,*]

[1]Interdisciplinary Research Institute, School of Mathematics, Statistics and Mechanics, Beijing University of Technology, Beijing, China
[2]College of Computer Science, Beijing University of Technology, Beijing, China

E-mail: leejing@bjut.edu.cn (J. Li), zhushaotao@bjut.edu.cn(S. T. Zhu).

Abstract. When appropriate parameters are selected, coupled SD oscillators can exhibit bistability, tristability, and even quadrastability. The presence of multiple potential wells introduces rich dynamical characteristics into these multi-stable systems. As a result, global dynamical analysis of quad-stable systems, particularly the study of subharmonic orbits, has garnered significant attention in recent years. This paper aims to investigate subharmonic motion in the quad-stable system of coupled SD oscillators and to establish a theoretical framework for Melnikov analysis of Type I subharmonic orbits. Based on the derived Melnikov function, the existence of Type I subharmonic orbits is verified. Finally, numerical simulation methods are employed to validate the proposed approach by plotting phase portrait, the time history diagram $t - x$ and spectrum diagram.

Keywords: Coupled SD oscillator; Subharmonic orbits; Quadruple-well; Melnikov function

1 Introduction

Piecewise smooth dynamical systems play a crucial role in various scientific and engineering disciplines due to their non-smooth characteristics. These systems can exhibit rich dynamic behaviors, including bifurcations and multiple steady states. Foundational theories by Filippov [1], Feigin [2], and others have provided crucial insights into the stability and response of such systems under varying conditions. Recently, extending the concepts and methodologies of smooth dynamical system theory to address non-smooth dynamics has emerged as a significant area of research.

Among the various dynamical phenomena observed in piecewise smooth systems, subharmonic orbits have garnered considerable attention due to their unique periodic properties [3, 4].The classical Melnikov method offers an analytical framework for assessing the dynamics of smooth systems after perturbation. It calculates the normal distance between the stable and unstable manifolds of hyperbolic fixed points to derive the Melnikov function. This method determines the persistence of periodic orbits under non-autonomous perturbations, with persistence guaranteed by subharmonic Melnikov functions and their simple zeros. Du et al. [5] and Li et al. [6] explored Type I and II periodic motions in nonlinear impact oscillators, identifying the conditions for subharmonic orbit generation using analytical methods. Recent advancements include Guo et al. [7], who extended the concept of subharmonic Melnikov functions to higher-dimensional non-smooth systems with unilateral impacts, such as double pendulums. Li and Sun [8] applied Melnikov analysis to investigate three types of subharmonic motions in bistable vibro-impact oscillators, emphasizing their significance in complex vibration systems. Fan and Zhang [9] explored multiple subharmonic solutions in Hamiltonian systems with symmetries, proposing a relationship between

Content from this work may be used under the terms of the Creative Commons Attribution 4.0 licence. Any further distribution of this work must maintain attribution to the author(s) and the title of the work, journal citation and DOI.
Published under licence by IOP Publishing Ltd

subharmonic solutions and system symmetries. Motonaga [10] examined the subharmonic Melnikov function in single-degree-of-freedom Hamiltonian systems, introducing new methodologies for characterizing subharmonic orbits under various perturbations. In summary, research on subharmonic orbits has significantly advanced our understanding of nonlinear dynamical systems and provided valuable theoretical frameworks for controlling and optimizing these systems.

In the field of non-smooth dynamics, the study of oscillator complexity and multi-stability has gained significant attention. Recent advancements, particularly those related to the SD oscillator, have revealed complex phenomena associated with various stable states. The Smooth and Discontinuous Oscillator (SD Oscillator), introduced by Cao et al. [11] and later developed by Han et al. [12], consists of a pair of inclined elastic springs connected to a mass m, with each spring having a stiffness k and capable of withstanding both tensile and compressive forces. Due to the loss of local hyperbolicity, the oscillator undergoes inter-well jumping in the discontinuous limit. Under damping and external excitation, the system exhibits coexisting attractors and chaotic saddle points, providing a foundation for exploring transitions from smooth to non-smooth behavior. Despite these advancements, research has primarily focused on single- and double-stable systems, with less emphasis on four-stable systems. Liu et al. [13] investigated the global bifurcations of a multi-stable nonlinear oscillator, highlighting its complex equilibrium configurations and bifurcation phenomena using the generalized cell mapping method. Sun et al. [14] compared the dynamics of a geometrically nonlinear oscillator, showing how variations in the horizontal distance ratio lead to different well configurations, each exhibiting distinct stiffness characteristics and dynamics. However, these studies have mainly concentrated on global bifurcations, with limited investigation into periodic orbits, particularly subharmonic orbits.

The structure of this article is organized as follows: Section 2 analyzes the dynamic equations governing the system, utilizing linear approximation method to approximate the complex restoring force function with piecewise linear segments. Section 3 derives the Melnikov function for subharmonic orbits within Type I homoclinic orbit and explores the conditions for the existence of Type I subharmonic orbits using pertinent theoretical frameworks. Section 4 validates the theoretical findings through numerical simulations. Finally, Section 5 summarizes the research results and conclusions.

2 A class of coupled SD oscillator with four-well structure

According to [15], after dimensionless transformation, the dynamic equation for governing the coupled SD oscillator is

$$\ddot{x} + F(x) = \varepsilon f\cos(\omega t) - \varepsilon\delta\dot{x},$$

where the restoring force $F(x)$ detailed as

$$F(x) = x\left(1 - \frac{1}{\sqrt{x^2+\alpha^2}}\right) + (x+\beta)\left(1 - \frac{1}{\sqrt{(x+\beta)^2+\gamma^2}}\right) + (x-\beta)\left(1 - \frac{1}{\sqrt{(x-\beta)^2+\gamma^2}}\right).$$

To facilitate subsequent analysis, the nonlinear equation $F(x)$ is approximated linearly as follows

$$\tilde{F}(x) = K_i(x - x_i^*), i = 1, 2, ...7,$$

among them, $x_i^*, i = 1, 2, ...7$, represents the abscissa of the seven equilibrium points.

Then, unperturbed equations of approximate system are

$$\begin{cases} \dot{x} = y, \\ \dot{y} = -\tilde{F}(x). \end{cases} \tag{1}$$

3 The Type I subharmonic Melnikov analysis

3.1 The unperturbed system

Due to the piecewise nature of the approximate restoring force $F(x)$, the phase plane is partitioned into seven distinct regions. For the purpose of analyzing Type I subharmonic orbits, the switching manifold is defined as

$$\Sigma_{c^+} = \Sigma_{c^+}^- \cup (x_c, 0) \cup \Sigma_{c^+}^+,$$

and where

$$\Sigma_{c^+}^+ = \{(x,y) \mid x = x_c, y > 0\}, \quad \Sigma_{c^+}^- = \{(x,y) \mid x = x_c, y < 0\}.$$

As illustrated in Figure 1(a), the Type I periodic orbit for the unperturbed system is defined as follows

$$\gamma_{\mathrm{I}}(t)=\begin{cases}\gamma_{\mathrm{I},1}(t)=(x_{\mathrm{I},1}(t),y_{\mathrm{I},1}(t)), & t\in[0,t_1),\\ \gamma_{\mathrm{I},2}(t)=(x_{\mathrm{I},2}(t),y_{\mathrm{I},2}(t)), & t\in[t_1,T_{\mathrm{I}}].\end{cases}$$

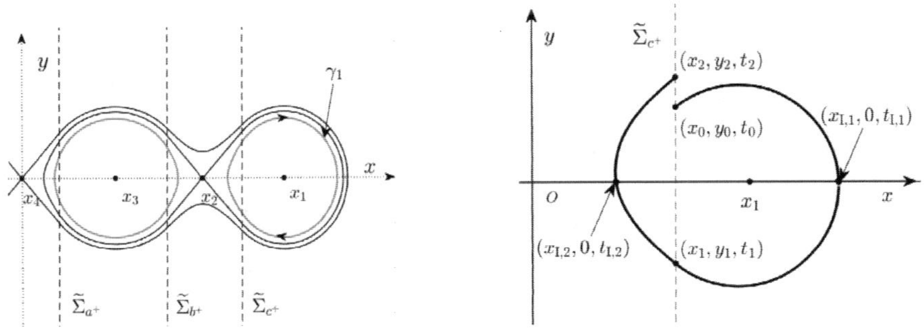

Figure 1: The geometrical structure of the unperturbed system: (a) Sections of the unperturbed manifold. Blue is used to mark the periodic orbits inside the homoclinic orbit; (b) Type I periodic orbit of the unperturbed system. The red dashed line represents the switching manifold Σ_c and the position indicated by the arrow is the intersection of the orbit and the x-axis

3.2 Melnikov processes of the Type I subharmonic orbits

It is apparent that

$$\begin{aligned}&(x_{\mathrm{I},1}(\frac{1}{2}t_1),y_{\mathrm{I},1}(\frac{1}{2}t_1))=(x_{\mathrm{I},1},0),\\&(x_{\mathrm{I},2}(\frac{1}{2}T_{\mathrm{I}}+\frac{1}{2}t_1),y_{\mathrm{I},1}(\frac{1}{2}T_{\mathrm{I}}+\frac{1}{2}t_1))=(x_{\mathrm{I},2},0),\end{aligned}\tag{2}$$

and the Hamiltonian function for the system described by (1) is

$$H(x,y)=\frac{1}{2}y^2+\tilde{V}(x).$$

Substituting the points $(x_{\mathrm{I},1},0)$ and $(x_{\mathrm{I},2},0)$ obtained from equation (2) into the Hamiltonian function, then we have

$$H(x_{\mathrm{I},1},0)=H(x_{\mathrm{I},2},0)=h_{\mathrm{I}},$$

which can be expressed equivalently as

$$y=\pm\sqrt{2h_{\mathrm{I}}-2\tilde{V}(x)}.$$

When $\varepsilon=0$, the positions of the two points are illustrated in Figure 1(b). To facilitate calculation, the analysis is restricted to the case where $m=1$. Next, the following Melnikov function can be derived by calculating the Hamiltonian energy difference between points $(x_0^{\varepsilon},y_0^{\varepsilon})$ and $(x_1^{\varepsilon},y_1^{\varepsilon})$

$$\begin{aligned}M_{\mathrm{I}}(t_0)&=\int_0^{T_{\mathrm{I}}}JD_xH(\gamma_1(t))\wedge g(\gamma_1(t),t+t_0)\mathrm{d}t\\&=-\delta\underbrace{\int_0^{T_{\mathrm{I}}}y_{\mathrm{I}}^2(t)\mathrm{d}t}_{I_{\mathrm{I}1}}+f\underbrace{\int_0^{T_{\mathrm{I}}}y_{\mathrm{I}}(t)\cos\omega(t+t_0)\mathrm{d}t.}_{I_{\mathrm{I}2}}\end{aligned}$$

By utilizing the symmetry of $I_{\text{I}1}$, the following transformation can be applied

$$I_{\text{I}1} = -2\int_{x_{\text{I},1}}^{x_c}\sqrt{2h_{\text{I}} - 2\tilde{V}_1(s)}\mathrm{d}s - 2\int_{x_c}^{x_{\text{I},2}}\sqrt{2h_{\text{I}} - 2\tilde{V}_2(s)}\mathrm{d}s.$$

Next, the calculation of $I_{\text{I}2}$ proceeds as follows:

$$\begin{aligned} I_{\text{I}2} &= \int_{-\frac{1}{2}T_{\text{I}}}^{\frac{1}{2}T_{\text{I}}} y_{\text{I}}(t + \frac{1}{2}T_{\text{I}})\cos(\omega(t + t_0 + \frac{1}{2}T_{\text{I}})\mathrm{d}t \\ &= -\sin(\omega(t_0 + \frac{1}{2}T_{\text{I}}))\int_0^{T_{\text{I}}} y_{\text{I}}(t)\sin\omega(t - \frac{1}{2}T_{\text{I}})\mathrm{d}t. \end{aligned}$$

Since $T_{\text{I}} = n\hat{T}$ and $\hat{T} = \dfrac{2\pi}{\omega}$, namely $T_{\text{I}} = \dfrac{2\pi n}{\omega}$, $I_{\text{I}2}$ can be simplified to

$$I_{\text{I}2} = 2\cdot(-1)^{n+1}\sin(\omega t_0)I_{\text{I}3},$$

where

$$I_{\text{I}3} = \int_{\frac{1}{2}t_1}^{t_1} y_{\text{I},1}(t)\sin\omega(t - \frac{1}{2}T_{\text{I}})\mathrm{d}t + \int_{t_1}^{\frac{1}{2}t_1 + \frac{1}{2}T_{\text{I}}} y_{\text{I},2}(t)\sin\omega(t - \frac{1}{2}T_{\text{I}})\mathrm{d}t.$$

As a result, a pair of subharmonics with order $m = 1$ will appear along the bifurcation curve. It is apparent that

$$y = \frac{\mathrm{d}x}{\mathrm{d}t} = \pm\sqrt{2h_{\text{I}} - 2\tilde{V}(x)} \Rightarrow \mathrm{d}t = \pm\frac{1}{\sqrt{2h_{\text{I}} - 2\tilde{V}(x)}}\mathrm{d}x,$$

hence, when $t \in (\frac{1}{2}t_1, t_1)$,

$$t = t_1 + \int_x^{x_c}\frac{1}{\sqrt{2h_{\text{I}} - 2\tilde{V}_1(s)}}\mathrm{d}s,$$

and for $t \in (t_1, \frac{1}{2}t_1 + \frac{1}{2}T_{\text{I}})$,

$$t = \frac{1}{2}t_1 + \frac{1}{2}T_{\text{I}} + \int_x^{x_{\text{I},2}}\frac{1}{\sqrt{2h_{\text{I}} - 2\tilde{V}_2(s)}}\mathrm{d}s.$$

Clearly

$$I_{\text{I}3} = \int_{x_{\text{I},1}}^{x_c}\sin\omega(\int_{x_{\text{I},1}}^{x}\frac{1}{\sqrt{2h_{\text{I}} - 2\tilde{V}_1(s)}}\mathrm{d}s + \frac{1}{2}t_1 - \frac{1}{2}T_{\text{I}})\mathrm{d}x + \int_{x_c}^{x_{\text{I},2}}\sin\omega(\frac{1}{2}t_1 - \int_x^{x_{\text{I},2}}\frac{1}{\sqrt{2h_{\text{I}} - 2\tilde{V}_2(s)}}\mathrm{d}s)\mathrm{d}x.$$

In summary, it can be concluded that

$$M_{\text{I}}^{n,1}(y_0, t_0) = 2\cdot(-1)^{n+1}fI_{\text{I}3}\sin(\omega t_0) - \delta I_{\text{I}1}. \tag{3}$$

Following the methodology of Li et al. [8], it is crucial to determine $(\bar{y}_0, \bar{t}_0)$ in the subsequent analysis, such that

$$\frac{\partial M_{\text{I}}^{n,1}(\bar{y}_0, \bar{t}_0)}{\partial t_0} \neq 0, \quad M_{\text{I}}^{n,1}(y_0, t_0) = 0.$$

Finally, based on equation (3), the threshold function is given by

$$f > \frac{\delta I_{\text{I}1}}{2|I_{\text{I}3}|}. \tag{4}$$

4 Numerical simulation

To evaluate the feasibility and effectiveness of the proposed theoretical framework, a series of numerical simulations focusing on the case $n = 1$ was conducted. The simulations used parameters consistent with the threshold function specified in equation (4). These parameter choices were intended to assess the robustness of the theoretical predictions and validate the applicability of the framework in practical scenarios.

Figure 2(a) and (b) illustrate the numerical simulation results for a specific set of parameters: $\varepsilon = 0.5, \delta = 0.5, \omega = 2.7$ and $f = 2.98$. In these figures, the blue dashed lines represent the positions of the switching manifolds. The time history plot confirms the periodic nature of the motion, demonstrating that the system exhibits regular oscillatory behavior over time. The spectrum diagram reveals that the observed motion is indeed subharmonic, with a ratio of $n/m = 1$, indicating that the response occurs at a subharmonic frequency, which aligns with theoretical expectations. Additionally, the phase plane diagram in Figure 2(a) classifies the observed subharmonic orbit as Type I. Furthermore, Figure 2(c) and (d) displays another Type I subharmonic orbit for the parameters $\varepsilon = 0.5, \delta = 0.5, \omega = 2.78$ and $f = 3.2$. This result further corroborates the findings, demonstrating that the theoretical framework accurately predicts Type I subharmonic orbits across a range of parameter values. The successful identification and characterization of these subharmonic orbits validate the proposed theoretical framework, confirming its ability to analyze and predict subharmonic behaviors within dynamical systems. These results not only support the theoretical predictions but also underscore the practical applicability of the framework.

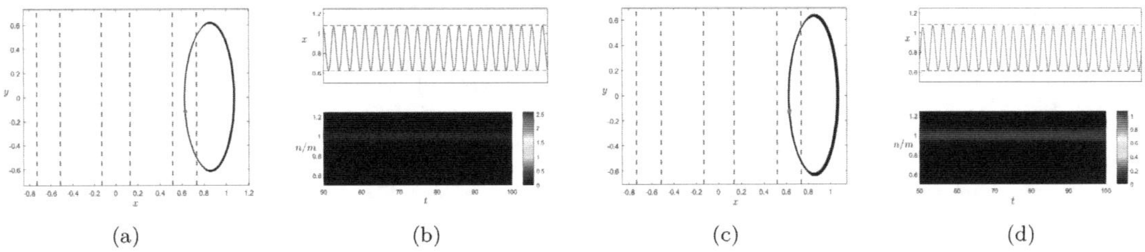

Figure 2: Periodic motions of the system for $n : 1 = 1 : 1$: (a) periodic orbit in projected phase plane (x, y) with $\varepsilon = 0.5, \delta = 0.5, \omega = 2.7$, $f = 2.98$; (b) the time history diagram $t - x$ and spectrum diagram with $\varepsilon = 0.5, \delta = 0.5, \omega = 2.7$; (c) periodic orbit in projected phase plane (x, y) with $\varepsilon = 0.5, \delta = 0.5, \omega = 2.78$, $f = 3.2$; (d) the time history diagram $t-x$ and spectrum diagram with $\varepsilon = 0.5, \delta = 0.5, \omega = 2.78$, $f = 3.2$

5 Conclusion

This article explores Type I subharmonic motion using Melnikov analysis, significantly advancing the understanding of global dynamics in coupled SD oscillators. A comprehensive theoretical framework has been established for applying the Melnikov method to Type I subharmonic orbits in piecewise smooth quad-stable systems. This framework includes a rigorous derivation of the threshold function based on the Melnikov function, providing a solid theoretical foundation for proving the existence of such subharmonic orbits. Due to the inherent nonlinearity of piecewise smooth systems, integrating the Melnikov function is relatively complex. Consequently, numerical integration techniques are employed to handle this complexity effectively. This approach successfully manages the intricate integrals involved and facilitates the identification of subharmonic orbits. The results demonstrate that Melnikov analysis is a powerful tool for detecting and characterizing subharmonic orbits in piecewise smooth systems. Future research will extend to more complex dynamical systems, particularly quad-stable systems, to explore additional types of subharmonic orbits.

Acknowledgments

This research was supported by the National Key R&D Program of China(Grant No.2022YFB3806000). This research was supported by the National Natural Science Foundation of China(Grant Nos.12272011 and 12332001).

Author contributions

Conceptualization, Jing Li and Ran Sun; methodology, Ran Sun and Shaotao Zhu; software, Ran Sun; validation, Ran Sun; formal analysis, Ran Sun and Jing Li; investigation, Ran Sun and Shaotao Zhu;

writingoriginal draft preparation, Ran Sun; writingreview and editing, Shaotao Zhu; supervision, Jing Li; funding acquisition, Jing Li. All authors have read and agreed to the published version of the manuscript.

Data Availability Statements:
No data was used for the research described in the article.

Declaration of competing interest:
The authors declare that they have no known competing financial interests or personal relationships that could have appeared to influence the work reported in this paper.

References

[1] Filippov A F 1988 Differential Equations with Discontinuous Right-Hand Sides: Mathematics and Its Applications Kluwer Academic *Dordrecht.*

[2] Feigin M I 1994 Forced Vibrations of Nonlinear Systems with Discontinuities Nauka *Moscow.*

[3] Yagasaki K 2013 Application of the subharmonic Melnikov method to piecewise-smooth systems *Journal of Dynamical and Control Systems* **33**(5), 2189-2209.

[4] Li S, Zhao S 2019 The analytical method of studying subharmonic periodic orbits for planar piecewise-smooth systems with two switching manifolds *International Journal of Dynamics and Control* **7**, 23-35.

[5] Du Z, Li Y 2007 Type I periodic motions for nonlinear impact oscillators *Nonlinear Analysis* **67**(5), 1344C1358.

[6] Li Y, Du Z, Zhang, W. 2008 Asymmetric type II periodic motions for nonlinear impact oscillators *Nonlinear Analysis* **68**(9), 2681-2696.

[7] Guo X Y, Zhang G, Tian R L 2022 Subharmonic Bifurcation for a Non-smooth Double Pendulum with Unilateral Impact *Journal of Nonlinear Mathematical Physics* **29**(2), 349-367.

[8] Li S, Sun R 2023 Melnikov analysis of subharmonic motions for a class of bistable vibro-impact oscillators *Nonlinear Dynamics* **111**(2), 1047-1069.

[9] Fan Z, Zhang D 2023 Multiple subharmonic solutions in Hamiltonian system with symmetries *Journal of Differential Equations* **351**, 1-48.

[10] Motonaga S 2024 Subharmonic Melnikov functions and nonintegrability for autonomous and non-autonomous perturbations of single-degree-of-freedom Hamiltonian systems near periodic orbits *Physica D: Nonlinear Phenomena* **460**, 134088.

[11] Cao Q J, Wiercigroch M, Pavlovskaia E E, Grebogi C T Thompson J M 2006 Archetypal oscillator for smooth and discontinuous dynamics *Physical Review E-Statistical, Nonlinear, and Soft Matter Physics* **74**(4), 046218.

[12] Han Y W, Cao Q J, Chen Y S, Wiercigroch M 2012 A novel mass-spring based smooth and discontinuous oscillator with strongly irrational nonlinearities *Science China Physics, Mechanics Astronomy* **55**, 1832-1843.

[13] Liu C, Jiang W A, Chen L 2023 Global bifurcations of a multi-stable nonlinear oscillator *Archive of Applied Mechanics* **93**(3), 1149-1165.

[14] Sun H, Shang H 2024 Comparisons for Global Dynamics of a Geometrically Nonlinear Oscillator among Single-, Double- and Quadruple-Well Configurations *Fractal and Fractional* **8**(4), 202.

[15] Han Y, Cao Q, Ji J 2015 Nonlinear Dynamics of a Smooth and Discontinuous Oscillator with Multiple Stability *International Journal of Bifurcation and Chaos* **25**(13), 1530038.

doi:10.1088/1742-6596/2910/1/012006

Analysis of the Goodwin business cycle model with discrete delay and distributed delay in induced investment

A.O. Antonova

Air Transportation Management Department, National Aviation University, Kyiv, Ukraine

*E-mail: anna.antonova@npp.nau.edu.ua

Abstract. In this paper, a Goodwin model of business cycle with both discrete and distributed delays is considered. With the corresponding characteristic equation analyzed, the linear stability of the equilibrium is investigated. Numerical simulations are carried out to study the excitation conditions of the long-periodic Goodwin oscillations.

1. Introduction

The differential equations with discrete or distributed time delay are the most adequate tools to model the endogenous theories of the business cycle Kalecki [1], Kaldor [2] and Goodwin [3]. These cycle models are still of interest today due to the complex nonlinear dynamics of the processes they describe [4-7].

In this paper we will revise Goodwin's model [3]. This model is based on multiplier-accelerator interaction and heavily influenced with the form of the induced investment function φ. According to Goodwin it depends only on the derivative of income y over time t, $\varphi = \varphi\left(\frac{dy(t)}{dt}\right)$ and is an S-shaped function.

The other necessary element of Goodwin's business cycle model is the time delay between investment decisions and the resulting outlays. This delay creates the limit cycles. In general, the Goodwin accelerator with delay can be written as [8]

$$\varphi_e(t) = \int_{-\infty}^{t} f\left(t-\xi\right)\varphi\left(\frac{dy(\xi)}{d\xi}\right)d\xi, \tag{1}$$

where $f(\xi)$ is a nonnegative continuous delay kernel defined and integrable on $[0,+\infty)$, which reflects the influence of the past states on the current dynamics, and $\int_0^\infty f(\xi)d\xi = 1$. Then the Goodwin business cycle model is written as follows

$$\varepsilon\frac{dy(t)}{dt} + y(t) = \alpha y(t) + \varphi_e(t) + A(t). \tag{2}$$

Here $\varepsilon > 0$ is time-lag of the dynamic multiplier, α is marginal propensity to consume, $0 \le \alpha \le 1$, A is the autonomous investment.

In general, the integro-differential equation (2) is difficult to consider even numerically. Therefore, the particular forms of $f(\xi)$ are chosen for which Eq. (2) is written as an ODE or as a differential equation with a fixed delay. The original Goodwin equation [3] corresponds to the

Content from this work may be used under the terms of the Creative Commons Attribution 4.0 licence. Any further distribution of this work must maintain attribution to the author(s) and the title of the work, journal citation and DOI.
Published under licence by IOP Publishing Ltd

case of exponential delay kernel [8-10], $f(\xi)=T^{-1}e^{-T^{-1}\xi}$, $T>0$, and has the form of the second order nonlinear differential equation

$$\varepsilon T\frac{d^2y(t)}{dt^2}+(\varepsilon+sT)\frac{dy(t)}{dt}+sy(t)=\varphi\left(\frac{dy(t)}{dt}\right)+T\frac{dA(t)}{dt}+A(t), \tag{3}$$

where $s=1-\alpha$. For the parameters chosen by Goodwin in [3] the solution to this equation is the limit cycle with a period of approximately 10 years. It is possible to prove an existence and uniqueness theorem for Eq. (3) [11].

However, if we choose another type of function $f(\xi)$, namely the Dirac function, $f(\xi)=\delta(\xi-\theta)$, the situation with the uniqueness of solutions changes. In this case we get Goodwin's model with a fixed delay in the form of the first order neutral delay differential equation

$$\varepsilon\frac{dy(t)}{dt}+y(t)=\alpha y(t)+\varphi\left(\frac{dy(t-\theta)}{dt}\right)+A(t). \tag{4}$$

As shown in [12-19], Eq. (4) describes the existence of an infinite number of limit cycles (multiplicity of solutions). One of them is the long-periodic cycle with period of about 10 years (we will call it mode 1), the rest are short-periodic cycles with periods $\frac{\theta}{m}$, $m=1, 2,\ldots$ and discontinuous derivative $\frac{dy(t)}{dt}$, (mode 2, 3, etc.). These higher modes are not simply superimposed upon the first mode, but each are an alternative to it. What will be the steady-state solutions - long-periodic Goodwin's cycle or periodic relaxation solutions depends on the type of initial function, $y(t),\ t\le 0$.

This nature of the solutions due to the fact that the Delta distribution has zero width. Taking into account that even small deviations from the delta function lead to a limited number of excited modes [7]. Limit cycle with period $\frac{\theta}{m}$ cannot exist if the distribution width is greater than $\frac{\theta}{m}$. On the other hand, the width effects cannot influence the long-periodic mode 1. As shown in [7], if $f(\xi)$ is the gamma distribution or the continuous uniform distribution the number of modes depends on the dimensionless half-width V

$$V=\frac{\sigma_f}{T_f},\quad T_f=\int_0^\infty \xi f(\xi)d\xi,\quad \sigma_f^2=\int_0^\infty(\xi-T_f)^2 f(\xi)d\xi,$$

where T_f is average delay time and σ_f^2 is its variance. For the parameters chosen by Goodwin it follows that if $V>0.3$ only mode 1 exists. For $0.1<V<0.3$ there may be modes 1 and 2.

We can generalize in this paper the Goodwin cycle model to the more general form of delay in induced investment. Suppose that in the absence of delay they can be represented as two terms

$$\varphi\left(\frac{dy(t)}{dt}\right)=\varphi_1\left(\frac{dy(t)}{dt}\right)+\varphi_2\left(\frac{dy(t)}{dt}\right),$$

and the delay in these terms is described by different kernels $f_1(\xi)$ and $f_2(\xi)$. Then we can write the modified Goodwin equation in the form

$$\varepsilon\frac{dy(t)}{dt}+y(t)=\alpha y(t)+\varphi_{e1}(t)+\varphi_{e2}(t)+A(t), \tag{5}$$

where

$$\varphi_{ek}(t) = \int_{-\infty}^{t} f_k\left(t-\tau\right)\varphi_k\left(\frac{dy(\tau)}{d\tau}\right)d\tau,\ k=1,2.$$

In this work we consider the accelerator with the following delay kernels

$$f_1(\xi) = \delta(\xi-\theta),\ f_2(\xi) = T^{-1}e^{-\xi/T}, \tag{6}$$

and

$$\varphi\left(\frac{dy(t)}{dt}\right) = \mu\varphi_1\left(\frac{dy(t)}{dt}\right) + (1-\mu)\varphi_2\left(\frac{dy(t)}{dt}\right),$$

where $0 \le \mu < 1$. By changing μ we significantly change the properties of the accelerator delay. We study the influence of the μ parameter on the stability of the stationary state and the conditions for the excitation of the long-periodic cycle and the short-periodic cycles for the Goodwin parameter values [3].

2. Model

If we assume that the delay kernels are given by Eq. (6), then it is possible to write the modified Goodwin model (5) in this form

$$\varepsilon\frac{dy(t)}{dt} = -sy(t) + \mu\varphi\left(\frac{dy_\theta(t)}{dt}\right) + \frac{(1-\mu)}{T}\int_{-\infty}^{t} e^{-(t-\tau)/T}\varphi\left(\frac{dy(\tau)}{d\tau}\right)d\tau + A(t),$$

$$y_\theta(t) = y(t-\theta).$$

This equation can be reduced to the ODE system with a fixed and continuous delay

$$\begin{aligned}
&\varepsilon\frac{dy(t)}{dt} + sy(t) = \mu\varphi\left(\frac{dy_\theta(t)}{dt}\right) + (1-\mu)I(t) + A(t),\\
&T\frac{dI(t)}{dt} + I(t) = \varphi\left(\frac{dy(t)}{dt}\right).
\end{aligned} \tag{7}$$

In Eq. (7) it is assumed that the time t is measured in years, the income $y(t)$, φ, I, A - in billions of dollars per year, ε, r and T, θ - in years. Eq. (7) can be reduced to the 2nd order nonlinear neutral delay differential equation

$$\varepsilon T\frac{d^2y}{dt^2} + \left(\varepsilon + sT\right)\frac{dy}{dt} + sy - \left(1-\mu\right)\varphi\left(\frac{dy}{dt}\right) - \mu T\varphi'\left(\frac{dy_\theta}{dt}\right)\frac{d^2y_\theta}{dt^2} - \mu\varphi\left(\frac{dy_\theta}{dt}\right) = T\frac{dA}{dt} + A. \tag{8}$$

3. Analysis of linearized equation

Let A=A_0=const. Then $y_s = A_0/s$, $I_s = 0$ is stationary solution. If the deviations from the stationary value $\delta y(t) = y(t) - y_s$ and $\delta I(t)$ are small, after linearization of Eq. (7) we obtain

$$\varepsilon\frac{d\delta y(t)}{dt} + s\delta y(t) = \mu r\frac{d\delta y_\theta(t)}{dt} + \left(1-\mu\right)\delta I(t),$$

$$T\frac{d\delta I(t)}{dt} + \delta I(t) = r\frac{d\delta y(t)}{dt}.$$

Here $r = \varphi'(0)$ is the acceleration coefficient. This system reduces to the linear neutral delay differential equation

$$\varepsilon T \frac{d^2\delta y(t)}{dt^2} - \mu r T \frac{d^2\delta y_\theta(t)}{dt^2} + \left(\varepsilon + sT - r + \mu r\right)\frac{d\delta y(t)}{dt} - \mu r \frac{d\delta y_\theta(t)}{dt} + s\delta y(t) = 0. \quad (9)$$

In Eq. (9), we make the ansatz that $\delta y(t) = Ce^{\lambda t}$, $\delta y_\theta(t) = Ce^{\lambda(t-\theta)} = e^{-\lambda\theta}\delta y(t)$ to obtain the associated characteristic equation, which is given by

$$\lambda^2 + \frac{\left(\varepsilon + sT - (1-\mu)r\right)}{\varepsilon T}\lambda + \frac{s}{\varepsilon T} - \frac{\mu r}{\varepsilon}\left(\lambda^2 + \frac{\lambda}{T}\right)e^{-\lambda\theta} = 0. \quad (10)$$

The stability analysis of Eq. (10) is equivalent to the problem of determining conditions under which all roots of its characteristic equation have negative real parts, $\operatorname{Re}\lambda < 0$. We will consider ε, s, T and r as fixed and study the dependence $\lambda(\theta)$.

For the value of λ with largest modulus, $|\lambda| \gg 1$ the characteristic equation (10) may be replaced by the approximate equation [20]

$$\lambda^2(\varepsilon - \mu r e^{-\lambda\theta}) = 0. \quad (11)$$

Then, from Eq. (11) we obtain $e^{\lambda\theta} = \dfrac{\mu r}{\varepsilon}$, therefore

$$\theta\lambda_n \approx \ln\frac{\mu r}{\varepsilon} + 2\pi(n-1)i, \; n = 2, 3, \ldots$$

If $\mu r > \varepsilon$, then for all $\theta > 0$, there is an infinite number of roots of Eq. (10) whose real parts $\operatorname{Re}\lambda_n(\theta)$ are positive and the zero solution δy of Eq. (9) is unstable for any $\theta > 0$.

The case $\mu r < \varepsilon$ is analyzed in detail in [20]. Using the results of [20], it is possible to verify that only the mode 1 will be excited for the Goodwin parameters $\varepsilon = 0.5$, $r = 2$, $T = \theta = 1$ and $s = 0.4$. Numerically found roots of Eq. (10) for these parameters are given in Table 1.

Table 1. Roots of the Eq. (10) for different values of μ for $\varepsilon = 0.5$, $r = 2$, $T = \theta = 1$ and $s = 0.4$.

μ	$\lambda_{1,1}$	$\lambda_{1,2}$	λ_2	λ_3	λ_4
0.0	0.46	1.74	-	-	-
0.1	0.47	1.75	-1.00±5.82 i	0.94±12.34i	0.93±18.70i
0.15	0.48	1.38	-0.55±5.84 i	0.52±12.36i	-0.52±18.71i
0.2	0.486	1.28	-0.230±5.875 i	-0.225±12.372 i	-0.224±18.721i
0.25	0.495	1.195	0.017 ±5.911 i	0.004 ±12.389 i	0.002±18.732 i
0.5	0.567	0.834	0.750 ±6.100 i	0.707 ±12.472 i	0.699±18.786 i
0.578		0.67 ±0.01 i	0.89 ±6.15 i	0.851 ±12.497 i	0.844±18.803 i
0.75		0.61±0.16 i	1.125 ±6.263 i	1.106±12.552 i	1.102±18.839 i
1.0		0.55 ±0.22 i	1.355 ±6.494 i	1.378 ±12.629 i	1.382±18.892 i

From Table 1it follows that if $0<\mu<0.25$, then the characteristic numbers with $\mathrm{Re}\,\lambda>0$ have small imaginary parts. Therefore, for $0<\mu<0.25$ the short-periodic cycles with periods $1, 1/2, 1/3, \ldots$ does not exist. For $0.25<\mu<0.5$ the increment of the mode 1 exceeds the increments of the higher modes and one might think that the mode 1 can be excited more easily than the higher modes.

4. Numerical modeling results

In this section, we present the results of numerical modeling of the behavior of solutions of Eq. (7) for some values of μ. The time dependences modelling of $y(t,\mu)$ were carried out for the Goodwin parameters and $\theta=1$. For the accelerator $\varphi(x)$ we use the following function [9]

$$\varphi(x)=\frac{\varphi_c|\varphi_f|\left(e^{r\gamma x}-1\right)}{|\varphi_f|e^{r\gamma x}+\varphi_c}, \qquad \gamma=\frac{1}{\varphi_c}+\frac{1}{|\varphi_f|}.$$

As in original Goodwin's paper [3] we use $r=2$, $\varphi_c=9$ and $\varphi_f=-3$.

We investigate the dependence $y(t,\mu)$ assuming that the initial function $y(t)$, $t\le 0$ is zero, initial induced investment is zero, $I(0)=0$ and the oscillations of income are excited by the autonomous investments $A(t)$ that have the form $A(t)=A_0te^{-t}$.

For this shape $A(t)$ and given μ, there exists a value $A_*(\mu)$ such that the sawtooth cycles with period 1 will be excited for all $A_0<A_*(\mu)$, and the long-period Goodwin cycles for $A_0>A_*(\mu)$. The dependence $A_*(\mu)$ is shown in Figure 1. It can be shown that with such choice of autonomous investments limit cycles with periods $1/2, 1/3, \ldots$ do not excited [16].

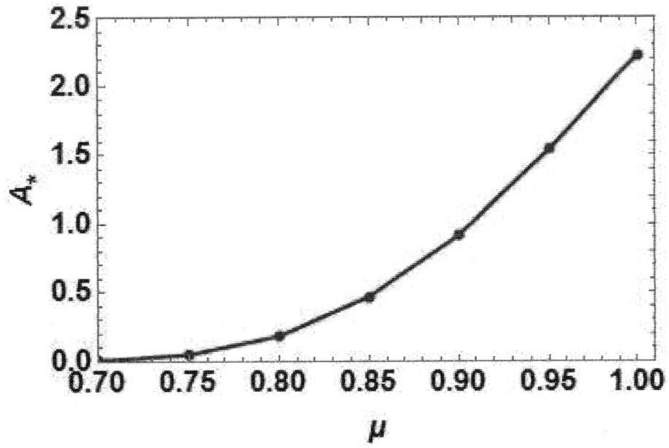

Figure 1. The dependence $A_*(\mu)$.

Figure 2 shows the time behavior $y(t)$ for two very close to $A_*(\mu)$ values of $A_-<A_*$ and $A_+>A_*$ for some μ. Blue lines correspond to the values $A_0=A_-$, green lines $A_0=A_+$.

When $0<\mu<0.6$ only long-period cycles are practically excited by using the very small A_0

.

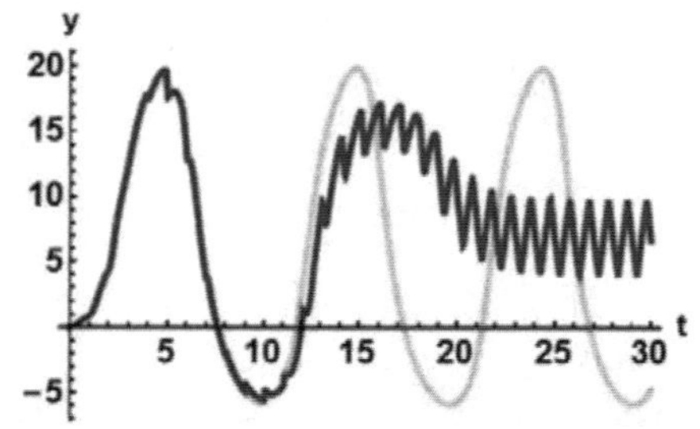

μ=1, A_{-} = 2.22, A_{+}=2.24

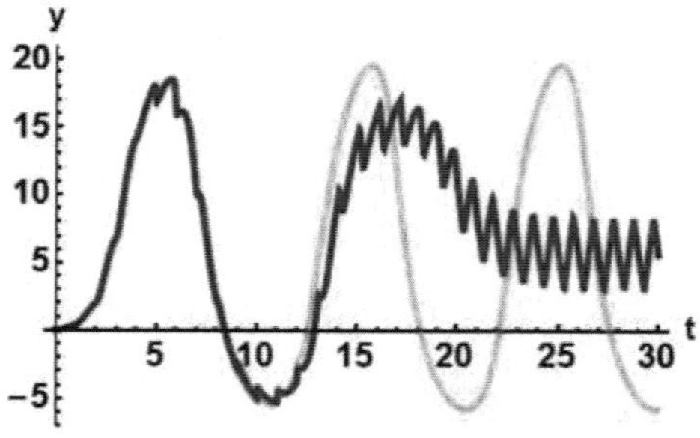

μ=0.9, $A_{-} = 0.91$, $A_{+} = 0.92$

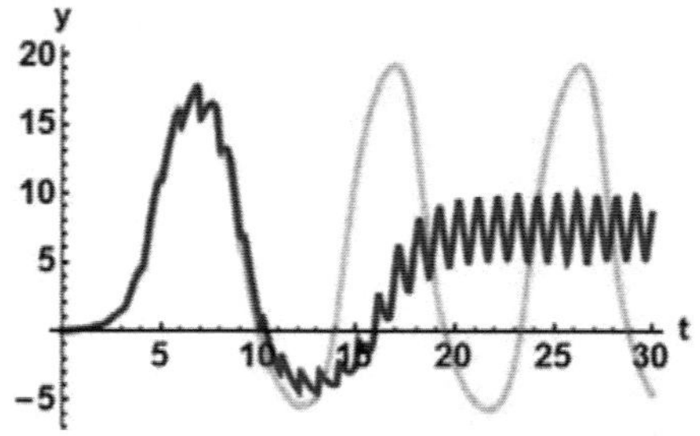

μ=0.8, $A_{-} = 0.18$, $A_{+} = 0.19$

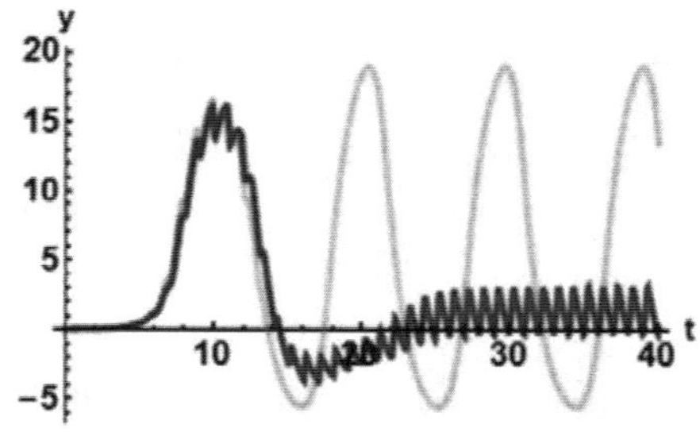

μ=0.7, $A_{-} = 0.008$, $A_{+} = 0.009$

Figure 2. Functions $y(t)$ for different values of μ for $A_{-} < A_{*}$ (blue lines) and for $A_{+} > A_{*}$ (green lines).

Figures 3,4 and 5 show the time dependences of the income $y(t)$, its time derivative $\frac{dy(t)}{dt}$ and phase plot $\frac{dy}{dt}(y)$ for three values of μ : μ =0 (red curve), μ =0.5 (green curve) and μ =1 (blue curve) . It is clear from the figures that the greater μ, the greater the range of oscillations $y(t)$ and $\frac{dy(t)}{dt}$.

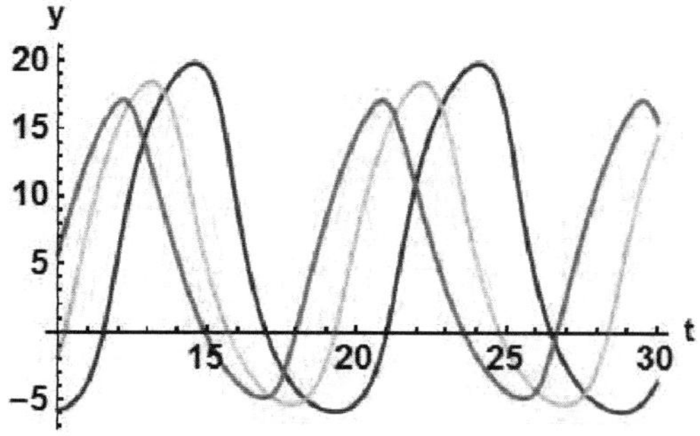

Figure 3. Dependencies $y(t)$. The red line corresponds to μ =0, green line - μ =0.5, blue line - μ =1.

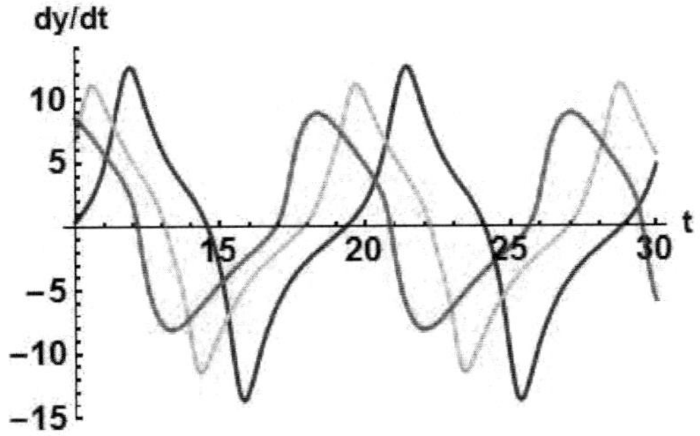

Figure 4. Dependencies $\frac{dy(t)}{dt}$. The red line corresponds to μ =0, green line - μ =0.5, blue line - μ =1.

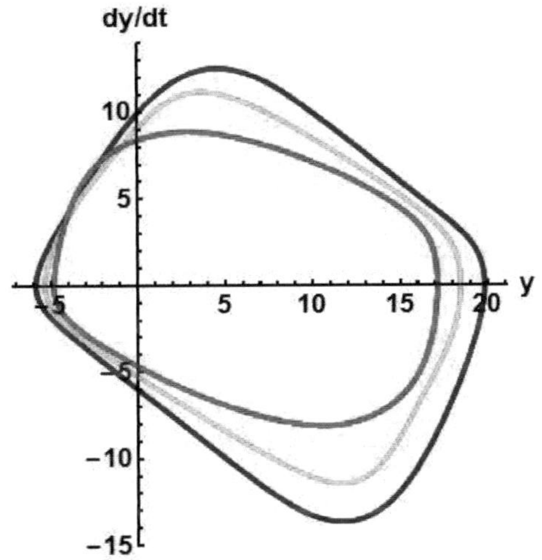

Figure 5. Phase curves $\frac{dy}{dt}(y)$. The red line corresponds to μ =0, green line - μ =0.5, blue line - μ =1.

Also, with the increase of μ, the period of mode 1 T increases almost linearly (Figure 6).

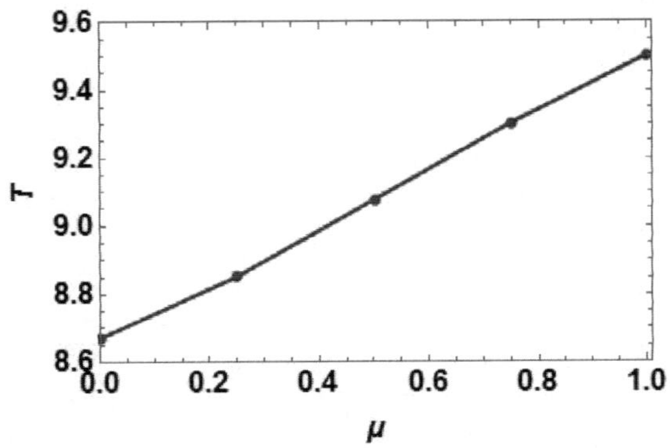

Figure 6. Dependence on μ the period of long period cycles.

5. Conclusion

We formulate a business cycle model by introducing the time delay in Goodwin's accelerator as follows

$$\varphi\left(\frac{dy(t)}{dt}\right)=\mu\varphi_1\left(\frac{dy(t)}{dt}\right)+(1-\mu)\varphi_2\left(\frac{dy(t)}{dt}\right).$$

It combines models with fixed delay φ_1 in induced investment and with exponentially distributed delay φ_2. Taking the time delay in investment as a bifurcation parameter we have investigated the necessary conditions for the excitation of the limit cycles for the different values of μ.

It has been shown that for $\mu < \varepsilon / r$ the short-period cycles do not excited.

We carry out the numerical analysis of the model and found both long-period limit cycles and short-period limit cycles. It is shown that Goodwin's cycles with a period of approximately 10 years can be maintained by choosing the time dependence of autonomous investments.

Acknowledgement

The author would like to thank Dr. S. Reznik for stimulating discussion and his assistance in carrying out the computations.

References

[1] Kalecki M. 1935 *Econometrica* **3**(3), 327–344.
[2] Kaldor N. 1940 *Econ. J.* **50**(197), 78–92 (1940).
[3] Goodwin R. M. 1951 *Econometrica* **19**(1) 1–17.
[4] Yu J., Peng M. 2016 *Phys. A* **460**, 66–75.
[5] Guerrini L., Krawiec A., Szydłowski M. 2020 *Nonlinear Dyn* **101**1263–1279
[6] Matsumoto A., Merlone U. and Szidarovszky F. 2018. *Communications in Nonlinear Science and Numerical Simulation* **58** 233-248.
[7] Antonova A. O., Reznik S. N. and Todorov M. D. 2022 High order ODE's approximations for Goodwin's business cycle model, in AMiTaNS-21, American Institute of Physics, AIP CP **2522** 070001-1, Ed. M. D. Todorov.
[8] Chiarella C. 1990 *The Elements of a Nonlinear Theory of Economic Dynamics* (Berlin: Springer-Verlag).
[9] Sordi S. 2006 'Floors' and/or 'ceilings' and the persistence of business cycles, in: Business Cycle Dynamics. Models and Tools, Ed. T. Puu and I. Sushko (Berlin: Springer-Verlag) 277-298.
[10] Matsumoto A. 2009 *J Econ Dyn Control* **33** 832–42.
[11] Sasakura K. 1996 *J Econ Dyn Control* **20** (9-10) 1763-1773.
[12] Bothwell F. E. 1952 *Econometrica* **20**(2) 269-284.
[13] Strotz R. H., McAnulty J.C., and Naines J.B. 1953 *Econometrica* **21**(3) 390-411.
[14] Antonova A. O., Reznik S. N. and Todorov M. D. 2011 Distributed time delay Goodwin's models of the business cycle, in AMiTaNS-11, American Institute of Physics, AIP CP **1404** 333-339, Ed. M. D. Todorov.
[15] Antonova A. O., Reznik S. N. and Todorov M. D. 2012 Dimensionless form for Goodwin's equation of business cycle, in AMiTaNS-12, American Institute of Physics, AIP CP **1487** 168-175, Ed. M. D. Todorov.
[16] Antonova A., Reznik S. and M. Todorov M. 2013 *Int. J. Computational Economics and Econometrics* **3** (3/4) 146-163.
[17] Antonova A. O., Reznik S. N. and Todorov M. D. 2014 The effect of consumption delay on the excitation of Goodwin's oscillations, in AMiTaNS-14, American Institute of Physics, AIP CP **1629** 247–252, Ed. M. D. Todorov.
[18] Matsumoto A., Szidarovszky F. 2015 *Struct Change Econ Dyn* **33** 1–9.
[19] Matsumoto A., Nakayama K. and Szidarovszky F. 2018 *Advances in Pure Mathematics* **8** 178-217.
[20] Kuang Y. 1993 *Delay Differential Equations with Applications in Population Dynamics* (NY: Academic Press).

A new application of the kinetic type theory in immunology

Mikhail Kolev[1,2], Iveta Nikolova[1,2], Meglena Lazarova[3] and Stanislava Stoilova[1]

[1]Department of Mathematics, University of Architecture, Civil Engineering and Geodesy (UACEG), Sofia, Bulgaria
[2]Department of Mathematics, University of Chemical Technology and Metallurgy, Sofia, Bulgaria
[3]Department of Mathematical modelling and numerical methods, Technical University, Sofia, Bulgaria

E-mail: mkolev_fte@uacg.bg

Abstract. The kinetic type theory for active particles is a methodology actively used for modeling processes and phenomena in biology and life sciences. One of the fields of its application is immunology, in particular the processes observed in the competition between immune system and foreign antigens. In this paper we present a new mathematical model describing such a complex system and possible occurrence of contemporary diseases. Preliminary qualitative analysis of the model is presented. The role of some of the main model parameters is studied using simulations.

1 Introduction

Knowing the regularities of the functioning of the immune system and the mechanisms of healing in viral infections at a time of continuous growth of chronic viral infections such as hepatitis B and C, cancers and AIDS, is one of the fundamental problems of modern medicine and immunology. Along with the developed theoretical, experimental and clinical immunological methods, an effective tool for studying these regularities is the mathematical modeling of the processes and systems [11, 12].

In recent decades, there has been a significant increase in the amount of work devoted to the modeling of immune processes. A large part of them have a rather formal character, which is why they cause a negative attitude of doctors and biologists. In these models, either the specificity of the studied phenomena is poorly reflected, or they are crowded with unjustified assumptions, the truth of which is quite doubtful. In the first case, the obtained results are trivial from the point of view of biology, and in the second case, it is difficult to separate the results that have a relation to reality from those that do not. Adequate mathematical models, quantitative and qualitative analysis, with an appropriate choice of initial conditions and parameter values, make it possible to obtain new information about the system, in particular, under conditions that have not been experimentally realized. It is also possible to predict different situations.

The problem of studying the occurrence and treatment of malignant neoplasms in the human body is not only medical. Methods from fundamental sciences such as physics, chemistry, mathematics, etc. are applied to study the growth and spread of cancer formations in the body [14].

By the middle of the 20th century, clinicians and immunologists had accumulated a huge amount of observational material on the course of various infectious diseases. As a result of the analyzes of this

Content from this work may be used under the terms of the Creative Commons Attribution 4.0 licence. Any further distribution of this work must maintain attribution to the author(s) and the title of the work, journal citation and DOI.
Published under licence by IOP Publishing Ltd

material, fundamental results were obtained concerning the interaction between antigens and immunocompetent cells at different levels: macroscopic, intracellular and genetic. These results have made it possible to build numerous mathematical models of immune processes.

The immune system is a highly adaptive defense system that has evolved in vertebrates to protect against invading pathogens and cancer. The most important characteristics of an immune system are:

(i) the ability to distinguish one's own cells from foreign cells, molecules, particles, etc., which may enter the host's body;

(ii) ability to eliminate foreign invaders, some of which are very dangerous;

(iii) ability to recognize and destroy own cells, changed for some reasons such as injuries or diseases such as cancer.

The actions of the immune system against foreign pathogens and modified own cells are defined as an immune response.

Our environment is filled with numerous infectious agents of different shapes, sizes, structures and destructive power. In the absence of protective mechanisms, the human organism can be used as an environment for the development and reproduction of these agents.

The immune system functions as: a physical barrier to foreign cells, viruses and other particles, innate immunity and acquired immunity.

The simplest way to prevent infection is to prevent infectious agents from entering the body. This is accomplished by the physical barrier - the skin and the secretions released by the sweat and sebaceous glands and the mucus released by the internal organs [8]. The body's next defense against pathogens that have crossed the physical barrier is innate immunity. It has two components: the complement system - a group of over 20 proteins that provide the non-specific humoral protection of the body and the cellular protection that is carried out by phagocytes (macrophages and neutrophils). If phagocytosis is not possible (due to the large size of the pathogens, etc.), the so-called NK-cells or natural killers are activated. In the body's defense, there are two different mechanisms for extracellular destruction. For the own cells, which do not have thick envelopes and are capable of apoptosis, it is appropriate to be destroyed by cytotoxicity, which maximally spares the surrounding tissue. Degranulation, on the other hand, releases toxic compounds into the extracellular space. They damage not only the object to be destroyed, but also the surrounding tissue. Therefore, degranulation is used against parasites to which cytotoxicity is not applicable [8]. Another major part of innate non-specific immunity are interferons, which play an important role in protecting the body from foreign pathogens – especially viruses and cancer cells. Interferons are a group of proteins that act as warning signals for viral infection. After infecting the cell with a pathogen, it produces interferon, secretes it outside, and after binding it to the interferon receptors of the neighboring cells, the latter receive the signal of viral danger and react to this signal by stopping the production of proteins and degrading all the RNA available in the cell. When a virus enters such a cell, it cannot develop. Hepatitis B and C viruses do not allow infected cells to produce interferon and neutralize already produced, which is why it has to be injected from the outside. Interferon injections are part of the standard treatment for chronic hepatitis C. Innate immunity is non-specific because it is directed against pathogens in general and not against a specific pathogen. Moreover, it does not improve during the life of the individual, but acts equally on the first and all subsequent encounters with the pathogen. Innate immunity is very useful because its reaction is very fast (up to a few hours after infection), but it is not always effective enough. It acts against all possible pathogens and is often insufficiently adapted to one or another specific antigen. An individual approach to each causative agent can only be carried out by acquired immunity. Acquired immunity is carried out by lymphocytes, called B- and T-lymphocytes according to whether they differentiate in the bone marrow or in the thymus.

Acquired immunity arose after a long evolution in humans and higher mammals to counteract pathogens that manage to overcome innate defenses and enter the body. It is highly specialized and is based on recognition between molecules. When first infected with the relevant antigen, it takes some time (usually one or two days) for this type of immune reaction to occur. If the immune system is already familiar with this type of antigen, the acquired immunity is much faster and stronger. In this subsection, the functions of only the cells that are included in the mathematical model proposed in the following section are considered.

During their reactions against foreign pathogens, which are necessary for the protection of the body, defects sometimes occur that lead to the so-called autoimmune diseases. These are diseases in which the body's defense mechanisms attack its own uninfected elements. The prevalence of these diseases has increased in recent decades. It is estimated that more than one-fifth of the population in developed countries suffers from some autoimmune disease. One of the leading organizations in this field, the American Autoimmune Disease Association (AARDA) cites data that more than 50 million US citizens have suffered from such diseases in recent years. To get a better idea of this amount, we can compare it to the number of Americans suffering from cancer, which is about 12 million, and those suffering from cardiovascular disease - about 25 million [3].

In researching diseases caused by a malfunctioning immune system against one's own tissues and organs, scientists have classified them into about 100 different diseases. Among them, some of the most widespread are insulin-dependent diabetes, Parkinson's disease, Alzheimer's disease, Meniere's disease, vitiligo, multiple sclerosis, psoriasis, psoriatic arthritis, Hashimoto's thyroiditis, and many others.

The causes of autoimmune diseases are diverse and not yet fully understood. Along with genetic factors, unfavorable environmental conditions (chemicals, pollutants, toxins), stress, infections, weakening of the immune system due to excessive use of drugs, as well as the relative cleanliness of the living environment play an important role [2, 10].

The role of nutrition is also important. On the one hand, the lack of important trace elements (vitamins and minerals) and on the other, the excessive use of gluten-containing foods can contribute to autoimmune diseases. Therefore, it is recommended to include high-quality meat and fish products, eggs, vegetables, fruits, nuts and seeds in the daily menu [3]. Among the factors that can help prevent or treat this type of diseases are stress control, normalizing sleep, avoiding a sedentary lifestyle, spending more time in nature, etc.

Recently mathematical modeling has been successfully used for the description and analysis of various diseases [1]. Among them are also autoimmune diseases [5, 6, 7]. During the process of modeling, information provided by immunology and medicine is used. Very important role plays the understanding and incorporation of the main components and functions of the body's systems of defense. As it is well known, they include physical barriers and the mechanisms of the immune system against foreign antigens [8].

In the long process of human evolution the defense system learned to produce specialized cells and molecules. Among them are lymphocytes, antibodies etc., which are able to destroy some of the foreign antigens and infected own cells. Usually, the immunological system must distinguish their own compartments from foreign agents. This feature of the immune system resulting in the fact that immunological response cannot be triggered against healthy own cells is known as immunological tolerance. Unfortunately, sometimes the mechanisms of tolerance fail and the immunological system attack the own compartments of the body. This is the main cause of autoimmune diseases [8].

The content of this paper is organized as follows:

We describe a kinetic model of autoimmune disease in Section 2. Further, in Section 3 we present a numerical scheme for solving approximately our model. In Section 4 results of numerical simulations are documented. Section 5 is devoted to conclusions.

2 Mathematical model of autoimmune diseases

In this Section we describe a new mathematical model of autoimmune disease. It belongs to the kinetic type models. Kinetic models are characterized by the fact that the populations involved in the interactions are not considered homogeneous, but heterogeneous, due to the different biological activity of their various elements. Historically, kinetic models are influenced by the nonequilibrium statistical mechanics and generalized kinetic theory, see [4] for reference. We use a special variable, denoted by u, which belongs to the interval $[0,1]$ to describe the biological activity of some of the interacting populations. Using this variable, the distribution function of the population denoted by the index i at time t is represented as:

$$g_i(t,u), \quad g_i : [0,\infty) \times [0,1] \to R^+.$$

The functions:

$$n_i(t) = \int_0^1 g_i(t,u)du, \quad n_i : [0,\infty) \to R^+. \tag{1}$$

represent the concentrations of the respective populations.

We consider the following four interacting populations with respective subscripts:

(1) healthy or target cells, labeled by 1;

(2) damaged cells, labeled by 2;

(3) immune cells, labeled by 3;

(4) viral particles, labeled by 4.

The following mechanisms of autoimmune disease development are taken into account. Some healthy cells may be damaged as result of viral infection. This can activate mechanisms of immune defense, which may results in attacks of target cells by the immune cells.

In our model, the biological activity of the immunological cells represents their ability to destroy healthy cells.

In order to keep the model as simple as possible, we assume that the remaining populations corresponding to subscripts $i = 1$, $i = 2$ and $i = 4$ are homogeneous with respect to their biological activaty. Therefore, we assume that for the distribution functions $g_1(t, u)$, $g_2(t, u)$ and $g_4(t, u)$ the following identities hold:

$$g_i(t, u) = n_i(t), \quad i = 1, 2, 4 \quad \forall u \in [0, 1], \quad t \geq 0.$$

As a result, our kinetic model of autoimmune disease has the form:

$$\begin{aligned} \frac{d}{dt}n_1(t) &= P_1(t) + p_{11}^{(1)} n_1(t) - \frac{p_{11}^{(1)}}{K_M} n_1^2(t) \\ &- d_{11}n_1(t) - d_{13}n_1(t) \int_0^1 v g_3(t, v) dv \end{aligned} \tag{2}$$

$$\frac{d}{dt}n_2(t) = d_{13}n_1(t) \int_0^1 v g_3(t, v) dv - d_{22}n_2(t) \tag{3}$$

$$\frac{\partial g_3}{\partial t}(t, u) = p_{23}^{(3)} n_2(t) + p_{34}^{(3)} n_4(t) - d_{33}g_3(t, u) \tag{4}$$

$$\frac{d}{dt}n_4(t) = p_{44}^{(4)} n_4(t) - d_{44}n_4(t) - d_{34}n_3(t)n_4(t) \tag{5}$$

with nonnegative initial conditions

$$n_i(0) = n_i^{(0)}, \quad i = 1, 2, 4, \quad g_3(0, u) = g_3^{(0)}(u).$$

The parameters of the system (2)-(5) are supposed to be nonnegative.

The equation (2) describes the time evolution of the concentration $n_1(t)$ of the healthy cells. The parameters in this equation have the following biological meaning:

- The function $P_1(t)$ characterizes the rate of production of healthy cells from inner sources;
- $p_{11}^{(1)}$ characterizes the maximal rate of proliferation of the healthy cells;
- K_M characterizes the concentration of the the healthy cells at which proliferation shuts off;
- d_{11} characterizes the rate of natural mortality of the healthy cells;
- d_{13} describes the rate of destruction of the healthy cells by the immune cells.

Equation (3) describes the time evolution of the concentration $n_2(t)$ of the damaged cells. The parameter d_{22} characterizes the rate of natural mortality of the damaged cells.

Equation (4) describes the temporary evolution of the distribution density $g_3(t, u)$ of the immune cells. The meaning of the parameters in this equation is the following:

- $p_{23}^{(3)}$ characterizes the rate of production of immunological cells due to self-antigens presented by damaged cells;
- $p_{34}^{(3)}$ characterizes the rate of production of immunological cells due to viral particles;
- d_{33} characterizes the rate of natural mortality of immune cells.

Equation (5) describes the evolution of the concentration $n_4(t)$ of the viral particles. The meaning of the parameters in this equation is the following:

- $p_{44}^{(4)}$ characterizes the rate of production of viruses;
- d_{34} characterizes the rate of destruction of the virusal particles to immune response;
- d_{44} describes the rate of natural mortality of the viruses.

Our model is more general version of our previous model [9]. In this previous model we did not consider the population of viruses. However, there are evidence that they can have important role in the process of development of autoimmune diseases [3].

3 Numerical scheme for solution

Here we present a numerical algorithm for finding the approximate solution for the unknown concentrations of the interacting populations $n_i(t)$, $i = 1,\ldots,4$ for any time $t \geq 0$. The concentrations $n_1(t), n_2(t), n_4(t)$ will be computed from eqs. (2), (3) and (5), while the concentration $n_3(t)$ - from eq. (1) using the function $g_3(t,u)$. To obtain the approximate values of the functions $n_1(t), n_2(t), n_4(t)$ and $g_3(t,u)$, we carry out a discretization of the modeled problem (2) - (5) with regard to the activity $u \in [0,1]$ by introducing a uniform grid using the grid-points

$$u_i = i\Delta u, \quad i = 0,\ldots,N,$$

where N is a positive integer and the step $\Delta u = 1/N$.

By using the constructed grid, we replace the values $g_3(t,u)$ in the modeled problem (2) - (5) with their approximations

$$g_3(t,u_i) \approx g_{3,i}(t), \tag{6}$$

at the state grid-points $u_i \in [0,1]$.

For each $t > 0$ and each $u_i \in [0,1]$ with $i = 0,\ldots,N$, we apply the approximations (6) for appropriate quadrature formulae in order to obtain approximations for the integrals:

$$\begin{aligned} \int_0^1 g_3(t,v)dv &\approx Q_0^N\Big[g_3(t,v)\Big], \\ \int_0^1 vg_3(t,v)dv &\approx Q_0^N\Big[vg_3(t,v)\Big]. \end{aligned} \tag{7}$$

In the approximation formulae (7) arbitrary quadratures could be used. For instance, in Section 4, the values $Q_0^N\Big[g_3(t,v)\Big]$ and $Q_0^N\Big[vg_3(t,v)\Big]$ are computed using the composite Simpson's rule [15].

After application of the approximations (6) and (7) to the modeled problem (2) - (5), we obtain a system of ODEs.

The results of the solution of this system are presented in Section 4. As it is mentioned above, the obtained approximate solutions $g_{3,i}(t)$, for $i = 0,\ldots,N$, are used for computing the needed concentrations $n_3(t)$. For this aim we used the formulae

$$n_3(t) \approx Q_0^N\Big[g_3(t,v)\Big]. \tag{8}$$

4 Results of simulations

After performing the procedure described in the previous Section, we obtained a system of ordinary differential equations. This system was solved by the use of the solver `ode15s` from the Matlab ODE suite [13], where we set the parameters $RelTol = 10^{-4}$ and $AbsTol = 10^{-5}$. This solver has been used instead of `ode45` because for some parameter values the corresponding system occurred to be sharp.

The obtained approximate solutions for the function $g_{3,i}(t)$ for $i = 0,\ldots,N$, have been used in (8) for computing the needed concentrations $n_3(t)$.

As initial conditions we assume the presence of certain amount of target cells, small amount of damaged cells and very small amount of viruses and immune cells with low activation state:

$$n_1(0) = 100, \quad n_2(0) = 0.5, \quad n_4(0) = 0.01,$$

$$g_3(0,0) = 0.01, \quad g_3(0,u) = 0.0, \quad \forall u \in (0,1].$$

We used the following parameter values in our set of simulations:

$$P_1(t) = 10, \forall t \geq 0, \quad K_M = 100000,$$

$$p_{11}^{(1)} = 0.5, \quad d_{11} = 0.2, \quad d_{13} = 0.1, \quad d_{22} = 1.1,$$

$$p_{23}^{(3)} = 0.001, \quad p_{34}^{(3)} = 0.001, \quad d_{33} = 0.1,$$

$$d_{34} = 0.001, \quad d_{44} = 0.001$$

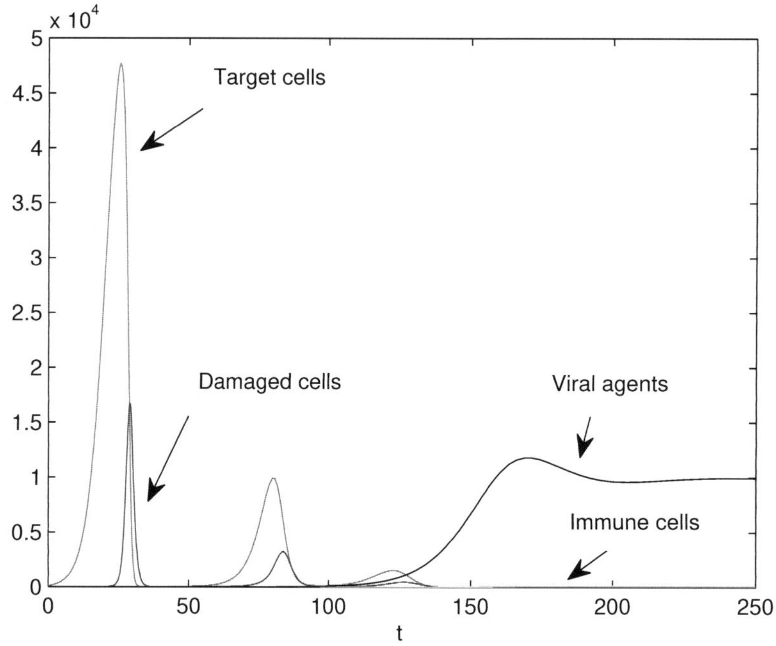

Figure 1: Autoimmune response with high aggressiveness: $p_{44}^{(4)} = 0.1$

The values of parameter $p_{44}^{(4)}$ were different in every case.

The goal of our simulations was to investigate the role of the viruses for the outcome of the interactions. Our numerical experiments show that in some cases, when the rate of production of viral particles $p_{44}^{(4)}$ is significantly bigger than the values of parameters of destruction of viruses d_{34} and d_{44}, the viruses can induce aggressive autoimmune response that destroys too many healthy cells, which can result even in death of a patient. Corresponding results are shown in Fig. 1, where $p_{44}^{(4)} = 0.1$. When the advantage of $p_{44}^{(4)}$ is not so large, more time is necessary for the autoimmune response to cause substantial damages to the organism (see Fig. 2, where $p_{44}^{(4)} = 0.02$). On the contrary, when the mentioned difference is even less, the autoimmune response is not very aggressive and the body tends to equilibrium state which can be considered as chronic autoimmune disease with mild symptoms or even as a healthy state (see Fig. 3, where $p_{44}^{(4)} = 0.001$).

For the new model with viruses (2))-(5)) we obtained the above scenarios for certain values of parameters. In addition, we studied the role of the viruses for the outcome of the interactions. Our numerical experiments show that in some cases, when the rate of production of viruses $p_{44}^{(4)}$ is significantly bigger than values of parameters of destruction of viruses d_{34} and d_{44}, the viral agents can induce strong autoimmune response that destroys too many healthy cells, which can lead even to death. Results for such case are presented in Fig. 1. When the advantage of $p_{44}^{(4)}$ is not so large, more time is necessary for the autoimmune response to cause substantial damages to the organism (see Fig. 2). On the contrary, when the mentioned difference is even less, the autoimmune response is not very strong and the organism tends to equilibrium state which can be considered as chronic autoimmune disease with mild symptoms or even as a healthy state (see Fig. 3).

5 Conclusions

The creation of adequate mathematical models reflecting the behavior of viral infections and tumors and the impact of the immune system on them is extremely useful.

A good mathematical model allows, without doing experiments, to study, analyze and even predict some new properties and changes in the behavior of the studied system.

Mathematical modeling of the object under study allows the often unsuccessful experimentation with the real object to be replaced by targeted research using a mathematical model, allowing the experiment

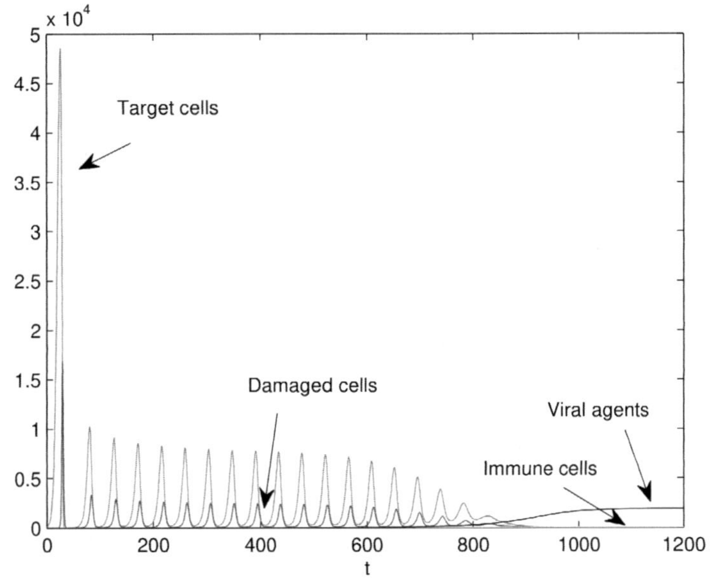

Figure 2: More prolonged aggressive autoimmune response: $p_{44}^{(4)} = 0.02$

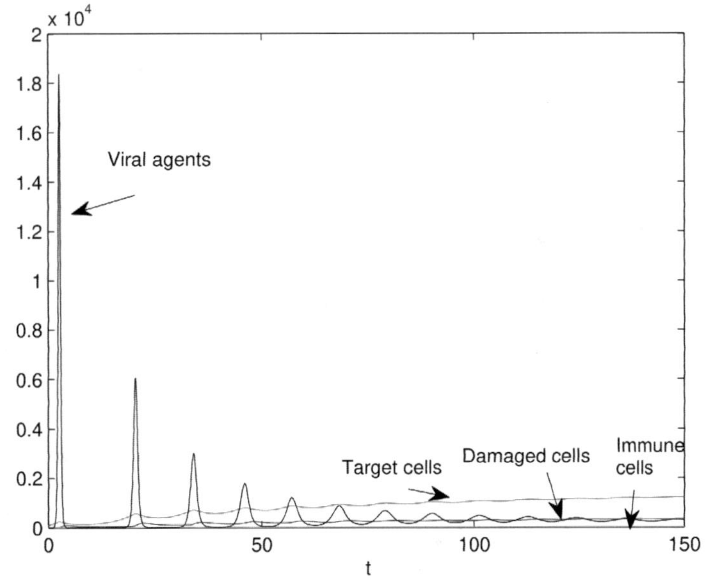

Figure 3: Weak autoimmune response: $p_{44}^{(4)} = 0.001$

to be performed on the model using a computer program.

Thus, mathematics from a means of calculation becomes a necessary method of modern research, and at certain stages it can be the only means of revealing the internal properties of the objects under consideration.

In our paper we propose a new mathematical model of autoimmune disease. The model takes into account the role of viral agents for occurrence and development of autoimmune disease. The modelled problem is solved numerically. The numerical solutions represent several typical dynamics of autoimmune disease.

Our future plans include analysis of the role of other model parameters as well as the development of the model through inclusion of activity for the populations of healthy cells, damaged cells and viral agents. In addition we plan to study the problem whether virus infections can lead to decrease of autoimmune reaction.

Acknowledgements

Mikhail Kolev and Iveta Nikolova thank to the financial support of European Union-NextGenerationEU, through the National Recovery and Resilience Plan of the Republic of Bulgaria, project № BG-RRP-2.004-0002-C01, "BiOrgaMCT".

CRediT roles: Conceptualization, M.L.; methodology, I.N. and M.L.; software, I.N.; validation, M.K. and I.N.; formal analysis, M.L.; investigation, S.S. and M.K.; resources, M.L. and S.S.; data curation, I.N. and S.S.; writing – original draft preparation, I.N. and M.K.; writing – review and editing, M.L. and I.N.; visualization, I.N. and M.K.; supervision, M.L.; project administration, S.S.; funding acquisition, S.S. and I.N.

References

[1] S. Andrew, C. Baker, G. Bocharov, *Rival approaches to mathematical modelling in immunology*, Journal of Computational and Applied Mathematics, **205** (2007), 669-686.

[2] J.F. Bach, *Protective role of infections and vaccinations on autoimmune diseases*, Journal of Autoimmunity, **16** (2001), 347-353.

[3] S. Ballantyne, *The Paleo Approach: Reverse Auto-immune Disease and Heal your Body*, Victory Belt Publishing, Las Vegas, 2013.

[4] N. Bellomo, *Modelling complex living systems*, Birkhäuser, Boston, 2008.

[5] K. Blyuss, L.S. Nicholson, *Understanding the roles of activation threshold and infections in the dynamics of autoimmune disease*, Journal of Theoretical Biology, **375** (2015), 13-20.

[6] S. Iwami, Y. Takeuchi, K. Iwamoto, Y. Naruo, M. Yasukawa, *A mathematical design of vector vaccine against autoimmune disease*, Journal of Theoretical Biology, **256** (2009), 382-392.

[7] S. Iwami, Y. Takeuchi, Miura, T. Sasaki, T. Kajiwara, *Dynamical properties of autoimmune disease models: Tolerance, flare-up, dormancy*, Journal of Theoretical Biology, **246** (2007), 646-659.

[8] C. Janeway, P. Travers, M. Walport, M. Shlomchik, *Immunobiology: The Immune System in Health and Disease*, Garland Science, New York, 2004.

[9] M. Kolev, I. Nikolova, *On a kinetic type model of autoimmune disease*, in Proceedings of the 23rd National Conference on Applicatons of Mathematics in Biology and Medicine, Silesian University of Technology, University of Warsaw, 101-107, 2017.

[10] K. Leon, J. Faro, A. Lage, J.S. Carneiro, *Inverse correlation between the incidences of autoimmune disease and infection predicted by a model of T cell mediated tolerance*, Journal of Autoimmunity, **22** (2004), 31-42.

[11] Marchuk, G. Mathematical modeling of Immune Response in Infections Diseases, Kluwer academic publishers, Dordrecht/ Boston/ London (1997)

[12] M. Nowak, R. May, *Virus Dynamics*, Oxford Univ. Press, Oxford , 2000.

[13] L. Shampine, M. Reichelt, *The Matlab ODE suite*, SIAM J. Sci. Comput. **18** (1997), 1-22.

Journal of Physics: Conference Series **2910** (2024) 012007 doi:10.1088/1742-6596/2910/1/012007

[14] Stetler-Stevenson, W.G.; Aznavoorian, S.; Liotta, L.A. Tumor cell interactions with the extracellular matrix during invasion and metastasis. *Annu. Rev. Cell Biol.* **1993**, *9*, 541–573.

[15] E. Volkov, *Numerical Methods*, Hemishere/Mir, New York/Moscow, 1990.

Forecasting Strategies in Retail: Utilizing Advanced Machine Learning Methods while Safeguarding Privacy

Irina Naskinova[1*], Mikhail Kolev[1,2] and Meglena Lazarova[3]

[1] University of Architecture, Civil Engineering and Geodesy (UACEG), Sofia, Bulgaria
[2] University of Chemical Technology and Metallurgy, Sofia, Bulgaria
[3] Technical University of Sofia, Sofia, Bulgaria

*E-mail: inaskinova_fte@uacg.bg

Abstract. Demand forecasting is critical to optimizing retail inventory management, pricing, and operations. This study compares several time series forecasting models on a publicly available retail dataset, including traditional stats like ARIMA and SARIMA, smoothing techniques, and new ones from Facebook, such as Prophet and Neural Prophet. We evaluate each model using metrics like Mean Absolute Error (MAE), Root Mean Squared Error (RMSE), and R^2. The results show that NeuralProphet and hybrid models outperform traditional models like ARIMA and SARIMA. We also look at the computational efficiency and practicality of the models and provide insights into how they can be used in real-world demand forecasting scenarios. The results show that traditional models are robust. Still, advanced models like NeuralProphet and hybrids, which combine machine learning with time series forecasting, have a lot of potential to improve forecast accuracy and operational decision-making.

1. Introduction

Accurate demand forecasting is one of the critical enablers for retail and other industry verticals. It forms the basis for making essential business decisions, directly impacting revenues, customer satisfaction, and operational efficiency. The accurate forecasting of demand in the retail industry optimizes inventory management. It makes it possible for a company to have the right products ready at the right time without either stockpiling them or facing a stockout. These help a firm arrive at the right balance to reduce carrying costs, minimize waste, and improve cash flow.

Demand forecasting helps decide pricing strategies, plan promotions, and manage the workforce, among other things. For example, knowing what the future will bring in-demand allows retailers to price competitively, plan promotional activity efficiently, and distribute the labor force optimally during peak hours. On the other hand, the demand forecasts can be grossly incorrect, leading to significant financial losses associated with unsold stocks, missed sales, or increased operating expenses. Demand forecasting is equally applied to manufacturing, logistics, supply chain management, capacity planning, production scheduling, and supply chain optimization. This will help manufacturers operate the production line efficiently without incurring the costs of either downtime or changing on short notice with rush orders.

Similarly, in logistics, better demand forecasting provides the opportunity for better route planning and fleet management, which saves on cost and increases service level. This paper aims

Content from this work may be used under the terms of the Creative Commons Attribution 4.0 licence. Any further distribution of this work must maintain attribution to the author(s) and the title of the work, journal citation and DOI.
Published under licence by IOP Publishing Ltd

to research and compare different forecasting models, including traditional and sophisticated ones, to propose insights into their effectiveness in real-world applications. With this knowledge, businesses can then adopt techniques that consider the strengths and limitations of these models in making appropriate decisions regarding applying a forecasting method to their strategic planning and operational performance.

Although demand forecasting is one of the most fundamental activities for retail and many other industries, selecting an effective forecasting model has been a big challenge because real-world data is usually complex and dynamic. Traditional time series models like ARIMA and SARIMA conventionally remain the competent choice for time series forecasting. However, these are restricted in capturing those complex patterns in data that may involve nonlinear relationships, seasonality, and other exogenous factors. Traditional models generally lead to incorrect forecasts, which show up as poor inventory management, inefficient pricing strategies, and missed revenue opportunities for the businesses concerned.

In the last couple of years, machine learning and advanced statistical models have been considered alternatives to traditional models, including Prophet, NeuralProphet, and hybrid versions. The above-mentioned new models give flexibility and accuracy in handling data complexity and non-linearity. However, the effectiveness of these new models compared with the traditional model has not been well established in practical use, especially for retail demand.

The research question to be addressed in this study is: "Which of the time series forecasting models, traditional or advanced, gives the best accurate yet computationally efficient predictions for retail industry demand?" The second part of the question involves examining if advanced models such as NeuralProphet and hybrid models could substantially outperform their traditional counterparts regarding forecast accuracy and applicability to operational settings.

The paper, therefore, systematically compares these models and provides clear guidance on the best approaches to improve demand forecasting in retail, with a view to helping businesses make more informed decisions. The paper compares different time series forecasting models against demand in the retail industry and finds the best. This work compares traditional methods with more advanced ones like Prophet, NeuralProphet, and hybrid machine learning models. The paper compares predictive accuracy with metrics like MAE, RMSE, and R^2. It further dwells on their efficiency during the processing concerning time and resource utilization. When adapting the models to real-world scenarios, essential factors such as seasonality and trends are considered. Throughout these objectives, the study seeks to confirm or rebuke other findings in the existing body of knowledge on demand forecast.

Results show that ARIMA and SARIMA models are less computationally intensive when compared with most advanced ones but possess limited applicability in non-linearity.

It underlines, above all, the potential of using such advanced methods as Prophet or NeuralProphet, which are designed to operate with complex nonlinear data patterns.

The final point may be interpreted as the comparisons between various models to bridge the theory of forecasting and its practical application. It also deals with some critical issues in this area that relate to the fact that choices regarding different forecasting methods should depend on the data characteristics selected. All these efforts are expected to substantially improve retail management by predicting future product sales with reasonable accuracy. Generally, the contribution of this research to developing the understanding of demand forecasting is related to two aspects: updating the framework of model performance evaluation and showing new opportunities to enhance the accuracy of retail demand forecasting models by applying machine learning techniques.

2. Related Work

Forecasting time series is tactile and constantly changes, but classical statistical and smart machine learning techniques have established the field since the first attempts. This literature review explores primary studies that exploit models like ARIMA, Prophet, NeuralProphet, and other new forecasting methods, and it examines the strengths and shortcomings of each, particularly in the context of demand forecasting.

2.1 Classical Statistical Models

The ARMA model is among the most elementary techniques applied to time series analysis. It could be appropriate for data featuring stationarity with no seasonal component. The models consider two crucial aspects simultaneously: one accounting for the autoregressive influence of past values of the time series on the current observation and another accounting for the moving average of past forecast errors. Therefore, an ARMA model, denoted as ARMA(p, q), can effectively capture the linear dependence of a time series. Its simplicity and robustness make the ARMA model applicable to various situations, ranging from financial market analysis to economic forecasting, represented by the seminal works of Box & Jenkins[1] and Brockwell and Davis [2]. Their research[3] showed that the performance for handling stationary data runs on a modest to large scale. Still, ARMA is not well adapted for series with either a trend or seasonality and may require employing higher-order models like ARIMA and SARIMA.

Table 1. Comparative performance of classical statistical models for demand forecasting models. This table presents a comparison of ARIMA and SARIMA time series models, highlighting their performance metrics

Model	Study (DOI)	MAE	RMSE	R^2	Typical Performance	Computational Efficiency
ARIMA (forecast package)	Hyndman & Khandakar (2008)	0.205	0.290	0.81	High	High
ARIMA	Brockwell & Davis (2016)	3.10	4.60	0.78	Moderate to High	Moderate to High
SARIMA	Hyndman et al. (2008)	0.190	0.280	0.85	High	Moderate
SARIMA	Brockwell & Davis (2016)	3.00	4.40	0.80	Moderate to High	Moderate to High
SARIMA (seasonal model)	Contreras et al. (2003)	0.180	0.270	0.87	High	Moderate
ARMA	Box & Jenkins (1976)	0.220	0.315	0.79	Moderate	High
ARMA	Brockwell & Davis (1991)	3.15	4.70	0.76	Moderate to High	Moderate to High

The ARMA models are usually efficient computationally, as the structure is not overly complicated, though it shows poorer performance in the case of seasonality of series. Thus, the ARMA will be an excellent model for the non-seasonal data series, providing good forecasts and reasonable computational burdens.

The traditional statistical model ARIMA (AutoRegressive Integrated Moving Average) is the basis of time series forecasting models. [3] provided a complete implementation of ARIMA in the R programming environment, which was endowed with the ability of automatic model selection and parameter tuning, thus allowing us to get accuracy and effectiveness. Their work, published in the Journal of Statistical Software, remains a benchmark for time series analysis, particularly for its balance of precision and computational efficiency.

Also, Brockwell and Davis [4] made ARIMA details and other classic methods available in their book "Introduction to Time Series and Forecasting," which is widely cited because it is informed by theory and supported by practice.

The most common extension is the Seasonal ARIMA-SARIMA-that, which extends the basic ARIMA model by including regular seasonal patterns, especially in many time series data with periodic trends.

In addition, SARIMA can represent the seasonality of the data that the basic ARIMA model cannot. That is where SARIMA picked up some actual seasonal variations that were important to forecast, performing slightly better for Hyndman et al. [5-6] than the traditional ARIMA models. It is preferred when working with regular cyclic data, like monthly sales or rain precipitation.

Another impressive example is the work of Contreras et al. [5], who applied a seasonal ARIMA to a seasonal data set quite effectively. Their results stressed the model's high efficiency by balancing high levels of accuracy against moderate computational demands and realistic options in many real-life forecasting tasks.

2.2 Exponential Smoothing Models

Exponential Smoothing models developed by Rubio[6], Holt [7], and Gardner [8] are the crucial components of forecasting, especially for seasonality and trends. Holt's technique, which is a smoothed exponentially weighted trend average, and Gardner's substantial evaluation of exponential smoothing initiate the current development in this area. Kalman Filter, which is Hamilton's [9] topic, and structural time series models mentioned by Harvey [10] are the ones that have also been significant in improving the quality of time series forecasting through the increased understanding of the underlying data structures and dynamics. Further, Caivano & Harvey's [11] research on time-series models with the EGB2 distribution proposes a more forceful alternative to the common assumption of a normal distribution, thus adding another effective tool for demand forecasting.

Hoerl and Kennard's work on Ridge Regression[12] has constituted a significant milestone within the frame of statistical modeling in general and for treating multicollinearity problems in linear regression, in particular. Ridge regression is one sort of regularization that stabilizes the coefficients against highly correlated predictors. This problem leads to unstable, biased estimates in regular ordinary least squares regression.

In demand forecasting, many interrelated factors that normally influence future demand allow Ridge Regression to enhance the accuracy of the predictions by handling the problem brought about by multicollinearity. However, its linear nature may not fully capture the complex nonlinear relationships that generally come with such data. It is, therefore, rated as moderate for this application compared to other advanced forecasting models.

Zou and Hastie gave the Elastic Net model [14], which is influential for dealing with complex data containing correlated variables. Elastic Net combines the strengths of Lasso (L1 regularization) and Ridge (L2 regularization) when selecting essential features to deal with multicollinearity.

This, in particular, makes it effective in genetics, finance, or marketing, where predictors are usually correlated. With a high R^2 value of 0.81, a relatively low MAE of 0.210, and an RMSE of 0.303, the performance metrics show the model is highly accurate in its prediction, with a variance explanation of 81% in the dependent variable.

Efficiency-wise, Elastic Net is more computationally expensive than Ridge and Lasso stand-alone, as it optimizes two penalties simultaneously. However, this is a good trade-off because it provides a good balance between accuracy and interpretability. Although having a moderate

Table 2. Replace this text with the table caption. Adapt the template table below or delete it and add a new table using either Insert – Table or by pasting from another document. Copy and paste the whole text box to add more tables. Place tables close to where they are cited in the text.

Model	Study (DOI)	MAE	RMSE	R^2	Typical Performance	Computational Efficiency
Structural Time Series Model	Harvey (1990)	0.220	0.310	0.78	Moderate	High
ARIMA & Exponential Smoothing	Rubio et al. (2021)	2.95	4.50	0.80	High	High
Exponential Smoothing	Holt (2004)	0.220	0.310	0.78	High	Very High
Exponential Smoothing	Gardner (1985)	3.25	4.75	0.75	Moderate	Moderate
Kalman Filter (Time Series)	Hamilton (1994)	0.208	0.299	0.80	High	Moderate
Time-series models with an EGB2 distribution	Caivano & Harvey (2014)	0.210	0.298	0.80	High	Low
Ridge Regression	Hoerl & Kennard (1970)	3.50	5.00	0.70	Moderate	Moderate
Elastic Net	Zou & Hastie (2005)	0.210	0.303	0.81	High	Moderate

computational cost, Elastic Net handles large and complex data, making it very reliable for real-world applications in predictive modeling.

2.3 Prophet and NeuralProphet Models

Prophet became a landmark in the development of the forecasting trend line when Taylor & Letham in 2018 [13] introduced an easy-to-use and accurate model. Building on the success of Prophet, Triebe et al. [14] introduced NeuralProphet, a hybrid model that combines the strengths of Prophet with neural networks to enhance predictive power. NeuralProphet is particularly well-suited for capturing nonlinear dependencies and complex seasonality, outperforming traditional methods in various settings. Recent studies, such as the one by Kenyi and Yamamoto [15], have further expanded on this approach by integrating SARIMA with Prophet, resulting in a model that excels in capturing both short-term fluctuations and long-term trends in time series data.

Other models discussed in the literature are machine learning and hybrid models, such as Random Forest, CatBoost, and XGBoost, LSTM-XGBoost hybrid model, and deep learning models like Long Short-Term Memory networks. However, this paper will deal with the more traditional structural and advanced models based on them, such as Prophet and NeuralProphet.

2.4 Gap Analysis

Despite the voluminous literature on time series forecasting, this study aims to bridge the gaps. ARIMA, Exponential Smoothing, and Kalman Filters are traditional statistical models that have been well-documented and widely used in various applications. Nevertheless, these methods

Table 3. Replace this text with the table caption. Adapt the template table below or delete it and add a new table using either Insert – Table or by pasting from another document. Copy and paste the whole text box to add more tables. Place tables close to where they are cited in the text.

Model	Study (DOI)	MAE	RMSE	R^2	Typical Performance	Computational Efficiency
Prophet	Taylor & Letham (2018)	3.00	4.50	0.80	High	High
NeuralProphet	Triebe et al. (2021)	2.50	3.75	0.88	High	Moderate
Hybrid SARIMA-Prophet	Springer (2024)	2.20	3.40	0.92	Very High	High

are usually unable to deal with the complexities of accurate data, especially regarding nonlinear relationships, multiple seasonality, and external influencing factors. Advanced models such as machine learning-based methods, including tree models like Random Forest and XGBoost, can use deep learning approaches, such as LSTM, which have shown considerable results. Still, their application in practical business forecasts has been limited.

Prophet and NeuralProphet are two new developments that try to take the best of both worlds: traditional prophecies and machine learning. Nevertheless, the literature remains elusive in comprehensive, head-to-head comparisons of these models' performance, especially in the retail demand forecasting scenarios. While most studies offer their exclusive focus on traditional methods or advanced machine learning models, they do not lead to precise, comparative analysis of intermediate solutions like Prophet and NeuralProphet.

Moreover, the applied model's computational efficiency on large-scale, real-life data sets is a missing point in the literature. Although the performance of these models has been studied, hardly any exhaustive research evaluates their computational requirements and scalability, which are essential in data-intensive environments.

This research will cover these issues by focusing on the performance of traditional statistics methods, Prophet and NeuralProphet, in demand forecasting for retail. Hence, it sheds light on these models' accuracy, computational efficiency, and applicability. It sets up the basis of future work in which the comparison will be extended to include tree-based models, like Random Forest and XGBoost, and deep learning approaches, such as LSTM and hybrid models.

3. Methodology

3.1 Data Description

The study is based on the Kaggle dataset for demand forecasting, which is generally recognized and especially suited for evaluating time series forecasting models. The dataset includes daily sales of various stores and items, thus being an ideal candidate for forecasting demand in retail.

The main features in the dataset are as follows:

- Date: The day on which the sales transaction has taken place.
- Store: A unique identifier for each store.
- Item: A unique identifier for each item sold across the stores.
- Sales: The number of units sold for each item on each date at each store.

3.2 Dataset Characteristics

The temporal dimension would be pretty rich, given that the data covers several years. Sales from various stores and items can be included in that dataset, which may show temporal patterns and cross-sectional variations. Its multi-level nature would then be beneficial for investigating how different models concerning capturing seasonality and trends perform on various levels of time and product categories.

3.3 Processing Steps

To transform the dataset into an analytically feasible form, several pre-processing steps were performed:

1. Date Conversion: The feature 'date' has been converted to datetime to support the time series analysis efficiently. The column 'date' has also been set as an index of the dataset to make the temporal operations easier.

2. Feature Engineering: To enable the model to capture the relationships correctly based on time, some features were derived from the 'date' column:

Year: Extracted from the 'date' to identify annual trends.

Month: Extracted from the 'date' to capture monthly seasonality.

Day of the Year: Extracted to track day-specific patterns across the year.

Weekday: Extracted to capture weekly seasonality because sales can differ on a weekday compared to a weekend.

3. Handling of Missing Values: The analysts found no missing values in the set. The required imputations were carried out in such a way that any loss of data in the dataset was considerably low.

4. Seasonal Decomposition: In the paper, the trend, seasonal, and residual components of the time series of sales that were observed were separated after Du made a seasonal decomposition. This action was the foundation for modeling specifics, which were selected and adjusted by forecasting protocols.

5. Stationarity Check: The Augmented Dickey-Fuller test was utilized to confirm whether data can be applied using models like ARIMA. Differencing is done to series wherever necessary to make them stationary.

6. Data Split: The dataset was split into training and test sets. The training set includes data from past years, while the test set includes the most recent year. In this way, it will be possible to evaluate the out-of-sample one-period-ahead forecasting performance of the models.

3.4 Model Descriptions

This research focuses on time series analysis of various retail sales forecasting methods, including traditional statistical and machine learning approaches. The models employed in this study are Prophet, NeuralProphet, ARIMA, SARIMA, and Exponential Smoothing techniques. Below is a detailed discussion of each model and its configurations.

3.5 Prophet

Prophet is a Facebook-developed time series forecasting model designed explicitly for data with seasonal solid trends and missing values. Prophet is suited chiefly for seasonal data in business applications with holiday effects.

The default Prophet program deals with seasonality by identifying the data's yearly, weekly, and daily seasonal components. The model facilitates the inclusion of custom seasonal components, thus granting it great flexibility and business-cycle adaptability.

Also, Prophet allows the adjustment of holiday effects, ensuring that the model can consider the spikes in the demands linked with the respective holidays or events.

Moreover, Prophet is a tool that can manage changepoints, which are times in the time series when the trend changes. The model automatically pinpointed these changepoints and corrected them, thus improving its ability to align with the data changes.

Being configurable, we set up Prophet with yearly seasonality, and a custom monthly seasonality was added. The model was further modified to include the weekday effects, which displayed the inconsistency of the sales across the different days of the week.

3.6 Neural Prophet

While the original Prophet model heavily uses traditional time series decomposition methods, NeuralProphet includes key neural network components. This hybrid model merges the strengths of the conventional time series decomposition with flexibility given by the neural networks for better capture of nonlinear dependencies and complex seasonal patterns.

Key Features and Configurations:

Extended Seasonality and Trend Components: Similar to Prophet, NeuralProphet can model yearly, weekly, and daily seasonality. It will do this with the added advantage of neural network layers to model more intricate relationships within the data.

Lagged Regressors: NeuralProphet introduces the concept of lagged regressors, where the model can consider delayed consequences of input variables on the target variable. This is quite helpful in cases where the effect of the lag factors carries over into future values.

Neural Network Layers: Using a neural network layer, the model of NeuralProphet captures more complex patterns than simple linear models. This then makes it more efficient in adapting to changes in trends and seasonality.

Configuration: NeuralProphet-configured lagged regressors for the day of the week and month capture both short- and long-term seasonality.

3.7 ARIMA

ARIMA (AutoRegressive Otherwise Integrated Moving Average) and SARIMA (Seasonal ARIMA) are widely adopted and often used to predict time series models. These models assume that future time series values are a linear function of past observations and errors. ARIMA:

AutoRegressive (AR) Component: The AR constituent of ARIMA refers to regressing the variable on its past values' lagged (metaphorically) ones. It captures the effect of prior observations on the current observation.

Integrated Component: This part of ARIMA is the operation of differencing the raw observations to make the time series stationary (i.e., with constant mean and variance over time).

Moving Average (MA) Component: The MA part of ARIMA models the relationship between an observation and a residual error from a moving average model applied to lagged observations.

Configuration: In this study, the ARIMA model was automatically configured using the auto_arima function, which selects the best combination of AR, I, and MA parameters based on model performance.

3.8 SARIMA

SARIMA extends ARIMA by including seasonal components. It models the data's non-seasonal and seasonal patterns by adding seasonal autoregressive, differencing, and moving average components.

Seasonal Components: SARIMA includes parameters for seasonal autoregression (P), seasonal differencing (D), and seasonal moving average (Q), along with the seasonal period (m).

Configuration: The SARIMA model in this study was configured with a seasonal period of 12 months, capturing annual seasonal patterns in the sales data.

3.9 Exponential Smoothing Techniques

Exponential smoothing is a family of forecasting methods that assigns exponentially decreasing weights to past observations. These methods are beneficial for data with trends and seasonality. Simple Exponential Smoothing (SES):

SES is used for time series data without a trend or seasonality. It smooths the series by averaging the past values with exponentially decreasing weights.

The smoothing level (α) is a parameter that controls the rate at which the influence of past observations decays. In this study, α was set manually for illustrative purposes.

3.10 Holt's Linear Trend Model

Holt's model extends SES by adding a trend component. It is suitable for data with a linear trend but no seasonality.

The model includes two parameters: one for the level (α) and one for the trend (β), both optimized for this study.

3.11 Holt-Winters Seasonal Model:

The Holt-Winters model is used for data with both trend and seasonality. It includes parameters for the level, trend, and seasonality (γ). In this study, the model was configured with additive seasonality to capture seasonal patterns in sales data.

In the research presented, various models are applied, including the traditional methods such as ARIMA and Exponential Smoothing and the advanced models such as Prophet and NeuralProphet. Each model's unique requirements regarding the retail sales data are considered thoroughly. These models are arranged so that their forecasting accuracy and computational efficiency can be compared comprehensively. The insight obtained from this comparison will allow the practitioner to choose an appropriate model based on their needs.

4. Experiments

4.1 Data pre-processing

Data pre-processing is a pivotal step in this study for extracting the most benefit from these forecasting models. Appropriate feature transformation and selection were significant; they increased the model accuracy and let these generalizing models work well with unseen data. Different steps were taken, starting from loading the dataset, such as splitting the data and handling missing points and outliers.

The dataset was sourced from the Kaggle competition "Demand Forecasting," which was based on the historical figure of the sales data of multiple stores and items. It was downloaded directly into the Google Colab environment using the Kaggle API to ease the drag and drop of his study using the platform.

4.2 Feature engineering

Feature engineering here involves constructing new input features from the available data to capture better the temporal pattern, an essential attribute in time series forecasting. The first step undertaken here was converting the 'date' column into a datetime format through Pandas for efficient data manipulation of the features, and to this, pre-processing added further features, including year, month, and day features, which were helpful identifications of long-term trends and seasonal effects by the models. The 'day' feature was further broken down to 'day of the year' to capture seasonality on more granular levels. A 'weekday' feature was also created—from 0 for

date	store	item	sales	year	month	day	weekday
2013-01-01	1	1	13	2013	1	1	1
2013-01-02	1	1	11	2013	1	2	2
2013-01-03	1	1	14	2013	1	3	3
2013-01-04	1	1	13	2013	1	4	4
2013-01-05	1	1	10	2013	1	5	5
2013-01-06	1	1	12	2013	1	6	6
2013-01-07	1	1	10	2013	1	7	0
2013-01-08	1	1	9	2013	1	8	1
2013-01-09	1	1	12	2013	1	9	2
2013-01-10	1	1	9	2013	1	10	3

Figure 1. Sample of the retail dataset used for forecasting model training.

Monday to 6 for Sunday, where models could account for the standard intense weekly sales cycles in retail data.

To enhance this predictive capability of the models, lag features were created that combined information from previous time steps—the number of sales in the prior day or week—along with the current prediction so that the models can learn patterns across time. Otherwise, some rolling statistics were implemented to smooth out short-term fluctuations and sharpen longer-term trends.

Once the feature engineering was done, the dataset was split into training and testing sets to assess the models' generalization performance. Usually, the training set consisted of about 70% or 80% of the data in the dataset, while the rest were for testing. Allowing the model to learn from the historical data, the rest, 20% or 30%, is reserved as a test set to represent the most recent sales data to evaluate how well the model can predict sales in the future.

Another critical phase in preparing the data was handling missing values. Each missing value was imputed with different imputation techniques, such as filling gaps with the median or mean of sales value or carrying forward the last available observation. Detailed outliers are also treated so as not to affect the models' performance. Capping extreme values or replacing them with median values will ensure no distortion in the dataset.

Normalization was the operation that scaled the numerical features. In this, we guaranteed that the features had an average of zero and a standard deviation of one—the step significant to the services, sensitively oriented to the scale of input data, like ARIMA and neural networks.

This was a required step to structure data so that it had the dataset ready for the accurate time series forecasting task. Hence, several pre-processing steps were thoroughly worked out, followed by extraction of meaningful features that treated missing values and outliers and eventually set splitting. Therefore, the study set a good foundation for building reliable forecasting models. This rigorous data preparation was mandatory to get interpretable and correct results in the succeeding analysis.

4.3 Model Training

The training of models was necessary for developing the most accurate and reliable forecasting models to suit this study. These range from simple traditional statistical models to advanced neural network models, each setting parameters that best fit the dataset with varying computational requirements.

Prophet is an advanced time series forecasting model developed by Facebook, configured to handle missing data, outliers, and seasonal solid effects. The model was trained with yearly, weekly, and custom monthly seasonality settings and extra regressors such as 'day_of_week' and 'month.' The training also involved fitting the model to the dataset using the default settings in Prophet by automatically detecting seasonality, trends, and holiday effects. Cross-validation was used to tune the hyperparameters, and the model was trained in Google Colab using CPU resources, thus being computationally efficient and well-suited for environments with no high-performance computing.

It is an extension of Prophet that integrates neural network components for higher flexibility and accuracy. Like Prophet, NeuralProphet was trained with yearly and weekly seasonality and lagged regressors like 'day_of_week' and 'month' to capture temporal dependencies. The model also uses a neural network structure that captures nonlinear patterns in the data; automated learning rate and hyperparameter tuning are used. The training comprised multiple epochs with early stops to avoid overfitting. Due to the increased complexity brought by neural networks, the model was trained using free-tier GPU resources on Google Colab to speed up the training of a computationally intensive process, albeit feasible in the cloud using free-tier resources.

ARIMA and SARIMA also represent typical statistical models trained in this work. In the case of the ARIMA model, parameters were selected using an auto_arima function, which automated the process of choosing the best configuration. In addition to the parameters of the ARIMA model,

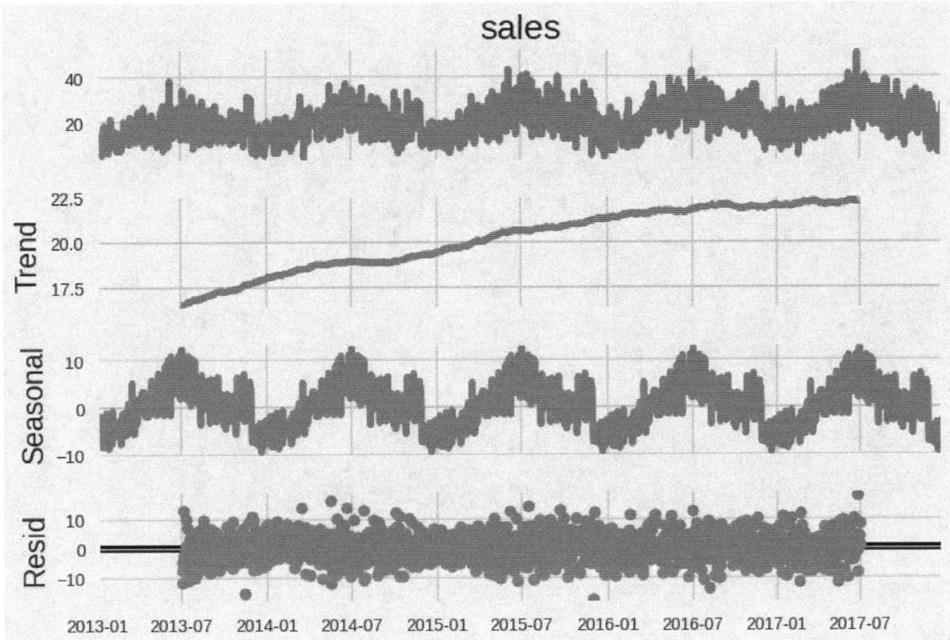

Figure 2. Seasonal Decomposition of the Sales Data

the SARIMA model had some additional seasonal parameters tuned to capture the weekly seasonality. Both models utilized training data differencing to ensure stationarity and were evaluated using cross-validation. While these methods are less resource-intensive than neural network-based methods, they may still use a good amount of resources when performing parameter tuning on larger datasets. The training was conducted in Google Colab using CPU resources.

Exponential Smoothing models represent one of the classic ways to forecast time series. These are applied in three variants: Simple Exponential Smoothing for series with neither trend nor seasonality, Holt's Linear Trend Model to capture linear trends, and the Holt-Winters Seasonal Model, which adds seasonal components. Each model was trained on the dataset and compared to the others concerning their forecast output against actual sales over a validation period. While the best parameters for smoothing levels and seasonal components were determined by grid search for the most computationally efficient models, training was done on CPU resources provided by Colab. In summary, each model in this study was trained with careful consideration of its unique configuration and requirements. Prophet and NeuralProphet were developed to capture seasonality and use additional regressors, while the latter benefited from Google Colab's GPU resources for neural network components. More classic models included ARIMA, SARIMA, and Exponential Smoothing, all trained using well-established statistical methods. Though the computational demands for some models were much more significant than others, careful attention in training meant models were set up for success in their demand forecasting performance within a retail context.

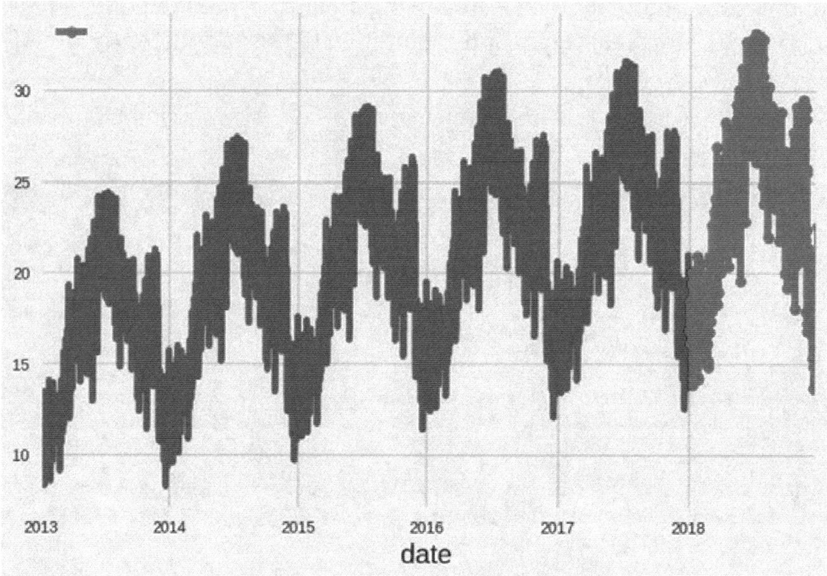

Figure 3. Results from the exponential smoothing forecasting.

4.4 Model Saving

Saving and reloading of models is a usual step in any machine learning workflow; it is especially true in the context of time series forecasting. In this way, the code can be reproducible, scalable, and efficient since it can reuse the models that have been trained previously and avoid retraining all the time. Because of these reasons, joblib has been used for the current study. Joblib is a Python library developed for object serialization and deserialization. It puts a strong emphasis on model serialization. The following describes the saving and reloading of the models, emphasizing how vital such a process is.

After training, each forecasting model was saved to disk using the joblib library in serialized format. As such, the model can easily be stored and retrieved whenever necessary, negating the need to retrain the model and thus not wasting time and computation resources. That involved using joblib.dump()to save the model in a format that could later be quickly loaded. This was important not only for efficiency in the running of predictions but also because of the reproducibility of the whole workflow.

We saved the trained models, allowing for a similar model version to be used through different project stages or even by other collaborators with no variations. Having model training archetypes further enhanced the scalability of the study in that models could be reused in a different environment or adapted for further analysis without the overhead of retraining. Since saving and seamless models would, therefore, be able to be performed utilizing joblib, reproducibility, a cornerstone of any reliable machine learning project, is thus guaranteed.

4.5 Model Evaluation

Different evaluation metrics are included in this study to measure the accuracy of the time series forecasting models. They give a complete explanation of how each of the models is in line with the accurate sales data. Prominent ones include Mean Absolute Error (MAE), Root Mean Squared Error (RMSE), and R-squared (R^2). Each of these metrics provides insight into the model's accuracy and reliability.

Mean Absolute Error (MAE) measures the average magnitude of errors in a set of predictions without considering their direction (that is, whether the error is positive or negative). It is the average of the absolute differences between prediction and actual observation over the test sample, where all individual differences have an equal weight. MAE is a simple metric to quantify the error, for which the lower, the better; it is instrumental when significant errors must be avoided since all the errors contribute equally to the metric.

Root Mean Squared Error (RMSE) is the square root of the average squared differences between prediction and actual observation. Significant errors are penalized much more than with MAE since errors are first squared before averaging.

RMSE is more sensitive to significant errors compared to MAE. It is a good metric when significant errors are especially unwanted. The lower the value of the RMSE, the better the model's performance. The RMSE is very useful when large deviations from actual values need to be kept low.

R-squared (R^2), also known as the coefficient of determination, measures the proportion of the variance in the dependent variable that is predictable from the independent variables. It indicates how well the model replicates the observed outcomes. R-squared values range from 0 to 1, with higher values indicating better model performance. An R^2 of 1 indicates that the model perfectly predicts the observed data, while an R^2 of 0 indicates that the model does not explain any of the variability in the data. Negative values of R^2 can occur if the model is worse than a horizontal line representing the mean of the data.

In some cases, other metrics like Mean Absolute Percentage Error (MAPE), Weighted Absolute Error (WAE), and Weighted Mean Absolute Percentage Error (WMAPE) are also used to provide further insights into model performance, especially when dealing with datasets where zero values or specific error distributions are of concern. Mean Absolute Percentage Error (MAPE) measures the accuracy of a forecasting method by calculating the average of the absolute percentage errors of the forecasts. It is expressed as a percentage.

Since MAPE expresses the error as a percentage, it is beneficial for comparing the accuracy of models between different data sets or periods. Percentage error calculations, in turn, can become misleading in time series where actual values are near zero since that would imply extremely high or undefined percentage errors.

Weighted Absolute Error (WAE) and Weighted Mean Absolute Percentage Error (WMAPE) are the weighted variants of MAE and MAPE, respectively. These include weights on observations, reflecting relative importance. In this sense, they provide a far more nuanced view of model performance, especially when some errors might be more consequential than others. These are useful when one needs to consider variability in the level of importance from one data point to another. For instance, in retail forecasting, the errors of high-volume items will be more critical compared to those of low volume.

All metrics together provide the overall performance of the models. MAE and RMSE give insight into average and extreme errors, respectively; R^2 measures how well the model explains the variance in the data. MAPE, WAE, and WMAPE go one step further to enlighten either conditions or error distributions of particular interest. Based on these metrics, the underlying analysis will conduct a comprehensive and balanced evaluation of each model's forecast performance.

4.6 Software and Tools

The current study has used some contemporary tools and programming environments available in time series forecasting. The preferred programming language for data analysis is Python, as it is flexible and has captured wide use by the community with extensive libraries related to data science, forecasting, and machine learning. Pandas and NumPy are used to handle and process large amounts of data, providing an essential basis for manipulation and analysis. Scikit-learn is used for model evaluation with metrics such as Mean Absolute Error and Root Mean Squared Error, whose results can be depicted on plots thanks to Matplotlib and Seaborn. The joblib was very helpful in saving and reloading trained models, which then, in turn, made resource-expensive machine learning models very efficient to use.

Prophet software, developed by Facebook, was used to manage complex time series data that exhibited strong seasonality and missing values.

NeuralProphet extended Prophet with neural network components and was thus able to extend this study to the advantages of blending these traditional time series techniques with deep learning methods.

The classical model side has been implemented using ARIMA and Exponential Smoothing with Statsmodels. The often complex procedure of choosing the best parameters for ARIMA models was partly automated with Pmdarima.

Google Colab platform makes easy collaboration as notebooks are accessible from any device online and shareable. Also, the free availability of GPUs and TPUs made tasks with complex model training possible. The integration of Google Drive means high-speed storage and retrieval of data to make the workflow smooth. These tools provided a robust, flexible, efficient environment for realizing the study's objectives with accuracy and computational efficiency.

5. Results

5.1 Models Evaluation

Each model has been assessed using key metrics: MAE, RMSE, and R^2. Below is a summary of the results from our experiments using different forecasting techniques by listing their predictive accuracies and computational efficiencies.

The performance of the NeuralProphet model was the best among the models, with the lowest MAE of 3.45 and an RMSE of 4.33, showing it was highly accurate in its predictions. It also showed a reasonably reasonable R^2 value of 0.37, showing that it captured a fair amount of variance in the data. Its computational efficiency was moderate due to the complexity of its neural network components. In contrast, ARIMA performed with an MAE of 22.20, RMSE at 23.35, and a negative R^2 of -10.31, proving it was ineffective in fitting the dataset. At the same time, SARIMA had significantly superior results compared to ARIMA. It still showed moderate error rates, with the R^2 value at -0.45, allowing room for further improvement. It is supported that the Smoothing Model performed exceptionally well, with the MAE being 4.29 and the RMSE being 5.49; hence, this model can be called an effective, simpler model, with the added advantage of high computational efficiency. Finally, the Prophet model, mostly known for handling seasonality and holidays, performed in the middle with a mean absolute error of 4.90 and an RMSE of 6.03, corresponding to the expectations set forth by similar studies in the literature.

Journal of Physics: Conference Series **2910** (2024) 012008 doi:10.1088/1742-6596/2910/1/012008

This can be seen by comparing the strengths and weaknesses of the models through the evaluation metrics. The MAE is a straightforward measure of the average prediction error, so the highest MAE for the NeuralProphet model was the most effective for this dataset. The RMSE emphasizes more significant errors and showed that NeuralProphet had the minimum number of significant deviations in its predictions. R^2 metric showed that models like ARIMA and SARIMA couldn't explain the variance in data at the same time when ARIMA was notably worse, and the robustness of Prophet and NeuralProphet was higher.

Insights from these experiments still hint that traditional smoothing techniques can be compelling in their simplicity. The Smoothing Model yields outstanding accuracy and low computational complexity, making it feasible for most forecasting tasks. However, the ARIMA and SARIMA models failed to produce the desired results; in fact, the ARIMA proved so inadequate that it did not seem suitable for this dataset. On the other hand, the SARIMA performed a little better than that but still poorly compared with both literature benchmarks and the different models tested in this work.

In contrast, the performance of the Prophet and NeuralProphet models was excellent, with the latter being slightly better than Prophet. The performance level is expected since Prophet and its variants are designed to handle seasonal patterns and other complex time series features. The neural network components incorporated in NeuralProphet seem to give it an increased boost in performance that turns it into a very effective model for forecasts in this context.

These results are competitive compared to results reported in the literature, and the MAE was also lower than usually reported in other works. That is why our NeuralProphet model seemed well-tuned for the particular dataset used in this research. At the same time, the ARIMA model had a poor performance compared to the usual outcomes provided by literature, as ARIMA models are often better in error metrics and R^2 value. As expected, though, the Prophet model gave moderate error rates, consistent with results from other works.

The study thus entails that, in one way or another, some time series forecasting models have consistently demonstrated superiority or inferiority in certain respects. The advanced models, like NeuralProphet, outperform conventional ones like ARIMA and SARIMA by a considerable margin of superiority in forecast accuracy, especially concerning complex data patterns. Contrarily, model interpretability and computational efficiency must be considered whether the methods are applied to large datasets or for a complex forecasting task.

Among these, the NeuralProphet model performed best in most of the metrics. This model had the most petite MAE of 3.45 and RMSE of 4.33, which means this model made the most negligible average errors and was least penalized for significant mistakes. Also, its R^2 value of 0.37 shows that this model would explain a much higher portion of the variance in data, hence being more suitable for more complex tasks such as prediction. Besides, NeuralProphet finds its application in moderately complex and sized datasets, as it is relatively above-average computationally efficient. In contrast with most other competitors, the Smoothing Model was relatively simple yet performed exceptionally well, making it the runner-up. Although it was the second-best model regarding predictive accuracy, with the MAE value coming out at 4.29 and the RMSE value at 5.49, it is decently good at capturing the variance within data, with its R-squared value at 0.38. This smoothing model is useful when forecasts must be made quickly and fairly accurately, especially considering that no high computational costs are associated.

Performance for the Prophet model was similarly consistent, although lagging behind both NeuralProphet and the Smoothing Model. It recorded an MAE of 4.90 and a root mean square error 6.03, still reasonably accurate forecasts. Its R^2 value is 0.25; therefore, it explains some variance, though not much compared to those two leading models. The Prophet shines where management of seasonality and incorporation of special events, such as holidays, is involved; hence, it is very effective in business applications where the integration of external factors is required.

On the contrary, the worst performance was recorded using ARIMA, with an MAE of 22.20 and RMSE of 23.35, coupled with an R^2 value of -10.31, which is an implication that the model prediction was worse than using the mean as a predictor. That would tend to indicate that ARIMA is not a good fit for this particular dataset or urgently needs serious tuning. Though the SARIMA model did better, its standing remained behind, with an MAE of 6.66, an RMSE of 8.37, and a negative R^2 value of -0.45. Similar to ARIMA, even SARIMA failed to capture the variance in the data, indicating that this traditional model has to be tuned carefully or the data has to be pre-processed further for better performance.

Table 4. Model Training Results

Model	MAE	RMSE	R^2
ARIMA	22.2	23.35	-10.31
SARIMA	6.66	8.37	-0.45
Smoothing Model	4.29	5.49	0.38
Prophet	4.9	6.03	0.25
NeuralProphet	3.45	4.33	0.37

Generally speaking, NeuralProphet is the best and most balanced model in predictive power and computational efficiency. This model can be a solid alternative for applications where simplicity and speed are of prime importance but do not compromise much on accuracy. Prophet is practical in contexts where external factors and seasonality play a significant role. Contrariwise, ARIMA and SARIMA models may be ineffective enough in datasets like this work under study unless they have been seriously tuned.

While advanced models such as NeuralProphet greatly enhance the forecasting accuracy of the benchmark, traditional models, including ARIMA and SARIMA, still have practical value with their computational efficiency and relatively good performance. This comparison underlines the importance of choosing an appropriate model concerning the specific requirements of the considered forecasting task.

6. Discussion

The comparison of different time series forecasting models has assisted in identifying insight into the performance and applicability of other models. Therefore, key performance metrics such as MAE, RMSE, and R^2 will go a long way in helping to draw several inferences relating to the relative effectiveness of each model besides highlighting reasons a few outperformed the others in the given context.

The Smoothing Model gave an excellent performance, yielding an MAE of 4.29 and an RMSE of 5.49 with an R^2 of 0.38. This model captured the overall underlying smoothness and seasonality exhibited in the sales data. It was great at capturing those, as it gave higher weights to recent observations; hence, it adapted quickly to changes happening in the time series. Focusing on the most recent data points helped the model to outperform the underlying patterns, though it struggled somewhat with the variances, as testified to by its RMSE.

On the contrary, the ARIMA model performed very poorly. It had a much higher MAE, amounting to 22.20, with an RMSE of 23.35 and a negative R^2 of -10.31. These results show that ARIMA struggled to capture the dynamics of the dataset. The model's performance may be due to an assumption of stationarity and linearity in the model that did not support such a level of complexity and seasonality of the data. The high RMSE, coupled with a negative R^2, indicates that ARIMA's predictions not only went off the chart concerning magnitude but also failed to capture the direction of actuals.

The best performance out of these, but still far from the performances of the Smoothing Model and Prophet-based models, was obtained with the SARIMA model that extends ARIMA to incorporate seasonality. The MAE was 6.66, the RMSE was 8.37, and the R^2 was -0.45. This model fared better but was still far from the best performance. Though the seasonal pattern is captured somehow, it is less capable of dealing with variability and complexity than other models, reflected in the negative R^2. The model training results suggest that the model has been overfitted or the best parameters were not selected during training.

Prophet was the best model, returning an MAE of 4.90 and an RMSE of 6.03 with an R^2 of 0.25. Even though its performance lagged a little behind the Smoothing Model regarding the RMSE, Prophet still produced good forecasts. Success factors included its ability to handle seasonal solid effects and robustness against missing data and outliers. However, the relatively low R^2 indicates that while Prophet effectively captured the seasonal components, it may have struggled with irregularities or noise in the data.

NeuralProphet, which is a model that incorporates elements of neural networks into the Prophet model, yielded an MAE of 4.89, an RMSE of 6.03, and a worst R^2 of 0.25, competing with the simple Prophet model. NeuralProphet has the power to handle challenging nonlinear relationships and flexibility in modeling lagged regressors. The fact that Prophet performed similarly suggests that the neural network elements provided little gain on this dataset, possibly because of the data's relatively tiny size or nature.

The conclusion derived from the above analysis is that the Smoothing Model was the best. Thus, more straightforward methods, which place a lot of emphasis on recent data points and account for seasonality concisely, work well on this particular dataset. On the other hand, more complex models such as SARIMA and NeuralProphet did not give very superior results, likely due to this dataset's nature or its possible overfitting problems. The R^2 values reveal severe model problems fit for ARIMA, and the negative values are poor fits-again, pointing, in this case, to the possibility that a parsimonious approach yields better results when either the seasonal or trend component is not well-represented in the data.

As mentioned above, the situation foresees improvements that may come from advanced models like NeuralProphet and SARIMA. Consequently, their advantages depend highly on dataset characteristics. Future studies can continue with more extensive and complex datasets, where advantages for these models would emerge. Further experiments with sophisticated tuning methods and hybrid models may further enhance predictive accuracy, at least for situations when traditional models perform inadequately.

Certain limitations establish uncertainties in the outcomes, such as the dataset scope, model selection and tuning, computation resources, and metrics adopted to assess performance. These limitations should be improved upon in a future study by utilizing more extensive and diverse datasets, improving model optimization, and utilizing more advanced computational resources. In this way, a deeper understanding of various model enabilities and shortcomings concerning the time series forecast could be acquired.

7. Implications

Models like Prophet and NeuralProphet are doing a pretty good job, considering how easy it is to implement and tune. Typically, they work well on data with explicitly stated seasonality and do not require heavy manual intervention. Thus, they will be convenient for an organization that does not possess the scarce resources of the data sciences.

On the other hand, when data shows strong autocorrelation and stationarity, the application's merit belongs to classical models like ARIMA and SARIMA. These models sometimes require more sensitive tuning and may be less computationally efficient from time to time, limiting their utilization for perhaps real-time applications or resource-constrained environments. Some good results were obtained using exponential smoothing methods for simple datasets with subtler trends and seasonality. In turn, their low computational overhead and robustness render them a workable choice for more straightforward forecasting tasks.

Another significant development of this research is the trade-off between model complexity and computational efficiency. This trade-off would, therefore, entail that in choosing the models, the practitioner has to consider the available computational resources. Yes, complex models, like the SARIMA and NeuralProphet, will generally require much computational processing power and time, mainly when applied to comprehensive datasets or when conditions include extensive hyperparameter tuning. Both Prophet and NeuralProphet can be scaled for organizations seeking to scale up their forecasting across a portfolio of products or regions in addition to seeking a good balance between performance and computational feasibility.

Results tell that model performance must be taken from a holistic perspective. The practitioner cannot base his decision on just one metric, such as MAE or RMSE, but on several other metrics like R^2 and computational efficiency. In this regard, it will ensure multi-dimensional model performance evaluations that enable business decisions relevant to the particular objective and constraints of the operation in question.

In particular, while the contributions mentioned above are essential, one crucial area of consideration for further research is the more advanced models such as LSTM and XGBoost and their hybrid models. Such models may perform generally better, being different from traditional ones, especially in some more dynamic or complicated situations. More importantly, investigating this deep learning model may shed more light on when models work, massive data models, and under what conditions these models fail.

Another promising line of research is improving techniques for hyperparameter tuning. Limitations identified here point toward future investigations exploring more advanced methods, such as automated machine learning, or more sophisticated wider search strategies like Bayesian optimization to find more promising model configurations.

Further confirmation may come through applying such models in a broader class of datasets: complex seasonality, nonlinear trends, or determined by exogenous variables themselves. In addition, such information would also give more details on where the specific models perform better precisely. Other than this, more sophisticated metrics in terms of MAPE or the development of new metrics will lend greater strength to performance assessments across various contexts.

That opens up another valuable area of exploration for scalability and real-time application of those models. Some of the models in this work are computationally expensive; hence, their optimization or approximation for real-time forecasting may be highly relevant in domains such as finance, supply chain management, and energy, where decisions must be made speedily with data-driven insight.

Therefore, this will provide practical guidance both to practitioners and researchers. While such insights will prompt the practitioners to choose models that best suit their needs and constraints, the researchers can build on those insights to address gaps and further push the pace of time series forecasting. Future research, if addressed, may help develop more robust, efficient, and accurate models that can enhance the decision-making processes across many industries.

8. Conclusion

The study reviewed numerous time series forecasting models, including traditional ones such as ARIMA, SARIMA, and Exponential Smoothing, with more modern approaches such as Prophet and NeuralProphet. Key takeaways of the research are therefore as follows. The model Prophet and its neural extension, NeuralProphet, performed well, notably when the data showed strong seasonal patterns. Thus, Hybrid SARIMA-Prophet performed better than the other models on all three metrics and should be a strong competitor for any demand forecasting.

Traditional models, like ARIMA and SARIMA, remain relevant when data shows strong autocorrelation with stationarity. However, such models required more meticulous tuning and were not computationally quite efficient compared to the Prophet models. While more straightforward, the Exponential Smoothing models yielded strong performance with less computational overhead.

Several evaluation metrics were applied in the study, such as MAE, RMSE, and R^2, to assess the models' performances. These highlighted the trade-off between accuracy and computational efficiency, entailing a nuanced understanding of the strengths and limitations of each model.

The more complex a model, the more computational power its operation requires. The computational power is a substantial requirement for practitioners, especially in real-time or large-scale forecasting. Future work Although the present study yields a lot of insight, there are several areas on which future work may focus to enhance the understanding and application of time series forecasting models. For instance, the exploration of machine learning and deep learning models in this current work could be extended into advanced machine learning models such as LSTM, XGBoost, and hybrid models, which could be considered alongside the models considered in the study. These models have given great promise for handling more prominent and complex datasets, and the inclusions of these may have provided a broader grasp of forecasting capabilities.

The study should be extended to a wide range of data, especially those exhibiting complex seasonality, nonlinear trends, or external influencing factors that could help establish the generalizability of the findings. Real-time Forecasting: Other research could be done to develop further models specifically for real-time forecasting, such that, in applications where decisions have to be made in the shortest time possible, the models would also do the same. Again, real-time forecasting has to do with enhancing computational efficiency by testing how the models refresh with new incoming data.

In the study, some of the provided metrics were problematic; examples include MAPE and WAE. The natural way to continue this work would be by refining these metrics or developing new ones that are better tailored to the various conditions arising in time series forecasting.

Based on this study, several practical recommendations may be derived and addressed to practitioners involved in demand forecasting and similar tasks.

A practitioner might use Prophet or NeuralProphet on a demand prediction task, especially when their datasets contain strong seasonal patterns. These models balance ease of use with solid performance, requiring less manual tuning than traditional models.

In cases where computations are an issue, a simple model, such as Exponential Smoothing, should be considered, especially with less complex datasets. On the other hand, when high accuracy is needed, and resources permit, then a hybrid model like SARIMA-Prophet has proven very effective.

In this respect, the practitioner should consider multiple metrics to evaluate model performance. Applying different metrics like MAE, RMSE, and R^2 provides a better understanding of how a model will perform in practice.

In real-world applications involving real-time data forecasting, the model selected should update efficiently with incoming data. Prophet and NeuralProphet are remarkably apt in such situations because of relatively faster training and lesser computational requirements when compared to complex models such as SARIMA.

Model monitoring is something that practitioners are giving serious attention to as they seek to project their models into dynamic environments continuously. Monitoring enhances refreshing or updating the models in time, with incoming data or trends that may emerge, to keep the forecasts accurate and reliable. By following such recommendations, the practitioner could make a more informed decision about deploying forecasting models, thus getting better results in accuracy and efficiency.

Acknowledgements

Mikhail Kolev thanks to the financial support of European Union-NextGenerationEU, through the National Recovery and Resilience Plan of the Republic of Bulgaria, project № BG-RRP-2.004-0002-C01, "BiOrgaMCT".

CRediT roles: Conceptualization, M.L.; methodology, I.N. and M.L.; software, I.N.; validation, M.K. and I.N.; formal analysis, M.L.; investigation, I.N. and M.K.; resources, M.L.; data curation, I.N. and M.K.; writing – original draft preparation, I.N. and M.K.; writing – review and editing, M.L. and I.N.; visualization, I.N. and M.K.; supervision, M.L.; project administration, M.K.; funding acquisition, M.K. and I.N.

References

[1] G. E. P. Box, G. M. Jenkins, and G. C. Reinsel, *Time series analysis: forecasting and control*, 3. ed. Englewood Cliffs, NJ: Prentice-Hall, 1994.

[2] P. J. Brockwell and R. A. Davis, *Time series: theory and methods*, 2nd ed. in Springer series in statistics. New York: Springer, 1996.

[3] G. Box, "Box and Jenkins: Time Series Analysis, Forecasting and Control," in *A Very British Affair*, London: Palgrave Macmillan UK, 2013, pp. 161–215. doi: 10.1057/9781137291264_6.

[4] P. J. Brockwell and R. A. Davis, *Introduction to Time Series and Forecasting*. in Springer Texts in Statistics. Cham: Springer International Publishing, 2016. doi: 10.1007/978-3-319-29854-2.

[5] J. Contreras, R. Espinola, F. J. Nogales, and A. J. Conejo, "ARIMA models to predict next-day electricity prices," *IEEE Trans. Power Syst.*, vol. 18, no. 3, pp. 1014–1020, Aug. 2003, doi: 10.1109/TPWRS.2002.804943.
[6] L. Rubio, A. J. Gutiérrez-Rodríguez, and M. G. Forero, "EBITDA Index Prediction Using Exponential Smoothing and ARIMA Model," *Mathematics*, vol. 9, no. 20, p. 2538, Oct. 2021, doi: 10.3390/math9202538.
[7] C. C. Holt, "Forecasting seasonals and trends by exponentially weighted moving averages," *Int. J. Forecast.*, vol. 20, no. 1, pp. 5–10, Jan. 2004, doi: 10.1016/j.ijforecast.2003.09.015.
[8] R. J. Hyndman and Y. Khandakar, "Automatic Time Series Forecasting: The **forecast** Package for *R*," *J. Stat. Softw.*, vol. 27, no. 3, 2008, doi: 10.18637/jss.v027.i03.
[9] J. D. Hamilton, *Time Series Analysis*. Princeton University Press, 1994. doi: 10.1515/9780691218632.
[10] A. C. Harvey, *Forecasting, Structural Time Series Models and the Kalman Filter*, 1st ed. Cambridge University Press, 1990. doi: 10.1017/CBO9781107049994.
[11] M. Caivano and A. Harvey, "Time-series models with an EGB2 conditional distribution," *J. Time Ser. Anal.*, vol. 35, no. 6, pp. 558–571, Nov. 2014, doi: 10.1111/jtsa.12081.
[12] A. E. Hoerl and R. W. Kennard, "Ridge Regression: Biased Estimation for Nonorthogonal Problems," *Technometrics*, vol. 12, no. 1, pp. 55–67, Feb. 1970, doi: 10.1080/00401706.1970.10488634.
[13] S. J. Taylor and B. Letham, "Forecasting at Scale," *Am. Stat.*, vol. 72, no. 1, pp. 37–45, Jan. 2018, doi: 10.1080/00031305.2017.1380080.
[14] O. Triebe, H. Hewamalage, P. Pilyugina, N. Laptev, C. Bergmeir, and R. Rajagopal, "NeuralProphet: Explainable Forecasting at Scale," Nov. 29, 2021, *arXiv*: arXiv:2111.15397. Accessed: Sep. 01, 2024. [Online]. Available: http://arxiv.org/abs/2111.15397
[15] M. G. S. Kenyi and K. Yamamoto, "A hybrid SARIMA-Prophet model for predicting historical streamflow time-series of the Sobat River in South Sudan," *Discov. Appl. Sci.*, vol. 6, no. 9, p. 457, Aug. 2024, doi: 10.1007/s42452-024-06083-x.

Road safety management system optimization methodology

Silvia Assenova[1], Plamena Zlateva[1,2] and Kamelia Assenova[2]

[1] Institute of Robotics, Bulgarian Academy of Sciences, Sofia, Bulgaria
[2] University of National and World Economy, Sofia, Bulgaria

*E-mail: silviaassenova@ir.bas.bg

Abstract. In this study, a traffic safety management system (TSMS) for asphalt surfaced roads is developed to improve safety on different classes of roads in Bulgaria. TSMS is a strategic and systematic procedure to improve traffic safety when funding is limited and it is important to determine the most appropriate combinations of traffic safety improvement projects to provide maximum benefits to the population in terms of reducing traffic accidents. The factors included in the developed optimization methodology are the annual report on traffic accidents of the National Institute of Statistics, a study of the roads, a functional classification of the same, historically established, countermeasures to improve safety, costs and determinants of crash reduction in relation to measures to positively modify safety and average daily traffic (ADT). This study demonstrates how the proposed model can determine the best combination of activities to increase safety in order to maximize the benefits in terms of reducing overall crash frequency. The proposed methodology has been applied to highways in Bulgaria. It can be modified for application to roads in a particular city, region and district.

1. Introduction

In 2022, a total of 31539 crashes occurred in Bulgaria, killing 531 people and injuring 8422. The damage in material terms has not been estimated and no such statistics are kept in the country. The State has and maintains a total of 19932.10 km of roads. In this study, only highways are considered. The main reason for not studying the first, second, third and fourth class roads is that there is lack of data collected for them. Although most European countries have their own traffic safety management system, Bulgaria does not. Therefore, this study focuses on the development of a TSMS for the aforementioned paved motorways. The collected data includes information about the roads: traffic data, roadway width, pavement condition. Some of this information would be useful for the development of the methodology (TSMS). Many of the main thoroughfares were constructed more than 35 years ago and are not consistently maintained. All this has led to the overall poor condition of the same. In addition, with the migration of the population to the larger cities, the number of cars and heavy goods vehicles travelling on roads of the types - highways, first and second class - has increased. All this leads to significant economic losses due to road accidents and there is a need to develop an innovative traffic management system that makes more efficient use of limited resources. The methodology developed, for the implementation of which a limited budget is foreseen, will ensure that the selected traffic safety tools will minimize the number of road accidents and the resulting fatalities and injuries. In the proposed methodology, the selection will show the improvement of traffic safety and it will depend not only on traffic but also on the number of crashes, the degree of reduction of the same and,

Content from this work may be used under the terms of the Creative Commons Attribution 4.0 licence. Any further distribution of this work must maintain attribution to the author(s) and the title of the work, journal citation and DOI.
Published under licence by IOP Publishing Ltd

consequently, on countermeasures. The improvement in safety will be depicted on the basis of the highest level of reduction in the number of crashes.

2. Literature Review

The literature review covering the recent research in the area of TSMS can be divided into three areas, viz:

- ✓ safety performance function,
- ✓ accident hotspots,
- ✓ safety management system optimization.

2.1 Safety performance function

To improve safety, it is important to understand the causes of accidents. There are a sufficient number of studies that model the occurrence of accidents [1, 2, 3, 6, 7, 12]. All of them investigate the modelling of the occurrence and involvement in road accidents. The results show that average daily traffic, speed, lane width, number of lanes and pavement width have statistically significant effects on crash occurrence and involvement. [4] performed a comprehensive analysis and developed models related to traffic accidents on two-lane roads in small cities. They found that the length of the stretch, the speed estimation and the number of speed differences of more than 10 km/h lead to an increase in the number of accidents to a large extent. The Handbook of Highway Safety /published in the United States/ lists the safety features of roads [16], which are divided: two-lane roads in small communities with two-way traffic, those with multi-lane roads, and urban and suburban primary roads. Safety performance functions can provide the expected total crash frequency for the baseline conditions of each section. The latter can be measured with certain metrics from the geometric design of the roadway network, its traffic flow, and its control functions [14]. Researchers have used different approaches to establish the relationship between crash frequency, geometric characteristics and traffic variables using statistical multiple linear regression models [5, 9]. [11] examined why multiple linear regression is not appropriate for modeling traffic accident frequency. The assumption in these models is that the frequency distribution of observations should be uniformly distributed. In most accident frequency data, the mean is lower than the standard deviation, i.e. we have overdispersion [17]. If there is overdispersion in the accident data, then it would be appropriate to use models that account for it.

2.2 Hot spots of accidents

There are 12 techniques for analysing accident hotspots. They mainly classify locations with potential safety problems. The criteria are: average, critical and predicted accident frequency, level of safety control. Methodologies exist in quite a few countries that have formulated them and complement these twelve techniques. A significant body of work on accident hotspot identification uses a variety of identification and selection methodologies, e.g., sliding scale analysis, empirical data analysis, basic data evaluation, Moran's Index I method, and [5, 8, 10, 15, 17]. The most accurate technique for finding the hot spots can be chosen based on the following considerations: accounting for regression to the mean bias and estimating the accident threshold level. The empirical research method in identifying accident hot spots is among the standard and available techniques.

2.3 Methology for optimising sifety sistem management
Selecting safety projects within a limited budget is an important element of vehicle traffic planning. Accident "hot spots" need to be identified for one main reason, namely that not all of these locations can be selected to implement safety measures due to rather limited resources. Within the budget, in order to prioritise projects, it is necessary to identify the best set of locations with many accidents. The traffic safety management system is a multi-objective optimization problem [9]. First, the total frequency of accidents within the limited resources should be minimized; Second, fatal and injury accidents should be minimized; Third, high-traffic roadways should be prioritized when selecting safety projects. Optimization in most cases involves minimizing the predicted crash frequency and involves a set of variables in selecting solutions, the same having different constraints - risk, budget, etc. [13, 18]. Optimization techniques in traffic safety management system include linear and integer programming.

3. **Modelling methodology**

The identification of accident-saturated locations requires an analysis of their data, which is followed by the identification of appropriate solution types (fig.1). The basic parameters of this crash identification model consist of two main steps: identification of critical crash points and potential countermeasures and funding allocation.

3.1 Identification of accident hotspots
Traffic accidents are random and infrequent with a tendency to occur in certain locations. The map of the same shows the clustering characteristics of crash occurrence. Weather and road conditions, light levels and the horizontal position of the lanes are factors that contribute to the occurrence of road accidents. The crash frequency in the study was calculated for individual sections using data for a two-year period (2021-2022). Crash frequency was reduced to one kilometre to allow comparison between sections. The empirical study method was used to identify accident "hot spots". Estimated crashes were calculated for two-lane two-way roads using a safety performance function. The study looked at reducing crash frequency and reducing crash-related fatalities to determine the best combination of designs to increase traffic safety.

3.2 Allocation of funding - strategy
Once the 'tipping points' of accidents have been identified, the location where they occur needs to be assessed to determine safety measures. Possible safety improvement measures with limited budgets and their associated one-off costs for asphalt surfaced primary roads are summarised in a list and apply to motorways only (Table 1).

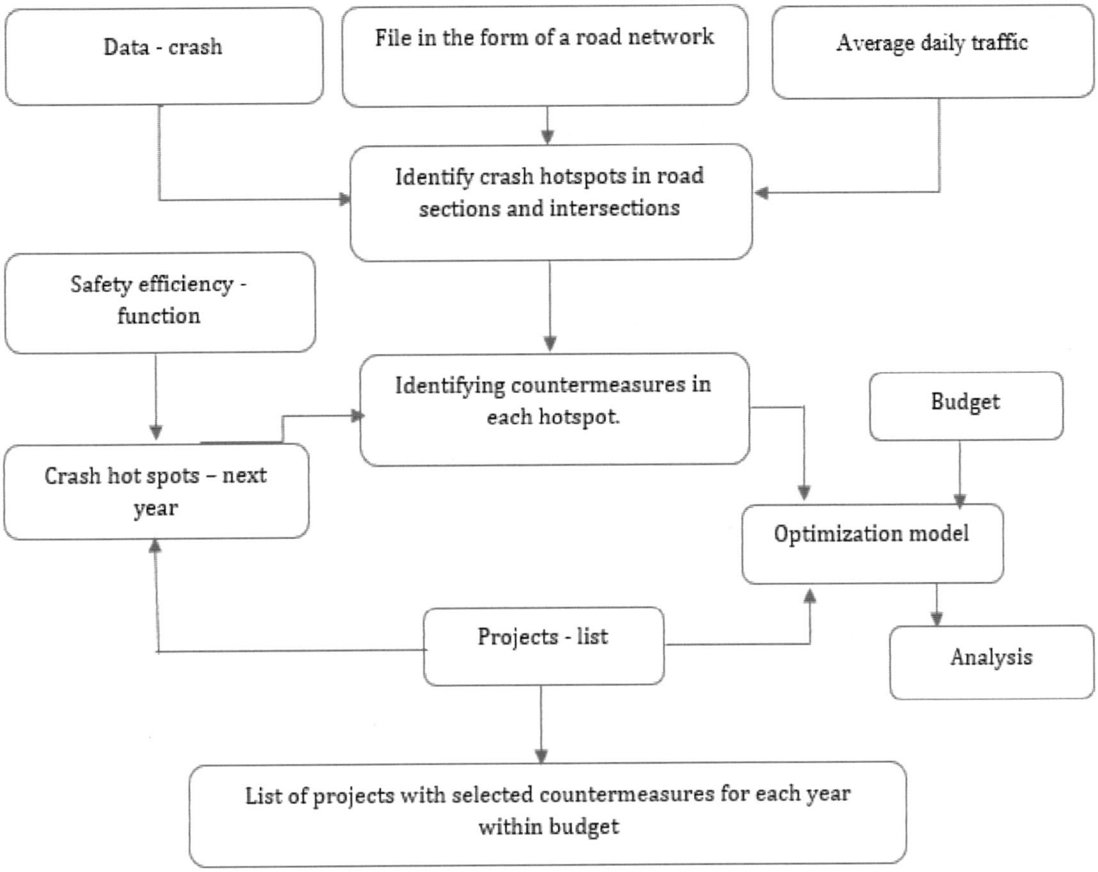

Fig. 1 Traffic safety management system research methodology

Table 1 Crash reduction factor (CRF) and cost of safety countermeasures for paved roads in one area

Countermeasures	Unit cost (lev)	CRF (%)
Install guide signs (general)	200	15
Install advance warning signs (positive guidance)	200	40
Install curve advance warning signs	200	30
Install delineators (general)	250	11
Install delineators (on bridges)	150	40
Install center line markings	7.2 per LM	33
Improve sight distance to intersection	15.5 per LM	56
Install guardrail (at bridge)	60 per LM	22
Install guardrail (outside curves)	30 per LM	63
Lengthen culvert	150 per LM	40

LM is linear meter

These measures are inexpensive in real terms and in most cases, in combination with other larger road rehabilitation projects, lead to significant improvements in traffic safety. For each individual section, the best measure suitable for that section is selected on the basis of cost, in accordance with the total project budget allocated to safety improvements. The specified optimization methodology is one in which the objective function minimizes the predicted crash frequency by selecting the best combination of road safety improvement measures within the selected project budgets, selecting the one with the highest average daily traffic value.

4. Optimization methodology

The proposed traffic safety management system for asphalt-surfaced roads meets the conditions in terms of accident frequency and average daily traffic. The objective of the developed methodology is to minimize the overall crash forecast in high traffic volume sections by prioritizing the sections where crash fatalities and injuries have occurred.

The mathematical model is described in the form of an equation.

$$\begin{cases} minimize\ \sum_{i=1}^{n} A_i \\ minimize\ \sum_{i=1}^{n} A_{f\&I_i} \end{cases} \quad (1)$$

where A_i and $A_{\mathrm{f\&I_i}}$ represent the predicted crashes and same with fatalities and injuries occurring on road *i*, respectively. This is a combinatorial optimization problem where a set of designs with minimum values must be chosen while satisfying some constraint. The predicted accidents are those of the section if it is selected for improvements multiplied by an accident reduction factor. The best combination to maximize traffic safety is selected using linear programming methods. Since the model is multistage, two objective functions are considered (Equation 2). The latter can be expressed by integer linear programming and is written as follows:

$$\begin{cases} minimize\ \sum_{i=1}^{n} A_i \\ minimize\ \sum_{i=1}^{n} A_{\mathrm{f\&I_i}} \\ complying\ with\ \left(\sum_{i=1}^{n} costs\ of\ improving\ safety\ _i * x_i\right) \leq budget \\ x_i \in \{0.1\} \end{cases} \quad (2)$$

For x_i is an integer equal to 1 if the project is selected and 0 if it is not selected.

5. Survey and data collection

Table 2 summarizes the type, number, and characteristics of data collected for the study. Figure 2 depicts the major paved roadways in the state. The total length of roads in the state is 19932.10 km. Of these, 9 highways are 879.10 km long, 9 primary roads are 2975 km long, 44 secondary roads are 4035 km long, and tertiary roads are 12051 km long. First class roads, second class roads, third class roads and municipal arteries are not considered as they are not the subject of the study.

5.1 Accident data

Data on road traffic accidents in the study area were obtained from the National Institute of Statistics, using basic mass data. The baseline data contain information about the accident: location, type of accident, road surface defects, accident severity, accident route, location. To solve the optimization problem, the period during which the accident occurred is needed. Accident data from January 2021 to December 2022 was used to ensure that there were no major changes in the roadway geometry in the study areas.

5.2 Classification of roadways

The functional classification of paved roads was obtained from the National Institute of Statistics and the road company. Roads are classified as highways, first, second, and third class, and these are primary roads, while municipal, forest, agricultural, and suburban roads are secondary in terms of their use.

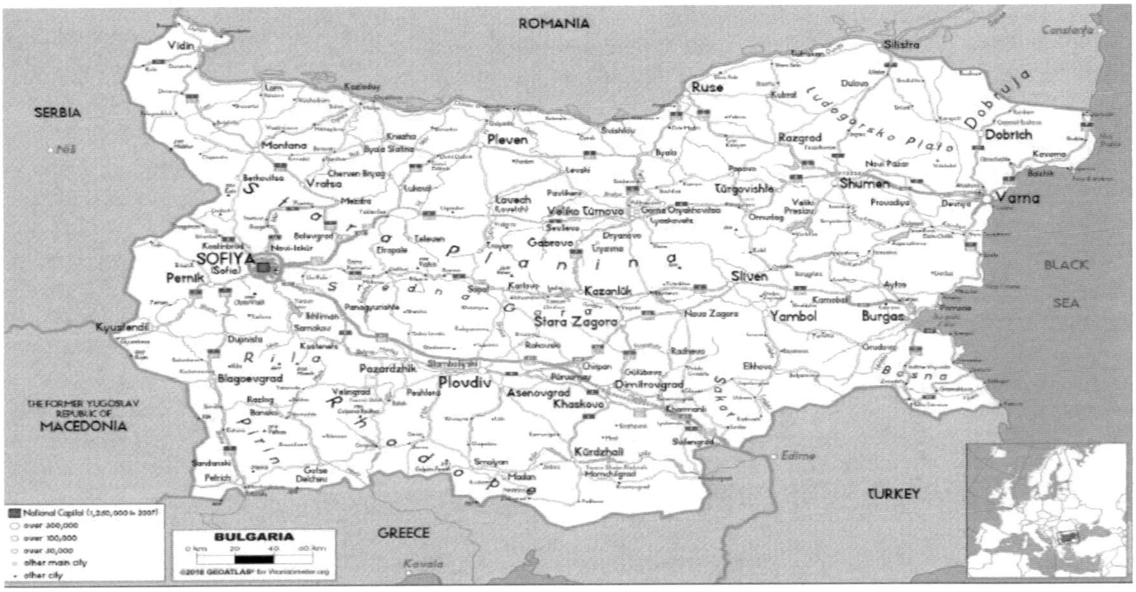

Fig. 2 Map of the main asphalt roads in Bulgaria

Table 2 Features and data collected for highways – basic

Feature	Data source	Quantity	Units	Data types
main asphalted roads	NSI	879.10	roads	layer of main paved roads
crash data	NSI	2	events	crash locations
functional classification	company Highways	7	segments	arterial
traffic counts	field	30	counts	average daily traffic

Table 3 shows the average daily traffic and Tables 4 and 5 below show that there is an increase in injury and fatal crashes in 2022 compared to 2021.

Table 3 Average daily traffic on selected segments from highways

Highways	Selected number of segments	ADT		
		maximum	minimum	average
A1-Trakia	6	2158	469	1912
A2-Hemus	5	3652	574	1045
A3-Struma	4	1886	293	1025
A4-Marica	4	1902	334	983
A5-Cherno more	4	985	308	678
A6-Evropa	4	1512	296	1017
A9-without name	3	785	188	532

Table 4 Crashes on the national road network – 2021 year

Class of road	2021						
	crash		killed		injured		Cost (lev)
	number	%	number	%	number	%	
Total	1936	100.0	351	100.0	2754	100.0	118100.00
highways	209	10.80	75	21.37	303	11.00	11700.00
first	601	31.04	122	34.76	847	30.76	55200.00
second	534	27.58	76	21.65	763	27.71	25100.00
third	592	30.58	78	22.22	841	30.53	25600.00

Table 5 Crashes on the national road network – 2022 year

Class of road	2022						
	crash		killed		injured		Cost (lev)
	number	%	number	%	number	%	
Total	2056	100.0	319	100.0	3074	100.0	227000.00
highways	202	9.82	32	10.03	314	10.22	13800.00
first	656	31.91	115	36.05	936	30.45	76300.00
second	596	28.99	88	27.59	961	31.26	70100.00
third	602	29.28	84	26.33	863	28.07	66800.00

Table 6 Deaths and injuries in crashes due to runway deficiencies – 2021 year

Violations	2021					
	RTA		killed		injured	
	number	%	number	%	number	%
Total	10	100.00	-	-	12	100.00
bumps in the roadway	1	10.00	-	-	1	8.30
road surface damage	1	10.00	-	-	1	8.30
damage to the banquettes/trotoirs	-	-	-	-	-	-
obstacles on the roadway	-	-	-	-	-	-
pavement not cleared of snow	-	-	-	-	-	-
pavement not cleared of ice	-	-	-	-	-	-
roadway not drained	-	-	-	-	-	-
other road deficiencies	8	80.00	-	-	10	83.40

Disaggregated Tables 6 and 7 focus on fatal and serious crashes due to pavement deficiencies in the respective two years.

Table 7 Deaths and injuries in crashes due to runway deficiencies – 2022 year

Violations	2022					
	RTA		killed		injured	
	number	%	number	%	number	%
Total	6	100.00	-	-	7	100.00
bumps in the roadway	1	16.70	-	-	1	14.30
road surface damage	1	16.70	-	-	1	14.30
damage to the banquettes/trotoirs	-	-	-	-	-	-
obstacles on the roadway	1	16.70	-	-	2	28.60
pavement not cleared of snow	-	-	-	-	-	-
pavement not cleared of ice	1	16.70	-	-	1	14.30
roadway not drained	-	-	-	-	-	-
other road deficiencies	2	33.20	-	-	2	28.50

Tables 8 and 9 show the road lighting and the time of day at which the accidents occurred. Based on available data for 2022 (no data for 2021) and it is evident that the greatest number of accidents occur during daylight hours when traffic on paved roads is most intense.

Table 8 Road crashes, deaths and injuries by road lighting – 2022 year (n.)

Lighting the way	RTA	killed	injured
Total	6609	531	8422
daylight	4954	332	6202
dim light	170	21	207
dark time of day	729	123	1038
artificial light	756	55	975

Table 9 Road accidents, deaths and injuries by hours – 2022 year (n.)

Time of day	RTA	killed	injured
Total	6609	531	8422
0-2	173	20	254
2-4	113	11	161
4-6	121	21	172
6-8	427	42	561
8-10	664	42	792
10-12	842	49	1031
12-14	831	52	1037
14-16	813	75	1053
16-18	1020	63	1271
18-20	811	67	1026
20-22	462	51	589
22-24	332	38	475

Traffic counts are not available for all roads, but are for major paved roads. It has been found that there is a significant difference in average daily traffic between the motorways and the primary road network. From selected data at thirty counts it is evident that there is a 6.2% increase in traffic accidents on paved roadways, comparing 2021 to 2022. To determine the best mix of projects resulting in improved traffic safety it is important to determine for each segment, along its length, which is saturated with accidents. To achieve this goal, the 7 highways and or segments currently in operation were used. The study focused on fatal and injury crashes.

6. Data analysis

In the data analysis section, the analysis is summarized in three parts crash hotspots, optimization process, and sensitivity. The optimization process then identifies projects among the selected crash hotspots within the estimated budget that could be allocated for highway safety improvements. Finally, a sensitivity analysis is performed to identify the critical budget that provides the greatest benefit to the public.

6.1 Critical points of crashes

Table 10 Low-cost safety countermeasures

ID Countermeasure	Countermeasures	Crash reduction factors (%)	Cost (lev)
1	install guide signs (general)	15	25400
2	install advance warning signs (positive guidance)	40	24600
3	install chevron signs on horizontal curves	35	31500
4	install curve advance warning signs	30	15400
5	install delineators (general)	11	16400
6	install delineators (on bridges)	40	2700

The table 10 shows the cheapest systems (according to the prices of materials and work) for improving safety for motorways in Bulgaria. In order to identify potential safety improvements in the future it will be necessary to visit the critical points on site. The method empirical investigation was applied to identify accident locations. Table 11 shows a list of the points with the most crashes on the highways. The expected crashes for this were calculated using SPF (protection factor) obtained from HSM (high speed memory) for a two-lane two-way road. In table 12, the last column is the performance index, which represents the increase in actual crashes compared to expected crashes and is greater than 1. There are a total of 21 hot spots identified along the entire length of the arterial roads due to their same currently being 879.10 km. Of this number, 7 have been investigated. Table 12 indicates which measures have been applied to certain critical points on the roads considered.

Table 11 Combined dataset for implementing TSMS model

Route	Beg. km.post	End km.post	Crash freq.	Crash frequency per km	Functional class	Countermeasure	CRF	Cost (lev)
A1-Trakia	16.000	16.999	5	48	0	3	0.30	60000
A2-Hemus	70.000	70.999	6	37	0	5	0.40	60000
A3-Struma	27.000	27.999	3	19	0	2	0.40	80000
A4-Marica	10.000	10.999	3	16	0	4	0.11	27500
A5-Cherno more	6.000	6.999	1	11	0	6	0.33	2376
A6-Evropa	36.000	36.999	3	8	0	1	0.15	30000
A9-without name	3.000	3.999	2	5	0	6	0.33	2376

The table 10 shows the cheapest systems (according to the prices of materials and work) for improving safety for motorways in Bulgaria. In order to identify potential safety improvements in the future it will be necessary to visit the critical points on site. The method empirical investigation was applied to identify accident locations. Table 11 provides a list of the locations with the most highways crashes, with their functional class noted as 0. The expected crashes for this were calculated using SPF (protection factor) obtained from HSM (high speed memory) for a two-lane two-way road. In table 12, the last column is the performance index, which represents the increase in actual crashes compared to expected crashes and is greater than 1. There are a total of 21 hot spots identified along the entire length of the arterial roads due to their same currently being 879.10 km. Of this number, 7 have been investigated. Table 12 indicates which measures have been applied to certain critical points on the roads considered.

Table 12 Selected critical crash spots using empirical investigation method

Route	Beg. km.post	End km.post	Crash freq.	Crash frequency km	Fatal and injury	ADT	Expected crashes (π)	Index of effectiveness (σ)
A1-Trakia	16.000	16.999	5	48	1	1912	2.20	1.52
A2-Hemus	70.000	70.999	6	37	1	1045	2.03	1.22
A3-Struma	27.000	27.999	3	19	0	1025	1.73	1.96
A4-Marica	10.000	10.999	3	16	0	983	1.49	2.29
A5-Cherno more	6.000	6.999	1	11	0	678	1.23	2.01
A6-Evropa	36.000	36.999	3	8	0	1017	1.12	2.18
A9-without name	3.000	3.999	2	5	0	532	0.99	2.31

The optimization model proposed in this study was used to select the best combination of safety improvement designs. For each critical point, the expected accidents were determined by multiplying the CRF and the accidents that occurred. The objective was to minimize the total expected number of accidents by selecting one of the projects after applying the safety countermeasures within a minimum budget.

Table 13 Selected safety enhancement projects for which a cost of 175000 leva is foreseen

Route	Beg. km.post	End km.post	Crash freq.	Fatal and injury	ADT	Countermeasure	CRF	Cost (lev)
A1-Trakia	$16^{.000}$	$16^{.999}$	5	1	1912	5	0.11	16400
A2-Hemus	$70^{.000}$	$70^{.999}$	6	1	1045	4	0.30	15400
A3-Struma	$27^{.000}$	$27^{.999}$	3	0	1025	4	0.30	15400
A4-Marica	$10^{.000}$	$10^{.999}$	3	0	983	2	0.40	24600
A5-Cherno more	$6^{.000}$	$6^{.999}$	1	0	678	5	0.11	16400
A6-Evropa	$36^{.000}$	$36^{.999}$	3	0	1017	2	0.40	24600
A9-without name	$3^{.000}$	$3^{.999}$	2	0	532	5	0.11	16400

At present, funds below the critical minimum of 100000 leva are allocated to improve traffic safety. We assume that the minimum budget that can be spent on motorways is BGN 120000. The optimization will produce the list of projects shown in Table 10. These countermeasures will lead to a reduction in crashes from 23 to 6, i.e. 18 fewer for the sections of motorway under consideration.

6.2 Analysis

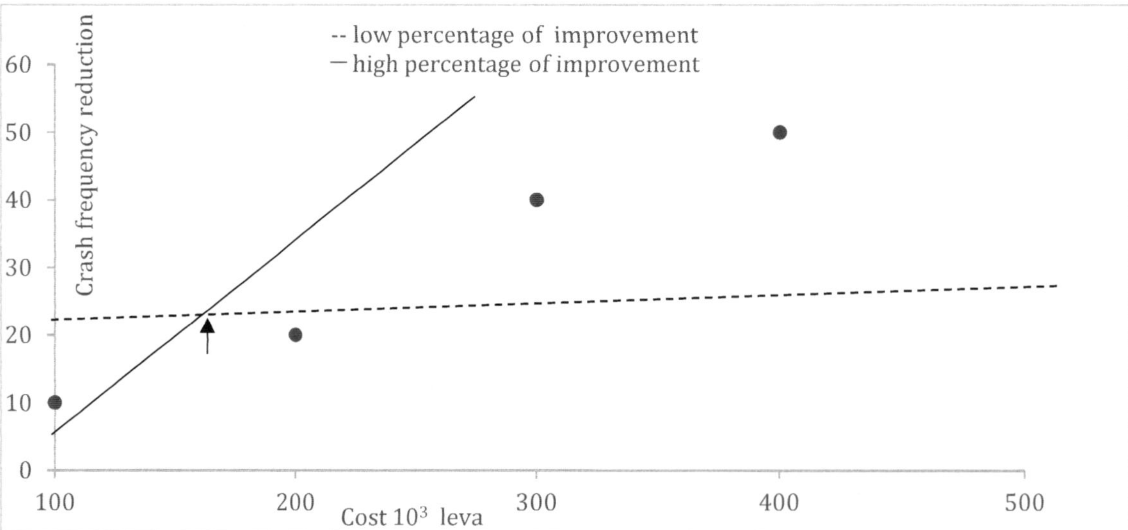

Fig. 3 TSMS submissions for a different budgets

When decisions are made, the budget should be set in such a way that it brings maximum benefit to the population. The study identified an appropriate budget depending on the financial conditions in the country and the expected reduction in the number of crashes. The optimization model was tested at different budget values (between 100000 and 425000 BGN). It can be seen from Figure 3 that the slope of the expected reduction in the number of crashes is greater when the budget is between 100000 and 175000 BGN than between 175000 and 425000 BGN. It follows that 175000 is the appropriate funding based on the optimization model applied.

7. Conclusion

The study developed an optimization methodology by determining the best combination of measures, under limited resources, to improve traffic safety on asphalt roads. The developed methodology is applied to the 879.10 km of highways in Bulgaria. The applied methodology minimized the total number of crashes expected by selecting a combination of measures to increase traffic safety. The sensitivity analysis identifies the most appropriate budget to ensure maximum utility for the population.

The developed methodology can be highlighted as follows:

- ✓ Suitable for arterial roadways;
- ✓ CRF countermeasures, their cost and budget for implementation, functional classification of roads are considered;
- ✓ The same prioritizes roads with high ADT and functional classification;
- ✓ It identifies the best set of projects leading to increased traffic safety;
- ✓ It minimizes expected crashes when a specific budget is limited;
- ✓ Determines the minimum budget needed to maximize human benefits in terms of crash reduction;
- ✓ Identifies critical crash points and appropriate safety improvement measures.

This paper demonstrates that limited budgets can be allocated in an economical and efficient method, to justify spending to increase safety and as a consequence to maximize crash reduction.

Acknowledgments

The scientific publication is in fulfillment of the objectives of the National Scientific Program "Security and Defense", financed by the Ministry of Education and Science of the Republic of Bulgaria in implementation of the Decision of the Council of Ministers No. 731 of October 21, 2021 and according to Agreement No. D01-74 /19.05.2022.

References

[1] Abdel-Aty, M.A., Radwan, E.A., 2000. Modeling traffic accident occurrence and involvement. Accident Analysis & Prevention 32 (5), 633-642.

[2] Ahmed, M., Huang, H., Abdel-Aty, M., et al., 2011. Exploring a Bayesian hierarchical approach for developing safety performance functions for a mountainous freeway. Accident Analysis & Prevention 43 (4), 1581-1589.

[3] Anderson, T., 2009. Kernel density estimation and K-means clustering to profile road accident hotspots. Accident Analysis & Prevention 41 (3), 359-364.

[4] Cafiso, S., Di Draziano, A., Di Silvestro, G., et al., 2010. Development of comprehensive accident models for two lane rural highways using exposure, geometry, consistency and context variables. Accident Analysis & Prevention 42 (4), 1072-1079.

[5] Cheng, W., Washington, S., 2008. New criteria for evaluating methods of identifying hot spots. Transportation Research Record 2083, 76-85.
[6] Chin, H., Quddus, M.A., 2003. Applying the random effect negative binomial model to examine traffic accident occurrence at signalized intersections. Accident Analysis & Prevention 35(2), 253-259.
[7] Elvik, R., 2008. Comparative analysis of techniques for identifying locations of hazardous roads. Transportation Research Record 2083, 72-75.
[8] ESRI, 2010. How Spatial Autocorrelation: Moran's I (Spatial Statistics) Works Available at: http://resources.esri.com/help/9.3/ArcGISengine/java/Gp_ToolRef/Spatial_Statistics_tools/how_spatial_autocorrelation_colon_moran_s_i_spatial_ statistics_works.htm (Accessed 25 January 2012).
[9] Getis, A., Ord, J.K., 1992. The analysis of spatial association by use of distance statistics. Geographical Analysis 24 (3), 189-206.
[10] Hauer, E., Allery, B.K., Kononov, J., et al., 2004. How best to rank sites with promise. Transportation Research Record 1897, 48-54.
[11] Jovanis, P., Chang, H., 1986. Modeling the relationship of accidents to miles traveled. Transportation Research Record 1068, 42-51.
[12] Miaou, S.P., Lord, D., 2012. Modeling traffic crash-flow relationships for intersections. Transportation Research Record 1840, 31-40.
[13] Mihova, V., Georgiev, I., Raeva, E., Georgiev, S., Pavlov, V., 2023. Validation of Stock Price Prediction Models in the Conditions of Financial Crisis, MDPI 12(1), 1-18.
[14] Mishra, S., Golias, M.M., Sharma, S., et al., 2015. Optimal funding allocation strategies for safety improvements on urban intersections. Transportation Research Part A: Policy and Practice 75, 113-133.
[15] Montella, A., 2010. A comparative analysis of hotspot identification methods. Accident Analysis & Prevention 42(2), 571-581.
[16] Persuad, B., Lyon, C., Nguyen, T., 1999. Empirical Bayes procedure for ranking sites for safety investigation by potential for safety improvement. Transportation Research Record 1665, 7-12.
[17] Saha, P., Ksaibati, K., 2015. A risk-based optimization methodology for managing county paved roads. International Journal of Pavement Engineering 17 (10), 913-923.
[18] Todorov, V., Georgiev, Sl., Dimov, I., 2023. New "Walk on Equations" Monte Carlo Algorithm for Linear Systems. Journal of Physics Conference Series 2675(1).

Group Classification and Numerical Study of Gross-Pitaevskii Systems

Yuri Bozhkov[1], **Stylianos Dimas**[2] **and Michail Todorov**[3]

[1]Institute of Mathematics, Statistics and Scientific Computing, University of Campinas – UNICAMP, 651 Sergio Buarque de Holanda str., CEP 13083-859, Campinas-SP, Brazil
[2]Department of Mathematics, Technological Institute of Aeronautics – ITA, 50 Marechal Eduardo Gomes Sq., Vila das Acacias, CEP 12228-900, Sao Jose dos Campos-SP, Brazil
[3]Institute of Mathematics and Informatics – Bulgarian Academy of Sciences, Akad. G. Bonchev str., bl.8, 1113 Sofia, Bulgaria

E-mail: bozhkov@unicamp.br; sdimas@ita.br; mdtodorov@math.bas.bg

Abstract. We carry out a preliminary group classification for the 3+1 Gross-Pitaevskii system and the complete group classification for the 1 + 1 integrable component of the Gross-Pitaevskii system. For the exceptional potentials that the group classification sheds light a numerical study was performed.

1 Introduction

We begin our study considering the Gross-Pitaevskii system in 3 + 1 dimensions,

$$\begin{cases} \mathrm{i}\psi_t = \beta\Delta\psi + \left[\alpha_1|\psi|^2 + (\alpha_1 + 2\alpha_2)|\phi|^2\right]\psi + V(x,y,z)\psi, \\ \mathrm{i}\phi_t = \beta\Delta\phi + \left[(\alpha_1 + 2\alpha_2)|\psi|^2 + \alpha_1|\phi|^2\right]\phi + V(x,y,z)\phi, \end{cases} \tag{1}$$

where x, y, z and t are the independent real variables, $\psi = \psi(x,y,z,t), \phi = \phi(x,y,z,t)$ are the dependent complex-valued variables, $V = V(x,y,z)$ is the real-valued potential, β is the dispersion coefficient, α_1 describes the self-focusing of a signal for pulses in birefringent media and α_2 — called the cross-modulation parameter— governs the nonlinear coupling between the equations. When $V(x,y,z) \equiv 0$ the system is called the *nonlinearly coupled Schrödinger equations* (CNLSE) or the *vector Schrödinger equation.* When $\alpha_2 = 0$, there is no nonlinear coupling despite the fact that "cross-terms" proportional to α_1 appear in the equations, additionally when $V \equiv 0$ as well, system (1) turns into the *Manakov system.* Be that as it may, when $\alpha_2 \neq 0$ the system is nonintegrable [5, 16, 26]. Finally, when $\phi = \psi$ the system "collapses" to the renowned *nonlinear Schrödinger equation* (NLSE),

$$\mathrm{i}\psi_t = \beta\Delta\psi + \alpha|\psi|^2\psi + V(x,y,z)\psi. \tag{2}$$

The system was initially derived in the three-dimensional space independently by Gross [8] and Pitaevskii [21] to describe the behavior of Bose-Einstein condensates as well as optic pulse propagation. However, the majority of works in literature concerns the 1+1 case where potentials in the terms of elliptic or other periodic functions were typically used. Due to the big number of results concerning this model we only point out the works [6], on singular instability of exact stationary solutions, and [14, 15]

Content from this work may be used under the terms of the Creative Commons Attribution 4.0 licence. Any further distribution of this work must maintain attribution to the author(s) and the title of the work, journal citation and DOI.
Published under licence by IOP Publishing Ltd

Journal of Physics: Conference Series **2910** (2024) 012010 doi:10.1088/1742-6596/2910/1/012010

where both exact and numerical solutions to the Gross-Pitaevskii equation with periodic potentials have been studied.

Manakov, in [17], solved analytically the $1+1$ version of system (1) for the case $\alpha_2 = 0$, $\beta = 1/2$, $V(x) \equiv 0$ via *Inverse Scattering Transform*, hence generalizing an earlier result by Zakharov and Shabat [27] for the scalar cubic NLSE. However, system (1), as previously mentioned, in its generality loses its integrability, a fact that diminishes the list of analytical tools available for its study, while numerically-wise it presents unique challenges that one needs to overcome.

On the analytical shore, the *de facto* all-weather tool for studying differential equations is *modern group analysis* with Lie's group theory of differential equations as its beating heart. Sophus Lie during the second half of the 19th century developed the general theory of Lie transformation groups with the specific intent of analyzing differential equations. After a hibernation period of limited use throughout the first half of the 20th century, over the past seven decades, Lie group techniques have been remarkably successful in studying nonlinear ordinary and partial differential equations, independently of their integrability status. There is now an extensive body of work in this field, and many of the fundamental methods and applications can be found in the now classical books [9, 19] as well as in the three volume handbook [10].

The present work is shaped as follows: in Section 2 we introduce the main tools used in our analytical study — that is the group classification of system (1) — adapted to the problem at hand, while the next 3 sections contain the main body of results — both analytical and numerical. Finally, we close the present work with some concluding remarks.

2 Methodology

In the heart of our analysis resides the concept of *symmetry*. Symmetry, loosely put, is a transformation — a *diffeomorphism* — between elements of a differential equation that leaves it invariant. These symmetries form a *group*, and for our purposes we will restrict ourselves to its connected component, which forms a *Lie group*. This means that the kind of symmetries that we shall use will depend on a continuous variable, ε, and we shall identify the symmetry that corresponds to $\varepsilon = 0$ with the identity transformation — which is a symmetry of any differential equation. But this is not the only reason for making this assumption: we can now elegantly represent any such symmetry with an element of a *Lie algebra*, that is, with a special type of *vector space*.

There is one more restriction to make: we shall assume that the transformations, our symmetries, involve only the independent and dependent variables. In other words we shall work with *Lie point symmetries*. In particular, for the study of (1) — where the dependent functions are complex functions — we shall work with symmetries of the form

$$\mathcal{X} = \sum_{i=1}^{3} \xi^i(\mathbf{x}, \mathbf{u}, \mathbf{u}^*)\frac{\partial}{\partial x^i} + \sum_{i=1}^{2} \eta_{u^i}(\mathbf{x}, \mathbf{u}, \mathbf{u}^*)\frac{\partial}{\partial u^i} + \sum_{i=1}^{2} \eta_{u^i}(\mathbf{x}, \mathbf{u}, \mathbf{u}^*)^*\frac{\partial}{\partial u^{i^*}}, \tag{3}$$

where $\mathbf{x} = (x^1, x^2, x^3) = (x, y, z)$, $\mathbf{u} = (u^1, u^2) = (\psi, \phi)$ and $\mathbf{u}^* = (u^{1^*}, u^{2^*}) = (\psi^*, \phi^*)$ the conjugates of $\mathbf{u}$. Having a symmetry as a element of a Lie algebra of the form (3) we can retrieve the corresponding continuous transformation, an element of a (local) Lie group, by solving the following Cauchy problem

$$\begin{cases} \dfrac{\mathrm{d}}{\mathrm{d}\varepsilon}\bar{x}^i = \xi^i(\bar{\mathbf{x}}, \bar{\mathbf{u}}, \bar{\mathbf{u}}^*), & \bar{x}^i(0) = x^i,\ i = 1, 2, 3, \\ \dfrac{\mathrm{d}}{\mathrm{d}\varepsilon}\bar{u}^j = \eta_{\bar{u}^j}(\bar{\mathbf{x}}, \bar{\mathbf{u}}, \bar{\mathbf{u}}^*), & \bar{u}^j(0) = u^j,\ j = 1, 2. \end{cases}$$

For this reason, $\mathcal{X}$ is also called *infinitesimal generator*. We call this procedure *exponentiation*.

But how we shall obtain the symmetries of a differential equation? Under certain conditions that the differential equation must satisfy — and for the case of (1) there are indeed satisfied — $\mathcal{X}$ is a symmetry of (1) if, and only if, it satisfies the linearised symmetry conditions:

$$\begin{aligned} \mathcal{X}^{(2)}[\mathrm{i}\psi_t - \beta\Delta\psi - [\alpha_1\psi\psi^* + (\alpha_1 + 2\alpha_2)\phi\phi^*]\psi - V(x,y,z)\psi]\Big|_{(1)} &= 0, \\ \mathcal{X}^{(2)}[\mathrm{i}\phi_t - \beta\Delta\phi - [(\alpha_1 + 2\alpha_2)\psi\psi^* + \alpha_1\phi\phi^*]\phi - V(x,y,z)\phi]\Big|_{(1)} &= 0, \end{aligned} \tag{4}$$

where $\mathcal{X}^{(2)}$ is the second *prolongation* of the vector $\mathcal{X}$,

$$\begin{aligned}\mathcal{X}^{(2)} =&\mathcal{X} + \sum_{j=1}^{2}\sum_{i=1}^{3}\eta_{u^j}^{(i)}(\mathbf{x},\mathbf{u},\mathbf{u}^*)\frac{\partial}{\partial u^j_{x^i}} + \sum_{j=1}^{2}\sum_{i=1}^{3}\eta_{u^j}^{(i)}(\mathbf{x},\mathbf{u},\mathbf{u}^*)^*\frac{\partial}{\partial u^{j\,*}_{x^i}} + \sum_{k=1}^{2}\sum_{i=1}^{3}\sum_{j=1}^{3}\eta_{u^k}^{(i,j)}(\mathbf{x},\mathbf{u},\mathbf{u}^*)\frac{\partial}{\partial u^k_{x^ix^j}}\\ &+\sum_{k=1}^{2}\sum_{i=1}^{3}\sum_{j=1}^{3}\eta_{u^k}^{(i,j)}(\mathbf{x},\mathbf{u},\mathbf{u}^*)^*\frac{\partial}{\partial u^{k\,*}_{x^ix^j}},\end{aligned}$$

where

$$\begin{aligned}\eta_\alpha^{(i)}(\mathbf{x},\mathbf{u},\mathbf{u}^*) =&D_{x^i}\eta_\alpha - \sum_{j=1}^{3}\alpha_{x^j}D_{x^i}\xi^j, \ \ \alpha=\psi,\psi^*,\phi,\phi^*,\\ \eta_\alpha^{(i,j)}(\mathbf{x},\mathbf{u},\mathbf{u}^*) =&D_{x^j}\eta_\alpha^{(i)} - \sum_{k=1}^{3}\alpha_{x^ix^k}D_{x^j}\xi^k, \ \ \alpha=\psi,\psi^*,\phi,\phi^*,\end{aligned}$$

with D_r the total derivative with respect to r.

Considering $\mathbf{x},\mathbf{u}$ and the first order derivatives of $\mathbf{u}$ as independent variables, (4) breaks down to an overdetermined — in our case — system of (linear) partial differential equations which are called the *determining equations.* Its general solution will provide us with the general form of the Lie algebra of Lie point symmetries of (1).

Getting the Lie point symmetries of a differential equations is only the beginning. Like the DNA for a living organism, symmetries help us identify and unearth important properties, reduce the order and even obtain solutions. We have just skimmed the surface of this remarkably beautiful theory, for more details see also [2–4, 9, 10, 19, 20, 24].

Evidently, getting the Lie algebra from (4) on the one hand involves a great deal of copious and error-prone calculations and on the other hand is a completely algorithmic procedure. These facts render the use of *computer algebra systems* essential. For our purposes we employed the symbolic package SYM for Wolfram Mathematica^{TM}, [7].

2.1 Equivalence transformations

When the differential equation involves arbitrary elements, for instance parameters, we deal in fact with a family, or a set, of differential equations. For each element of this family, we can associate the Lie algebra of its Lie point symmetries. Due to the structural importance of symmetries, it merits our attention to try to classify the family with respect to the possible Lie algebras that can be admitted: enter the *group classification*!

An efficient way to perform the group classification of a family of differential equations is by employing its *equivalence transformations.* Equivalence transformations can be considered as *meta-symmetries* of the family, being defined as *transformations* that leave invariant this set of differential equations as a whole, and as such, they contain the Lie point symmetries of the family.

Hence, following the same reasoning as in the previous section we shall restrict ourselves to continuous equivalence transformations that involve only the independent and dependent variables. Therefore, the algorithmic procedure previously described remains the same with the only difference being that now the infinitesimal generator has the form

$$\begin{aligned}\hat{\mathcal{X}} =&\sum_{i=1}^{3}\xi^i(\mathbf{x},\mathbf{u},\mathbf{u}^*;\alpha_1,\alpha_2,\beta,V)\frac{\partial}{\partial x^i} + \sum_{j=1}^{2}\eta_{u^j}(\mathbf{x},\mathbf{u},\mathbf{u}^*;\alpha_1,\alpha_2,\beta,V)\frac{\partial}{\partial u^j}\\ &+\sum_{j=1}^{2}\eta_{u^j}(\mathbf{x},\mathbf{u},\mathbf{u}^*;\alpha_1,\alpha_2,\beta,V)^*\frac{\partial}{\partial u^{j*}} + \omega_{\alpha_1}(\mathbf{x},\mathbf{u},\mathbf{u}^*;\alpha_1,\alpha_2,\beta,V)\frac{\partial}{\partial\alpha_1}\\ &+\omega_{\alpha_2}(\mathbf{x},\mathbf{u},\mathbf{u}^*;\alpha_1,\alpha_2,\beta,V)\frac{\partial}{\partial\alpha_2} + \omega_{\beta}(\mathbf{x},\mathbf{u},\mathbf{u}^*;\alpha_1,\alpha_2,\beta,V)\frac{\partial}{\partial\beta}\\ &+\omega_{V}(\mathbf{x},\mathbf{u},\mathbf{u}^*;\alpha_1,\alpha_2,\beta,V)\frac{\partial}{\partial V},\end{aligned} \tag{5}$$

where $\alpha = \alpha(\mathbf{x}, \mathbf{u}, \mathbf{u}^*)$, $\alpha = \alpha_1, \alpha_2, \beta, V$, and the first prolongations of ω_α follow the formulas [11],

$$\begin{aligned}
\omega_\alpha^{(i)}(\mathbf{x}, \mathbf{u}, \mathbf{u}^*; \alpha_1, \alpha_2, \beta, V) =& D_{x^i}\omega_\alpha - \sum_{j=1}^{3} \alpha_{x^j} D_{x^i}\xi^j - \sum_{j=1}^{2} \alpha_{u^j} D_{x^i}\eta_{u^j} - \sum_{j=1}^{2} \alpha_{u^{j*}} D_{x^i}\eta^*_{u^j},\\
\omega_\alpha^{u^i}(\mathbf{x}, \mathbf{u}, \mathbf{u}^*; \alpha_1, \alpha_2, \beta, V) =& D_{u^i}\omega_\alpha - \sum_{j=1}^{3} \alpha_{x^j} D_{u^i}\xi^j - \sum_{j=1}^{2} \alpha_{u^j} D_{u^i}\eta_{u^j} - \sum_{j=1}^{2} \alpha_{u^{j*}} D_{u^i}\eta^*_{u^j},\\
\omega_\alpha^{u^{i*}}(\mathbf{x}, \mathbf{u}, \mathbf{u}^*; \alpha_1, \alpha_2, \beta, V) =& D_{u^{i*}}\omega_\alpha - \sum_{j=1}^{3} \alpha_{x^j} D_{u^{i*}}\xi^j - \sum_{j=1}^{2} \alpha_{u^j} D_{u^{i*}}\eta_{u^j} - \sum_{j=1}^{2} \alpha_{u^{j*}} D_{u^{i*}}\eta^*_{u^j}.
\end{aligned}$$

By construction, the equivalence transformations split the family of differential equations into *equivalence classes* and all the elements of a class will have the same Lie algebra of Lie point symmetries. As a result, through the equivalence classification of the family we achieve also a first take on its group classification. We choose as the *canonical representative* of each equivalence class the simplest possible differential equation — in form — of that class.

3 The Group Classification

We start the study of (1) by its group classification. The group classification method for differential equations or systems involving a number of arbitrary parameters and functions aims to find the Lie symmetry groups of these equations or systems and then determine specific subcases that can enlarge the symmetry group. In most cases, there is a geometrical or physical justification for considering these particular subcases - see [19], p.182. As previously mentioned, we shall first employ its equivalence transformations. The Lie algebra of equivalence transformations is

$$\begin{aligned}
&\left(\mathtt{f} + \mathtt{f}_4 y + \mathtt{f}_3 z - \frac{x}{2\alpha_1\beta}(\beta\mathtt{f}_1 - \alpha_1\mathtt{f}_2 - 2\beta\alpha_1\mathtt{f}_{11})\right)\frac{\partial}{\partial x}\\
&+\left(\mathtt{f}_8 - \mathtt{f}_4 x + \mathtt{f}_5 z - \frac{y}{2\alpha_1\beta}(\beta\mathtt{f}_1 - \alpha_1\mathtt{f}_2 - 2\beta\alpha_1\mathtt{f}_{11})\right)\frac{\partial}{\partial y}\\
&+\left(\mathtt{f}_9 - \mathtt{f}_3 x - \mathtt{f}_5 y - \frac{z}{2\alpha_1\beta}(\beta\mathtt{f}_1 - \alpha_1\mathtt{f}_2 - 2\beta\alpha_1\mathtt{f}_{11})\right)\frac{\partial}{\partial z}\\
&+\left(\mathtt{f}_7 - \frac{t}{\alpha_1}(\mathtt{f}_1 + 2\alpha_1\mathtt{f}_{11})\right)\frac{\partial}{\partial t} + \left(\mathtt{f}_{11} + \mathrm{i}\mathtt{f}_{12} - \frac{\mathrm{i}t}{2\alpha_2}\mathtt{f}_{10}\right)\psi\frac{\partial}{\partial\psi}\\
&+\left(\mathtt{f}_{11} - \mathrm{i}\mathtt{f}_{12} + \frac{\mathrm{i}t}{2\alpha_2}\mathtt{f}_{10}\right)\psi^*\frac{\partial}{\partial\psi^*} + \left(\mathtt{f}_{11} - \frac{\mathrm{i}t}{2\alpha_2}\mathtt{f}_{10} + \frac{\mathrm{i}}{\alpha_1}(\mathtt{f}_{13} + (\alpha_1 + 2\alpha_2)\mathtt{f}_{12})\right)\phi\frac{\partial}{\partial\phi}\\
&+\left(\mathtt{f}_{11} + \frac{\mathrm{i}t}{2\alpha_2}\mathtt{f}_{10} - \frac{\mathrm{i}}{\alpha_1}(\mathtt{f}_{13} + (\alpha_1 + 2\alpha_2)\mathtt{f}_{12})\right)\phi^*\frac{\partial}{\partial\phi^*} + \mathtt{f}_1\frac{\partial}{\partial\alpha_1} + \frac{\alpha_2}{\alpha_1}\mathtt{f}_1\frac{\partial}{\partial\alpha_2} + \mathtt{f}_2\frac{\partial}{\partial\beta}\\
&+\left(\frac{1}{2\alpha_2}\mathtt{f}_{10} + \frac{V}{\alpha_1}(\mathtt{f}_1 + 2\mathtt{f}_{11})\right)\frac{\partial}{\partial V},
\end{aligned} \tag{6}$$

where $\mathtt{f}_i = \mathtt{f}_i(\alpha_1, \alpha_2, \beta)$, $i = 1, 2, \ldots, 12, 13$ are arbitrary complex functions. To obtain the *principal algebra*, that is the maximal Lie algebra that all the possible cases will include, we need to apply the Lie algebra of equivalence transformations found to the relations $\alpha_1 = \text{const}$, $\alpha_2 = \text{const}$, $\beta = \text{const}$ and $V = \mathcal{F}(x, y, z)$:

$$\begin{aligned}
&\mathtt{f}_1 = 0,\\
&\mathtt{f}_2 = 0,\\
&\frac{1}{2\alpha_2}\mathtt{f}_{10} + 2\mathtt{f}_{11}\mathcal{F} - \mathtt{f}_6\mathcal{F}_x - \mathtt{f}_8\mathcal{F}_y - \mathtt{f}_9\mathcal{F}_z + x\left(\mathtt{f}_{11}\mathcal{F}_x + \mathtt{f}_4\mathcal{F}_y + \mathtt{f}_3\mathcal{F}_z\right)\\
&+ y\left(-\mathtt{f}_4\mathcal{F}_x + \mathtt{f}_{11}\mathcal{F}_y + \mathtt{f}_5\mathcal{F}_z\right) + z\left(-\mathtt{f}_3\mathcal{F}_x - \mathtt{f}_5\mathcal{F}_y + \mathtt{f}_{11}\mathcal{F}_z\right) = 0.
\end{aligned}$$

These equations for any given constant and function $\mathcal{F}$, yield

$$\mathtt{f}_1 = \mathtt{f}_2 = \mathtt{f}_3 = \mathtt{f}_4 = \mathtt{f}_5 = \mathtt{f}_6 = \mathtt{f}_8 = \mathtt{f}_9 = \mathtt{f}_{10} = \mathtt{f}_{11} = 0.$$

Thus, the principal algebra for equation (1) is spanned by the vectors

$$\hat{\mathcal{X}}_1 = \partial_t, \qquad \hat{\mathcal{X}}_2 = \mathrm{i}\partial_t, \qquad \hat{\mathcal{X}}_3 = \mathrm{i}\phi\partial_\phi - \mathrm{i}\phi^*\partial_{\phi^*}, \qquad \hat{\mathcal{X}}_4 = \mathrm{i}\psi\partial_\psi - \mathrm{i}\psi^*\partial_{\psi^*}.$$

Exponentiating now the part of the Lie algebra of equivalence transformations (5) that involves only the arbitrary functions $\mathtt{f}_1, \mathtt{f}_2, \mathtt{f}_{10}$ and $\mathtt{f}_{11}$ we get that under the equivalence transformation

$$x \mapsto \sqrt{\left|\frac{\alpha_1}{\beta}\right|}x, \qquad y \mapsto \sqrt{\left|\frac{\alpha_1}{\beta}\right|}y, \qquad z \mapsto \sqrt{\left|\frac{\alpha_1}{\beta}\right|}z, \qquad t \mapsto \alpha_1 t,\ \alpha_1 \neq 0$$

system (1) is equivalent to the system

$$\begin{cases} \mathrm{i}\psi_t = \operatorname{sign}\left(\frac{\alpha_1}{\beta}\right)\Delta\psi + \left[|\psi|^2 + \alpha|\phi|^2\right]\psi + \tilde{V}(x,y,z)\psi, \\ \mathrm{i}\phi_t = \operatorname{sign}\left(\frac{\alpha_1}{\beta}\right)\Delta\phi + \left[\alpha|\psi|^2 + |\phi|^2\right]\phi + \tilde{V}(x,y,z)\phi, \end{cases} \tag{7}$$

where $\alpha = 1 + 2\dfrac{\alpha_2}{\alpha_1}$ and $\tilde{V} = \dfrac{V}{\alpha_1}$. Similarly, if $\alpha_1 = 0$ we get that under the equivalence transformation

$$x \mapsto \sqrt{\left|\frac{2\alpha_2}{\beta}\right|}x, \qquad y \mapsto \sqrt{\left|\frac{2\alpha_2}{\beta}\right|}y, \qquad z \mapsto \sqrt{\left|\frac{2\alpha_2}{\beta}\right|}z, \qquad t \mapsto 2\alpha_2 t,\ \alpha_2 \neq 0$$

system (1) with $\alpha_1 = 0$ is equivalent to the system

$$\begin{cases} \mathrm{i}\psi_t = \operatorname{sign}\left(\frac{\alpha_2}{\beta}\right)\Delta\psi + |\phi|^2\psi + \tilde{V}(x,y,z)\psi, \\ \mathrm{i}\phi_t = \operatorname{sign}\left(\frac{\alpha_2}{\beta}\right)\Delta\phi + |\psi|^2\phi + \tilde{V}(x,y,z)\phi, \end{cases} \tag{8}$$

where $\tilde{V} = \dfrac{V}{2\alpha_2}$. When $\alpha_1 = \alpha_2 = 0$ we arrive to a decoupled system of Schrodinger equations.

Even though the use of equivalence transformations reduced the number of arbitrary constants we had to consider, unfortunately it did not help in classifying the function V. At this point is worth mentioning that due to its high dimensionality, in conjunction with the fact that for $\alpha \neq 1$ system (7) is nonintegrable, its numerical study is challenging since one must take care of the issues of computational complexity, stability and convergence, among others, that such high dimensional system gives rise to. For that reason, we shall carry on with the rest of the group classification by considering the positive $1+1$ case with $\alpha = 1$, that is, we shall consider the following Gross-Pitaevskii system:

$$\begin{cases} \mathrm{i}\psi_t = \psi_{xx} + \left(|\psi|^2 + |\phi|^2 + V(x)\right)\psi, \\ \mathrm{i}\phi_t = \phi_{xx} + \left(|\psi|^2 + |\phi|^2 + V(x)\right)\phi. \end{cases} \tag{9}$$

Similarly to the previous process, by obtaining the equivalence transformations of (9), we can find its principal algebra. Since we choose $\alpha = 1$ that symmetrized the system we expect an increase in the dimension of the principal algebra. The next step is to construct the system of determining equations of (9) using the linearized symmetry condition (4) and investigate particular forms of V that expand its solution set. We realize that by resolving the determining equations that do not involve the function V in order to arrive to the following *classification equations*:

$$c_5 V' + \frac{c_6}{2} = 0,$$
$$f_5' + (2c_6 t + c_7)V + f_6 V' + x\left((c_6 t + \frac{c_7}{2})V' - \frac{f_6''}{2}\right) = 0,$$

where $c_6, c_7, f_5(t), f_6(t)$ constants and functions of integration that arise during the resolution of the determining equations. By solving the above system we obtain the complete group classification of system (9). The result of this process can be expressed in the following theorem:

Theorem. *The principal symmetry algebra of system (9) is generated by the following vector fields:*

$$\mathcal{X}_1 = \partial_t, \mathcal{X}_2 = \mathrm{i}\{\psi\partial_\psi - \psi^*\partial_{\psi^*}\},\ \mathcal{X}_3 = \mathrm{i}\{\phi\partial_\phi - \phi^*\partial_{\phi^*}\},\ \mathcal{X}_4 = \phi\partial_\psi - \psi\partial_\phi + \phi^*\partial_{\psi^*} - \psi^*\partial_{\phi^*}, \\ \mathcal{X}_5 = \mathrm{i}\{\psi\partial_\phi - \psi^*\partial_{\phi^*} + \phi\partial_\psi - \phi^*\partial_{\psi^*}\}. \quad (10)$$

Furthermore, we have three special classes of real functions $V(x)$ that enlarge the principal algebra (10):

$\mathbf{V(x) = cx + d}$: *where c and d are real constants. The symmetry algebra is spanned by the principal algebra and the additional operators*

$$\mathcal{R}_1 = -\sqrt{2}t\partial_x + \mathrm{i}(ct^2 + \sqrt{2}x)\{\psi\partial_\psi - \psi^*\partial_{\psi^*}\} + \mathrm{i}(ct^2 + \sqrt{2}x)\{\phi\partial_\phi - \phi^*\partial_{\phi^*}\},$$

$$\mathcal{R}_2 = 4t\partial_t + (3\sqrt{2}ct^2 + 2x)\partial_x - 2(\psi\partial_\psi + \psi^*\partial_{\psi^*}) - \mathrm{i}(4dt + c^2t^3 + 6ct)\{\psi\partial_\psi - \psi^*\partial_{\psi^*}\} \\ -2(\phi\partial_\phi + \phi^*\partial_{\phi^*}) - \mathrm{i}(4dt + c^2t^3 + 6ct)\{\phi\partial_\phi - \phi^*\partial_{\phi^*}\},$$

$$\mathcal{R}_3 = 4ct\partial_t + (-4\sqrt{2}d + 3\sqrt{2}c^2t^2 + 2cx)\partial_x - 2c(\psi\partial_\psi + \psi^*\partial_{\psi^*}) - \mathrm{i}c^2t(-ct + 3\sqrt{2}x)\{\psi\partial_\psi - \psi^*\partial_{\psi^*}\} \\ -2c(\phi\partial_\phi + \phi^*\partial_{\phi^*}) - \mathrm{i}c^2t(-ct + 3\sqrt{2}x)\{\phi\partial_\phi - \phi^*\partial_{\phi^*}\}.$$

$\mathbf{V(x) = ax^2 + bx + c}$: *where $a \neq 0, b, c$ are real constants. The symmetry algebra is spanned by the principal algebra and the additional operators*

$$\mathcal{Y}_1 = \mathrm{i}(b + \sqrt{2}ax)\exp[-\sqrt{2a}t]\{\psi\partial_\psi - \psi^*\partial_{\psi^*} + \phi\partial_\phi - \phi^*\partial_{\phi^*}\} + 2\sqrt{a}\exp[-\sqrt{2a}t]\partial_x,$$

$$\mathcal{Y}_2 = -\mathrm{i}(b + \sqrt{2}ax)\exp[\sqrt{2a}t]\{\psi\partial_\psi - \psi^*\partial_{\psi^*} + \phi\partial_\phi - \phi^*\partial_{\phi^*}\} + 2\sqrt{a}\exp[\sqrt{2a}t]\partial_x.$$

$\mathbf{V = a(x + b)^{-2} + c}$: *where $a \neq 0, b, c$ are real constants. The symmetry algebra is spanned by the principal algebra and the additional operators*

$$\mathcal{Z} = 2t\partial_t + (x + b)\partial_x - (\psi + 2\mathrm{i}ct)\partial_\psi - (\psi^* - 2\mathrm{i}ct)\partial_{\psi^*} - (\phi + 2\mathrm{i}ct)p_\phi - (\phi^* - 2\mathrm{i}ct)\partial_{\phi^*}.$$

To summarize, the principal algebra of (9) is 5-dimensional and for the particular forms of V we illustrate in the theorem the Lie algebra can be enlarged to a 6,7 and 8-dimensional Lie algebra. The maximal dimension 8 is achieved in the linear case, which contains Manakov's system.

4 Lie point symmetries of Nonlinear Schrödinger Equations

In this section we briefly review some results concerning the Lie point symmetries of nonlinear Schrödinger equations.

The group classification of the generalized scalar NLSE

$$\mathrm{i}\psi_t + \Delta\psi + F(\psi, \psi^*) = 0 \quad (11)$$

for a complex function $\psi(x, t)$ of $n+1$ independent real variables was established by Nikitin and Popovich in [18]. Then, in [12], Ivanova studied the symmetries of the following scalar NLSE with harmonic oscillator type potential:

$$\mathrm{i}\psi_t + \Delta\psi + k|x|^2\psi - f(|\psi|)\psi = 0, \quad (12)$$

where k is a constant. Further, Popovich, Ivanova and Eshraghi performed in [22] the complete group classification of the $(1 + 1)-$dimensional Schrödinger equation involving a power nonlinearity

$$\mathrm{i}\psi_t + \Delta\psi + |\psi|^\gamma\psi + V\psi = 0, \quad (13)$$

where the potential $V = V(x, t)$ is a smooth complex-valued function and $\gamma \neq 0$ a real constant.

In regard to the vector NLSE

$$\mathrm{i}\Psi_t + \Delta\Psi = |\Psi|^2\Psi, \quad (14)$$

where Ψ is a complex-valued vector function, its Lie point symmetries were obtained in [23] by Sciarrino and Winternitz. In [13], Kodama and Mikhailov show, amongst other things, that the vector NLSE possesses an infinite dimensional algebra of nonlocal noncommutative symmetries. In [1], J. Belmonte-Beitia, F. Gungor, and P.J. Torres construct explicit solutions with nontrivial phase of systems of type (1) with spatially inhomogeneous nonlinearities.

5 Numerical results

5.1 *Posing the problem*

The corresponding vector N-soliton configuration — soliton train — is determined by the equation (9) under the initial condition

$$\vec{u}(x,t=0) = \sum_{k=1}^{N} k\vec{u}_k(x,t=0) \tag{15}$$

where $\vec{u}_k(x,t) = 2\nu_k e^{i\varphi_k} \mathrm{sech} z_k \vec{n}_k$, $z_k = 2\nu_k(x - \xi_k(t))$, $\varphi_k = \frac{\mu_k}{\nu_k} z_k + \delta_k(t)$, $\delta_k(t) = z(\mu_k^2 + \nu_k^2)t + \delta_{k,0}$, $\vec{u} = (\psi, \phi)^{\mathrm{T}}$, μ_k – amplitudes, ν_k – velocities, ξ_k – soliton centers, δ_k – phase shifts, $k = 1, ..., N$, N – number of solitons. In our present numerical study we have selected $N = 3$, $\xi_k = -8 + 8(k-1)$, $\nu_{1,3} = 0.5 \mp 0.005$, $\nu_2 = 0.5$, $\mu_k = 0$, $\delta_{1,0} = 0, \delta_{2,0} = \frac{\pi}{2}$ and $\delta_{3,0} = \frac{\pi}{36}$.

5.2 *Criterium for perturbation*

The effect of the external potential on the soliton train can be viewed as a perturbation, keeping the initial soliton structure of the solution. The unspecified choice of parameters leads inevitably to a violation of the initial soliton structure and turbulence. Let us consider the Hamiltonians of the LHS of the system, Eq.(9) and the external potential $V(x)$. They are as follows

$$H_0 = \int_{-\infty}^{\infty} \left[|\psi_x|^2 + |\phi_x|^2 - \frac{1}{2}(|\psi|^2 + |\phi|^2) \right] \mathrm{d}x$$
$$H_V = \int_{-\infty}^{\infty} V(x)(|\psi|^2 + |\phi|^2)\mathrm{d}x.$$

If the ratio $|H_V| \ll |H_0|$ then the initial soliton dynamics are determined.

Remark: H_0 and H_V depend on the soliton parameters. The leading term of H_0, however, depends only on the phase velocities ν_k and amplitudes μ_k, while H_V depends substantially also on the positions ξ_k of the solitons and their number.

5.3 *Effects of external potentials*

The numerical predictions for Eq.(9) for the six potentials of theorem 1 are implemented by a fully implicit difference scheme of Crank-Nicolson type, which conserves the energy, the mass, and the pseudomomentum. The scheme is implemented in the field of complex numbers [25].

(i) **Manakov model, $\mathbf{V(x) \equiv 0}$:** First we compute the 3-soliton Manakov model by the above described set of parameters. In Figures 1 and 2 are plotted trajectories of soliton envelopes up to times of 960 dimensionless units.

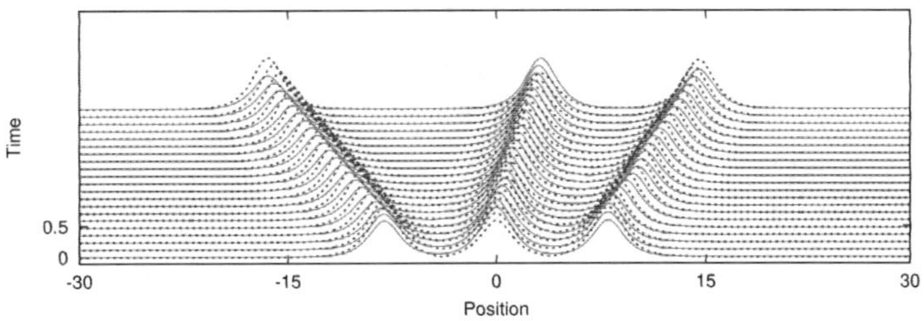

Figure 1: Dynamics of 3-soliton Manakov envelopes from Figure 2: red – $|\psi|$, blue – $|\phi|$.

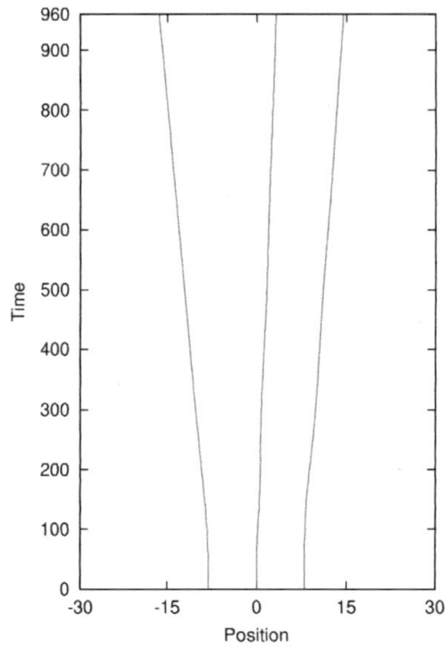

Figure 2: Trajectories of the centers of 3-soliton Manakov solution.

(ii) **Constant external potential $\mathbf{V(x)} = \text{const} \neq \mathbf{0}$:** We implemented two experiments - for $V(x) = 1$ and $V(x) = 10$. Because the differences with the Manakov system are insignificant — about 1% for the trajectories and 8% for the amplitudes — their graphs are qualitatively indistinguishable and will not be presented on separate graphs.

(iii) **Linear external potential $\mathbf{V(x)} = \mathbf{ax} + \mathbf{b}, \mathbf{a} \neq \mathbf{0}, \mathbf{b} \neq \mathbf{0}$:** Starting with $a = b = 1$ we established that these magnitudes of parameters cause rapidly a tearing down of the initial soliton structure. In order to keep the structure we repeated the experiment with $a = b = 0.0001$. Results are present in Figures 3 and 4 that follow.

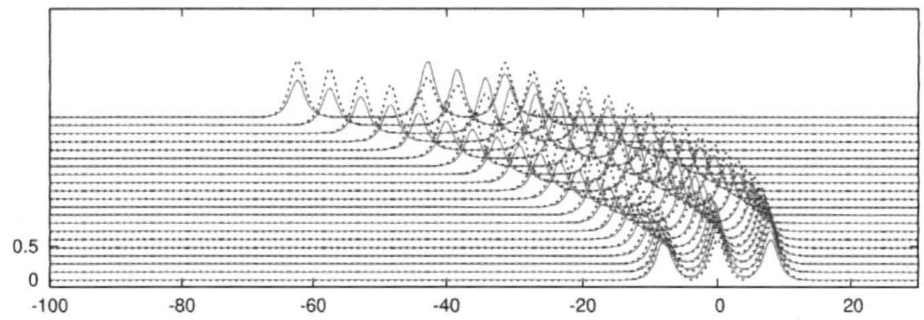

Figure 3: Dynamics of 3-soliton envelopes from Figure 4: : red – $|\psi|$, blue – $|\phi|$.

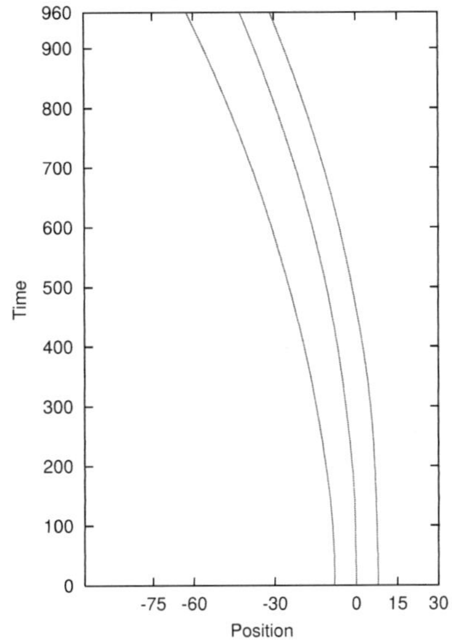

Figure 4: Trajectories of the centers of 3-soliton configuration with an external potential $V(x) = 0.0001(x+1)$ up to 960 dimensionless time units.

It is seen that the linear external potential preserves the amplitudes of the solitons and the polarization angle $\theta = \arctan\left(|\phi|_{\max}/|\psi|_{\max}\right)$. On the other hand, it changes the phase velocities whilst keeping the bound asymptotic state of the soliton train.

(iv) **External potential** $\mathbf{V(x) = a^2(x+b)^2, a \neq 0, b \neq 0}$**:** For this case, we choose $a = 0.01$ and $b = 1$. If a is small enough the initial 3-soliton structure is conserved keeping the amplitudes and polarization. The trajectories start to oscillate maintaining the bound asymptotic state (see Figures 5 and 6).

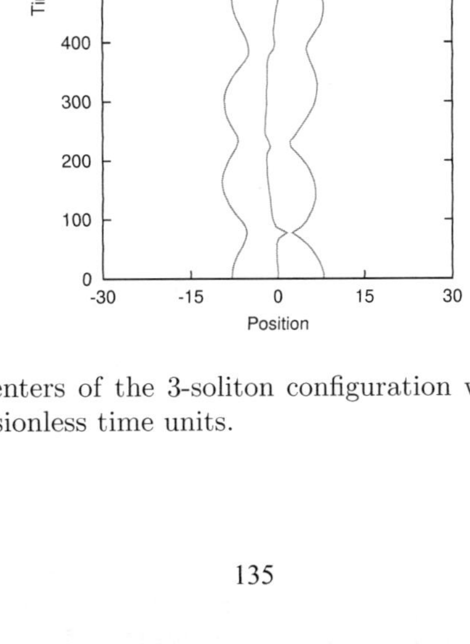

Figure 5: Trajectories of the centers of the 3-soliton configuration with an external potential $V(x) = 0.0001(x+1)^2$ up to 960 dimensionless time units.

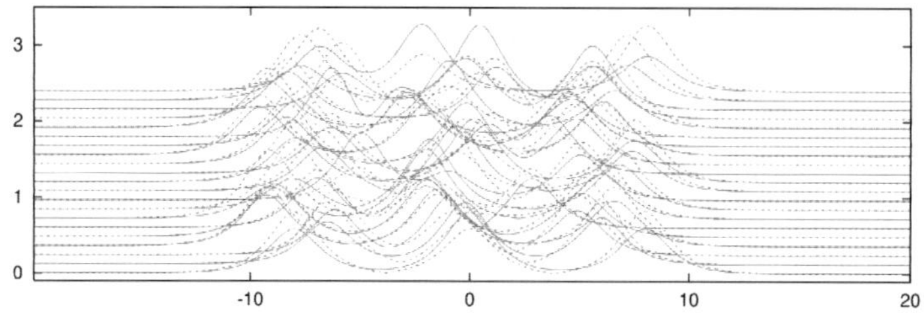

Figure 6: Dynamics of the 3-soliton configuration from Figure 5.

(v) **External potential $\mathbf{V(x) = -a^2(x+b)^2, a \neq 0, b \neq 0}$:** Similarly to the previous case we again choose $a = 0.01$ and $b = 1$. In this case the dynamics are more complicated. The solitons are freely propagating until reflected elastically in the left and right border at $x = -500$ and $x = 500$ (Figures 7 and 8).

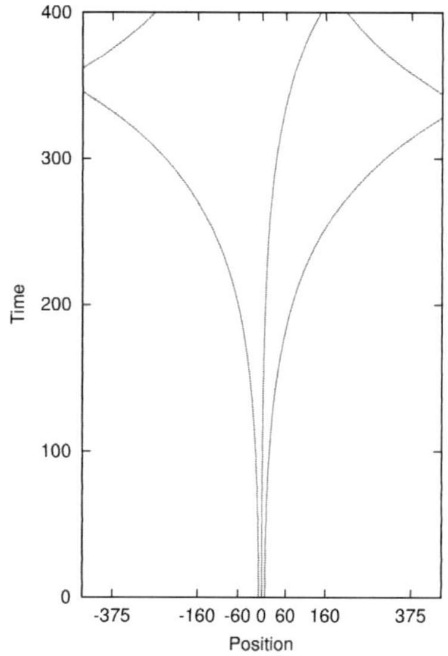

Figure 7: Trajectories of the centers of the 3-soliton configuration with an external potential $V(x) = -0.0001(x+1)^2$ up to 450 dimensionless time units.

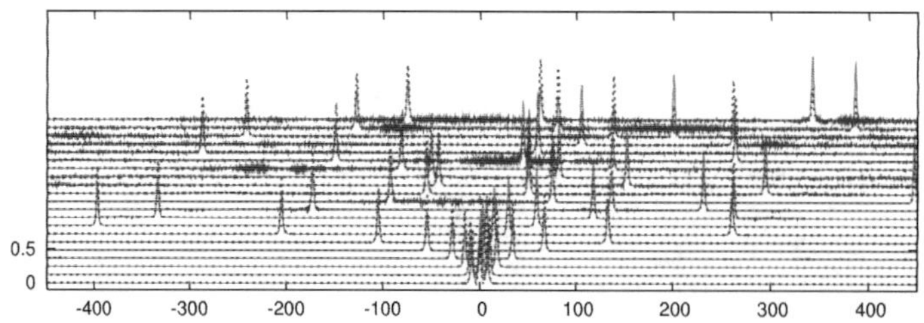

Figure 8: Dynamics of the 3-soliton configuration from Figure 7

(vi) **Rational external potential $\mathbf{V(x) = \frac{1}{(ax+b)^2}, a \neq 0, b \neq 0}$:** We choose $a = 1$, $b = 1.51$. The pattern is similar to the case *(iv)* if the point $x = -\frac{b}{a}$ does not belong to the computation interval.

Journal of Physics: Conference Series **2910** (2024) 012010 doi:10.1088/1742-6596/2910/1/012010

If it belongs to it (our consideration), the potential forms a peakon and functions as a barrier that elastically reflects the envelopes (see Figure 9). After reflection at the right boundary, the right reflex soliton interacts with itself (head-on collision).

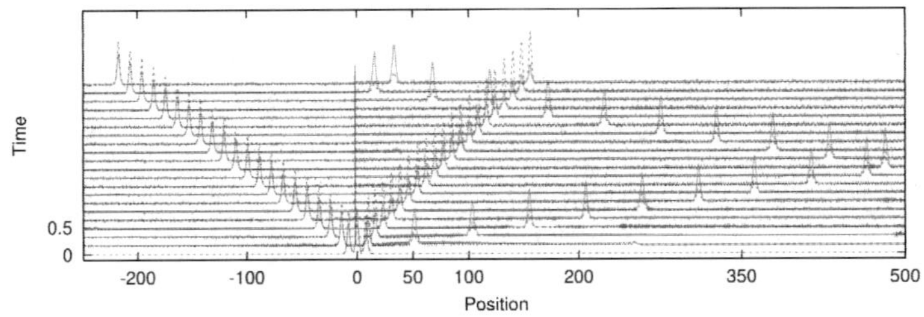

Figure 9: Dynamics of 3-soliton configuration with an external potential $V(x) = \frac{1}{(x+1.51)^2}$ and a barrier at $x = 1.5$.

As a supplement to the considered 6 external potentials, we illustrate 2 more 5-soliton configurations with quadratic potentials. They are in asymptotic bound states. In the second case they are oscillating as well.

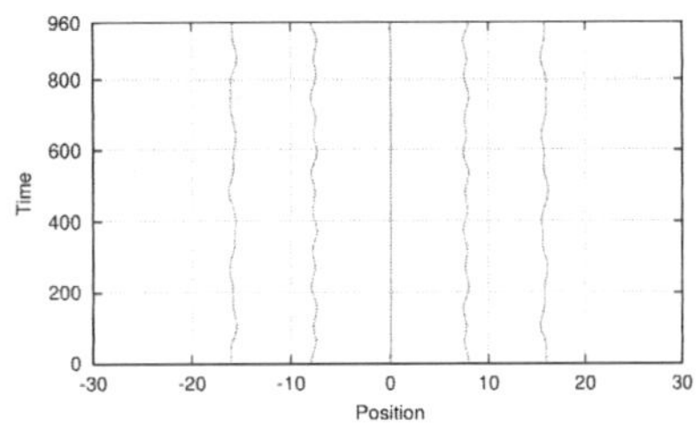

Figure 10: Five-soliton train with initial parameters $r_0 = 8$, $\xi_k(0) = -24+8k$, $k = 1, \ldots, 5$ with potential $V(x) = 0.000036x^2$.

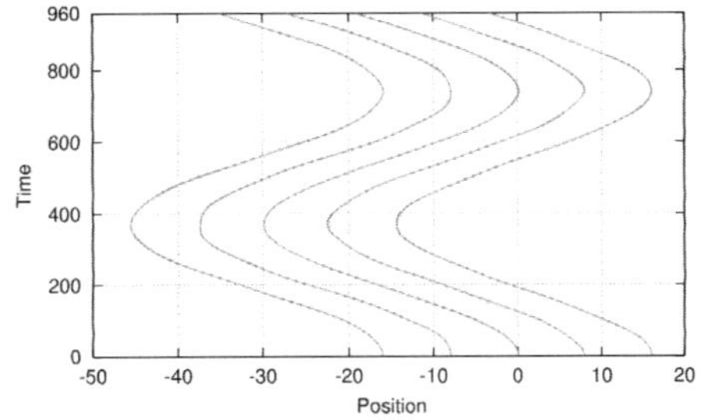

Figure 11: Five-soliton train with initial parameters $r_0 = 8$, $\xi_k(0) = -24 + 8k$, $k = 1, \ldots, 5$, $V(x) = 0.000036(x + 15)^2$.

6 Conclusion

In the present work we initiated the group classification of the Gross-Pitaevskii system in $3+1$ dimensions and we concluded the group classification in $1+1$ dimensions for the integrable cases only, its integrable component. Then, the exceptional cases found were studied numerically employing an implicit difference scheme of Crank-Nicolson type.

The reason for this transition was the inherent difficulties in studying high-dimensional nonintegrable systems numerically. However, it will be interesting, and helpful for numerical integration, to study the transition from integrable to nonintegrable, that is, $\alpha_2 = 0 \to \alpha_2 \ll 1$ using *approximate symmetries*. The success of this endeavor has the potential to lead to improved numerical schemes and will be explored in a future work.

Furthermore, in a second future work we shall complete the classification for $V(x, y, z)$ and we shall study how its symmetries can reduce the order of the system hence rendering it more tractable for a numerical study.

Acknowledgements

We would like to thank the São Paulo Research Foundation - FAPESP, Grant No. 2022/10525-2, for the financial support, which has given M. Todorov the opportunity to visit UNICAMP, where this work was initiated. Y. Bozhkov would also like to thank the São Paulo Research Foundation - FAPESP, Grant No. 2024/02615-7.

References

[1] J. Belmonte-Beitia, F. Güngör, and P.J. Torres, Explicit solutions with non-trivial phase of the inhomogeneous coupled two-component NLS system, J. Phys. A: Math. Theor. 53 (2020), 015201.

[2] G. W. Bluman and S. Anco, Symmetry and Integration Methods for Differential Equations, (Springer, New York, 2002).

[3] G. W. Bluman, A. F. Cheviakov, and S. C. Anco, Applications of Symmetry Methods to Partial Differential Equations (Springer, New York, 2010).

[4] G. W. Bluman and S. Kumei, Symmetries and Differential Equations, Applied Mathematical Sciences 81, (Springer, New York, 1989).

[5] C. I. Christov, S. Dost, G. A. Maugin, Inelasticity of soliton collisions in systems of coupled nls equations, Physica Scripta 50 (1994), 449–54.

[6] B. Deconinck and T. N. Kutz, Singular instability of exact stationary solutions to the non-local Gross-Pitaevskii equation, Phys. Lett. A 319 (2003), 97–103.

[7] S. Dimas and D. Tsoubelis, SYM: A new symmetry-finding package for mathematica, in: The 10th International Conference on Modern Group Analysis, Nicosia, (2005), 64–70.

[8] E. P. Gross, Structure of a quantized vortex in boson systems, Il Nuovo Cimento 20 (1961) 454–457, doi: 10.1007/BF02731494.

[9] N. H. Ibragimov, Transformation groups applied to mathematical physics, Translated from the Russian Mathematics and its Applications (Soviet Series), (D. Reidel Publishing Co., Dordrecht, 1985).

[10] N. H. Ibragimov (ed.), CRC Handbook of Lie Group Analysis of Differential Equations, vols. 1–3, CRC Press, Boca Raton, Fl., 1994–6.

[11] N. H. Ibragimov, Selected Works, volume IV, pp. 57 – 58, ALGA publications, Blekinge Institute of Technology, S-371 79 Karlskrona, Sweden., 2009.

[12] N. Ivanova, Symmetry of nonlinear Schrödinger equations with harmonic oscillator type potential, Proc. Inst. Math. NAS Ukraine 43 (2002), 149–150.

[13] Y. Kodama and A.V. Mikhailov, Symmetry and perturbation of the vector nonlinear Schrödinger equation, Physica D 152-153 (2001), 171–177.

Journal of Physics: Conference Series **2910** (2024) 012010 doi:10.1088/1742-6596/2910/1/012010

[14] N. A. Kostov. Exact solutions of the Manakov system. In: Stancho Dimiev and Kouei Sekigawa (eds), Topics in Contemporary Differential Geometry, Complex Analysis and Mathematical Physics, Proceedings of the 8th International Workshop on Complex Structures and Vector Fields, (2007), pp. 158–167.

[15] N. A. Kostov, V. Z. Enolskii, V. S. Gerdjikov, V. V. Konotop, and M. Salerno, Two-component Bose-Einstein condensates in periodic potential, Phys. Rev. E, 70:056617, 2004.

[16] B. A. Malomed and S. Kivshar, Dynamics of solitons in nearly integrable systems, Rev. Mod. Phys. 61:763, 1989.

[17] S. V. Manakov, On the theory of two-dimensional stationary self-focusing of electromagnetic waves, Sov. Phys. JETP 38 (1974) 248–253.

[18] A. G. Nikitin and R. O. Popovych, Group classification of nonlinear Schrödinger equations, Ukrainian Math. J. 53 (2001), 1255–1265.

[19] P. J. Olver, Applications of Lie Groups to Differential Equations (Springer, New York, 1986).

[20] L. V. Ovsiannikov, Group Analysis of Differential Equations (Academic Press, London 1982).

[21] L. P. Pitaevskii, Vortex lines in an imperfect Bose gas, Sov. Phys. JETP 13(2) (1961) 451–454.

[22] R. O. Popovych, N. Ivanova and H. Eshraghi, Group classification of 1+1-dimensional Schrödinger equations with potentials and power nonlinearities, J. Math. Phys. 45 (2004), 3049–3057.

[23] A. Sciarrino and P. Winternitz, Symmetries and solutions of the vector nonlinear Schrödinger equation, Il Nuovo Cimento 112B (1997), 853–871.

[24] H. Stephani, Differential Equations: Their Solutions Using Symmetries (Cambridge Univ. Press, Cambridge, 1989).

[25] M. D. Todorov and C. I. Christov, Conservative numerical scheme in complex arithmetic for Coupled Nonlinear Schrödinger Equations, Discrete and Continuous Dynamical Systems, Suppl. 2007, AIMS, 982–992.

[26] V. E. Zakharov and E. I. Schulman, To the integrability of the system of two coupled nonlinear nonlinear Schrödinger equations, Physica D, 4 (1982), 270–274.

[27] V. E. Zakharov, A.B. Shabat, Exact theory of two-dimensional self-focusing and one-dimensional self-modulation of waves in nonlinear media, Journal of Experimental and Theoretical Physics 34(1) (1972) 62–69.

ϕ^4 model under Dirichlet-Neumann boundary conditions

Daniel Dantchev[1,2], **Vassil Vassilev**[1]

[1] Institute of Mechanics, Bulgarian Academy of Sciences, Academic Georgy Bonchev St. building 4, 1113 Sofia, Bulgaria

[2] Max-Planck-Institut für Intelligente Systeme, Heisenbergstrasse 3, D-70569 Stuttgart, Germany

E-mail: danieldantchev@gmail.com, vasilvas53@gmail.com

Abstract. The minimization of the Ginsburg-Landau ϕ^4 functional is at core of the so-called ϕ^4 model, which is one of the basic models of statistical mechanics. The minimization leads to a second order nonlinear differential equation that has to be solved under specific boundary conditions. In the current article we consider a system with a film geometry with thickness L under Dirichlet-Neumann boundary conditions applied along the finite-direction. We study the modifications of the bulk phase diagram for the finite system, as well as the field-temperature behavior of the characterizing the system order parameter profile, as well as the connected to it corresponding response functions (local and total susceptibilities).

1. Introduction

Currently, there is a substantial interest in the behavior of low-dimensional systems undergoing phase transitions. This is both due to the internal needs of theory to be developed for such systems, but also due to the experimental interest of scaled down systems. One considers systems with finite geometry like fully finite systems, chains, films, etc.

In the current article we consider a system with a film geometry of width L at temperature T and exposed to some external ordering field h. In statistical mechanics the systems are described via some order parameter, which can be scalar, vector, tensor, etc. We specifically envisage a system described by scalar, i.e., Ising type mean-field order parameter under Dirichlet-Neumann boundary conditions. Let us assume that along the finite direction of length L the corresponding Cartesian coordinate is z. For such a system in the grand canonical ensemble the finite-size scaling theory [1–3] predicts:

- For the order parameter ϕ (e.g., magnetization) profile

$$\phi(z|T,h,L) \simeq a_h L^{-\beta/\nu} X_m\left(z|L,x_t,x_h\right) \tag{1}$$

where

$$x_t = a_t \tau L^{1/\nu}, \qquad x_h = a_h h L^{\Delta/\nu}. \tag{2}$$

Here $\tau = (T-T_c)/T_c$ is the dimensionless distance from the bulk critical point T_c of the infinite system. Under Dirichlet-Neumann boundary conditions in the current article we

Content from this work may be used under the terms of the Creative Commons Attribution 4.0 licence. Any further distribution of this work must maintain attribution to the author(s) and the title of the work, journal citation and DOI.
Published under licence by IOP Publishing Ltd

mean that

$$\phi(z=0|T,h,L)=0 \quad \text{and} \quad \frac{\partial}{\partial z}\phi(z|T,h,L)\Big|_{z=1}=0. \tag{3}$$

Obviously, this is a specific case of mixed boundary conditions problem in which the solution of a given differential equation is required to satisfy different boundary conditions on disjoint parts $\mathcal{B}_1$ and $\mathcal{B}_2$ of the boundary $\partial\Omega = \mathcal{B}_1 \bigcup \mathcal{B}_2$ of the domain Ω where the condition is stated

$$\phi_{|\mathcal{B}_1} = f_1 \quad \text{and} \quad \frac{\partial}{\partial n}\phi_{|\mathcal{B}_2} = f_2. \tag{4}$$

Here ∂n means a derivative in a direction normal to the boundary $\mathcal{B}_2$. In the current article the case $f_1 = f_2 = 0$ has been chosen and, because of the film geometry, the normal derivative is along the z axis.

A basic characteristic of the order parameter profile is its response to the external influence on the system described in the form of some external filed h, say a magnetic one. Thus, one defines the local and total response functions (susceptibilities). For them the finite-size scaling theory predicts:

- For the local (layer) susceptibility profile

$$k_BT\,\chi_l(z|T,h,L) \equiv \frac{\partial}{\partial h}\phi(z|T,h,L) \simeq a_h^2 L^{\gamma/\nu} X_\chi\left(z|L,x_t,x_h\right) \tag{5}$$

with

$$X_\chi\left(z|L,x_t,x_h\right) = \frac{\partial}{\partial x_h} X_m\left(z|L,x_t,x_h\right); \tag{6}$$

- For the total susceptibility $\chi(T,h,L) \equiv L^{-1}\int_0^L \chi_l(z|T,h,L)dz$ one has

$$k_BT\,\chi(T,h,L) \simeq a_h^2 L^{\gamma/\nu} X\left(x_t,x_h\right). \tag{7}$$

In the above equations (1) – (7), k_B is the Boltzmann constant, β and γ are the critical exponents for the order parameter and the susceptibility (compressibility), the quantities a_t and a_h are nonuniversal metric factors that can be fixed, for a given system, by taking them to be, e.g., $a_t = 1/\left[\xi_0^+\right]^{1/\nu}$, and $a_h = 1/\left[\xi_{0,h}\right]^{\Delta/\nu}$, where ξ_0^+ and $\xi_{0,h}$ are the respective amplitudes of the correlation length along the τ and h axes. In addition, ν is a critical exponent characterizing the behavior of the correlation length, while Δ is another exponent related to the behavior of, say, order parameter as a function of the external field h.

Let us recall that the Ising system with a film geometry $\infty^2 \times L$ possesses a critical point $T_{c,L}$ of its own with coordinates $(x_t^{(c)}, x_h^{(c)})$. The scaling functions X_m, X_χ and X will exhibit singularities near this point. For example

$$X(x_t,x_h^{(c)}) \simeq X_{c,t}\left(x_t - x_t^{(c)}\right)^{-\gamma_2}, \qquad x_t \to x_t^{(c)}, \tag{8}$$

where the subscript in γ_2 reminds that γ_2 is the critical exponent of the two-dimensional infinite system that is to be distinguished from the corresponding exponent γ for the three dimensional bulk system.

In the current article we will derive new exact analytical results for the scaling functions X_m, X_χ and X for the Ginzburg-Landau Ising type mean-field model under Dirichlet-Neumann boundary conditions. Let us recall that in the mean-field approximation $\beta = \nu = 1/2$, $\Delta = 3/2$ and $\gamma = \gamma_2 = 1$. For the version of the model considered here $\tau = (T - T_c)/T_c\,(\xi_0^+)^{-2}$, $z \in [0, L]$, $\xi_0^+ = 1$, $\xi_{0,h} = 1/\sqrt[3]{3}$, $a_t = 1$ and $a_h = 3$ [4–6].

2. The Ginzburg-Landau functional

In the present work we consider the standard ϕ^4 Ginzburg–Landau functional

$$\mathcal{F}[\phi|\tau, h, L] = \int_0^L \mathcal{L}(\phi, \phi'|\tau, h)\,\mathrm{d}z \tag{9}$$

with

$$\mathcal{L}(\phi, \phi'|\tau, h) = \frac{1}{2}\phi'^2 + \frac{1}{2}\tau\phi^2 + \frac{1}{4}g\phi^4 - h\phi. \tag{10}$$

Here L, $g \in \mathbb{R}^+$, while $\tau, h \in \mathbb{R}$, $z \in (0, L)$ and $\phi = \phi(z)$ are the independent and dependent variables, respectively, and the prime indicates differentiation with respect to the variable z.

This functional describes a critical system of Ising type in a film geometry $\infty^2 \times L$, where the film thickness L is supposed to be along the z axis. In Eq. (9), $\phi(z; \tau, h, L)$ is the order parameter of the system, g is the bare coupling constant, and, as before, τ is the bare reduced temperature and h is the external ordering field.

The physical state of the regarded system is described by the minimizers of the Ginzburg–Landau functional $\mathcal{F}[\phi; \tau, h, L]$ given above whose extremals are determined by the solutions of the corresponding Euler-Lagrange equation

$$\frac{d}{dz}\frac{\partial \mathcal{L}}{\partial \phi'} - \frac{\partial \mathcal{L}}{\partial \phi} = 0, \tag{11}$$

which, on account of Eq. (10), reads

$$\phi'' - \phi\left[\tau + g\,\phi^2\right] + h = 0. \tag{12}$$

Multiplying equation (12) by ϕ' and integrating once over z one obtains the following first integral of the system

$$\frac{1}{2}\phi'^2 - \frac{1}{2}\tau\phi^2 - \frac{1}{4}g\phi^4 + h\phi = c, \tag{13}$$

where c is the constant of integration. Another quantities of interest are the local $\chi_l(z, T, h, L)$ and total $\chi(z, T, h, L)$ susceptibilities

$$\chi_l(z|T, h, L) \equiv \frac{\partial}{\partial h}\phi(z|T, h, L) \quad \text{and} \quad L^{-1}\chi_l(z|T, h, L) = \int_0^L \chi_l(z|T, h, L)dz. \tag{14}$$

From Eq. (12) for χ_l one has

$$-\chi_l'' + \chi_l\left[\tau + 3g\,\phi^2\right] = 1. \tag{15}$$

It is convenient to introduce the following notations

$$\zeta = z/L, \quad x_t = \tau L^{1/\nu}, \quad \bar{x}_h = \sqrt{2g}hL^{\Delta/\nu}, \quad \nu = 1/2, \quad \Delta = 3/2 \tag{16}$$

and

$$\phi(z) = \sqrt{\frac{2}{g}}\,L^{-\beta/\nu}X_m(\zeta|x_t, \bar{x}_h). \tag{17}$$

In terms of the above variables the energy functional Eq. (9) and its Lagrangian density Eq. (10) read

$$\mathcal{F}[X_m|x_t, \bar{x}_h] = \frac{1}{gL^4}\int_0^1 \mathcal{L}[X_m, X_m'|x_t, \bar{x}_h]d\zeta, \tag{18}$$

with

$$\mathcal{L}[X_m, X_m'|x_t, \bar{x}_h] = X_m'^2(\zeta) + X_m^4(\zeta) + x_t X_m^2(\zeta) - \bar{x}_h X_m(\zeta). \tag{19}$$

The primes here and hereafter indicate differentiation with respect to the variable $\zeta \in [0, 1]$. Accordingly, Eq. (12) and its first integral (13) become

$$X''_m(\zeta) = X_m(\zeta)\left[x_t + 2X_m^2(\zeta)\right] - \frac{\bar{x}_h}{2}. \tag{20}$$

and

$$X'^2_m(\zeta) = P[X_m], \qquad \text{with} \qquad P[X_m] = X_m^4(\zeta) + x_t X_m^2(\zeta) - \bar{x}_h X_m(\zeta) + \varepsilon, \tag{21}$$

respectively, where ε denotes the respective constant of integration.

We are looking for such a solution of Eq. (19) that

$$X_m(\zeta = 0|x_t, \bar{x}_h) = 0 \qquad \text{and} \qquad X'_m(\zeta = 1|x_t, \bar{x}_h) = 0, \tag{22}$$

i.e., we are looking for a solution of the problem under Dirichlet-Neumann boundary conditions.

3. Expression for the order parameter profile

Below we derive the corresponding result within two different approaches.

3.1. Results for the order parameter profiles in the case of zero field

In this case the expression for the order parameter profiles can be expressed in terms of Jacobi elliptic functions.

Let us look for a solution of Eq. (20) with $\bar{x}_h = 0$ of the type

$$X_m(\zeta|a, b, k) = a\,\mathrm{sn}\,(b\zeta\,|k\,), \tag{23}$$

where $\mathrm{sn}(.|k)$ is the corresponding Jacobi elliptic function with $0 \le k \le 1$. Here a, b and k are parameters that are to be determined. Let us first check what restrictions the boundary conditions Eq. (22) impose on these parameters. First, it is clear that $X_m(\zeta = 0|a, b, k) = 0$ by the very form in which we have chosen to look for a solution. Next, the condition $X'_m(\zeta = 1|a, b, k) = 0$ leads to

$$ab\,\mathrm{cn}\,(b\,|k\,)\,\mathrm{dn}\,(b\,|k\,) = 0. \tag{24}$$

Solving for b, renouncing the trivial solution $b = 0$, we obtain

$$b = K\,(k)\,. \tag{25}$$

Plugging $a\,\mathrm{sn}\,(b\zeta\,|K\,(k))$ into Eq. (20) (with $\bar{x}_h = 0$) and solving for x_t, we derive

$$x_t = \left(k^2 - 1\right) K\,(k)^2 - 2a^2\,\mathrm{sn}\,(\zeta\,K\,(k)\,|k)^2\,. \tag{26}$$

The above shall be valid for any ζ. Thus, from $\zeta = 0$ we conclude that

$$x_t(k) = -\left(k^2 + 1\right) K\,(k)^2\,, \tag{27}$$

while from $\zeta = 1$, with x_t given by Eq. (27), it follows that

$$a = \pm k\,K\,(k)\,. \tag{28}$$

Summarizing, we conclude that

$$X_m(\zeta) = \pm\,k\,K\,(k)\,\mathrm{sn}\left(\zeta K\,(k)\,|k\right), \quad \text{with} \quad x_t = -\left(k^2 + 1\right) K\,(k)^2\,, \quad 0 \le k \le 1. \tag{29}$$

The phase diagram of the considered system follows from the above relations and its details are given in Figure 1 - on the left panel for the bulk system and for the finite film with Dirichlet-Neumann boundary conditions - on the right panel. From Eq. (27) it follows that the critical temperature of the finite system, below which the order parameter profile is nonzero, is

$$x_{t,\,\mathrm{crit}} = x_t(k = 0) = -\frac{\pi^2}{4}. \tag{30}$$

The last means that for $x_t \ge -\pi^2/4$ one has $X_m = 0$ and $X_m \ne 0$ for $x_t < -\pi^2/4 \simeq -2.4674$. Several order parameter profiles for different values of x_t are shown in Fig. 2.

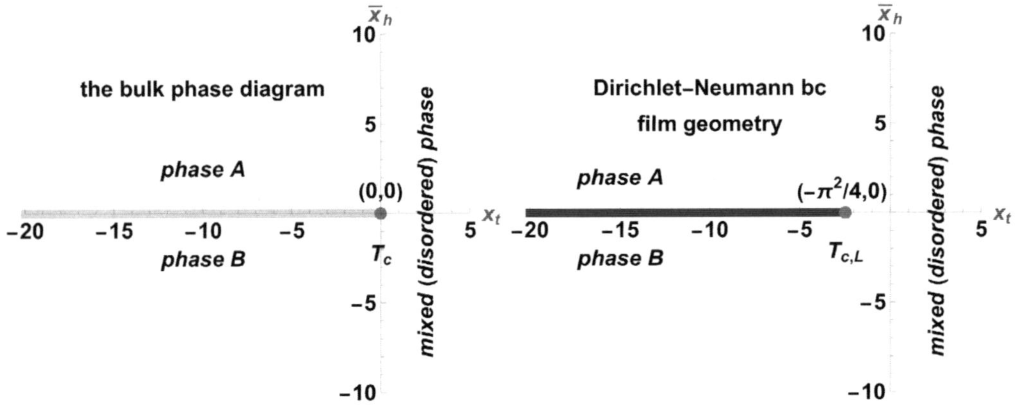

Figure 1. Left panel: The phase diagram of the bulk system (i.e., of the system in which the limit $L \to \infty$ has been performed, keeping all other parameters, T and h, fixed). **Right panel**: The phase diagram of the finite system with Dirichlet-Neumann boundary conditions. In the bulk system a phase transition of first order happens when crossing the phase coexistence line that is at $\bar{x}_h = 0$ and spans for $T \in (0, T = T_c)$. At $T = T_c$ the system exhibits a second order phase transition. In the finite system the coexistence line is at $\bar{x}_h = 0$ and spans for $T \in (0, T = T_{c,L})$. The second order phase transition happens again at $\bar{x}_h = 0$ but at, see Eq. (30), $T = T_{c,L} \equiv T_c(1 - x_{t,\text{crit}} L^{-2}) = T_c(1 - (\pi^2/4)L^{-2})$. Note the change with Dirichlet-Dirichlet boundary conditions where the critical point is at $T_{c,L} = T_c(1 - \pi^2 L^{-2})$. We emphasize that in the finite system the second order phase transition does *not* happen at $T = T_c$ but at some other, L-dependent temperature $T_{c,L}$, such that $T_{c,L} \to \infty = T_c$.

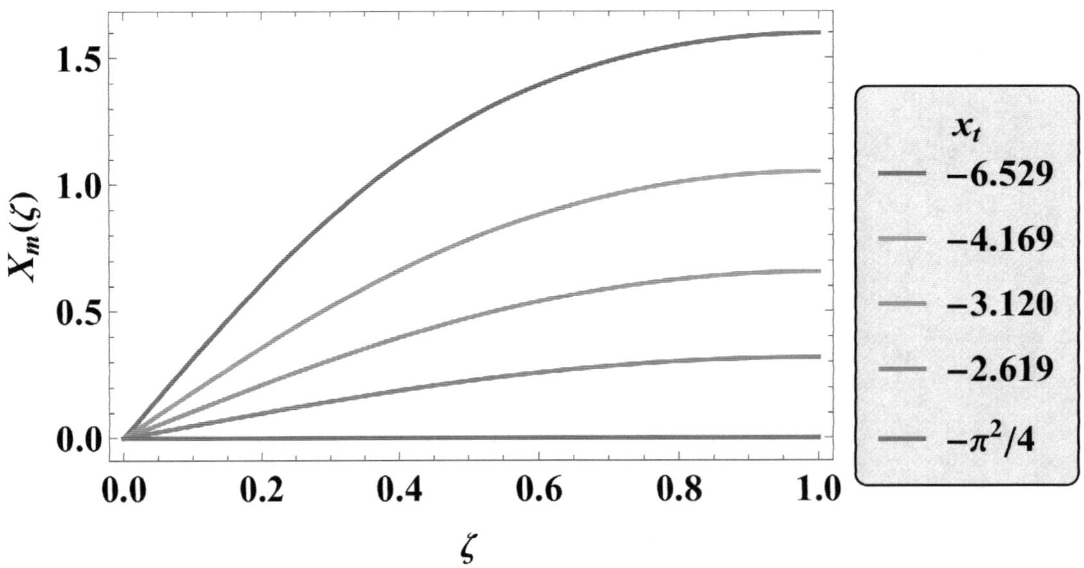

Figure 2. The behavior of the order parameter profile for $x_t = -6.529, -4.169, -3.120, -2.619$, and $x_t = -\pi^2/4$.

3.2. Results for the order parameter profile in the case of nonzero field

3.2.1. Approach I: Using the first integral. Let $X_m(\zeta = 1) = X_{m,r}$. Then, from Eq. (21) and Eq. (22) we obtain $P[X_{m,r}] = 0$, and thus

$$P[X_m] = X_m^4(\zeta) - X_{m,r}^4 + x_t\left[X_m^2(\zeta) - X_{m,r}^2\right] - \bar{x}_h\left[X_m(\zeta) - X_{m,r}\right]. \tag{31}$$

In addition Eq. (21) delivers

$$\zeta = \int_0^{X_m(\zeta)} \frac{dX_m}{\sqrt{X_m^4 - X_{m,r}^4 + x_t\left[X_m^2 - X_{m,r}^2\right] - \bar{x}_h\left[X_m - X_{m,r}\right]}}, \tag{32}$$

and

$$1 = \int_0^{X_{m,r}} \frac{dX_m}{\sqrt{X_m^4 - X_{m,r}^4 + x_t\left[X_m^2 - X_{m,r}^2\right] - \bar{x}_h\left[X_m - X_{m,r}\right]}}, \tag{33}$$

The algorithm for solving numerically the problem for finding the order parameter profile under Dirichlet-Neumann boundary conditions is now clear. From Eq. (33) one can determine $X_{m,r}$ as a function of x_t and $\bar{x}_h$. Then, from Eq. (32) one determines ζ as a function of X_m. Inverting this dependence, one shall arrive at the desired function $X_m(\zeta)$. If there is more than one solution obtained under the procedure described above, one shall choose this profile $X_m(\zeta)$ that provides the minimum of the free energy $\mathcal{F}$, see Eq. (18).

For $\bar{x}_h = 0$ one has $P[X_m] = P[-X_m]$. Thus, in this case there are two equally probable possibilities $X_m'(\zeta) > 0$ and $X_m'(\zeta) < 0$ which will lead to two order parameter profiles with the same free energy. If $h > 0$ since the physical solution has to minimize the functional Eq. (18) it is clear that $\bar{x}_h X_m(\zeta) > 0$, when $h > 0$. Next, the problem is symmetric under the *simultaneous* change of the signs of X_m and h. Therefore, it is enough to study the case $h \geq 0$ with $X_m(\zeta) \geq 0$.

Let us assume that $X_m'(\zeta) > 0$, where $X_m(\zeta)$ starts from $X_m(\zeta = 0) = 0$ and reaches its maximum at $X_m(\zeta = 1) = X_{m,r}$. Then, from Eq. (32) one obtains

$$\zeta = X_{m,r}^{-1} \int_0^{X_m(\zeta)/X_{m,r}} \frac{dy}{\sqrt{y^4 - 1 + x_{t,r}\left[y^2 - 1\right] - \bar{x}_{h,r}\left[y - 1\right]}}, \tag{34}$$

and

$$X_{m,r} = \int_0^1 \frac{dy}{\sqrt{y^4 - 1 + x_{t,r}\left[y^2 - 1\right] - \bar{x}_{h,r}\left[y - 1\right]}}, \tag{35}$$

where

$$y = X_m/X_{m,r}, \qquad x_{t,r} = x_t/X_{m,r}^2 \quad \text{and} \quad \bar{x}_{h,r} = \bar{x}_h/X_{m,r}^3. \tag{36}$$

Eqs. (34) - (36) provide an even simpler, than the described above, numerical algorithm for solving the problem about the determination of the order parameter profile. First, for any given set of values $\{x_{t,r}, \bar{x}_{h,r} \geq 0\} \in \mathbb{R}^2$, Eq. (35) delivers a *single* value of $X_{m,r} \in \mathbb{Z}$. Of course, we are only interested in $X_{m,r} \geq 0$. When this is the case, Eq. (34) delivers numerically $X_m(\zeta)$. Actually, we have shown that if for a given set of values $\{x_t, \bar{x}_h \geq 0\}$ the order parameter for the Dirichlet-Neumann boundary conditions exist, it is only one. Thus, we can prove the following:

Proposition *If the external magnetic field is nonzero, i.e.,* $\bar{x}_h \neq 0$, *in the case when for given* x_t *and* $\bar{x}_h$ *the order parameter profile* $X_m(\zeta)$ *which minimizes the free energy exist under the Dirichlet-Neumann boundary conditions* $X_m(0) = 0$, $X_m'(1) = 0$, *it is a single one, monotonic on the interval* $(0, 1)$, *and for it* $\bar{x}_h X_m(\zeta) \geq 0$.

Several order parameter profiles for different values of x_t are shown in Fig. 2.

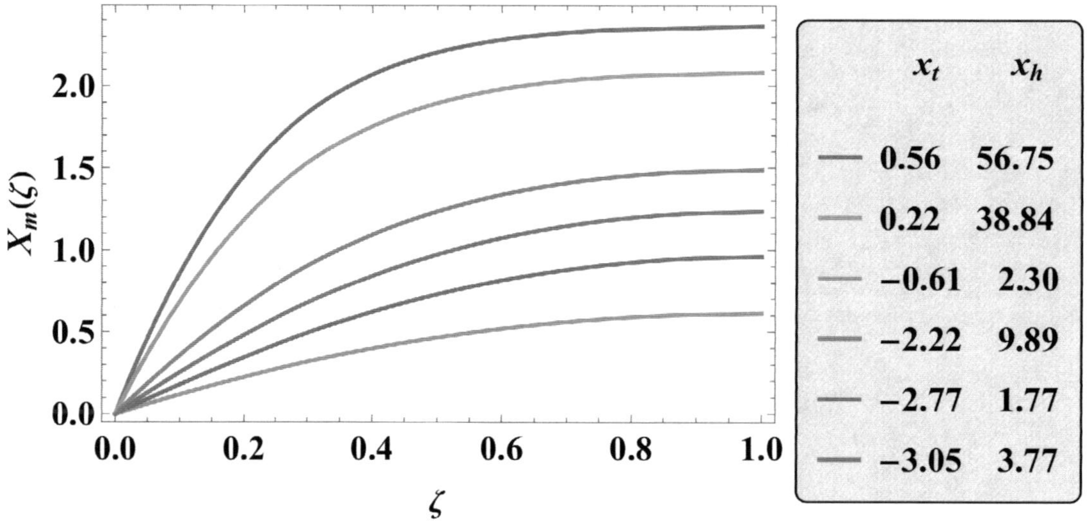

Figure 3. The behavior of the order parameter profile for $\{x_t = 0.56, x_h = 56.75\}$, $\{x_t = 0.22, x_h = 38.84\}$, $\{x_t = -0.61, x_h = 2.30\}$, $\{x_t = -2.22, x_h = 9.89\}$, $\{x_t = -2.77, x_h = 1.77\}$, $\{x_t = -3.05, x_h = 3.77\}$.

Let us demonstrate how from above expressions one can obtain the ones for the specific case of $\bar{x}_h = 0$. Then the integrals Eq. (34) and Eq. (35) can be calculated explicitly in terms of elliptic functions. From Eq. (35) one obtains

$$X_{m,r} = \frac{K\left[\left(-x_t/X_{m,r}^2 - 1\right)^{-1/2}\right]}{\sqrt{-x_t/X_{m,r}^2 - 1}} \tag{37}$$

while Eq. (34) delivers

$$\zeta X_{m,r} = \frac{F\left[\sin^{-1}\left(X_m(\zeta)/X_{m,r}\right) \Big| \left(-x_t/X_{m,r}^2 - 1\right)^{-1/2}\right]}{\sqrt{-x_t/X_{m,r}^2 - 1}}, \tag{38}$$

or, reverting the last result

$$X_m(\zeta)/X_{m,r} = \mathrm{sn}\left(\zeta\; X_{m,r}\sqrt{-x_t/X_{m,r}^2 - 1} \Big| \left(-x_t/X_{m,r}^2 - 1\right)^{-1/2}\right), \tag{39}$$

where $K(z)$ is the complete elliptic integral of the first kind, $F(z|m)$ gives the elliptic integral of the first kind, and $\mathrm{sn}(z|m)$ — the Jacobi elliptic function.

From Eq. (37) and Eq. (39) one derives

$$X_m(\zeta) = X_{m,r}\, \mathrm{sn}\left\{\zeta\; K\left[\left(-x_t/X_{m,r}^2 - 1\right)^{-1/2}\right] \Big| \left(-x_t/X_{m,r}^2 - 1\right)^{-1/2}\right\}. \tag{40}$$

Setting

$$k = \left(-x_t/X_{m,r}^2 - 1\right)^{-1/2} \tag{41}$$

from Eq. (40) one arrives at

$$X_m(\zeta) = X_{m,r}\, \mathrm{sn}(\zeta\; K(k)|k), \tag{42}$$

and from Eq. (37) it follows

$$X_{m,r} = k\,K(k). \tag{43}$$

Eq. (41) then delivers Eq. (27) from the previous approach. The inspection of the equations of the two approaches ensures as that they lead to the same result for the order parameter profile, as it shall be expected.

3.2.2. Approach II: Exact explicit results for the order parameter profile in the case of nonzero field via Weierstrass function. Following [7, p. 454] one obtains that the order parameter for given values of the temperature and the field reads

$$X_m\left(\zeta|x_t,\bar{x}_h,X_{m,r}\right) = X_{m,r} + \frac{6X_{m,r}\left(x_t+2X_{m,r}^2\right)-3\bar{x}_h}{12\wp\left(\zeta-1;g_2,g_3\right)-\left(x_t+6X_{m,r}^2\right)}, \tag{44}$$

where $X_{m,r} = X_{m,r}(x_t,x_h)$ is the value of the order parameter profile at the right end of the system. Note that $X_{m,r}$ is to be determined by the boundary conditions. Here $\wp\left(v;g_2,g_3\right)$ is the Weierstrass elliptic function whose invariants g_2 and g_3 read

$$\begin{aligned} g_2 &= \frac{1}{12}x_t^2 - X_{m,r}\left(X_{m,r}^3+x_tX_{m,r}-\bar{x}_h\right), \\ g_3 &= -\frac{1}{432}\left[27\bar{x}_h^2+2x_t^3+72x_tX_{m,r}\left(X_{m,r}^3+x_tX_{m,r}-\bar{x}_h\right)\right]. \end{aligned} \tag{45}$$

The procedure for the explicit numerical determination of $X_m\left(\zeta|x_t,\bar{x}_h,X_{m,r}\right)$ is now clear. First, using that (see, e.g., [8]) $\wp(\zeta\to 0)\simeq\zeta^{-2}$, we immediately conclude that $X_m\left(1|x_t,\bar{x}_h,X_{m,r}\right) = X_{m,r}$, as it ought to be. It is also easy to show that $X_m'\left(\zeta\to 1|x_t,\bar{x}_h,X_{m,r}\right)=0$. The only remaining requirement that $X_m\left(\zeta\to 0|x_t,\bar{x}_h,X_{m,r}\right)=0$ leads to an transcendental equation from where we have to determine $X_{m,r}$. Normally, one obtains more than one solution of this equation. The one, that corresponds to the physical reality is the one which minimizes the energy given by Eq. (19).

4. Expression for the susceptibility

In terms of the notations of Eq. (16), for the local susceptibility (6) in the model under consideration (with $\nu=1/2$, $\gamma=1$) one derives

$$X_\chi(\zeta|x_t,\bar{x}_h) = 2\frac{\partial}{\partial\bar{x}_h}X_m(\zeta|x_t,\bar{x}_h). \tag{46}$$

From Eq. (20) one derives

$$-X_\chi'' + (x_t+6X_m^2)X_\chi = 1. \tag{47}$$

This equation does not depend explicitly on x_h. It is possible to obtain its solution for $x_h=0$ in terms of Jacobi elliptic functions.

Below we show that if one knows $X_m\left(\zeta|x_t,\bar{x}_h,X_{m,r}\right)$ one can determine the scaling function of the total susceptibility. Indeed, for a fixed x_t Eq. (46) implies

$$\frac{1}{2}X_\chi\left(\zeta|x_t,\bar{x}_h,X_{m,r}\right) = \frac{\partial}{\partial\bar{x}_h}X_m\left(\zeta|x_t,\bar{x}_h,X_{m,r}\right)+\frac{\partial}{\partial\bar{x}_h}X_{m,r}(\bar{x}_h)\frac{\partial}{\partial X_{m,r}}X_m\left(\zeta|x_t,\bar{x}_h,X_{m,r}\right). \tag{48}$$

Since for the local susceptibility at the end of the system one has

$$X_\chi(0|x_t,\bar{x}_h,,X_{m,r}) = 0, \tag{49}$$

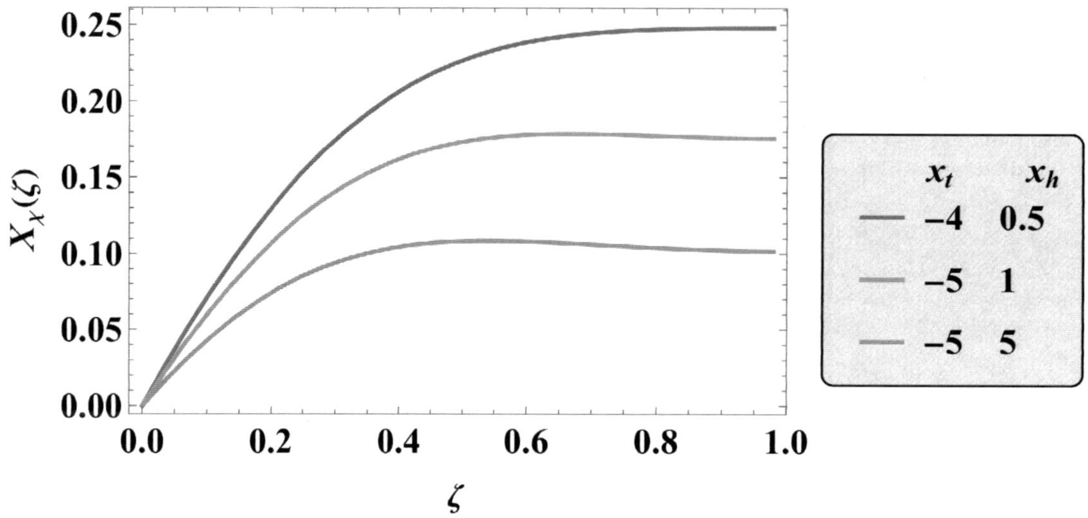

Figure 4. The behavior of the local susceptibility for $\{x_t = -4, x_h = 0.5\}$, $\{x_t = -5, x_h = 1\}$, $\{x_t = -1, x_h = 5\}$. Please note that for strong field and low temperatures the profile is *not* a monotonic function of ζ but becomes such for higher x_t and smaller x_h.

one derives

$$\frac{\partial}{\partial \bar{x}_h} X_{m,r}(\bar{x}_h) = -\frac{\frac{\partial}{\partial \bar{x}_h} X_m\left(0|x_t, \bar{x}_h, X_{m,r}\right)}{\frac{\partial}{\partial X_{m,r}} X_m\left(0|x_t, \bar{x}_h, X_{m,r}\right)}. \tag{50}$$

Thus, the total susceptibility

$$X(x_t, \bar{x}_h, X_{m,r}) = \int_0^1 X_\chi\left(\zeta|x_t, \bar{x}_h, X_{m,r}\right) d\zeta, \tag{51}$$

is determined by Eqs. (48) and (50) in terms of the scaling function of the order parameter profile $X_m(\zeta|x_t, \bar{x}_h, X_{m,r})$, given in analytic form by Eq. (44), and its partial derivatives with respect to $\bar{x}_h$ and $X_{m,r}$.

Using the procedure described above can determine the behavior of the local and total susceptibilities for given values of x_t and x_h The behavior of $X_\chi(\zeta|x_t, x_h)$ is shown in Fig. 4, while the behavior of the total susceptibility $X(x_t, x_h)$ is demonstrated in Fig. 5 One clearly observes the divergence of X for $x_h = 0$ when $x_t \to -\pi^2/4$.

4.1. On the analytical solution for the susceptibility in zero external field

In the current section we solve Eq. (47) for $X_\chi(\zeta|x_t) \equiv X_\chi(\zeta|x_t, \bar{x}_h = 0)$. When $\bar{x}_h = 0$ the order parameter profile X_m is given by Eq. (29) (see also Eq. (42)). Eq. (47) is a second order differential equation. According to the general theory [9] of the differential equations of the second order, the solution of such an equation is given by

$$X_\chi(\zeta|x_t) = c_1 y_1(\zeta|x_t) + c_2 y_2(\zeta|x_t) + c_i y_i(\zeta|x_t), \tag{52}$$

where $y_1(\zeta|x_t)$ and $y_2(\zeta|x_t)$ are linearly independent solutions of the homogeneous equation

$$-X_\chi'' + (x_t + 6X_m^2)X_\chi = 0, \tag{53}$$

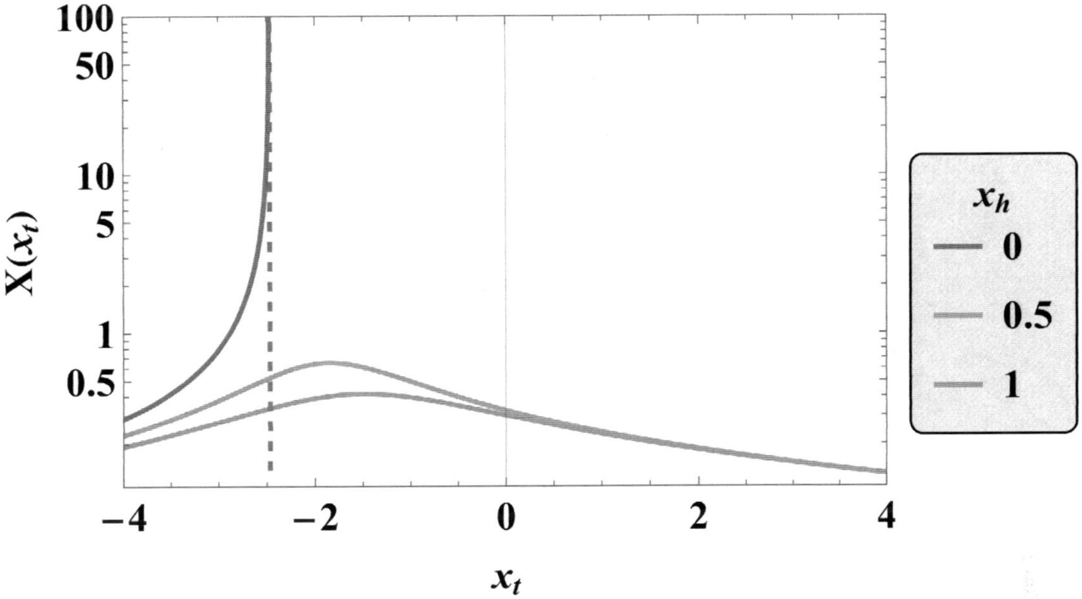

Figure 5. The behavior of the total susceptibility for $\{x_h = 0\}$, $\{x_h = 0.5\}$, and $\{x_h = 1\}$. Note that, as expected, larger the field, smaller the total susceptibility. It diverges near the critical point $\{x_t = -\pi^2/4, 0\}$ of the finite system.

while $y_i(\zeta|x_t)$ is a particular solution of the inhomogeneous Eq. (47); c_1, c_2 and c_i are constants.

We start with the determination of y_1. One can check that

$$y_1(\zeta|x_t) = \frac{d}{d\zeta} X_m(\zeta|x_t) \tag{54}$$

is a solution of Eq. (53). Performing the calculations, one obtains the explicit form for $y_1(\zeta|x_t)$

$$y_1(\zeta|x_t) = kK(k)^2 \,\mathrm{cn}\,[\zeta K(k)\,|k]\,\mathrm{dn}\,[\zeta K(k)\,|k]\,. \tag{55}$$

The behavior of $y_1(\zeta|x_t)$ for several values of k (i.e., of x_t) is shown in Fig. 6. Knowing y_1 one finds y_2 via [9] the construction:

$$y_2(\zeta|x_t) = y_1(\zeta|x_t) \int \frac{d\zeta}{[y_1(\zeta|x_t)]^2}. \tag{56}$$

From Eq. (56) it follows that

$$y_2''(\zeta|x_t) = y_1''(\zeta|x_t) \int \frac{d\zeta}{[y_1(\zeta|x_t)]^2}. \tag{57}$$

Thus, since $y_1(\zeta|x_t)$ is a solution of Eq. (53), it follows that $y_2(\zeta|x_t)$ is also a solution of Eq. (53). Having $y_1(\zeta|x_t)$ and $y_2(\zeta|x_t)$, one can determine the corresponding Wronskian

$$W = y_1(\zeta|x_t)y_2'(\zeta|x_t) - y_2(\zeta, x_t)y_1'(\zeta, x_t). \tag{58}$$

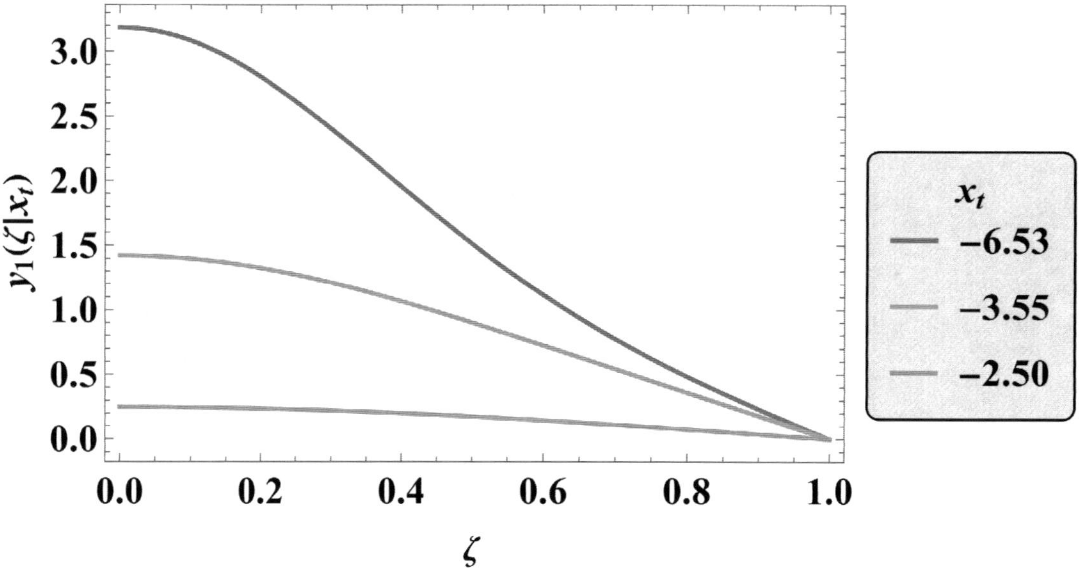

Figure 6. The behavior of $y_1(\zeta|x_t)$ as a function of ζ for $x_t = -6.53, -3.55$, and $x_t = -2.50$. We observe that it is a monotonically decreasing function equal to zero for $\zeta = 1$.

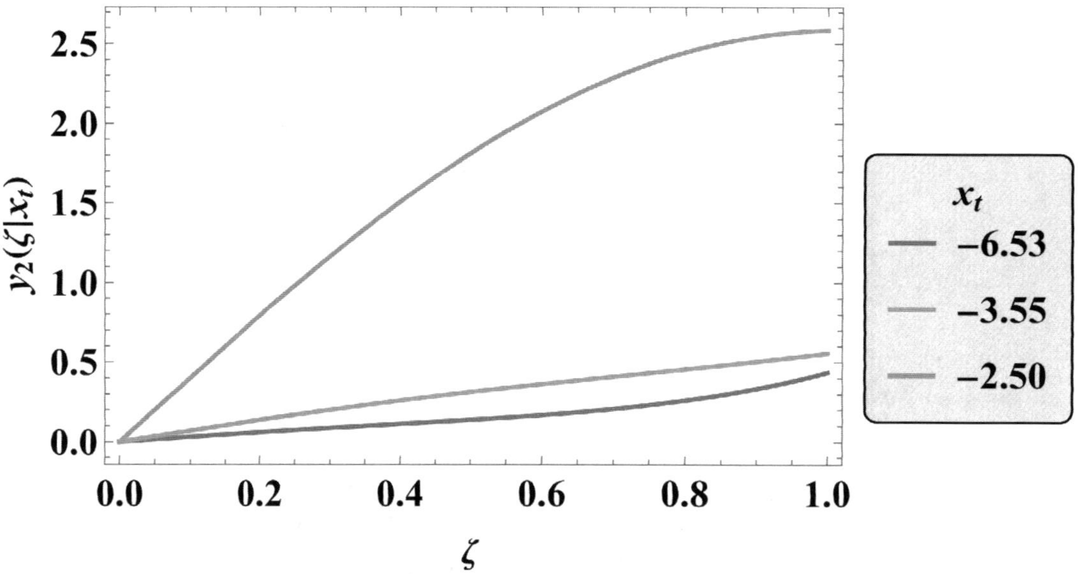

Figure 7. The behavior of $y_2(\zeta|x_t)$ as a function of ζ for $x_t = -6.53, -3.55$, and $x_t = -2.50$. We observe that it is a monotonically increasing function equal to zero for $\zeta = 0$.

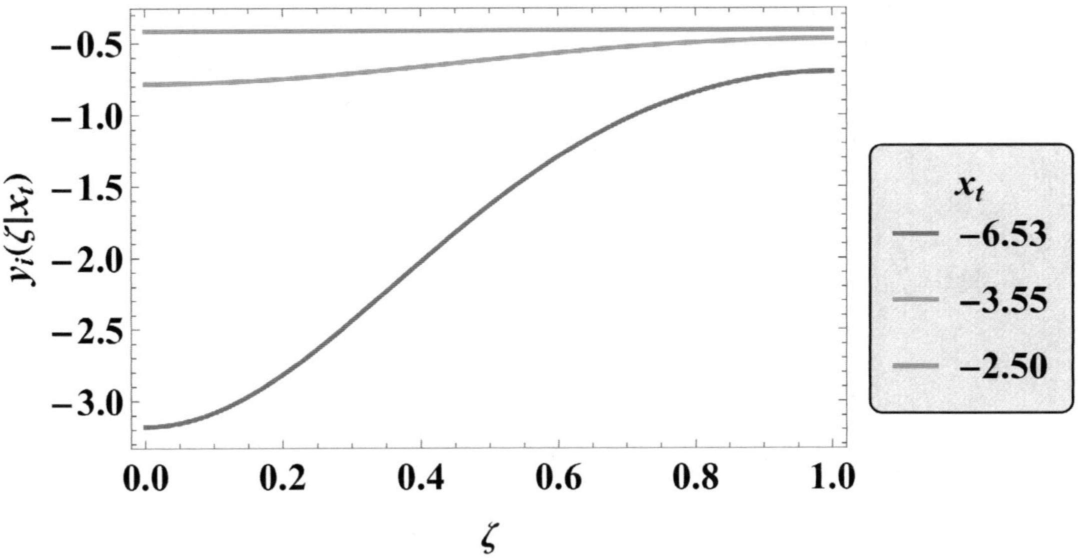

Figure 8. The behavior of $y_i(\zeta|x_t)$ as a function of ζ for $x_t = -6.53, -3.55$, and $x_t = -2.50$. We observe that it is a negative and monotonically increasing function of $\zeta = 0$.

From Eq. (56) it immediately follows that $W = 1$, which means that y_1 and y_2 are linearly independent solutions. The standard behavior of $y_2(\zeta)$ for several selected values of x_t is shown in Fig. 7.

According to the general recipe [9], for the particular solution of the homogeneous equation one has

$$y_i(\zeta|x_t) = y_1(\zeta|x_t) \int y_2(\zeta|x_t) d\zeta - y_2(\zeta|x_t) \int y_1(\zeta|x_t) d\zeta. \tag{59}$$

Having in mind Eq. (54) the above expression simplifies to

$$y_i(\zeta|x_t) = -y_1(\zeta|x_t) \int \frac{X_m(\zeta|x_t)}{[y_1(\zeta|x_t)]^2} d\zeta. \tag{60}$$

Performing the integration one obtains the explicit expression

$$y_i(\zeta|x_t) = \frac{1 - k^2 - 2\mathrm{dn}\left(\zeta K\left(k^2\right)|k^2\right)^2}{\left(k^2-1\right)^2 K\left(k^2\right)^2}. \tag{61}$$

The behavior of $y_i(\zeta)$ for selected values of x_t is shown in Fig. 8.

It is easy to check that $c_i y_i(\zeta|x_t)$ with $c_i = 1$ is a solution of Eq. (47). Requiring the the solution for X_χ preserves the symmetru due to the Dirichlet - Neumann boundary conditions we obtain that

$$c_1(k) = \frac{k^2+1}{k\left(k^2-1\right)^2 K\left(k^2\right)^4} \tag{62}$$

and

$$c_2(k) = \frac{k\left(k^2+1\right) K\left(k^2\right)}{\left(k^2-1\right) K\left(k^2\right) + \left(k^2+1\right) E\left(k^2\right)}. \tag{63}$$

The behavior of $X_\chi(\zeta|x_t)$ (shown in logarithmic scale) as a function of ζ for several values of x_t is shown in Fig. 9. The behavior of the total susceptibility is given in Fig. 10.

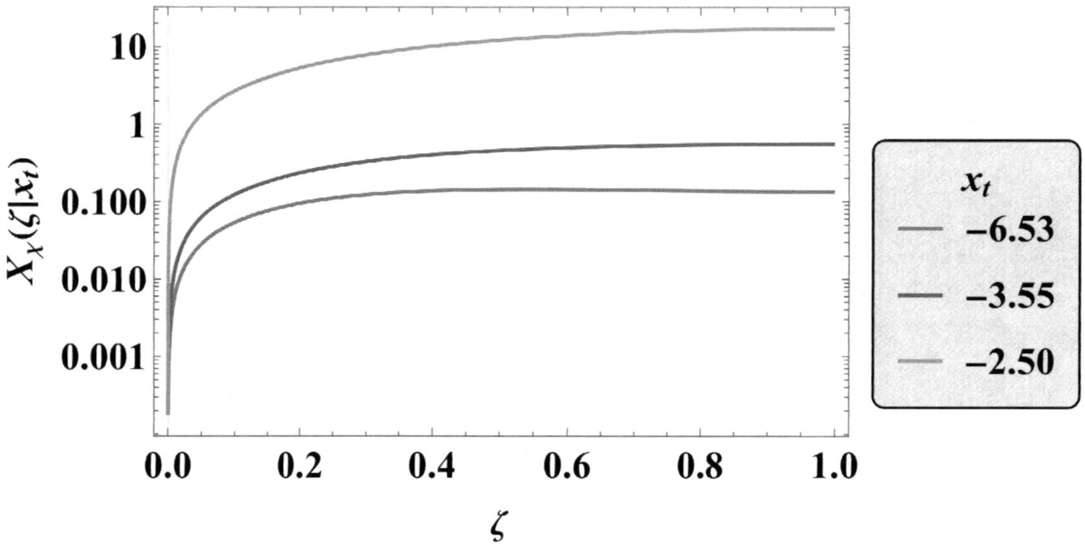

Figure 9. The behavior of $X_\chi(\zeta|x_t)$ (in logarithmic scale) as a function of ζ for $x_t = -6.53, -3.55$, and $x_t = -2.50$. We observe that the function sharply increases on approaching the critical point.

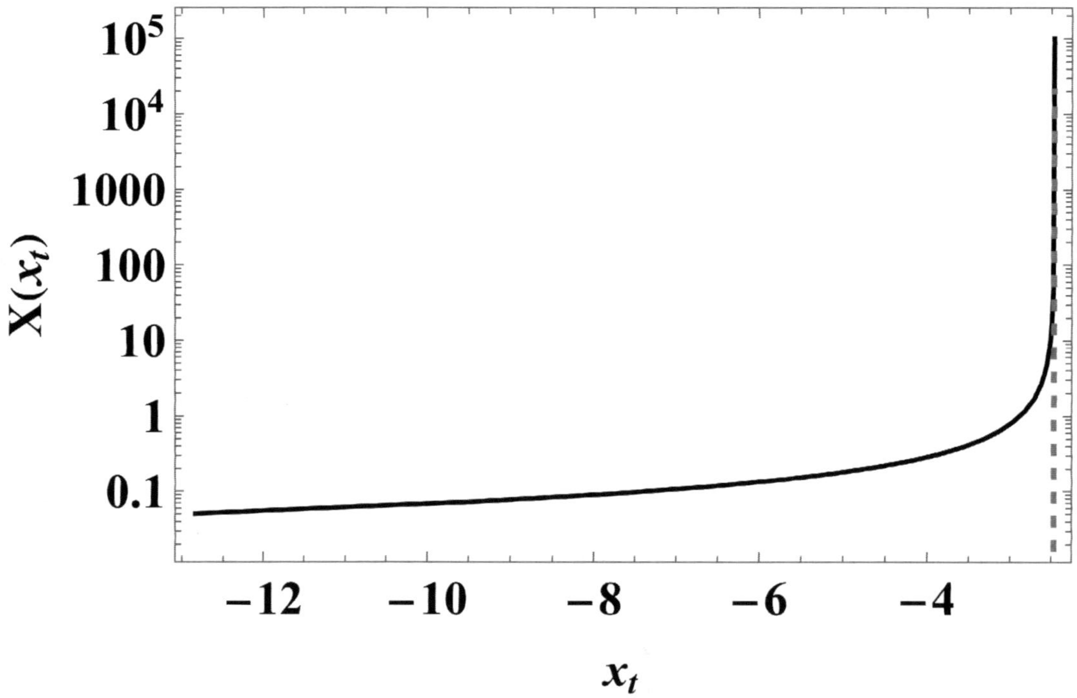

Figure 10. The behavior of the total susceptibility$X(x_t)$ as a function of x_t. We observe that the function dramatically increases upon approaching the critical point. The red vertical dashed line marks the position of the critical temperature $T_{c,L} = -\pi^2/4$.

5. Concluding Comments and Discussion

In the current article we study a system with a film geometry describing a critical fluid under Dirichlet-Neumann boundary conditions. We determine the order parameter profile, local and total susceptibilities as a function both of temperature T and the external ordering field h. As a result we obtained the phase diagram of the finite system shown in Fig. 1. We observe that the finite system possesses a critical point of its own at a temperature $T_{c,L} = -\pi^2/4$ *below* the bulk critical temperature T_c. The last implies that, for a zero field, the order parameter is zero for $T > T_{c,L}$. Its behavior is explicitly given, in terms of the Jacobi elliptic functions, in Eq. (29) (see also Eq. (42)). The behavior of the order parameter for several selected values of the scaled temperature x_t is shown in Fig. 2. The behavior of the order parameter profiles for non-zero external field h is determined in two independent ways - *i)* using the first integral of the system - see Eq. (34) — Eq. (36), and *ii)* via the via Weierstrass function - see Eq. (44) — Eq. (45). A visualization of the order parameter profile for the case of nonzero field for specific values of x_t and x_h is shown in Fig. 3. The response of the order parameter profile to the change of the xternal influence on the system, i.e., the external field, is reflected by the local and total susceptibilities. General analytic expression for the susceptibilities are given in Eq. (48) and Eq. (51). From these expressions it is clear, that if one knows analytically the behavior of the order parameter as a function of x_t and x_h, one, in principle, can also determine the behavior of the local $X_\chi(\zeta|x_t, x_h)$ and total $X(x_t, x_h)$ susceptibilities. The behavior of the local susceptibility is derived in Sec. 4.1. It is given as a linear combination of three components y_1, y_2 and y_i — see Eq. (55), Eq. (56) and Eq. (61), correspondingly. The coefficients of the linear combinations c_1 and c_2 are given in Eq. (62) and Eq. (63), while $c_i = 1$. The behavior of $X_\chi(\zeta|x_t, x_h = 0)$ as a function of ζ for several selected values of x_t i shown in Fig. 9. We observe that the function sharply increases on approaching the critical point. The behavior of $X(x_t, x_h = 0)$ as a function of x_t is given in Fig. 10. This function has a singularity at the critical temperature $T_{c,L}$ and tends to infinity upon approaching this point from below.

From the above it is clear that mathematically the problem for finding the order parameter profile and the susceptibilities reduces itself to a problem for solving a system of two nonlinear differential equations of second order in which the second equation involves the solution of the first one as an input. In the current article we have demonstrated how one can tackle this problem and obtain analytical solutions for these two equations.

Acknowledgments

Partial financial support via Grant No KP-06-H72/5 of the Bulgarian NSF is gratefully acknowledged.

Appendix A. Some details for derivation of the expression with Weierstrass function

Let us explain some details of the derivation of Eq. (44).

In Ref. [7, 20.6] one considered integrals of the type

$$z = \int_a^x \frac{1}{\sqrt{f(t)}} dt, \tag{A.1}$$

with $f(t)$ a quartic polynomial with no repeated factors

$$f(t) = a_0 t^4 + 4a_1 t^3 + 6a_2 t^2 + 4a_3 t + a_4. \tag{A.2}$$

Then it is shown that the integral Eq. (A.1) can be inverted to represent x as a function of z.

The corresponding result is [7, 20.6, Example 2, p. 454]

$$x(z) = a + \frac{\sqrt{f(a)}\,\wp'[z] + \frac{1}{2}f'(a)\left[\wp(z) - \frac{1}{24}f''(a)\right] + \frac{1}{24}f(a)f'''(a)}{2\left[\wp(z) - \frac{1}{24}f''(a)\right]^2 - \frac{1}{48}f(a)f^{iv}(a)}. \tag{A.3}$$

Here, a is an arbitrary constant and $\wp(z) = \wp(z; g_2, g_3)$ is the Weierstrass elliptic function with invariants

$$\begin{aligned} g_2 &= a_0a_4 - 4a_1a_3 + 3a_2^2, \\ g_3 &= a_0a_2a_4 + 2a_1a_2a_3 - a_2^3 - a_0a_3^2 - a_1^2a_4. \end{aligned} \tag{A.4}$$

If a is a root of the polynomial $f(t)$, then (A.3) simplifies and reads

$$x(z) = a + \frac{f'(a)}{4\left[\wp(z) - \frac{1}{24}f''(a)\right]}. \tag{A.5}$$

Thus, when $a = X_{m,r}$ and $f = P[X_m]$, (A.5) leads to the expression (44) since $X_{m,r}$ is a zero of the polynomial $P[X_m]$, while the invariants (A.4) take the form (45).

Acknowledgments

The partial financial support via Grant No KP-06-H72/5 of Bulgarian NSF is gratefully acknowledged.

References

[1] Brankov J G, Dantchev D M and Tonchev N S 2000 *The Theory of Critical Phenomena in Finite-Size Systems - Scaling and Quantum Effects* (World Scientific, Singapore)

[2] Barber M N 1983 *Phase Transitions and Critical Phenomena* vol 8 ed Domb C and Lebowitz J L (Academic, London) chap 2, pp 146–266

[3] Privman V 1990 *Finite Size Scaling and Numerical Simulations of Statistical Systems* ed Privman V (World Scientific, Singapore) pp 1–98

[4] Schlesener F, Hanke A and Dietrich S 2003 *J. Stat. Phys.* **110** 981 – 1013

[5] Dantchev D, Schlesener F and Dietrich S 2007 *Phys. Rev. E* **76** 011121 (pages 24) URL `http://link.aps.org/abstract/PRE/v76/e011121`

[6] Privman V, Hohenberg P C and Aharony A 1991 *Phase ransitions and critical phenomena* vol 14 ed Domb C and Lebowitz J L (Academic, New York) pp 1–134

[7] Whittaker E T and Watson G N 2027 *A Course of Modern Analysis* 4th ed (Cambridge University Press, London)

[8] 2010 *NIST Handbook of Mathematical Functions* (NIST and Cambridge University Press, Cambridge)

[9] Boyce W E, DiPrima R C and Meade D B 2021 *Elementary differential equations and boundary value problems* (John Wiley & Sons)

Journal of Physics: Conference Series **2910** (2024) 012012 doi:10.1088/1742-6596/2910/1/012012

Minimization Principle for Analytical Solution of Turbulent Flow in Channel

Alex Fedoseyev

Ultra Quantum Inc., Huntsville, Alabama, USA

E-mail: af@ultraquantum.com

Abstract. The analytical solution for turbulent flow in a channel, presented in Fedoseyev (2023), described the mean turbulent flow velocity as a superposition of the parabolic and superexponential solutions. In this study, the coefficients of superposition are obtained through the minimization principle, the principle of minimum viscous dissipation. The laminar flow showed a significantly larger dissipation rate. The obtained analytical solutions agreed well with turbulent flow experiments.

1 Introduction

The turbulent flow in a channel remains the subject of experimental investigations to provide data that can be used to validate numerical simulations, to improve theoretical models, and to bring new insights into the fundamental mechanisms of turbulence that may not be captured by existing theoretical models Van Doorne (2007), Pasch (2023), Wei (1989).

The theoretical models fall under Hilbertś 6th problem Hilbert (1902), which addresses the transition from the "atomistic view to the laws of motion of continua". It includes at least two steps: (i) from mechanics to kinetics (from Newton to Boltzmann), and (ii) from kinetics to mechanics and nonequilibrium thermodynamics of continua (from Boltzmann to Euler and Navier, Stokes, and Fourier), Gorban (2014). The results of theoretical works show that the well-known Euler and Navier-Stokes equations are valid only at "a limit of very slow flows with very small gradients of all fields, i.e., almost no flow at all", Gorban (2014). Examples from laminar flow experiments, such as lid-driven cavity flow and backward-facing step flow, show that the Navier-Stokes results are far from the experimental data Armaly (1983), Koseff (1984). Therefore, the search for improved kinetic equation and their respective hydrodynamic equations continues.

In the analytical solution by Fedoseyev (2023), the Generalized Hydrodynamic Equations (GHE) have been employed for turbulent flow in a channel. The GHE were derived from the Generalized Boltzmann Equation (GBE), which takes into account finite particle size Alexeev (1994), while in the traditional Boltzmann equation, particles are treated as material points. So far, the improved kinetic equation was chosen, and the respective improved hydrodynamic equations have been used in Fedoseyev (2023).

The approximate analytical solution was obtained for a mean turbulent flow velocity U_{GHE} as a superposition of the laminar (parabolic) U_L and turbulent (superexponential) U_T solutions,

$$U_{GHE} = \gamma U_T + (1 - \gamma) U_L, \tag{1}$$

where the coefficients γ and $(1 - \gamma)$ were introduced. The expressions for U_T and U_L were explicitly provided giving

$$U_{GHE} = U_0 \left[\gamma \left(1 - e^{1 - e^{y/\delta}} \right) + (1 - \gamma) 4y(L - y)/L^2 \right] \tag{2}$$

Content from this work may be used under the terms of the Creative Commons Attribution 4.0 licence. Any further distribution of this work must maintain attribution to the author(s) and the title of the work, journal citation and DOI.
Published under licence by IOP Publishing Ltd

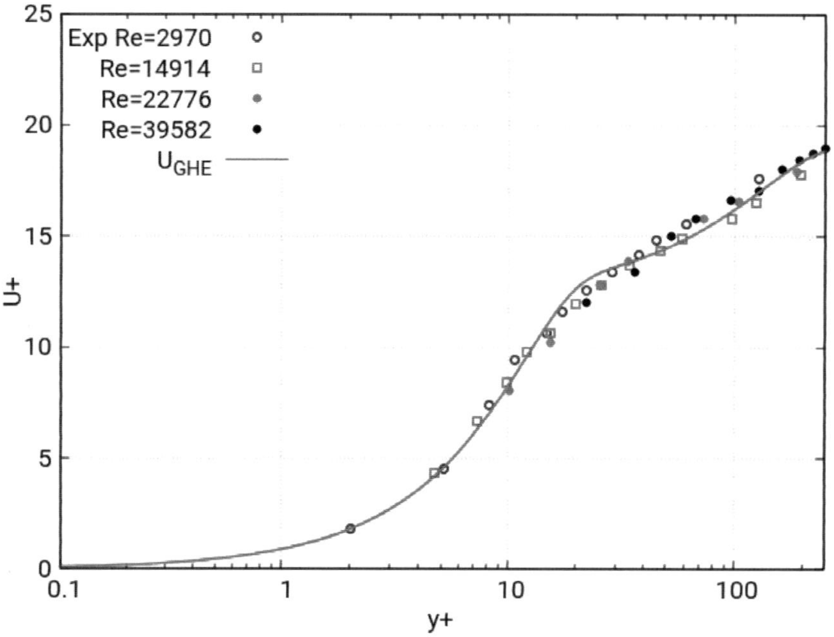

Figure 1: Mean velocity profiles in turbulent boundary layer from Wei (1989) experiment (distilled water), non-dimensionalized on inner variables, for the four Reynolds numbers: 2970 (circles); 14914 (squares); 22776 (red dots) and 39582 (black dots). The red line is an analytical GHE solution.

in 2D channel, where x is the coordinate along a channel, $y \leq L/2$ is the transversal coordinate, L is the width of a channel with a centerline velocity U_0. All parameters are nondimensional. The Eq. (2) presents the laminar flow velocity if $\gamma = 0$, and the turbulent flow if $\gamma > 0$. The parameters δ is

$$\delta = \sqrt{\tau^* \nu}/L_0, \tag{3}$$

where τ^* is the relaxation time, or timescale, a material property for particular liquid or gas used in the experiments, ν is the kinematic viscosity, and L_0 is the hydrodynamic scale. The nondimensional τ, a timescale coefficient for the fluctuation terms in GHE, is expressed as

$$\tau = \tau^* L_0^{-1} U_0 = \delta^2 Re, \tag{4}$$

where $Re = U_0 L_0/\nu$ denotes the Reynolds number. The analytical solution to Eq. (2) can also be used for the turbulent flow in circular pipe, if $\delta \ll 1$.

As an example, the analytical solution U_{GHE} (red line) for the experiments of Wei (1989) is shown in Figure 1 in (U^+, y^+) coordinates. The experimental velocity is shown as points for four Reynolds numbers. The parameter $y^+ = yu_\tau/\nu$ where u_τ is so called friction velocity, y is the absolute distance from the wall, and ν is the kinematic viscosity. One can interpret y^+ as a local Reynolds number. The friction velocity u_τ is defined as

$$u_\tau = \sqrt{\tau_w/\rho}, \tag{5}$$

where wall shear stress τ_w, $\tau_w = \rho\nu\frac{dU}{dy}$ at y=0, and the dimensionless velocity is given by $U^+ = u/u_\tau$. The Figure 1 demonstrates that the superposition U_{GHE} provides an excellent fit to the experimental mean velocity profile for γ=0.65 and δ=0.052, Fedoseyev (2023).

As to parameter γ in Eq.(1), the method proposed in Fedoseyev (2023) to obtain γ from the momentum equation was approximate. In this study, a different approach is proposed, that is to obtain the parameter γ using the minimum viscous dissipation principle.

The contents of the paper is the following. Section 2 presents the GHE and its simplified form for turbulent flow in channel to which the minimization principle to be applied. Section 3 shows a general analytical solution of the

GHE, and Section 4 (i) formulates a minimization principle, that was chosen from several candidates, (ii) applies the principle to several problems, and (iii) compares the obtained analytical solution with the experimental data for turbulent flows. Section 5 provides discussion of the obtained results, which is summarized by the Conclusions.

2 Governing Equation

To proceed with the minimization principle, we have to present the governing equation used, the Generalized Hydrodynamic Equations (GHE). The GHE are obtained from Generalized Boltzmann Transport Equation, Alexeev (2004), by multiplying the latter by the standard collision invariants (mass, momentum, and energy), and integrating the result in the velocity space. The particles of finite size are considered. The obtained equations below are valid for incompressible viscous flow, and have the following non-dimensional form, Fedoseyev (2012):

$$\frac{\partial \mathbf{V}}{\partial t} + (\mathbf{V}\nabla)\mathbf{V} - Re^{-1}\nabla^2\mathbf{V} + \nabla p - \mathbf{F} = \tau\left\{2\frac{\partial}{\partial t}(\nabla p) + \nabla^2(p\mathbf{V}) + \nabla(\nabla\cdot(p\mathbf{V}))\right\}, \tag{6}$$

$$\nabla\cdot\mathbf{V} = \tau\left\{2\frac{\partial}{\partial \mathbf{t}}(\nabla\cdot\mathbf{V}) + \nabla\cdot(\mathbf{V}\nabla)\mathbf{V} + \nabla^2 p - \nabla\cdot\mathbf{F}\right\}, \tag{7}$$

where $\mathbf{V}$ and p are nondimensional velocity and pressure respectively, $Re = U_0L_0/\nu$ - the Reynolds number, U_0 - velocity scale, L_0 - hydrodynamic length scale, ν - kinematic viscosity, $\mathbf{F}$ is nondimensional body force and nondimensional timescale $\tau = \tau^* L_0^{-1} U_0$. Terms containing τ are called the fluctuations (temporal and spatial) by Alexeev (1994). One can see the equations become the Navier-Stokes equations if $\tau = 0$.

The following assumptions were made by deriving Eq. (6, 7):

(i) τ is assumed to be constant.

(ii) The nonlinear terms of the third order in the fluctuations, and the terms of order τ/Re, are neglected. The focus is on large Re numbers.

(iii) Assumed slow flow variation, so second derivatives in time are neglected.

Additional boundary condition was set for pressure on walls:

$$(\nabla p - \mathbf{F})\cdot\mathbf{n} = 0, \tag{8}$$

where $\mathbf{n}$ is a wall normal.

The GHE is not a turbulence model, and no additional equations are introduced. The solution time of the GHE on a computer is the same as that of the Navier-Stokes equations. By setting $\tau = 0$ in the GHE, one actually solves the Navier-Stokes equations.

2.1 GHE for 2D Stationary Incompressible Flow

The case of 2D incompressible fluid flow in channel is considered with the flow direction in x. For the stationary analytical solution in Fedoseyev (2023), GHE (6), (7) were simplified by: (a) dropping all temporal derivatives, (b) dropping all the terms (with coefficient τ) in the momentum equations, (c) the nonlinear terms were neglected in the fluctuations, (d) all the derivatives in x were neglected, except for the pressure gradient p_x=const, so the Laplacian of pressure was $\nabla^2 p = p_{yy}$.

The resulting continuity equation of GHE model is as follows:

$$v_y = \tau p_{yy} \tag{9}$$

while the momentum equations are :

$$v\,u_y + p_x = Re^{-1}u_{yy} \tag{10}$$

$$v\,v_y + p_y = Re^{-1}v_{yy} \tag{11}$$

where Re is Reynolds number, τ is given by Eq.(4). The boundary conditions are as follows: $u = 0$, $v = 0$ and the normal derivative of pressure $p_n = 0$ at the wall $y = 0$; $u = U_0$, and $v = 0$, $\frac{\partial p}{\partial y} = 0$ (the symmetry conditions) at $y = L/2$. A symmetry about the centerline $y = L/2$ is assumed, and the problem is solved in half of the domain.

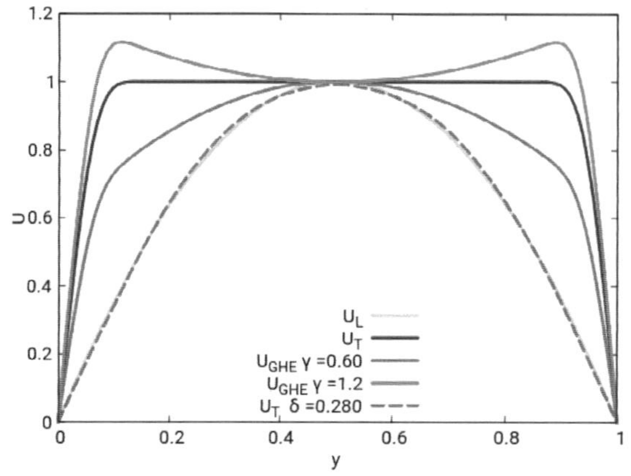

Figure 2: Examples of laminar and turbulent solutions, that is U_L, Eq.(12) (parabolic solution, green line), and U_T, Eq.(13) (super exponential solution, blue line, $\delta = 0.060$) and their superposition U_{GHE} for different γ. It is interesting to note that U_T for $\delta = 0.280$ (red dashed line) coincides well with the parabolic solution.

3 General Analytical Solution of GHE for Turbulent Channel Flow

It was shown in Fedoseyev (2023) that the stationary channel flow problem Eq. (9),(10),(11) has two solutions. The first is the parabolic velocity $u(y)$ for laminar flow profile :

$$U_L = 4U_0 y(L-y)/L^2, \tag{12}$$

where $U_0 = -\frac{1}{8} Re\ p_x L^2$ and $v = 0$ everywhere. This solution is also a solution of the Navier-Stokes equations, as $\nabla^2 p = 0$ in Eq.(9).

However, equations (9),(10), and (11) have a second solution $u(y)$, which was called a turbulent solution. It was obtained analytically by Fedoseyev (2023), and it is a super exponential function:

$$U_T = U_0 \left(1 - e^{1-e^{y/\delta}}\right), \tag{13}$$

and $v(y) = \frac{1}{\delta Re}(1 - e^{y/\delta})$, where the parameter δ is defined by Eq.(3) above. The Navier-Stokes equations do not have such a solution.

Figure 2 shows examples of both U_L (Eq.(12), parabolic, green line) and U_T (Eq.(13), super exponential, blue line) solutions for laminar and turbulent flows respectively, and U_{GHE} solutions for $\gamma = 0.6$ (red) and $\gamma = 1.2$ (pink). It is interesting to note that U_T for $\delta = 0.280$ (red dashed line) coincides well with the parabolic solution U_L (green).

The general solution for turbulent flow is proposed as a linear superposition of laminar and turbulent solutions, as shown in Eq.(2):

$$U_{GHE} = U_0 \left[\gamma \left(1 - e^{1-e^{y/\delta}}\right) + (1-\gamma) 4y(L-y)/L^2\right].$$

The governing equations are nonlinear, and the linear combination above need to be justified. The first term in brackets, γU_T, grows superexponentially in the boundary layer, whereas the second term containing $(1-\gamma)U_L$ is nearly zero, Figure 3. Outside the boundary layer, the first term is constant and the second term starts to grow.

Therefore U_{GHE} becomes a function that is approximated by γU_T in the boundary layer and by $\gamma\cdot$ const + $(1-\gamma)U_L$ outside the boundary layer. As in Eq. (10), (11) and (9) the function $u = U_{GHE}$ enters only as a derivative in y (all derivatives in x are zero), and the constant disappears. As a result, the derivatives of u are approximated by the derivatives of γU_T in the boundary layer and by the derivatives of $(1-\gamma)U_L$ outside the boundary layer, and the proposed linear superposition of solutions is valid.

The minimization principle will be applied to the analytical solution U_{GHE} for turbulent flow.

4 Minimization Principles for Viscous Fluid Dynamics

Several papers provided minimization principles for viscous fluid flow: principles of minimum pressure gradient (PMPG), Taha (2023), minimum kinetic energy dissipation, Lyu'lka (2001), the principle of minimal viscous

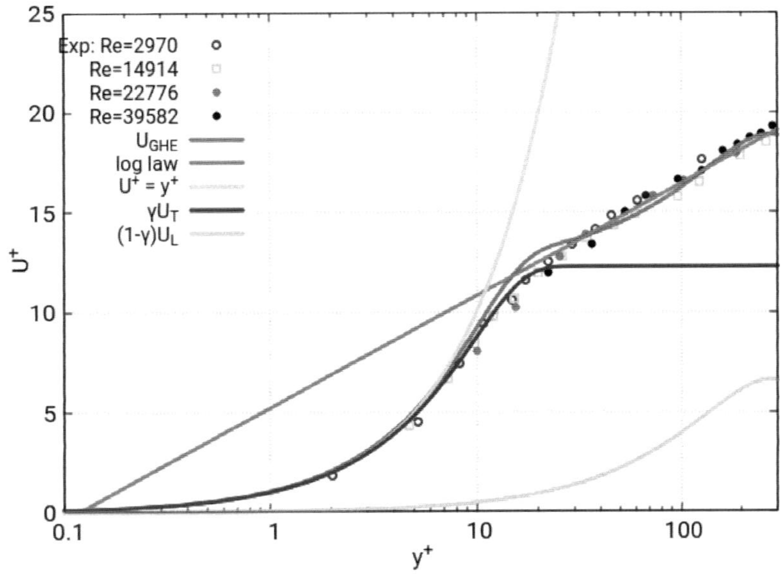

Figure 3: Velocity profiles for turbulent flow in channel: experimental data (dots) by Wei (1989), analytical solution U_{GHE} (red line) and its constituents, the turbulent solution γU_T and the laminar solution $(1-\gamma)U_L$; log law by von Karman U^+ = 1/k log y^+ + B, k = 0.41, B = 5.2 (log law line), and linear law $U^+ = y^+$ (cyan line).

dissipation, Ruangkriengsin (2022), and Helmholtz-Korteweg minimization principle, Borisov (1998), confirming their ideas by the respective examples. Lyu'lka (2001) considered the case of cylindrical pipe with obstacles inside, Taha (2023) examples included the unsteady laminar flow in a channel and the flow from harmonically oscillating plate, Ruangkriengsin (2022) considered the pressure driven Stokes flow in channel of different cross-sections to confirm the proposed principles. Talon (2021) considered the minimization of dissipation as a general principle in physics, and used it for the flow of inelastic non-Newtonian fluids in macroscopic heterogeneous porous medium. Borisov (1998) stated that the Helmholtz-Korteweg minimization principle is widely known principle of viscous incompressible fluid mechanics, analyzed the kinetic energy dissipation functional for internal flows, and found that the stationary point corresponds to the Stokes equations.

The principle of minimal viscous dissipation was selected in this study. The dissipation function of a Newtonian fluid with viscosity μ in 2D channel is

$$E = 2\mu\left[\left(\frac{\partial U}{\partial x}\right)^2 + \left(\frac{\partial V}{\partial y}\right)^2 - \frac{1}{3}(\nabla\cdot\mathbf{V})^2\right] + \mu\left[\frac{\partial V}{\partial x} + \frac{\partial U}{\partial y}\right]^2, \tag{14}$$

$$\nabla\cdot\mathbf{V} = \frac{\partial U}{\partial x} + \frac{\partial V}{\partial y} \tag{15}$$

where the U and V are the components of the dimensional velocity vector, $\mathbf{V}$ is the velocity vector, and x, y are streamwise and transversal coordinates.

In the case of a stationary flow in channel all derivatives with respect to x are zero, and the term $\frac{\partial V}{\partial y}$ is small and can be neglected. Therefore, the remaining term for dissipation is the last term in Eq.(14), and in our notations, it is as follows:

$$\varepsilon(y) = \left(\frac{u_y(y)}{u_y(0)}\right)^2, \tag{16}$$

where the term is made nondimensional using $u_y(0)$, the velocity derivative on the wall, in the denominator.

4.1 Obtaining parameter γ by minimization principle

Parameter γ is obtained using the principle of minimal total viscous dissipation. To obtain the total viscous dissipation, one needs to identify the volume to integrate Eq.(16). The chosen volume is formed by a channel

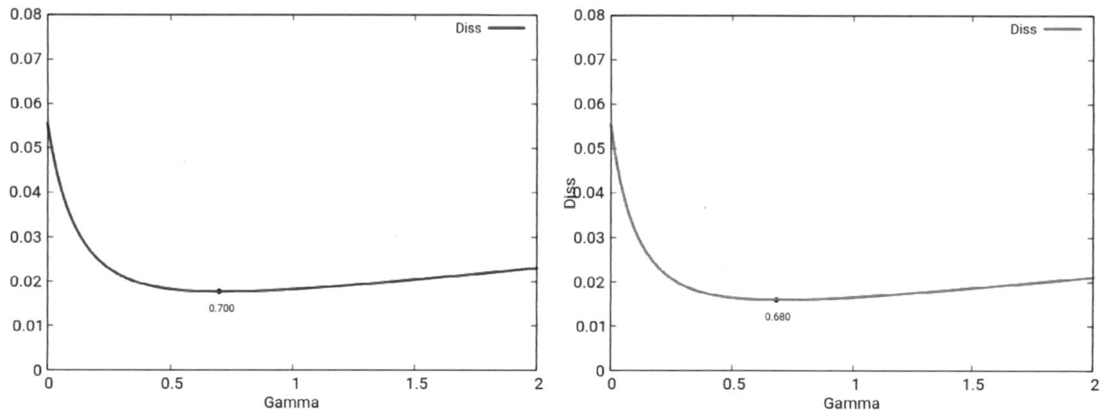

(a) (b)

Figure 4: (a) Total viscous dissipation for the Wei (1989) experiment with $Re = 39582$ versus γ, where γ has an extended range [0,2]. The dissipation has a minimum at $\gamma = 0.70$. Laminar flow ($\gamma = 0$) exhibits significantly larger dissipation than turbulent flow. (b) Total viscous dissipation for Van Doorne (2007) experiment in circular pipe, $Re = 7200$, the minimum is at $\gamma = 0.68$.

cross-section line, channel walls, and the line where cross-section line will be in a unit of time, at a distance of the average velocity of the fluid. Therefore, the integral of Eq.(16) across the channel is multiplied by the distance (average velocity, an integral of the velocity divided by L). Total energy dissipation per unit time is expressed as:

$$\varepsilon_T = \frac{1}{u_y(0)^2} \int_0^L u_y^2 dy \cdot \frac{1}{L} \int_0^L u \, dy, \tag{17}$$

where $u(y) = U_{GHE}$ that depend on γ, and is given by Eq. (2). A similar equation was derived in Horne (1986) (p.6, Eq.(15)).

The dependence of total viscous dissipation on γ calculated for the Wei (1989) experiment is shown in Figure 4(a) at $\delta = 0.052$. The parameter δ is related to the material properties τ^*, ν for distilled water as working fluid, Eq.(3).

The dissipation has minimum $\varepsilon_T = 0.019$ at $\gamma = 0.70$. This is close to the experimental fit $\gamma = 0.65$ in Fedoseyev (2023). The dissipation plot demonstrates that the laminar flow ($\gamma = 0$) has significantly larger dissipation $\varepsilon_T = 0.056$ than the turbulent flow ($\gamma > 0$), and the minimum is almost flat in a range of $\gamma = [0.5, 1.0]$.

4.2 *Comparison with experiments*

Figure 4(b) shows γ calculated for the Van Doorne (2007) experiment, with $Re = 7200$ and $\delta = 0.047$. The minimum of dissipation is at $\gamma = 0.68$, while the best experimental fit is $\gamma = 0.65$ in Fedoseyev (2023).

Figure 5(a) shows the experimental data digitized from Van Doorne (2007), along with several plots: (i) laminar (parabolic) flow profile (green line), (ii) turbulent (superexponential) solution (blue line) and (iii) GHE analytical solution (red line). The left part of the GHE plot is for $\gamma = 0.68$ (minimal dissipation), and the right part is for $\gamma = 0.65$ from Fedoseyev (2023). The figure demonstrates that neither the laminar nor turbulent solution fit the data, but the superposition U_{GHE} provides a good comparison to the experimental data.

Figure 5(b) shows the streamwise velocity in a turbulent channel experiment by Pasch (2023), with $Re = 14000$, along with the analytical solution U_{GHE} with coefficient $\gamma = 0.62$ obtained by the minimum of viscous dissipation principle. Here, $\delta = 0.033$, the working fluid is air.

5 Discussion

It was shown that the parameter γ can be obtained using the minimization principle. Comparison with several experiments have been provided demonstrating good agreement of the analytical solution with the experimental data. The GHE solution for turbulent flows depends on two similarity parameters: the Reynolds number Re, and the parameter $\delta = \sqrt{\tau^* \nu}/L_0$, which is related to the material properties τ^* and ν.

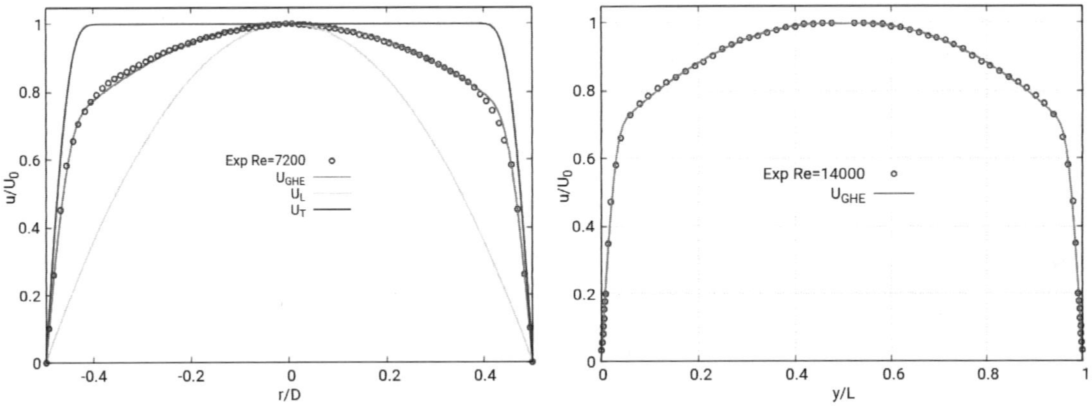

(a) (b)

Figure 5: *(a)* A comparison of the experimental data for the streamwise velocity $U = u/U_0$ versus radius r/D (D-diameter) by Van Doorne (2007), (blue dots) at $Re = 7200$, with the GHE solution (red line). The right part of the U_{GHE} plot is for $\gamma = 0.65$ from Fedoseyev (2023), and the left part is for $\gamma = 0.68$, obtained from the minimum viscous dissipation principle. Both GHE analytical solutions fit the experimental velocity profile well. Also shown are the laminar solution (green line) and the turbulent solution (blue line). *(b)* Comparison of the streamwise velocity in the turbulent channel experiment by Pasch (2023), $Re = 14000$, (dots) and the analytical solution U_{GHE} with coefficient $\gamma = 0.62$ (red line) found by the minimization principle.

5.1 Similarity Parameter δ

The similarity parameter δ does not depend on the Reynolds number as shown in the case of Wei & Willmarth experiment, Wei (1989), where the experimental data for different Reynolds numbers are falling into the same curve (Figure 3, red line U_{GHE}) defined by the parameter δ. Knowing δ one can find the timescale coefficient τ^*, which is the material property.

5.2 Timescale Coefficient τ

Analyzing several experiments and simulations, Wei (1989), Koseff (1984), Van Doorne (2007), and Fedoseyev (2010), Fedoseyev (2012), Fedoseyev (2023) we have found that the dimensional timescale coefficient $\tau^* = \delta^2 L_0^2/\nu$ for distilled water is $\tau^* = 0.40 \pm 0.05$ s, and for tap water is $\tau^* \approx 0.80$ s. There is no estimation of τ^* for air, as the viscosity, temperature and pressure were not provided in Pasch (2023).

5.3 Turbulent Boundary Layer

One can see that the analytical solution of GHE presents well the turbulent boundary layer in all the regions. The linear law is in the range $0 < y^+ < 5$, where the parabolic profile $(1 - \gamma)U_L$ (laminar solution) is very small, and the analytical solution for small y/δ becomes

$$U_{GHE} \;\; = \;\; \gamma U_T = U_0 \gamma \left(1 - e^{1-e^{y/\delta}}\right) = U_0 \gamma y/\delta,$$

that is a linear law.

The near-middle (buffer) boundary layer region is the range $5 < y^+ < 30$, a strictly nonlinear region, and the analytical solution fits the experiment quite satisfactory. In the far-middle (inner) boundary layer region, the range of $30 < y^+ < 200$, the superexponential part of γU_T becomes nearly constant ≈ 12.5 (Figure 3, blue line), and the GHE analytical solution (red line) changes due to the growth of the laminar part of solution $(1 - \gamma)U_L$ (Figure 3, cyan line). In this region the GHE solution is $U_{GHE} = (12.5 + (1 - \gamma)U_L)$ and fits well to logarithmic von Karman law, and the experimental data. The outer (non-linear, essentially inviscid) region starts at $y^+ > 200$ and continues to the center line, where the analytical solution fit the experiment well too.

Conclusions

The minimum viscous dissipation principle as a method for calculating the coefficients of an analytical solution for turbulent flow in a channel has been presented. The obtained analytical solution was compared with experimental data from several turbulent flow experiments, demonstrating a good agreement with the mean experimental velocity data, successfully capturing the correct velocity behavior across the entire turbulent boundary layer and into the external flow, spanning from the inner viscous sublayer to the outer layer of the boundary layer.

References

Alexeev B.V. 1994 The generalized Boltzmann equation, generalized hydrodynamic equations and their applications. Phil. Trans. Roy. Soc. London, A. 349 (1994), 417-443.

Alexeev B.V. 2004 Generalized Boltzmann Physical Kinetics, Elsevier.

Armaly B.F., Durst F., Pereira J.C.F., Schonung B. 1983 Experimental and theoretical investigation of backward-facing step flow, J. Fluid Mech. (1983), vol.127, pp. 473-496.

Borisov V.M., Lyul'ka V.A.1998 Some viscous flows that implement the principle of minimum dissipation of kinetic energy. Comput. Math. and Math. Phys., v.39, iss.12,pp.2015-2019

Van Doorne C.W.H., Westerweel J. 2007 Measurement of laminar, transitional and turbulent pipe flow using Stereoscopic-PIV, February 2007, Experiments in Fluids 42(2), DOI: 10.1007/s00348-006-0235-5.

Fedoseyev A. 2023 Approximate Analytical Solution for Turbulent Flow in Channel, J. Phys.: Conf. Ser. 2675 012011.

Fedoseyev A.I., Alexeev, B.V. 2010 Simulation of viscous flows with boundary layers within multiscale model using generalized hydrodynamics equations, Procedia Computer Science, 1, 665-672.

Fedoseyev A., Alexeev B.V. 2012 Generalized hydrodynamic equationsfor viscous flows-simulation versus experimental data,in AMiTaNS-12, American Institute of Physics AIP CP 1487, pp.241-247.

Gorban A.N., Karlin I. 2014 Hilberts 6th problem: exact and approximate hydrodynamic manifolds for kinetic equations, American Mathematical Society, V. 51, N. 2, April 2014, 187-246, S0273-0979(2013)01439-3

Hilbert, D. 1902 Mathematical problems, Bull. Amer. Math. Soc. 8 (1902), no. 10, 437-479.

Horne C.F., Karamcheti K. 1986 Vortical Dissipation in Two-Dimensional Shear Flows, NASA Technical Memorandum 89408, Dec. 1986.

Koseff R., Street R. L. 1984 The Lid-Driven Cavity Flow: A Synthesis of Qualitative and Quantitative ObservationsTrans, ASME J. Fluids Eng. 106, 390-398.

Lyul'ka V.A. 2001 On the Principle of Minimum Kinetic Energy Dissipation in the Nonlinear Dynamics of Viscous Fluid, Technical Physics, Vol. 46, No. 12, 2001, pp. 1501-1503.

Pasch S., Leister R., Gatti D. et al. 2023 Measurements in a Turbulent Channel Flow by Means of an LDV Profile Sensor, Flow, Turbulence and Combustion, https://doi.org/10.1007/s10494-023-00469-4.

Ruangkriengsin T., Roper M. 2022 Approximating pressure-driven Stokes flow using the principle of minimal excess dissipation, arXiv:2204.07240v [physics.flu-dyn]

Taha H., Gonzalez C., Shorbagy M. 2023 A minimization principle for incompressible fluid mechanics, Phys.Fluids 35, 127110.

Talon L. 2021 Minimum principle for the flow of inelastic non-Newtonian fluids in macroscopic heterogeneous porous media, Phys. Rev. Fluids 7, L042101.

Wei T., Willmarth W.W. 1989 Reynolds-number effects on the structure of a turbulent channel flow, J. Fluid Mech., vol 204, pp. 57-95.

Journal of Physics: Conference Series **2910** (2024) 012013 doi:10.1088/1742-6596/2910/1/012013

Analytical Solution for Transverse Velocity of Turbulent Flow in Channel

Alex Fedoseyev

Ultra Quantum Inc., Huntsville, Alabama, USA

E-mail: af@ultraquantum.com

Abstract. The analytical solution for the streamwise mean velocity of turbulent flow of incompressible fluid in a channel was presented in Fedoseyev (2023) paper. In this work, we present both the stationary and transient solutions for the transverse velocity component. The obtained transient solution is a product of a series of periodic functions in space and time, multiplied by a growing exponential function of time. This analytical solution reveals the instability and imminence of oscillations in the turbulent flow. The obtained solutions are compared with numerical simulation results and experimental data, demonstrating a close agreement or strong similarity.

1 Introduction

Generalized Hydrodynamic Equations (GHE) have been proposed by Boris Alexeev in 1994 [1], [2], more details are provided in [3]. They were used for the simulations of incompressible viscous flows for a wide range of problems and flow parameters [5, 6, 7, 8, 10], resulting in good agreement with experiments [14], [13] and other works.

The approximate analytical solution for a mean streamwise velocity for turbulent flow in channel was obtained in [11] as a superposition of parabolic and superexponential functions:

$$U = V_0 \left[\gamma \left(1 - e^{1-e^{y/\delta}} \right) + (1-\gamma) 4y(L-y)/L^2 \right], \tag{1}$$

where $y \leq L/2$ is the transverse coordinate, L is the width of a channel with a centerline velocity V_0, and γ and $(1-\gamma)$ are the coefficients of superposition. All parameters are nondimensional. The parameter δ, the boundary layer scale, is

$$\delta = \sqrt{\tau^* \nu}/L_0, \tag{2}$$

where τ^* is the relaxation time, or timescale, a material property for particular liquid or gas used in the experiments, ν is the kinematic viscosity, and L_0 is the hydrodynamic scale. The nondimensional τ, a timescale coefficient for the fluctuation terms in GHE, is expressed as

$$\tau = \tau^* L_0^{-1} U_0 = \delta^2 Re, \tag{3}$$

where $Re = U_0 L_0/\nu$ denotes the Reynolds number.

This analytical solution demonstrated good comparison with experiments, Figure 1 shows a comparison with experiment of [16] for four Reynolds numbers, the values are non-dimensionalized on inner variables, (y^+, U^+) coordinates.

Content from this work may be used under the terms of the Creative Commons Attribution 4.0 licence. Any further distribution of this work must maintain attribution to the author(s) and the title of the work, journal citation and DOI.
Published under licence by IOP Publishing Ltd

Journal of Physics: Conference Series **2910** (2024) 012013 doi:10.1088/1742-6596/2910/1/012013

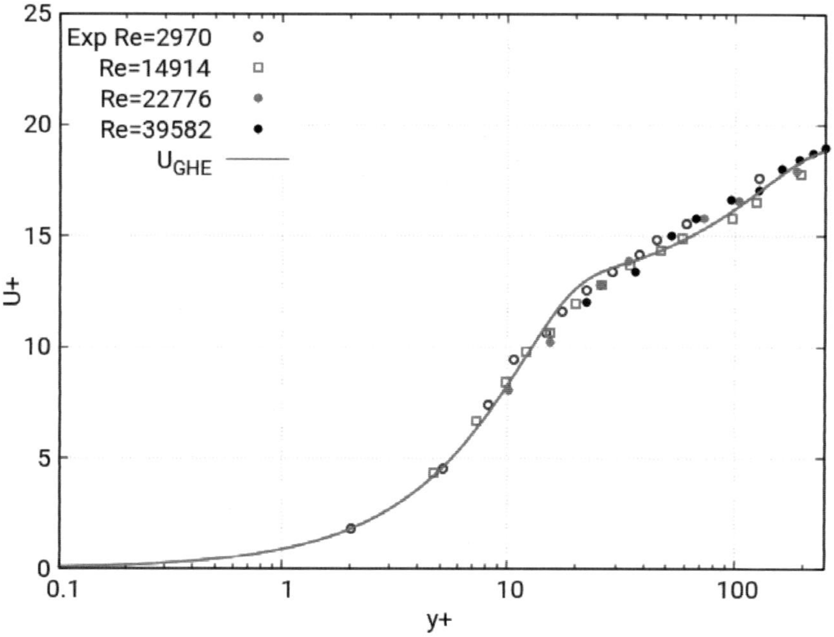

Figure 1: Mean velocity profiles in turbulent boundary layer from [16] experiment (distilled water), non-dimensionalized on inner variables, for the four Reynolds numbers: 2970 (green circles); 14914 (blue squares); 22776 (red dots) and 39582 (black dots). The red line is an analytical GHE solution.

A better way to obtain the coefficients of superposition was presented in [12] where the principle of minimal viscous dissipation for turbulent flow in channel was used.

The estimation for transverse mean velocity V in turbulent boundary layer was provided in [11] as

$$V \sim \frac{1}{\delta^2 Re^2}(1 - e^{y/\delta}). \tag{4}$$

In this work we present the exact and approximate stationary solutions and transient solution for transverse velocity of turbulent flow in channel and compare it with the experimental data. For the laminar flow, the transverse velocity is zero.

The Generalized Hydrodynamic Equations will be employed for turbulent flow in a channel. The GHE were derived from the Generalized Boltzmann Equation (GBE), which takes into account finite particle size [1], while in the traditional Boltzmann equation, particles are treated as material points. So far, the improved kinetic equation was chosen, and the respective improved hydrodynamic equations have been used in [11]. A brief outline of the basic ideas of GBE and derivation of governing Generalized Hydrodynamic Equations is presented in Section 1.1.

1.1 Generalized Boltzmann Transport Equation (GBE)

Physical derivation of the standard Navier-Stokes equations (NS) can be obtained from the kinetic theory of gases, which is based on the solution to the Boltzmann transport equation for space-time evolution of particle velocity distribution function, f, written in the form

$$\frac{Df}{Dt} = J, \tag{5}$$

where

$$\frac{D}{Dt} = \frac{\partial}{\partial t} + \mathbf{v} \cdot \frac{\partial}{\partial \mathbf{r}} + \mathbf{F} \cdot \frac{\partial}{\partial \mathbf{v}}$$

represents material derivative in space, velocity space and time and J is the collision integral ([3], p.11), where $\mathbf{v}$ and $\mathbf{r}$ are the velocity and the radius-vector of the particle, respectively, and $\mathbf{F}$ is the force acting on the particle. The standard Boltzmann transport equation takes into account the changes in distribution function f on hydrodynamic and mean time between collision scales of infinitesimal particles. Accounting for a third time scale associated with finite dimensions of interacting particles gives rise to an additional term in the Boltzmann transport equation resulting in a generalization of the form as

$$\frac{Df}{Dt} - \frac{D}{Dt}(\tau^* \frac{Df}{Dt}) = J, \tag{6}$$

where τ^* is the mean time between particle collisions. The new term is thermodynamically consistent and is proportional to the Knudsen number, $Kn = \lambda/L_0$ (λ is the mean free path between collisions, and L_0 is the hydrodynamic scale), and therefore in the hydrodynamic limit, to viscosity. More details on the GBE are provided in Alexeev's book [3].

2 Generalized Hydrodynamic Equations

Hydrodynamic equations can be obtained from Eq. (6) by multiplying the latter by the standard collision invariants (mass, momentum, energy) and integrating the result in the velocity space. These equation are for incompressible viscous flow, and are the following:

$$\frac{\partial \mathbf{V}}{\partial \mathrm{t}} + (\mathbf{V}\nabla)\mathbf{V} - Re^{-1}\nabla^2\mathbf{V} + \nabla \mathrm{p} - \mathbf{F} = \tau \left\{ \frac{\partial^2 \mathbf{V}}{\partial \mathrm{t}^2} + 2\frac{\partial}{\partial \mathrm{t}}(\nabla \mathrm{p}) + \nabla^2(\mathrm{p}\mathbf{V}) + \nabla(\nabla \cdot (\mathrm{p}\mathbf{V})) \right\} \tag{7}$$

while continuity equation is

$$\nabla \cdot \mathbf{V} = \tau \left\{ \mathbf{2}\frac{\partial}{\partial \mathbf{t}}(\nabla \cdot \mathbf{V}) + \nabla \cdot (\mathbf{V}\nabla)\mathbf{V} + \nabla^2 \mathrm{p} - \nabla \cdot \mathbf{F} \right\} \tag{8}$$

where $\mathbf{V}$ and p are nondimensional velocity and pressure, $Re = V_0 L_0/\nu$ - the Reynolds number, V_0 - velocity scale, L_0 - hydrodynamic length scale, ν - kinematic viscosity, $\mathbf{F}$ is a body force and a nondimensional τ is

$$\tau = \tau^* L_0^{-1} V_0 = l^2/L_0^2 Re = \delta^2 Re. \tag{9}$$

Note that the right-hand side of Eq.(8) is the divergence of the fluctuation part of flow velocity, expressed explicitly through original primitive variables according to [1]. The parameter τ is assumed to be constant. No assumption of slow flow variation is made, so the second derivatives in time are kept.

Additional boundary conditions on walls are for fluctuations to be zero. The following boundary condition for pressure on walls is

$$(\nabla \mathrm{p} - \mathbf{F}) \cdot \mathbf{n} = 0, \tag{10}$$

where $\mathbf{n}$ is a wall normal.

Note, that the dimension of a product $\tau^*\nu$ is a square of length. We introduce fluctuation length scale l, $l^2 = \tau^*\nu$ and rewrite nondimensional τ as $\tau = l^2 L_0^{-2} Re = \delta^2 \cdot Re$, where $\delta = l/L_0$. The parameter τ^* is a material property, the value is known for a distilled water [11].

In this paper, to obtain the analytical solution, we further simplify the GHE equations: the nonlinear terms are neglected in the fluctuations, so we consider a case of small velocities.

2.1 *Governing Equations for 2D Transient Incompressible Flow*

The case of 2D incompressible fluid flow is considered. The equations are taken from [10] where they are presented explicitly, and further simplification is done by dropping the nonlinear terms from fluctuations (terms with coefficient τ) in the momentum and continuity equations.

The resulting continuity equation of GHE model is the following:

$$u_x + v_y = \tau \left[2(u_{xt} + v_{yt}) + \nabla^2 p\right] \tag{11}$$

while the momentum equations are :

$$u_t + u\,u_x + v\,u_y + p_x = Re^{-1}\nabla^2 u + \tau(u_{tt} + 2p_{xt}) \tag{12}$$

$$v_t + u\,v_x + v\,v_y + p_y = Re^{-1}\nabla^2 v + \tau(v_{tt} + 2p_{yt}) \tag{13}$$

where the subscript at the variable means a derivative, $Re = L_0V_0/\nu$ is Reynolds number, L is the length scale, V_0 is velocity scale, ν is kinematic viscosity, $\delta^2 = \tau^*\nu/L_0^2$.

One may point out that the continuity Eq. (11) is not correct, and the mass is not conserved for incompressible fluid (may have sources and sinks). This problem is discussed in [11], [12], and the validity of the continuity equation is justified .

3 Governing Equations for 2D Channel Flow

In this case we take the equations (11,12,13) for 2D channel flow and assume $u_x = v_x = 0$, and $p_x = const$. We get

$$v_y = \tau(2v_{yt} + p_{yy}) \tag{14}$$

$$u_t + vu_y + p_x = Re^{-1}u_{yy} + \tau \cdot u_{tt} \tag{15}$$

$$v_t + vv_y + p_y = Re^{-1}v_{yy} + \tau \cdot (v_{tt} + 2p_{yt}) \tag{16}$$

The analytical solution for streamwise velocity in a stationary case (mean velocity) has been obtained in [11]. In this work we will obtain the stationary and transient analytical solutions for the transverse velocity.

3.1 Equation for Transverse Velocity in 2D Channel Flow

Integrating Eq.(14) in y one obtains

$$\tau \cdot p_y = v - 2\tau \cdot v_t + C \tag{17}$$

As $p_y = 0, v = 0$ at the wall, then $C = 0$. Then

$$p_y = 1/\tau \cdot v - 2v_t \tag{18}$$

and differentiating in t one obtains

$$p_{yt} = 1/\tau \cdot v_t - 2v_{tt}. \tag{19}$$

Substituting Eq.(18, 19) into Eq.(16), the following equation is obtained:

$$v_t + vv_y + 1/\tau \cdot v - 2v_t = Re^{-1}v_{yy} + \tau \cdot (v_{tt} + 2/\tau \cdot v_t - 4v_{tt}) \tag{20}$$

or

$$-v_t + vv_y + 1/\tau \cdot v = Re^{-1}v_{yy} - 3\tau \cdot v_{tt} + 2v_t. \tag{21}$$

Moving the time derivatives into the left hand side,

$$3\tau \cdot v_{tt} - 3v_t + 1/\tau \cdot v = Re^{-1}v_{yy} - vv_y, \tag{22}$$

resulting in

$$v_{tt} - 1/\tau \cdot v_t = 1/(3\tau Re)v_{yy} - 1/(3\tau) \cdot vv_y - 1/(3\tau^2) \cdot v. \tag{23}$$

4 Exact Stationary Solution

In the stationary case the Eq. (27) reduces to

$$\alpha \cdot v_{yy} - vv_y - 1/\tau \cdot v = 0, \tag{24}$$

where $\alpha = 1/Re$. The analytical solution of Eq.(24) is available in [15], Sec.2.2.3-2, Eq.2, and Sec.1.3.1-2 Eq.1, in implicit form.

The alternative analytical solution is available from WolframAlpha.com in implicit form as Eq. (25),

$$y + C_2 = -\int_1^{v(y)} \tau/(W(-e^{-\tau(\xi^2+2\alpha\tau C_1)/(2\alpha)-1}) + 1)d\xi \tag{25}$$

where $W(z) = W[k, z]$ is Lambert W function (or the Product Log function), the solution for w in $z = w\exp(w)$. Constants C_1 and C_2 are chosen to satisfy boundary conditions. There is the $k - th$ solution of the Eq.(24) corresponding to $W[k, z]$ in Eq.(25), $k = 0, -1$ for real values of $W[k, z]$. There are multiple solutions of $v(y)$, and then multiple solutions $u(y)$ can be found by substituting $v(y)$ into Eq.(15). The pressure can be found using Eq.(14). This can not be done analytically to obtain the explicit formulas for $u(y)$ and $p(y)$.

4.1 Limits of Stationary Solution
Recalling that $\tau = \delta^2 Re$, the Eq.(24) becomes

$$\frac{1}{Re} v_{yy} - v v_y - \frac{1}{\delta^2 Re} v = 0. \tag{26}$$

In the limit of $Re \to \infty$ the Eq.(24) becomes

$$v v_y = 0,$$

with the solution $v^2 = C_1$. As the boundary condition is $v = 0$ at $y = 0$, then $C_1 = 0$, and the solution limit is $v = 0$ for $Re \to \infty$.
In the limit of $Re \to 0$ the Eq.(24) becomes

$$v_{yy} - 1/\delta^2 \cdot v = 0,$$

with the non-zero solution $v = C_1 exp(y/\delta)$. As the boundary condition is $v = 0$ at $y = 0$, then $C_1 = 0$. Therefore, the solution limit is $v = 0$ for $Re \to 0$.

5 Transient Solution by Method of Separation of Variables

We drop the nonlinear term in Eq.(23), assuming it is small:

$$v_{tt} - 1/\tau \cdot v_t = 1/(3\tau Re) v_{yy} - 1/(3\tau^2) \cdot v. \tag{27}$$

A separation of variables is used to find $v(y,t)$ solution:

$$v(y,t) = g(y) \cdot h(t). \tag{28}$$

Substituting Eq. (28) into Eq.(27) we have the following

$$g \cdot h_{tt} - 1/\tau \cdot g \cdot h_t + 1/(3\tau^2) \cdot g \cdot h = 1/(3\tau Re) h \cdot g_{yy}. \tag{29}$$

Dividing by $g \cdot h$ we get

$$(h_{tt} - 1/\tau \cdot h_t + 1/(3\tau^2) \cdot h)/h = 1/(3\tau Re) g_{yy}/g = -C. \tag{30}$$

where C is a constant (to be found).
As a result, we get two separate equations for $g(y)$ and $h(t)$:

$$h_{tt} - 1/\tau \cdot h_t + (C + 1/(3\tau^2)) \cdot h = 0 \tag{31}$$
$$g_{yy} + C(3\tau Re) \cdot g = 0. \tag{32}$$

5.1 Solution for G-equation
The solution for $g(y)$ in Eq.(32) is $g(y) = C_1 \sin(ky)$ chosen to satisfy boundary conditions $v(0) = v(L) = 0$, and so $g(0) = g(L) = 0$. Therefore $k = \sqrt{3C\tau Re}$ has to satisfy $kL = n\pi$:

$$\sqrt{C3\tau Re} L = n\pi, \tag{33}$$
$$\sqrt{C} = \pi n/(\sqrt{3\tau Re} L), \tag{34}$$
$$C = \pi^2 n^2/(3L^2 \tau Re). \tag{35}$$

Then the solution for $g(y)$ will be the following sum

$$g(y) = \sum_{n=1}^{\infty} g_n \sin(\pi n y/L). \tag{36}$$

5.2 Solution for H-equation

Eq.(31) has a solution following [15](Section 2.1.2-2) for $h_{tt} + ah_t + bh = 0$:

$$\begin{aligned}
\lambda^2 &= a^2 - 4b, \\
h &= exp(-\frac{1}{2}at)|C_1 exp(\frac{1}{2}\lambda t) + C_2 exp(-\frac{1}{2}\lambda t)|, \lambda^2 > 0, \\
h &= exp(-\frac{1}{2}at)|C_1 sin(\frac{1}{2}\lambda t) + C_2 cos(\frac{1}{2}\lambda t)|, \lambda^2 < 0, \\
h &= exp(-\frac{1}{2}at)(C_1 t + C_2), \lambda^2 = 0.
\end{aligned}$$

Taking coefficients of Eq.(31) we get:

$$\begin{aligned}
a &= -1/\tau, \\
b &= C + 1/(3\tau^2) = \pi^2 n^2/(3L^2\tau Re) + 1/(3\tau^2), \\
\lambda^2 &= a^2 - 4b.
\end{aligned}$$

In our case $\lambda^2 < 0$, $\lambda = \sqrt{4b - a^2} = \sqrt{\pi^2 n^2/(3L^2\delta^2 Re^2) + 1/(3\delta^4 Re^2)}$, and the solution $h(t)$ is:

$$h(t) = exp(\frac{t}{2\tau})|C_1 sin(\frac{1}{2}\lambda t) + C_2 cos(\frac{1}{2}\lambda t)|.$$

The solution of H equation is an exponential function of time multiplied by a periodic function of time.

5.3 Solution for Velocity V(y,t) and Comparison with Experiment

The solution for $v(y, t)$ is

$$v(y,t) = exp(\frac{t}{2\delta^2 Re}) \sum_{n=1}^{\infty} |C_n^1 sin(\frac{1}{2}\lambda_n t) + C_n^2 cos(\frac{1}{2}\lambda_n t)| \cdot g_n \sin(\pi n y/L), \quad (37)$$

$$\lambda_n = \frac{1}{\delta Re\sqrt{3}} \sqrt{\pi^2 n^2/L^2 + 1/\delta^2}. \quad (38)$$

where τ was replaced by $\tau = \delta^2 Re$. The coefficients C_n^1, C_n^2 and g_n are defined by the initial condition.

6 Discussions

A qualitative comparison of the transient analytical solution given by Eq.(37) (four terms) with the experiment [16] for the turbulent velocity $v(t)$ in the time interval $\Delta T = 0.1s$ in the boundary layer at $y^+ = 45.3$ is shown in Figure 2. A overall change of amplitude is not visible, as the the dimensional time scale $2\delta^2 Re$ in the multiplier $exp(t/(2\delta^2 Re))$ is rather large.

A growing exponential multiplier in the solution reveals the imminence of oscillations in the turbulent flow in the linear approximation. Any perturbation will grow in the amplitude to some extent, then the presented linear analysis will loose the validity. Apparently, the nonlinear effects then will come into play, limiting the oscillation amplitude of $v(y, t)$ (typically the scale of v is much less then the scale of streamwise velocity V_0).

The space-dependent part of $v(y, t)$ solution, the function $g(y)$, is compared with the simulation result, where the stationary solution of GHE, Eq.(7) and Eq.(8), was obtained by the finite-element method in a channel of length $L_{ch} = 50$, and channel width $L = 1$. Figure 3(a) presents the simulation result (blue line), analytical solution $g(y)$ (red line), Eq.(36) where only the second term is present, $n = 2$, $g_2 = 0.00026$ at $Re = 14914$ (case of Wei (1989) experiment), and the approximate analytical solution for V from [11](dashed brown line marked as $'V(2023)'$). The analytical solution $g(y)$ and the simulation result $Vsym$ and are close enough. The solution for V from [11] is close to $Vsym$ only in the boundary layer. Figure 3(b) presents the analytical solution for the streamwise velocity U using the formula

$$U = V_0 \left[\gamma \left(1 - e^{1-e^{y/\delta}}\right) + (1-\gamma)4y(L-y)/L^2\right]$$

from Fedoseyev (2023) [11] and the experimental data from Wei (1989) [16] at different Reynolds numbers, for comparison of scales with Figure 3(a).

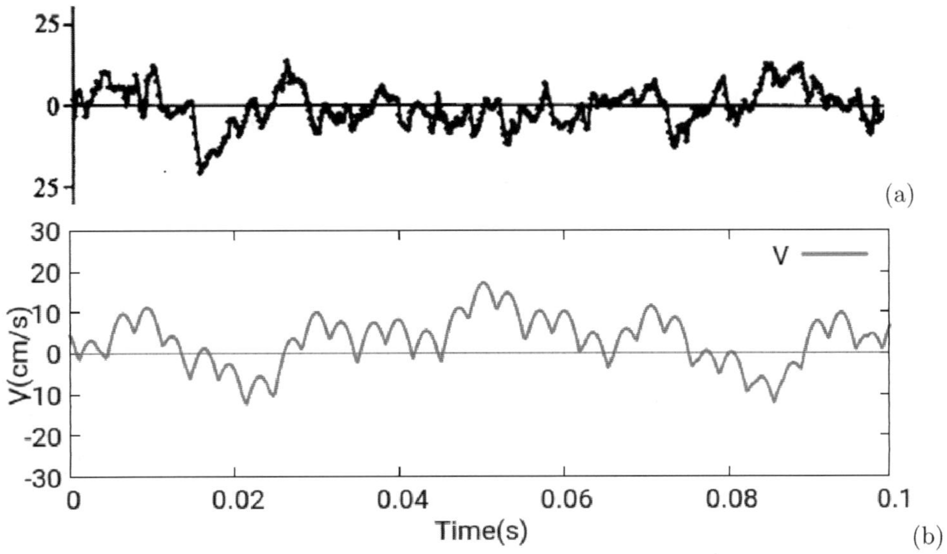

Figure 2: (a) Instantaneous turbulent velocity $v(t)$ in the time interval 0.1 s in the boundary layer at $y^+ = 45.3$ (see Figure 1(b) for location) from experiment [16], and (b) analytical solution, Eq.(38) (four terms), qualitative comparison, at $Re = 14914$.

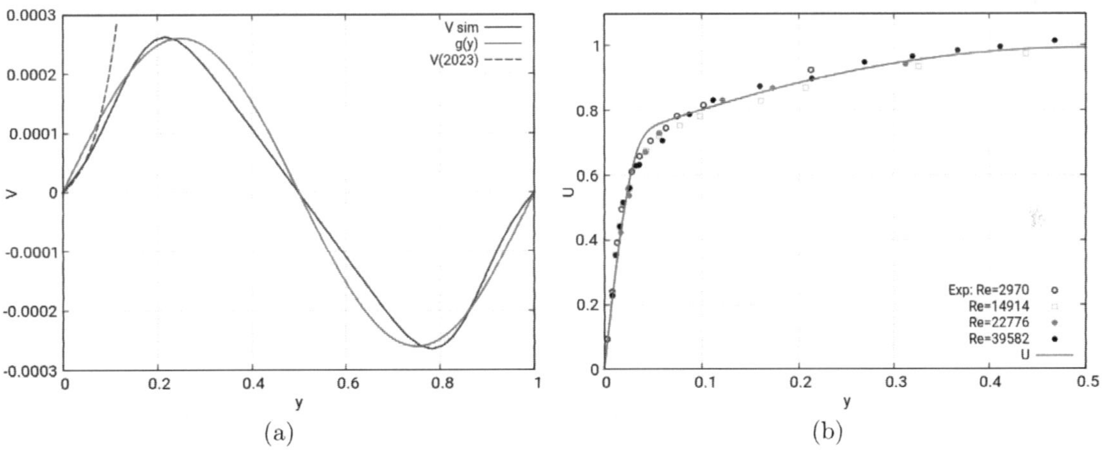

Figure 3: (a) Analytical solution for the transverse velocity V, the space-dependent part (red line marked as $g(y)$), Eq.(36) (term for $n = 2$), the simulation result, the stationary solution of GHE in a channel of length $L_{ch} = 50$ at $Re = 14914$ (blue line marked as V_{sim}) and the analytical solution for V in [11] (dashed brown line marked as V[11]). (b) Analytical solution for the streamwise velocity U using the formula from Fedoseyev (2023) [11] (red line) and the experimental data from Wei (1989) [16] at different Reynolds numbers (dots).

Conclusions

The stationary and transient analytical solutions of the Generalized Hydrodynamic Equations for the transverse velocity in turbulent channel flow are presented. The results are compared with numerical simulation and experimental data, demonstrating similarity for transient solution, and close agreement for spacial velocity distribution.

The transient solutions is a product of series of periodic functions in space and time multiplied by a slowly growing exponential function of time. Any perturbation will grow in the amplitude, then the presented linear analysis will loose the validity. Thus the analytical solution reveals the instability and imminence of oscillations in the turbulent flow.

Nomenclature and Units
Units are provided for dimensional parameters only.

L_0 hydrodynamic scale m
U_0 velocity scale m/s
L channel width
V_0 centerline velocity
U analytical solution for streamwise velocity
V analytical solution for transverse velocity
$\mathbf{V}$ velocity vector
$\mathbf{F}$ body force vector
$W(z)$ Lambert W function
f particle distribution function
Re Reynolds number
D symbol of material derivative
n integer number
p pressure
$g(y)$ space-dependent part of solution
$h(y)$ time-dependent part of solution
u streamwise velocity
v transverse velocity
x streamwise coordinate
y transverse coordinate
γ coefficient of superposition
ν kinematic viscosity m^2/s
τ^* relaxation time s
τ relaxation time
$z = wexp(w)$ w is solution, a Lambert W function
τ relaxation time
C, C_1, C_2 integration constants
δ boundary layer scale
k integer number
a coefficient of h-equation
b coefficient of h-equation
λ parameter of h-equation solution

References

[1] Alexeev B.V.,1994 The generalized Boltzmann equation, generalized hydrodynamic equations and their applications. Phil. Trans. Roy. Soc. London, A. 349, 417-443.

[2] Alekseev, B.V., Grushin, I.T., 1994 Protsessy Perenosa v Reagiruyushchikh Gazakh i Plazme (Transport Processes in Reacting Gases and Plasma), Energoatomizdat, Moscow.

[3] Alexeev B.V.,2004 Generalized Boltzmann Physical Kinetics, Elsevier.

[4] Van Doorne C.W.H., Jerry Westerweel, 2007, Measurement of laminar, transitional and turbulent pipe flow using Stereoscopic-PIV, February 2007, Experiments in Fluids 42(2), DOI: 10.1007/s00348-006-0235-5.

[5] Fedoseyev A.I., Alexeev, B.V. 1998 Mathematical model for viscous incompressible fluid flow using Alexeev equations and comparison with experimental data. In: Dey, S.K., Ziebarth, G., Ferrandiz (Eds.), Proceedings of Advances in Scientific Computing Modelling (Special Proceedings of IMACS'98). Alicante, Spain, 158-163.

[6] Fedoseyev A.I., Kansa E.J., C. Marin, M. Volz, and A. G. Ostrogorsky, 2000 AIAA Paper 2000-0698.

[7] Fedoseyev A., 2001a A regularization approach to solving the Navier-Stokes Equations for Problems with Boundary Layer, Comput. Fluid Dynamics J., 9, 317-324.

[8] Fedoseyev A.I., Kansa E.J., C. Marin, and A.G. Ostrogorsky, 2001b Japanese Computational Fluid Dynamics Journal, 10(3), 325-333.

[9] Fedoseyev A., Alexeev, B.V., 2010 Simulation of viscous flows with boundary layers within multiscale model using generalized hydrodynamics equations, Procedia Computer Science, 1, 665-672.

[10] Fedoseyev A., Alexeev B.V.,2012 Generalized hydrodynamic equationsfor viscous flows-simulation versus experimental data,in AMiTaNS-12, American Institute of Physics AIP CP 1487, pp.241-247.

[11] Fedoseyev A. 2023 Approximate Analytical Solution for Turbulent Flow in Channel, J. Phys.: Conf. Ser. 2675 012011.

[12] Fedoseyev A. 2024 Minimization Principle for Analytical Solution of Turbulent Flow in Channel, arXiv:2404.08130 [physics.flu-dyn]; Phys. Rev. Fluids (submitted).

[13] Kim J., S.J. Kline, and J. P. Johnston, 1980 ASME J. Fluids Engng. 102, 302-308.

[14] Koseff R., R. L. Street ,1984 Trans. ASME/Journal of Fluids Engineering 106, 390-398.

[15] Polyanin A.D., Zaitsev V.F., 2002 Handbook of Exact Solutions for Ordinary Differential Equations, Chapman Hall, CRC Press Company, Boca Raton, London.

[16] Wei T., Willmarth W.W., 1989 Reynolds-number effects on the structure of a turbulent channel flow, J. Fluid Mech., v. 204, pp. 57-95.

AMITANS-2024 IOP Publishing
Journal of Physics: Conference Series **2910** (2024) 012014 doi:10.1088/1742-6596/2910/1/012014

Reparametrization Invariant Scaling Symmetry and the Value of the Einstein Cosmological Constant Λ_E

Vesselin G. Gueorguiev[1,2,3]

[1]Institute for Advanced Physical Studies, Sofia, Bulgaria,
[2]Ronin Institute for Independent Scholarship, Montclair, NJ, USA
[3]National Coalition of Independent Scholars, Battleboro, VT, USA

E-mail: Vesselin at MailAPS.org

Abstract. The nature of Dark Energy (DE) remains an enigma. This work explores the non-zero Einstein Cosmological Constant (Λ_E) as a manifestation of DE – viewing it through the lens of Reparametrization Invariant Scaling Symmetry (RISS), where Λ_E is interpreted as a "kinetic energy" term within the energy density. The scale factor $\lambda(t)$ relevant for cosmic reparametrization removes Λ_E from the extended Einstein GR, its defining equations, and their relations to Λ_E are derived. The value for $\Lambda_E = 1.8 \times 10^{-52}$ m^{-2} is consistent with the current observations for Λ_E. The current approach avoids the puzzles associated with interpreting Λ_E as vacuum zero-point energy (often called dark energy). Further applications are related to the Dark Matter problem by insisting on a reparametrization symmetry for the equations of motion, where un-proper (non-co-moving) time parametrization gives rise to fictitious forces.

Introduction

Modern physics faces a captivating paradox. Quantum Field Theory (QFT) and General Relativity (GR), the foundation of our understanding of the universe, successfully describe phenomena within the Solar System and on Earth. However, at larger scales encompassing galaxies and the cosmos, models reveal a stark disparity. Ordinary matter, the observable building block of the universe, constitutes only $\approx 5\%$ of its energy content. Dark Energy (DE) dominates the total energy with $\approx 70\%$, while Dark Matter (DM) accounts for $\approx 25\%$ [1].

Numerous proposals exist regarding the nature of DE and DM, including new fields, particles, or modifications of GR [2, 3, 4]. However, the prevailing explanation for DE and DM has faced a detection deficit for over 40 years.

This work presents a potential resolution to the DE and DM puzzles through a symmetry extension of Einstein GR. While Weyl's 1918 proposal for such an extension was initially rejected by [5], subsequent reformulations addressed the initial concerns [6, 7], particularly via the Weyl Integrable Geometry (WIG) considered by [8, 9].

In a general Weyl Geometry there are two more objects besides the metric tensor $g_{\mu\nu}$; there is a vector of metrical connexion κ_μ that relates the change in magnitude of the length $dl = l\kappa_\mu dx^\mu$ and a gauge scale factor λ. In the Weyl Integrable Geometry (WIG) the connexion vector κ_μ is related to the scale factor λ i.e. $\kappa_\mu = -\partial_\mu \ln\lambda$.

The working hypothesis here is that the existence of problems at galactic and cosmological scales can be attributed to the deviation away from the Einstein GR, where the lengths are preserved ($dl = 0$) upon parallel transport while utilizing the metric compatible connection. That is, there could be a small and mostly negligible $dl \approx 0$ that accumulates over large distances and fool us that there are some new fields

Content from this work may be used under the terms of the Creative Commons Attribution 4.0 licence. Any further distribution of this work must maintain attribution to the author(s) and the title of the work, journal citation and DOI.
Published under licence by IOP Publishing Ltd

causing these dark phenomena. In this respect the Weyl Integrable Geometry is very relevant because nature could conspire so that observations of distant phenomena will have a constructive interference at the observer and by doing so will circumvent the Einstein original objection; that is, WIG will reinforce models consistent with our local understanding of nature, while non-integrable contributions will be subject to destructive interference - therefore harder to be understood.

The concept of scale invariance gained prominence through [9], later forming the basis of scale-covariant cosmology by [10]. Initially, these works relied on the Large Numbers Hypothesis introduced by [11], which faced skepticism. Subsequently, the notion of Scale-Invariant Vacuum (SIV) emerged in 2016 [12, 13, 14], further detailed in 2017 by [15]. Since its proposal, SIV has undergone rigorous testing against various phenomena [16, 17, 18]. Recent reviews include [19, 20]. The framework's fundamentals were revisited by [21], with its potential connection to DM and DE proposed in 2020 by [22]. The original skepticism about the scale-covariant cosmology may still be present, which may also be the reason why the modern reincarnation into the Scale-Invariant Vacuum (SIV) cosmology is yet under-appreciated despite its various successes outlined in [19]. However, the idea of reparametrization invariance do not seems to be subject to such concerns and allows for inhomogeneous and anisotropic Universe.

The idea of reparametrization invariance proves invaluable in geometrically justifying fundamental interaction fields for classical long-range forces [23] and understanding key properties of physical systems [24]. Unlike scale invariance, which is questionable in the presence of matter [25], RISS appears as fundamental as diffeomorphism invariance, offering an additional layer of symmetry to the observer. Thus, RISS aids us in the understanding of DE and DM phenomena and may even turnout to be the new fruitful way of thinking about DE, DM, and other natural phenomena.

Here, I present my current understanding of DE as the Einstein cosmological constant Λ_E. Relevant formulas and the numerical value of Λ_E will be derived and computed within the framework of Reparametrization Invariant Scaling Symmetry (RISS), equivalent to the SIV paradigm for a homogeneous and isotropic Universe. The relevant formulas and the numerical value of the MOND a_0 within RISS will be presented elsewhere.

Reparametrization Invariant Scaling Symmetry (RISS)

Systems obeying reparametrization invariance are important in physics. However, such systems usually are not easy to work with since their Hamiltonian is identically equal to zero. In general, these systems possess gauge symmetries, i.e. diffeomorphism invariance of the General Relativity, the gauge invariance of the electromagnetism, gauge symmetries in particle physics and in string theory, etc. [23, 24].

A gauge symmetry does not affect the underlining understanding of a system/process under study, just as the choice of a coordinate system should not influence the system/process under study. However, a suitably chosen coordinate system may significantly simplify the description and therefore the understanding of a system/process under study. In this respect reparametrization invariance could be viewed as a gauge symmetry, although one often uses the coordinate time or the proper time of the system itself, but upon reparametrization invariance one should be able to use any desired, but reasonable, process parametrization.

It sounds as if there is no new physics related to reparametrization invariance, and it may be so, but only if the equations utilized to study and model systems do obey reparametrization invariance. Would the following well-known equation be appropriate for a reparametrization invariant system?

$$\frac{d^2\vec{r}}{dt^2} = -\frac{GM}{r^2}\frac{\vec{r}}{r}. \tag{1}$$

This is the well-known Newtonian gravity equation that has been tested and verified many times over in our solar system. So, yes it is appropriate as long as the time t is the corresponding proper-time of the test object whose motion is to be studied. However, if one performs a reasonable time reparametrization, say by using $dt \to \lambda(t)\,dt$ then this equation will acquire extra terms on the left hand side of (1). So, the actual answer is that (1) is not reparametrization invariant and therefore one's conclusions upon utilizing this equations will be only valid within the comoving frame where the well-studied Newtonian physics applies. Thus, one needs to verify that the equations of motion used for the study of a system obey reparametrization invariant symmetry in order to claim that there will be no new physics due to this gauge symmetry. This will be discussed in more details elsewhere.

Another important property to be preserved upon the reparametrization transformation is the constancy of the speed of light since the modern view on the fundamental constants such as c, $\hbar$, and k_B is that these are conversion constants that can be fixed by the choice of units [26]. In this respect along with $dt \to \lambda(t)\,dt$ one should perform also rescaling $dr \to \lambda(t)\,dr$ in order to keep the speed of light c

unaltered when studying the light null-geodesics. Thus, this brings the scaling symmetry into place via the Weyl transformation:

$$g_{\mu\nu} \to \lambda^2 g_{\mu\nu} \tag{2}$$

The important point here is that one arrives at the Weyl transformation (2) upon considering the reparametrization gauge along with the desire to maintain the constancy of the speed of light. Otherwise concepts like tired light can enter the framework and by doing so would obscure the actual symmetry at place [27, 28]. In the RISS framework λ is a function of time only from the get go, while in the general Weyl transformation it could depend on all of the spacetime coordinates and one imposes time-only-dependence upon considering scale invariant cosmology for homogeneous and isotropic universe [22, 21].

At this point, as long as the equations of motion are reparametrization invariant, one cannot specify a functional form for $\lambda(t)$ but one can get some understanding of $\lambda(t)$ and the related terms in the equations of motion, which will be discussed in the next section. In order to arrive at some specific functional form for $\lambda(t)$ one will have to study a specific system and ints defining equations. This will be the result in the forthcoming sections on scale invariant cosmology, and in particular the SIV cosmology.

RISS as framework for scale invariant cosmology
As noted previously, the application of Weyl geometry as a foundation for scale-invariant cosmology has been explored by [9, 10]. Additionally, [29, 30] investigated potential astronomical applications, while more recent reviews have focused on highly abstract mathematical formulations [31]. However, it is worth noting that *nature-based considerations of phenomena with potential observational validations have been scarce*, with only a few examples existing, such as [30, 32, 19], while other studies have been misguided by varying speed of light considerations [27, 28], which according to our current view for defining the International System of Units [26] is unreasonable. Furthermore, even fewer studies have focused on the application in the context of particle physics, such as [33]. Here, I will highlight the key properties of the Ricci tensor and scalar that are crucial for understanding the current perspective and could be used later to identify a particular form of $\lambda(t)$ to be identified with the SIV gauge choice.

The importance of the RISS as only-time-dependent λ in the Weyl transformation (2) and its consequences for the equations of motion were already discussed above. A more formal discussions have been given along the lines of the Dirac co-calculus and its applications [19, 21]. Here we focus only on the minimal formalism needed to justify the $\lambda(t)$ dependence for the SIV gauge.

Upon a Weyl transformation of the metric [29, 30], one has the following Ricci tensor and scalar expressions:

$$g_{\mu\nu} \to \lambda^2 g_{\mu\nu} \Rightarrow R_{\mu\nu} \to R_{\mu\nu} + K_{\mu\nu}, \tag{3}$$

where $K_{\mu\nu}$ is given in terms of $\kappa_\mu = -\partial_\mu \ln \lambda$ as:

$$K_{\mu\nu} = g_{\mu\nu}\kappa^\rho\kappa_\rho + 2\kappa_\mu\kappa_\nu + \kappa_{\mu;\nu} + \kappa_{\nu;\mu} - 2g_{\mu\nu}\kappa^\rho_{;\rho}. \tag{4}$$

Such expressions have been discussed in Eq. (89.2) by [8] first and later by [9, 33] as well. After contracting (3) one obtains the Ricci scalar to be mapped so that $R \to (R+K)/\lambda^2$ along with:

$$K = 6\kappa^\rho\kappa_\rho - 6\kappa^\rho_{;\rho}, \tag{5}$$

In a general Weyl framework, λ may depend on space and time and so is the corresponding connexion vector κ_μ. The french spelling "connexion" is used to avoid confusion with the usual connection concept within EGR.

λ in the SIV gauge
The EGR equation with cosmological constant Λ_E is:

$$R_{\mu\nu} - \frac{1}{2}Rg_{\mu\nu} + \Lambda_E g_{\mu\nu} = \varkappa T_{\mu\nu}. \tag{6}$$

This formulation assumes a metric compatible connection, where the Einstein tensor has zero divergence. Similarly, the stress-energy tensor $T_{\mu\nu}$ satisfies zero divergence ensuring covariant conservation laws. As long as $\varkappa = 8\pi G/c^4$ and Λ_E are constants, the equation remains consistent. Otherwise, one has $\Lambda_{E,\mu} = \varkappa_{,\nu}T^\nu_\mu$. The LHS of (6) represents gravity through the metric tensor and its connection, while the right-hand side represents matter and its energy density, with G and c implying the constancy of Λ_E.

One might consider moving the term $\Lambda_E g_{\mu\nu}$ from the left-hand side to the right-hand side of equation (6) and to associate Λ_E with the zero-point/vacuum energy. However, this approach leads to the **cosmological constant problem**, characterized by the vast discrepancy between estimated values based on quantum field theory (QFT) arguments and observed/measured values [34, 35, 36, 1]. Furthermore, introducing a non-zero constant energy density due to matter fields (non-gravitational fields) in a homogeneous and isotropic universe implies the existence of a black hole event horizon at a sufficiently large distance R_S. This seemingly implies that our Universe resides within a black hole, as proposed in [37, 38]. Yet, such a scenario would predict a contracting flow of matter towards the center, where the highest concentration of matter should reside; however, one instead is observing an expanding Hubble flow towards the future event horizon.

Alternatively, considering the observed age of the Universe, $\tau_0 \approx 1/H_0$, along with c and G, provides an accurate estimate for Λ_E, albeit interpreted differently within the proposed paradigm. To explore this further, let's employ the Weyl transformation (2). This transformation relates the metric in the EGR frame, $g'_{\mu\nu}$, to a metric in a more general WIG frame, $g_{\mu\nu}$, via a λ factor: $g'_{\mu\nu} = \lambda^2 g_{\mu\nu}$. Applying equations (3), (4), and (5) within the Einstein equation (6), this Weyl transformation expresses the Einstein equation in the more general WIG framework, resulting into:

$$R_{\mu\nu} + K_{\mu\nu} - \frac{1}{2}\lambda^{-2}(R+K)\lambda^2 g_{\mu\nu} + \Lambda_E \lambda^2 g_{\mu\nu} = \varkappa T_{\mu\nu}. \tag{7}$$

Under Reparametrization Invariant Scaling Symmetry (RISS), the local time coordinate is not the only choice for parameterizing the evolution of the Universe (i.e., the process under observation). This allows us to split equation (7). One equation will contain $\Lambda = \lambda^2 \Lambda_E$, λ, and its derivatives expressed as $k_\mu = -\partial_\mu \ln \lambda$. The other equation will not include the Λ term nor λ directly:

$$R_{\mu\nu} - \frac{1}{2} R g_{\mu\nu} = \varkappa T_{\mu\nu}, \tag{8}$$

$$K_{\mu\nu} - \frac{1}{2} K g_{\mu\nu} + \Lambda g_{\mu\nu} = 0. \tag{9}$$

One can interpret the above separation of equation (7) as a prescient insight into choosing a special "gauge" for the factor λ. The second equation (9) defines λ, and when satisfied, the first equation (8) remains alone. Consequently, taking the trace of (9) yields $\Lambda = K/4$. Using only a time-dependent λ in (5), as needed for re-parameterization ($d\tau' = \lambda\, dt$), along with $\kappa = \kappa_0 = -\dot{\lambda}/\lambda$, one arrives at the key SIV equations:

$$\Lambda = \frac{3}{2}\left(\kappa^2 - \dot{\kappa}\right) \Rightarrow \Lambda = 3\kappa^2 = 3\left(\frac{\dot{\lambda}}{\lambda}\right)^2 \text{ if } \dot{\kappa} = -\kappa^2 \tag{10}$$

Usually, one arrives at this point within SIV by assuming homogeneous and isotropic Universe which for cosmology translates into dependence of λ only on time. Within the RISS paradigm, the dependence of λ only on time is a fundamental assumption which does not preclude the Universe from taking upon inhomogeneous and anisotropic state(s). *Thus, the question whether scale invariance is an actual symmetry of nature is of no concern for the RISS paradigm.*

By taking the time derivative of κ/λ, one can see that it will be equal to $(\dot{\kappa} + \kappa^2)/\lambda$, which will vanish if $\dot{\kappa} = -\kappa^2$; thus, Λ/λ^2 will be a constant; therefore; let's denote this constant also with Λ_E since these two constants will coincide eventually. Thus, for the case $\dot{\kappa} = -\kappa^2$ the solution for $\lambda(t)$ is very simple and one can use the constant $\Lambda_E = \Lambda/\lambda^2 = 3\dot{\lambda}^2/\lambda^4$ to characterize it:

$$\varepsilon(t_0 - t)\sqrt{\Lambda_E/3} = 1/\lambda - 1/\lambda_0. \tag{11}$$

That is, $\kappa^2 = \dot{\lambda}^2/\lambda^2 = \Lambda/3 = (\Lambda_E/3)\,\lambda^2$ along with

$$\lambda = \lambda_0 / \left(1 + \lambda_0 \varepsilon (t_0 - t)\sqrt{\Lambda_E/3}\right), \tag{12}$$

and $\kappa = -\dot{\lambda}/\lambda = -\varepsilon\lambda\sqrt{\Lambda_E/3}$ are key expressions of the new paradigm; therefore, $\dot{\kappa} \sim \dot{\lambda} = \varepsilon\lambda^2\sqrt{\Lambda_E/3}$ implies $\dot{\kappa} = -\varepsilon^2\,(\Lambda_E/3)\,\lambda^2 = -\kappa^2$ as imposed; here $\varepsilon = \pm 1$ and therefore $\varepsilon^2 = 1$. By setting $\lambda_0\sqrt{\Lambda_E/3} = 1/t_0$, one has $\lambda = \lambda_0 t_0/t$ for $\varepsilon = -1$; therefore, $\kappa = -\dot{\lambda}/\lambda = 1/t$ with $t \in [t_{\text{in}}, t_0]$, where t_{in} is the moment of the Big-Bang when $a(t_{\text{in}}) = 0$ [22].

It is often convenient to choose $\lambda_0 = 1$ along with SIV time units such that $t_0 = 1$. Thus, one is led here to the same expression of the scale factor λ as obtained by the fundamental SIV hypothesis, according to which the macroscopic empty space is scale invariant, homogeneous and isotropic [15].

Within the RISS paradigm, however, the dependence of λ on time only is a fundamental assumption which does not preclude the Universe from being in anisotropic and inhomogeneous states. Thus, the question whether scale invariance is an actual symmetry of nature is of no concern for the RISS paradigm. The RISS freedom also allows us to consider/choose parametrization such that $\dot{\kappa} = -\kappa^2$ which results in $\kappa^2 = \Lambda/3$ along with $\Lambda_E = \Lambda/\lambda^2$ being a constant.

Interpretation of Λ_E within the New Framework

The presence of a non-zero cosmological term Λ_E poses significant issues: (1) a mismatch between the observed value and zero-point energy estimates based on QFT, and (2) the puzzling notion that one is inside a black hole.

Within the present paradigm, one can circumvent these problems by showing that Λ_E should be a constant. As mentioned earlier the Einstein Cosmological Constant Λ_E must be a constant if one views the Newton Gravitational Constant as a true constant. This can be used to choose a new parametrization that removes the cosmological term Λ_E. This implies that the extra energy density due to the cosmological constant can be viewed as an observer effect, just as the kinetic energy of a system depends on the relative motion of the two systems; however, in the case for the cosmological constant this seems to be about the difference in time parameterizations.

That is, the metric $g_{\mu\nu}$ provides Λ-free EGR equations as seen in (8) while the presence of non-zero constant Λ_E-term is due to the choice of the EGR metric tensor $g'_{\mu\nu} = \lambda^2 g_{\mu\nu}$ where the factor $\lambda = t_0/t$ is defined via $\sqrt{\Lambda_E/3} = 1/t_0$ (12); thus, this relates the value of the cosmological constant Λ_E to the age of the Universe t_0, which is consistent with the simple dimensional estimates [39]. In the usual SI units where age of the Universe is $\tau_0 = 13.8$ billion years and speed of light is $c = 3 \times 10^8$ m/s, one obtains:

$$\Lambda_E = 3/(c\tau_0)^2 \approx 1.8 \times 10^{-52}\ \mathrm{m}^{-2}, \tag{13}$$

which is reasonably close to the measured value [39, 1]. Note that this measured value is based on model fitting using the ΛCDM

In the above considerations, one has settled for a time dependence of λ where upon a proper choice of parametrization via an equivalent conformal metric $g_{\mu\nu}$, the Λ_E term disappears and one has only (8) without the observer-related Cosmological Constant Λ_E given by (13). This shows that the Cosmological Constant Λ_E is an energy density related to the possibility that our time coordinate is not the proper time parametrization for our Universe. That is, the energy density induced by the relative motion through time of the observer with respect to the system studied acquires an extra energy component just as it is the case of the non-zero kinetic energy when the observer is not in a co-moving frame.

Conclusion

To summarize, this paper argues that the Einstein cosmological constant, Λ_E, is a true constant within the presented framework. In particular, an expression for the scale factor $\lambda(t)$ (12) can be derived based on (9), resulting in the condition (10), while the metric tensor satisfies the usual Einstein equation without cosmological constant (8).

The current approach avoids the puzzles associated with interpreting Λ_E as vacuum zero-point energy (often referred to as dark energy). These puzzles include residing inside a black hole and the vast discrepancy (123 orders of magnitude) between Λ_E and theoretical estimates from quantum field theory (QFT). In the current framework, a non-zero Λ_E within the ΛCDM model arises from the choice of time-parameterization. Analogous to kinetic energy associated with spatial motion, a difference in time-parameterization introduces "extra energy" due to relative temporal motion.

Acknowledgments

This research did not receive any specific grant from funding agencies in the public, commercial, or not-for-profit sectors. The author is extremely grateful to his wife and daughters for their understanding and family support during the various stages of the research presented, as well as to Prof. A. Maeder for the many years of fruitful and joyful explorations of the SIV paradigm.

References

[1] Planck Collaboration: Ade et al 2016 A&A **594** A13 (*Preprint* 1502.01589)

[2] Frieman J A, Turner M S and Huterer D 2008 ARA&A **46** 385–432 (*Preprint* 0803.0982)

[3] Li M, Li X D, Wang S and Wang Y 2011 *Communications in Theoretical Physics* **56** 525–604 (*Preprint* 1103.5870)

[4] Capozziello S and de Laurentis M 2011 Phys. Rep. **509** 167–321 (*Preprint* 1108.6266)

[5] Einstein A 1918 *Sitzungsberichte der Königlich, Preussischen Akademie der Wissenschaften zu Berlin* 478

[6] Weyl H 1922 *Space - Time - Matter* (Dutton)

[7] Weyl H 1970 *Raum, Zeit, Materie. Vorlesungen über allgemeine Relativitätstheorie.* (Berlin: Springer)

[8] Eddington A S 1923 *The mathematical theory of relativity* 2nd ed (Cambridge University Press) ISBN 978-052-109-165-7

[9] Dirac P A M 1973 *Proceedings of the Royal Society of London Series A* **333** 403–418

[10] Canuto V, Adams P J, Hsieh S H and Tsiang E 1977 Phys. Rev. D **16** 1643–1663

[11] Dirac P A M 1974 *Proceedings of the Royal Society of London Series A* **338** 439–446

[12] Maeder A 2016 *arXiv e-prints* arXiv:1605.06315 (*Preprint* 1605.06315)

[13] Maeder A 2016 *arXiv e-prints* arXiv:1605.06316 (*Preprint* 1605.06316)

[14] Maeder A 2016 *arXiv e-prints* arXiv:1605.06314 (*Preprint* 1605.06314)

[15] Maeder A 2017 ApJ **834** 194 (*Preprint* 1701.03964)

[16] Maeder A and Gueorguiev V G 2019 *Physics of the Dark Universe* **25** 100315 (*Preprint* 1811.03495)

[17] Maeder A and Gueorguiev V G 2020 MNRAS **492** 2698–2708 (*Preprint* 2001.04978)

[18] Maeder A and Gueorguiev V G 2021 MNRAS **504** 4005–4014 (*Preprint* 2104.09314)

[19] Gueorguiev V G and Maeder A 2022 *Universe* **8** 213 (*Preprint* 2202.08412)

[20] Gueorguiev V G and Maeder A 2024 *Symmetry* **16** 657 (*Preprint* 2311.14569)

[21] Maeder A and Gueorguiev V G 2023 *Symmetry* **15** 1966 (*Preprint* 2310.16913)

[22] Maeder A and Gueorguiev V G 2020 *Universe* **6** 46 (*Preprint* 2311.15391)

[23] Gueorguiev V G and Maeder A 2021 *Symmetry* **13** 379 (*Preprint* 1907.05248)

[24] Gueorguiev V G and Maeder A 2021 *Symmetry* **13** 522 (*Preprint* 1903.02483)

[25] Feynman R P, Leighton R B and Sands M 1963 *Feynman lectures on physics - Volume 1* 1st ed (Addinson-Wesley) ISBN 978-0-465-04085-8

[26] Newell D B, Tiesinga E *et al.* 2019 *NIST Special Publication* **330** 1–138

[27] Lee S 2021 *arXiv e-prints* arXiv:2101.09862 (*Preprint* 2101.09862)

[28] Gupta R P 2024 ApJ **964** 55 (*Preprint* 2401.09483)

[29] Bouvier P and Maeder A 1978 Ap&SS **54** 497–508

[30] Maeder A and Bouvier P 1979 A&A **73** 82–89

[31] Wheeler J T 2018 *General Relativity and Gravitation* **50** 80 (*Preprint* 1801.03178)

[32] Ferreira P G and Tattersall O J 2020 Phys. Rev. D **101** 024011 (*Preprint* 1910.04480)

Journal of Physics: Conference Series **2910** (2024) 012014 doi:10.1088/1742-6596/2910/1/012014

[33] Parker L 1973 Phys. Rev. D **7** 976–983

[34] Weinberg S 1989 *Reviews of Modern Physics* **61** 1–23

[35] Carroll S M, Press W H and Turner E L 1992 ARA&A **30** 499–542

[36] WMAP Colaboration: Hinshaw et al 2013 ApJS **208** 19 (*Preprint* 1212.5226)

[37] Pathria R K 1972 Nature **240** 298–299

[38] Gaztanaga E 2022 *Symmetry* **14** 1849

[39] Gueorguiev V G and Maeder A 2020 *Universe* **6** 108 (*Preprint* 2009.03866)

Combined electron and proton radiation effects on solar cells for earth orbit

Yana Gurimskaya, Alex Fedoseyev and Stanislau Herasimenka

Solestial Inc., Tempe, Arizona, USA

E-mail: yg@solestial.com

Abstract.

We present a computational approach for accurate calculations of the combined effects of electron and proton radiation in space, tailored to a specific orbit and accounting for the full range of particle energies. The radiation-induced defect formation is shown to depend on the NIEL (Non-Ionizing Energy Loss), which reflects the small fraction of particle energy (from electrons, protons or other ions) lost through the ionisation of the material. The model developed by the Naval Research Laboratory (NRL) for predicting solar cell degradation based on NIEL aligns closely with the results from the Jet Propulsion Laboratory (JPL). However, the JPL model involves a more comprehensive methodology, incorporating experimental validation.

To determine the energy dependence of NIEL, $S(E)$, for both electrons and protons, we use the ESA NIEL calculation tool. This computational process involves assessing the radiation environment and its effects with an assistance from SPENVIS — a simulation platform developed by ESA/NASA for generating spacecraft trajectories in LEO, MEO or GEO. Proton and electron fluxes are simulated using the International Radiation Environment Near Earth (IRENE) model within SPENVIS. This model offers various calculation options — such as mean, percentile, perturbed or Monte Carlo simulations — based on the mission duration and launch date.

The resulting proton and electron fluences, both integral and differential, are used to calculate the Displacement Damage Dose (DDD). This is achieved by integrating the product of NIEL with the differential fluences over the relevant particles energy range.

To enhance efficiency, specialized software tools have been developed to extract particle fluxes and fluences from SPENVIS, interpolate them and perform the necessary integration to obtain DDD values for protons, electrons and the combined DDD. Equivalent fluences for defined energy ranges are also provided, enabling detailed, mission-specific radiation analyses. These tools streamline the evaluation of the radiation environment and its impact on solar cells, determining the design and material choices for mission planning and providing the data to predict space power generation throughout the mission's duration.

1 Introduction

The increasing advancement of space technology has created a demand for more efficient and reliable solar cells to power satellites and spacecrafts in Earth's orbit. The reliability of solar cells, which serve as the primary power source, is critically impacted by the space radiation environment, especially the high-energy electrons and protons within the Van Allen belts. This study focuses on understanding how

Content from this work may be used under the terms of the Creative Commons Attribution 4.0 licence. Any further distribution of this work must maintain attribution to the author(s) and the title of the work, journal citation and DOI.
Published under licence by IOP Publishing Ltd

IOP Publishing
doi:10.1088/1742-6596/2910/1/012015

these particles degrade ultra-thin silicon solar cells, specifically through displacement damage caused by collisions with atomic nuclei within the solar cell material. The degradation processes are examined in terms of Non-Ionizing Energy Loss (NIEL) and the contributions of elastic and inelastic scattering, leading to efficiency loss of solar cells over the course of a mission. Our experimental approach incorporates ultra-thin silicon solar cells engineered to withstand high radiation levels, while the computational model integrates orbit-specific radiation environment calculations and Displacement Damage Dose (DDD) estimations. The results demonstrate the effectiveness of the developed models in predicting radiation-induced degradation, underlining the importance of mission-specific modelling to ensure accurate predictions of hardware lifespan.

This paper explores the mechanisms by which high-energy electrons and protons affect ultra-thin Silicon Hetero-Junction (SHJ) solar cells operating in space. By exploring these degradation pathways and developing precise computational models, we aim to predict long-term performance of solar cells throughout a mission. This understanding is vital for designing robust solar cells that can withstand the harsh space radiation environment while maintaining optimal performance for the duration of the mission.

2 Space radiation environment

During space missions, solar cells are subjected to extreme environmental conditions, including drastic temperature fluctuations and exposure to high-energy particles such as protons and electrons trapped within Earth's magnetosphere. In addition, solar events contribute to the radiation environment by emitting protons, neutrons, electrons, X- and γ-rays. The fluence and energy spectra of these high-energy particles can be predicted using models that combine observational data with spacecraft trajectories and orbital parameters. Fig. 1 illustrates the fluxes of electrons and protons in Earth's near-space environment, highlighting the complex radiation environment the spacecraft must endure [1].

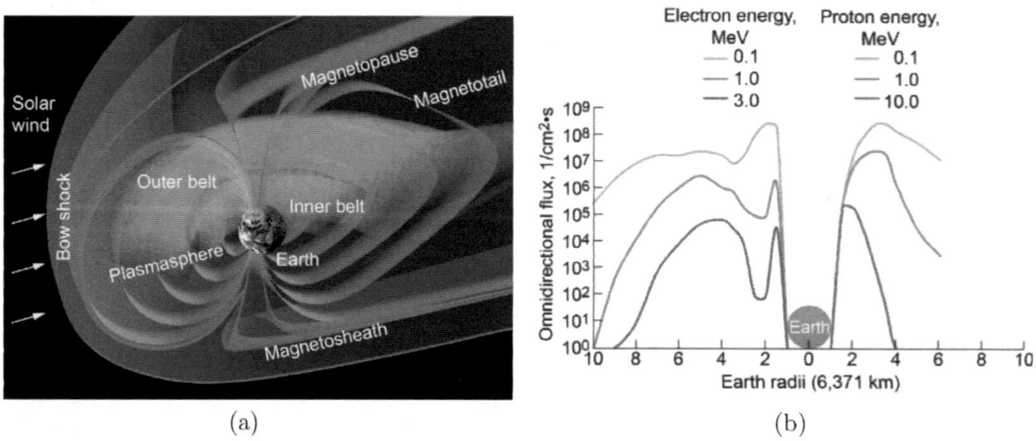

(a) (b)

Figure 1: (a) Schematic view of the Earth magnetosphere and Solar Cosmic Rays (SCRs). (b) Flux of electrons and protons trapped in the Van Allen belts as a function of altitude (Courtesy of NASA, *Space radiation and impact on instrumentation technologies, J.D. and S.Y. Wrbanek, 2020, NASA/TP-2020-220002).*

3 Radiation Effects in Space

3.1 Radiation damage mechanism

Particles interact with matter primarily through two mechanisms: ionizing and non-ionizing radiation. Non-ionizing radiation, in particular, has been identified as the dominant mechanism responsible for the degradation of the current space photovoltaic (PV) technologies. When energetic protons traverse a crystalline lattice, they may collide with atomic nuclei, causing atomic displacements and the formation of defect centers.

Electrons, being lighter and elementary charged particles, primarily interact with the material through continuous scattering with lattice electrons. These interactions result in the formation of electron-hole pairs and atomic displacements. As electrons travel through the material, they lose energy more uniformly due to frequent electron-electron and electron-phonon scattering. In contrast, protons tend to penetrate deeper into the bulk before losing significant energy. Consequently, the primary knock-on atoms (PKAs) generated by protons are concentrated along their path, resulting in localized displacement cascades.

A basic example of atomic displacement is the formation of a vacancy and an interstitial defect pair, known as a Frenkel pair. However, more complex interactions, involving multiple proton-nuclei collisions, can lead to the creation of defect clusters [2] or even amorphous zones [3, 4]. The degree of these interactions depends on the energy of the incident protons and the effective cross-section of the target nuclei.

Defects formed within the active regions of space solar cells are particularly harmful to their performance, as they introduce parasitic trap states and recombination centers, significantly reducing power output. The highest number of atomic displacements occurs when particle energy is sufficient to penetrate the encapsulation and conductive layers. Despite slowing down, these particles retain enough energy to surpass so-called displacement energy threshold of the host lattice. This threshold, often considered between 21–30 eV for silicon [6], is the minimum energy required to create a stable defect in the material [5]. As a result, protons typically stop within the active regions of the cell, provided the cell is sufficiently thick or the proton energy is relatively low. Conversely, high-energy protons are less likely to interact with the absorber regions due to a reduced effective cross-section. They pass through the lattice with fewer interactions, potentially stopping deeper in the substrate or bypassing thin layers entirely.

3.2 Radiation damage quantification and existing models

Two primary modeling approaches, developed by the Naval Research Laboratory (NRL) and the Jet Propulsion Laboratory (JPL) are used for predicting solar cell degradation [7–9].

The NIEL value is a key parameter that quantifies the rate of energy transfer from particles to the nuclei of the host lattice. NIEL measures the energy dissipated by a particle as it traverses silicon, primarily via non-ionizing processes, such as kinetic energy transfer during recoil events or energy conversion into heat within the crystal lattice.

Fig. 2 illustrates the NIEL for electrons and protons in silicon calculated using the SR-NIEL tool [10].

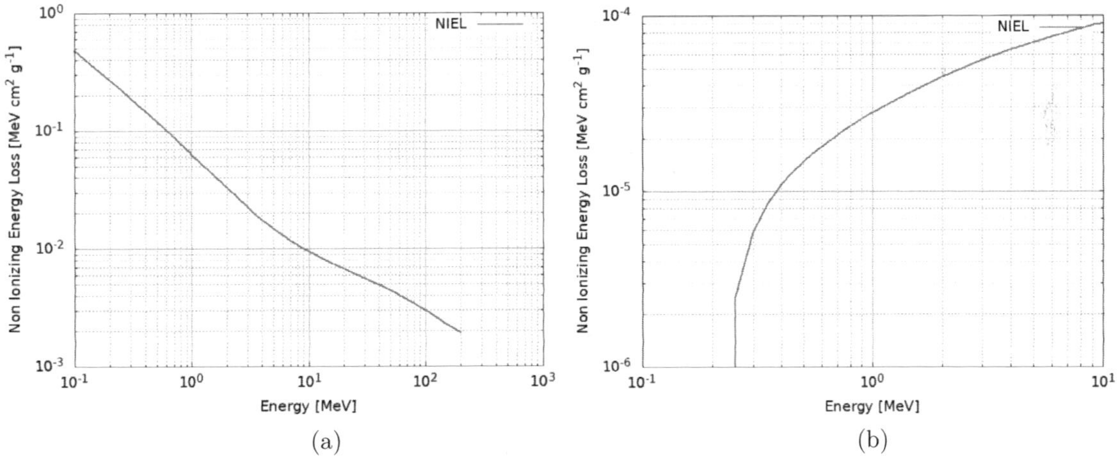

Figure 2: Calculation of (a) proton and (b) electron NIEL in silicon as a function of energy, using SR-NIEL calculator available in [10].

A fundamental concept in evaluating material damage is the correlation between NIEL and radiation damage, as described by the NIEL scaling model. This relationship is essential for predicting the lifespan of solar cells, detectors and other semiconductor devices operating in radiation-harsh environments. The NIEL scaling hypothesis links NIEL to bulk damage, highlighting a linear relationship between NIEL and experimental damage coefficients. The Displacement Damage Dose (DDD) serves as a common metric for predicting the operational life of semiconductor devices in space, where radiation exposure is a major concern [11, 12].

Simulation tools such as GEANT4 [13] and FLUKA [14] can be employed to calculate DDD values based on the NIEL database along with expected particle fluences. This NIEL database for electrons and protons, typically sourced from the SR-NIEL calculator [10], includes contributions from Coulomb scattering but also hadronic contribution in silicon employing Robinson and Akkerman partition functions. This integrated approach allows more accurate predictions of radiation-induced damage and helps in the development of more radiation-hard PV technologies for space missions.

The NRL [7, 15] and JPL [8, 16] have developed distinct approaches for modeling a degradation of solar cells of different architectures, each with unique strengths.

The NRL approach primarily focuses on the effects of protons and heavy ions with a strong emphasis on Single Event Effects (SEE) and Displacement Damage Dose (DDD). Grounded in empirical data from particle accelerator tests, this model is designed to predict solar cells degradation due to displacement damage, primarily from protons. This method ensures high accuracy in proton and heavy-ion environments, making it well-suited for specific missions dominated by these particles.

In the NRL model [7], DDD is calculated as the product of NIEL and particle fluence for mono-energetic particles. It serves as a key parameter for determining the durability of these devices under radiation:

$$DDD = \phi \cdot S, \tag{1}$$

where ϕ is the particle fluence and S is the NIEL for the particle.

For a range of particle energies, the DDD is calculated using the following equation:

$$DDD = \int_{E_1}^{E_2} \frac{d\phi}{dE} \cdot S(E)\, dE, \tag{2}$$

where $d\phi/dE$ is the differential fluence (for the selected orbit) and $S(E)$ is the energy-dependent NIEL for the particle.

However, the NRL model's narrow focus on protons and heavy ions may limit its applicability in scenarios where other particles, such as electrons, are more significant. The enhanced equation for electron DDD is provided in [9]:

$$DDD = E_{ep} \int_{E_1}^{E_2} \frac{d\Phi}{dE_e} \cdot NIEL(E_e) \cdot \left[\frac{NIEL(E_e)}{NIEL(1MeV)}\right]^{n-1} dE_e, \tag{3}$$

where $d\Phi/dE$ is the differential fluence, $NIEL(E_e)$ is the energy-dependent NIEL for the particle, $NIEL(1MeV)$ is the electron NIEL at 1 MeV and $n > 2$.

In contrast, the JPL approach adopts a broader perspective, incorporating both electrons and protons, and focuses on the dynamic trapped particle environments found in various orbits.

While the JPL approach is more comprehensive and adaptable to dynamic space conditions, it is also more complex and dependent on multiple data sources. This complexity can present challenges in implementation, particularly in missions with limited data [8].

4 Combined electron and proton radiation effects on silicon solar cells for Earth orbit

Building upon the established methodologies of NRL and JPL, we are developing a novel computational approach tailored to ultra-thin silicon solar cells for space applications. This model incorporates mission-specific factors such as radiation environment, illumination and temperature. By integrating empirical data from ground-based testing weighted by real-time mission conditions, this approach provides more accurate predictions of solar cell degradation. It balances defect formation and annealing rates, leveraging the unique features of Solestial's ultra-thin solar cells to self-repair radiation damage [17] achieved by defect engineering. The model requires inputs such as radiation environment, cell temperature, light intensity and operating conditions to predict the solar array's power output as a function of defect formation $D(t)$ and annealing $A(t)$ at any given time during a mission.

4.1 Computational model

4.1.1 Orbit Selection and Coordinate Calculation .

The model begins by selecting the appropriate orbit for the mission through SPENVIS (Space Environment Information System) [18]. Options include very Low Earth Orbit (vLEO), Low Earth Orbit (LEO), Medium Earth Orbit (MEO), Geosynchronous Earth Orbit (GEO) or transition orbits. Once the orbit is chosen, the corresponding radius and inclination are determined based on the particular trajectory. Mission-specific parameters, such as launch dates and mission duration, are then used as an input to finalize the orbit calculation, providing the spatial coordinates necessary for subsequent radiation

Journal of Physics: Conference Series **2910** (2024) 012015 doi:10.1088/1742-6596/2910/1/012015

environment analysis. Fig. 3 shows the trajectory of the spacecraft, providing key orbit parameters such as altitude, inclination and geographical coverage.

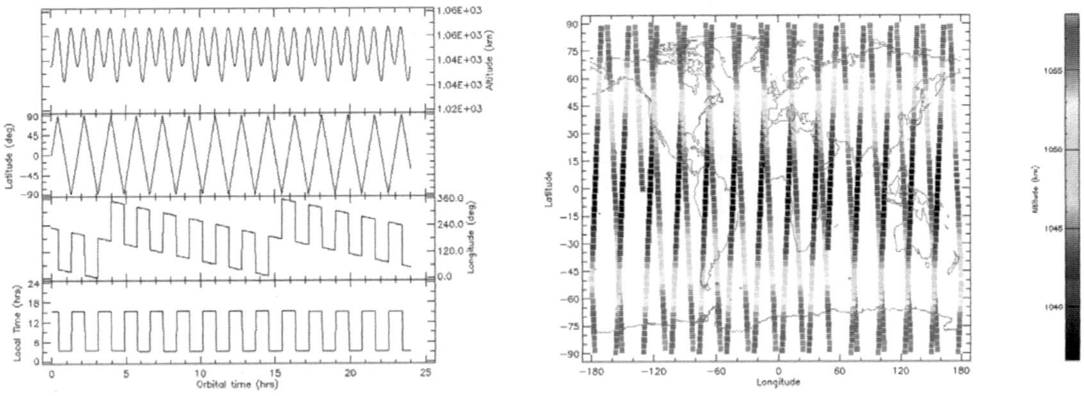

(a) (b)

Figure 3: (a) Orbit parameters as a function of time and (b) World map of the orbit obtained for 8 years mission in LEO at 1080 km orbit and 90° inclination, using SPENVIS.

4.1.2 Radiation Environment Calculation .

To assess the radiation environment, one of several established models is used, for example IRENE-AP9/AE9 model. The type of calculation is selected based on the required level of details and uncertainty management. Options include mean values (for average behavior), percentile-based estimates, perturbed models accounting for measurement uncertainties or Monte Carlo methods, which incorporate dynamic space weather variations. Once the model is selected, it is executed to generate the fluence data, which is crucial for understanding the potential radiation exposure in the orbit of interest. These results are saved for further processing and analysis.

4.1.3 Displacement Damage Dose (DDD) Calculation .

The next stage involves calculating the Displacement Damage Dose (DDD) for protons and electrons, accounting for particle-specific damage mechanism and the obtained differential fluences.

The fluence data is processed and plotted to ensure its accuracy and quality. After verification, the data is compiled into a comprehensive table and converted into equivalent fluences. These fluences are critical for evaluating the impact of radiation on materials and predicting their degradation over the mission's duration.

Fig. 4 shows the electron fluences in Low Earth Orbit (LEO), over an eight-year period for a 1080 km orbit with an inclination of 90°, obtained using SPENVIS software tools.

The Displacement Damage Dose (DDD) can be calculated by integrating the differential fluences with NIEL over the relevant energy range, as provided by SPENVIS data.

For the considered example of the 1080 km orbit, inclination 90°, the results are presented in Table 1. The equivalent fluences are obtained using Eq. (1), expressed as $\Phi = DDD/S$, where the NIEL scaling factor S is defined as $S = S_e(1\,\text{MeV})$ for electrons, and $S = S_p(3\,\text{MeV})$ for protons.

Particle	DDD (8 years) [MeV/g]	Equiv. Fluence (3 MeV proton)	Equiv. Fluence (1 MeV electron)
Proton	6.030E+11	2.41E+14	2.15E+16
Electron	1.922E+09	7.69E+11	6.86E+13
Total	6.050E+11	2.41E+14	2.15E+16

Table 1: Calculated displacement damage dose (DDD) and equivalent fluence in a 1080 km / 90° orbit for a silicon solar cell over 8 years.

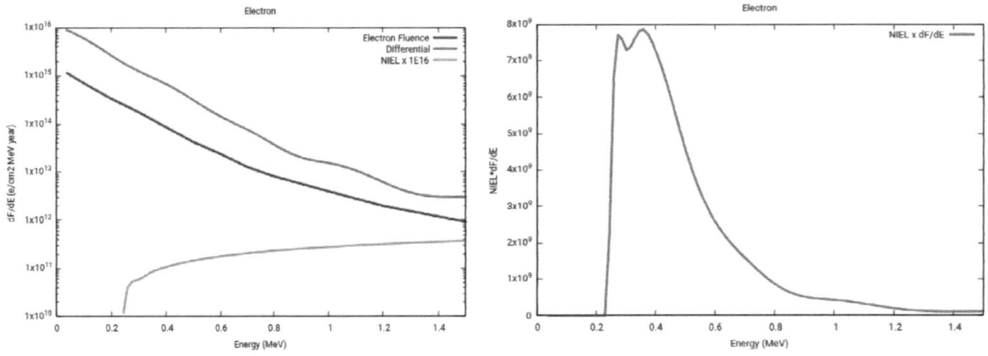

(a) (b)

Figure 4: (a) Example of electron fluence and differential fluence for 8 years mission in LEO at 1080 km orbit and 90° inclination. (b) The product $\frac{d\phi}{dE} \cdot S(E)$ under the integral in Eq. (2).

Employing experimental ground-based data for ultra-thin silicon solar cells, the model ensures realistic defect formation rates, derived from particle accelerator test results simulating space radiation environments. Using the calculated DDD values, the model predicts defect formation rates, taking into account proton and electron interactions within the ultra-thin silicon.

5 Conclusions

In this work, we introduced a computational approach for accurately calculating the combined effects of electron and proton radiation on solar cells in space, focusing on mission-specific orbits and incorporating real-time radiation data. By leveraging tools such as SPENVIS for orbit determination and radiation environment modeling, combined with the ESA NIEL tool for calculating the energy dependence of Non-Ionizing Energy Loss (NIEL), we can predict Displacement Damage Dose (DDD) for both electrons and protons.

The comparison of our method with the established NRL and JPL models demonstrates that our approach aligns with these standardised methodologies, while offering a more streamlined integration with real-time mission parameters. The ability to predict radiation effects on ultra-thin silicon solar cells, based on differential particle fluence and NIEL integration, offers new insights into the degradation processes and self-healing mechanisms achieved through defect engineering.

The computational tools developed in this work allow for a precise evaluation of radiation effects on solar cells, facilitating better planning for space missions. By predicting the power output of solar arrays throughout the mission duration, we can offer more accurate data for the design and material selection, ultimately improving the reliability and longevity of space power systems. These developments are crucial for long-duration missions and advancing the durability of space-based photovoltaic technologies.

6 Acknowledgments

This material is based upon work supported by the National Aeronautics and Space Administration under SBIR Phase II Contract №80NSSC22CA123.

CRediT roles: Conceptualization, A.F.; methodology, A.F.; software, A.F; validation, A.F. and Y.G.; writing – original draft preparation, Y.G.; writing – review and editing, A.F. and Y.G.; supervision, S.H.; project administration, S.H.; funding acquisition, S.H. and A.F.

References

[1] Wrbanek, D. J. and Wrbanek, Y. S., 2020, Space Radiation and Impact on Instrumentation Technologies, NASA/TP–2020-220002. Available: `https://ntrs.nasa.gov/api/citations/20200001895/downloads/20200001895.pdf`

[2] P. Partyka, Y. Zhong, K. Nordlund, R. S. Averback, I. M. Robinson and P. Ehrhart, 2001, *Phys. Rev. B*, vol. 64, no. 23, p. 235207. DOI: 10.1103/PhysRevB.64.235207.

[3] M. O. Ruault, J. Chaumont, J. M. Penisson and A. Bourret, 1984, *Philosophical Magazine A*, vol. 50, no. 5, pp. 667–675. DOI: 10.1080/01418618408237526.

[4] L. Pelaz, L. A. Marqués, M. Aboy, J. Barbolla and G. H. Gilmer, 2003, *Applied Physics Letters*, vol. 82, no. 13, pp. 2038–2040. DOI: 10.1063/1.1564296.

[5] R. E. Stoller, 2012, Primary Radiation Damage Formation, *Comprehensive Nuclear Materials*, R. J. M. Konings, Ed. Oxford: Elsevier, pp. 293–332. DOI: 10.1016/B978-0-08-056033-5.00027-6.

[6] E. Holmström, A. Kuronen and K. Nordlund, 2008, *Phys. Rev. B*, vol. 78, no. 4, p. 045202. DOI: 10.1103/PhysRevB.78.045202.

[7] Ch. Inguimbert and S. Messenger, 2012, Equivalent Displacement Damage Dose for On-Orbit Space Applications, *IEEE Trans. Nucl.Sci.*. DOI: 10.1109/TNS.2012.2221477.

[8] S. R. Messenger, G. P. Summers, E. A. Burke, R. J. Walters and M. A. Xapsos, 2001, *Progress in Photovoltaics: Research and Applications*, vol. 9, pp. 103–121, 2001.

[9] S. R. Messenger, E. A. Jackson, J. H. Warner and R. J. Walters, 2012, Advancements to SCREAM: Multiple Spectrum Input and Shield Dose Options in *IEEE Photovoltaic Specialists Conference*. DOI: 10.1109/PVSC.2012.6318276.

[10] M. J. Boschini, P. G. Rancoita and M. Tacconi, 2014, SR-NIEL–7 Calculator: Screened Relativistic (SR) Treatment for NIEL Dose, Nuclear and Electronic Stopping Power Calculator (version 10.16). Available: `https://www.sr-niel.org`.

[11] G. P. Summers, E. A. Burke and M. A. Xapsos, 1995, *Radiation Measurements*, vol. 24, no. 1, pp. 1–8. DOI: 10.1016/1350-4487(94)00093-G.

[12] C. Inguimbert and S. Messenger, 2012, *IEEE Transactions on Nuclear Science*, vol. 59, no. 6, pp. 3117–3125. DOI: 10.1109/TNS.2012.2221477.

[13] J. Allison et al., 2016, *Nuclear Instruments and Methods in Physics Research Section A: Accelerators, Spectrometers, Detectors and Associated Equipment*, vol. 835, pp. 186–225, 2016. DOI: 10.1016/j.nima.2016.06.125.

[14] T. T. Böhlen, F. Cerutti, M. P. W. Chin, A. Fassò, A. Ferrari, P. G. Ortega, A. Mairani, P. R. Sala, G. Smirnov and V. Vlachoudis, 2014, *Nuclear Data Sheets*, vol. 120, pp. 211–214. DOI: 10.1016/j.nds.2014.07.049.

[15] G. P. Summers, R. J. Walters, M. A. Xapsos, E. A. Burke, S. R. Messenger, P. Shapiro and R. L. Statler, 1994, *Proceedings of 1994 IEEE 1st World Conference on Photovoltaic Energy Conversion - WCPEC (A Joint Conference of PVSC, PVSEC and PSEC)*, vol. 2, pp. 2068–2075. DOI: 10.1109/WCPEC.1994.521828.

[16] B. E. Anspaugh and R. G. Downing, 1984, *JPL Publication*, pp. 61–84. Available: `https://ntrs.nasa.gov/api/citations/19840026722/downloads/19840026722.pdf`.

[17] Y. Gurimskaya, S. Herasimenka, A. Fedoseyev, M. Reginevich, S. Bowden, N. Honesty, M. Eyink, T. Moraca, R. Cariou and N. Enjalbert, 2023, *2023 13th European Space Power Conference (ESPC)*, pp. 1–5. DOI: 10.1109/ESPC59009.2023.10298179.

[18] R. Walters, G. P. Summers, J. H. Warner, S. Messenger, J. R. Lorentzen, T. Morton, S. J. Taylor, H. Evans, D. Heynderickx, B. Quaghebeur and F. Lei, 2007, NASA/CP-2007-214494. Available: `https://ntrs.nasa.gov/api/citations/20090022289/downloads/20090022289.pdf`

Journal of Physics: Conference Series **2910** (2024) 012016 doi:10.1088/1742-6596/2910/1/012016

Building Artificial Intelligence, Machine Learning, and Causal Models to Improve Cardiac Health

Roman Haynatzki[1, a)], Thomas A Windle[1], Ran Dai, PhD[1], Ronnie D Horner, PhD[1], James C McClay, MD[3], Peter Z Revesz, PhD[2], Gleb Haynatzki, PhD[1], John R Windle, MD[1]

[1]*University of Nebraska Medical Center,* 42nd and Emile St, Omaha, NE 68198, USA

[2]*University of Nebraska Lincoln,* 1144 T Street, Lincoln, NE 68588, USA

[3]*University of Missouri,* 5 Hospital Dr, Columbia, MO 65201, USA

[a)] Corresponding author: rhaynatzki@unmc.edu, rhaynatzki@gmail.com

Abstract. The prediction of myocardial infarction (MI) outcomes remains pivotal for advancing treatment strategies in cardiology. This study proposes to compare the variable selection efficacy and predictive power of LASSO logistic regression and XGBoost, incorporating a causal mediation approach to elaborate on the underlying mechanisms influencing MI risks. Leveraging the GUSTO-I dataset on MI patients, this research has employed LASSO logistic regression to facilitate variable selection, aiming to identify critical predictors with substantial impacts on MI outcomes. Alternatively, the XGBoost algorithm has been utilized to assess its approach to variable selection and prediction. Both methods have been evaluated based on their Area Under the Receiver Operating Characteristic Curve (AUC) to determine their predictive accuracy. Furthermore, causal mediation analysis has been integrated to explore the putative effects of key predictors, identified in previous studies. The analysis aimed to reveal which method offers superior predictive performance and how well each can be interpreted in a clinical context. This study can contribute valuable insights into the comparative advantages of using LASSO logistic regression versus XGBoost in the field of cardiovascular disease research. By integrating causal mediation analysis, we aim to extend beyond mere prediction to offer a deeper understanding of the causal relationships and mechanisms at play.

1. INTRODUCTION AND BACKGROUND

The overarching objective of this paper is to model the 30-day survival after MI with high precision that will leverage Artificial Intelligence and Machine Learning (AI/ML) advanced analytics to "Big Data", such as electronic health records (EHRs). We have utilized the GUSTO-1 large dataset [26], which contains clinical data from 40,830 patients with acute MI, providing an opportunity to develop predictive models for the main outcome in this study, MI survival within 30 days (denoted as DAY30 in the sequel). These models can leverage various clinical variables and biomarkers to aid in improving prediction of MI survival rate. The GUSTO-I includes demographics, clinical diagnosis, medications, patient reported measures and

Content from this work may be used under the terms of the Creative Commons Attribution 4.0 licence. Any further distribution of this work must maintain attribution to the author(s) and the title of the work, journal citation and DOI.
Published under licence by IOP Publishing Ltd

outcomes, and over 40 other variables per patient. Moreover, we are using models such as XGBoost (XGB) and the Peter-Clark (PC) Causal Discovery Algorithm on a dataset that has not been previously analyzed using these specific methods. We aim to identify the most optimal model for variable selection and efficiency, using AUC as a metric, by comparing XGBoost and LASSO Penalized Logistic Regression (LASSO), both of which are prominent methods for identifying key predictors in the data. We have noticed that Steyerberg et al. demonstrated that stepwise selection methods used in regular logistic regression introduce substantial bias, especially in small datasets [15]. We want to add interaction terms for any causal relationships found by the PC algorithm in the LASSO model to determine if this amendment can surpass XGBoost in AUC (Area Under the ROC Curve as defined below). Disclaimer: This study aims to investigate the application of AI/ML models in predicting MI outcomes. The methodologies and findings discussed are intended solely for research purposes and should not be construed as medical advice or used to guide clinical decision-making without the involvement of a qualified healthcare professional. While the GUSTO-I dataset is from the 1990s, and the medical outcomes in the dataset may not directly reflect current clinical practices, the primary aim of this study is to compare the performance of AI/ML models and their interpretations, not to draw modern medical conclusions. The GUSTO-I dataset provides a valuable test case for assessing the robustness and performance of these models due to its size and richness in clinical variables. Therefore, while the outcomes may not perfectly mirror present-day scenarios, the comparison of LASSO and XGBoost in handling high-dimensional, heterogeneous clinical data remains relevant for advancing AI/ML methodologies in healthcare.

Myocardial infarction (MI), commonly known as a heart attack, is a significant health concern in the United States. MI occurs when blood flow to a part of the heart is blocked for a long enough time that part of the heart muscle is damaged or dies. This blockage is most often caused by a buildup of plaque—a combination of fat, cholesterol, and other substances—on the walls of coronary arteries. MI is a major health problem, affecting over 800,000 Americans annually [23]. This staggering number highlights the widespread nature of the issue, placing millions at risk of severe health complications and death. Despite advancements in medical science, MI remains one of the leading causes of mortality, accounting for nearly 365,000 deaths [23] each year in the US. Several risk factors contribute to the prevalence of MI. These include hypertension (HT), high cholesterol, smoking, diabetes, obesity, and a sedentary lifestyle [3]. Age, gender, and genetic predisposition also play crucial roles. Men are generally at higher risk, but the incidence in women increases and often surpasses that in men post-menopause. MI disproportionately affects African Americans, Hispanic Americans, and older adults, with these groups experiencing higher rates of complications and mortality. The financial burden of MI on the healthcare system is substantial. The American Heart Association (AHA) estimates that the direct and indirect costs of cardiovascular diseases, including MI, amount to over $316 billion annually. This includes healthcare services, medications, and lost productivity. The personal impact is equally significant, affecting patients' quality of life and their ability to work and participate in daily activities. A challenge in managing MI is timely and accurate diagnosis. Many individuals at risk of MI are either undiagnosed or inadequately treated for underlying conditions like HT and high cholesterol. Early detection through regular medical check-ups and advanced diagnostic techniques is essential. Innovations in medical technology, including the assistance for medical experts through use of Artificial Intelligence (AI) and Machine Learning (ML) in analyzing electronic health records (EHRs), offer promising solutions to improve diagnosis and treatment outcomes.

1.1 Personalized Risk Assessment in Myocardial Infarction

Recent advances in the management of myocardial infarction (MI) have underscored the importance of personalized risk assessment. Traditional methods, while valuable, often overlook the complex interplay of multiple factors unique to each patient. By incorporating AI/ML models, we can better identify and analyze these factors, allowing for a more nuanced and individualized approach to predicting MI outcomes. Personalized risk assessment leverages detailed patient data, including lifestyle factors and specific clinical markers to enhance the predictive power of models. For example, the interaction between hypertension and comorbidities like diabetes or prior angina significantly influences MI risk. Customizing risk models to reflect these interactions can improve the accuracy of predictions and lead to more targeted interventions.

Machine learning, particularly through algorithms like XGB and LASSO, plays a crucial role in this personalized approach. These models can parse large datasets, such as electronic health records, to uncover novel risk factors and refine existing ones, offering deeper insights into patient-specific risks. By incorporating SHAP-based Feature Importance Values (FIV) [13], clinicians can gain a clearer understanding of how individual features contribute to a patient's risk profile. This enhances both the predictive accuracy and the interpretability of the model, making it a valuable tool in clinical decision-making.

In summary, integrating personalized risk assessment into MI management using advanced machine learning techniques aligns with the overarching goals of improving prediction accuracy and offering individualized patient care. This approach not only optimizes patient outcomes but also ensures the efficient use of healthcare resources.

2. METHODS

Machine learning (ML) and *artificial intelligence* (AI) are algorithmic analyses of *large* and *complex* data. For purposes of this paper, we consider ML to be a subset of AI used to model data, namely, *prediction/estimation* of MI survival probabilities. AI, in the form of mathematical or statistical algorithms for prediction, refers to the application of statistical techniques to analyzing data and making predictions or inferences. That involves using mathematical models and statistical methods to identify patterns, correlations, and trends in large datasets, and then using those patterns to make predictions about future or unseen data.

Thus, the focus is the correct development and implementation of the ML algorithm:

- tuning/finding suitable hyperparameters (for both LASSO and XGB);
- selecting the best set of hyperparameters (for both LASSO and XGB);
- running the algorithm in real time (for both LASSO and XGB);
- creating an ensemble model by combining multiple weak learners (for XGB);
- improving overall prediction accuracy by sequentially addressing the weaknesses of previous models (for XGB);
- learning complex patterns and relationships in data (for XGB);
- solving classification, regression, or other complex tasks (for both LASSO and XGB);
- adapting to new inputs through training and adjusting weights (for both LASSO and XGB);
- reducing overfitting (for both LASSO and XGB); and
- (in case of categorical response variable) classifying instances based on the likelihood of belonging to a particular class (for both LASSO and XGB).

2.1 SHAP (SHapley Additive exPlanations) Importance Values

Shapley values quantify the contribution of each feature (aka, variable) to the prediction of a machine learning model. Shapley values provide local (at patient level) explanations by estimating the contribution of each feature for a particular patient and considering all possible feature combinations and averaging their effects, allocating the total effect of a feature among all possible feature subsets. However, it's important to note that Shapley values are suited for explaining the contributions of features after models have been finalized, and not for model selection.

The Shapley values were originally proposed as a method for fairly distributing the "value" of a coalition (a group of players working together) among its members [4]. In our case, these Shapley values are being calculated for different features, or variables, instead of players. The larger the Shapley importance value for a given feature (e.g., age), the more this feature contributes to the mortality within 30 days of MI, denoted above as DAY30. Given the value function

$$val_x(S) = \int \hat{f}(x_1, \dots, x_p) d\mathbb{P}_{x \notin S} - E_X\left(\hat{f}(X)\right)$$

and a predictive model, such as a linear model:

$$\hat{f}(X) = \beta_0 + \beta_1 x_1 + \dots + \beta_p x_p$$

for MI risk where $x_1, x_2, \dots, x_p$ are the p features' values, β_j refers to the weight of feature j and S refers to a coalition, or subset of features, drawn from $\{x_1, x_2, \dots, x_p\}$. Additionally, $|S|$ denotes the number of features in subset S. The formula used to calculate the Shapley value for each feature j is as follows [11], [13]:

$$\varphi_j(val) = \sum_{S \subseteq \{1,\dots,p\} \setminus \{j\}} \frac{|S|!\,(p - |S| - 1)!}{p!} \left(val(S \cup \{j\}) - val(S)\right)$$

$$= \frac{1}{\text{number of features}} \sum_{\text{coalitions including feature } j} \frac{\text{marginal contribution of } j \text{ to coalition}}{\text{number of coalitions excluding } j \text{ of this size}}.$$

The SHAP value implements the Shapley value formulas [12]. SHAP incorporates computational optimizations and approximations to efficiently compute feature attributions for high-dimensional datasets and complex models. This scalability makes SHAP applicable to real-world machine learning tasks, where the number of features and instances can be significant. The number of features is important when presenting the number of ways to choose a subset of features from a set of features, and the resulting binomial coefficients contribute to the computational complexity of calculating the Shapley values using the original formula. The SHAP framework addresses this issue by introducing several techniques to reduce the computational burden while approximating the Shapley values efficiently. They include coalition reduction where SHAP employs sampling or other approximation methods to estimate the Shapley values without exhaustively evaluating all possible coalitions. This reduces the computational complexity significantly. Also, SHAP employs the consistency and linearity properties of Shapley values, thus ensuring that the feature attribution methods have desirable characteristics like local accuracy, consistency, (even when there is missing data), making SHAP importance values more interpretable and efficient [1].

In the context of our study, SHAP-based feature-importance values (FIV) are not only instrumental in providing transparent and interpretable model explanations but also serve as a powerful tool for enhancing our understanding of complex medical data. By leveraging SHAP values, we can view the intricacies of feature contributions, which is particularly beneficial in a clinical setting where the interpretability of model predictions is crucial for informed decision-making.

One of the novel aspects of using SHAP values in our analysis is their ability to handle interactions between features effectively. Unlike traditional feature importance metrics that often consider features in isolation, SHAP-based FIV values account for the interdependencies among features. This characteristic is especially valuable when dealing with medical data, where if present, interactions between different clinical variables can significantly influence patient outcomes. For instance, the interplay between age, comorbidities, and treatment protocols may jointly affect the risk of mortality within 30 days of MI (DAY30).

Furthermore, SHAP values facilitate individualized patient-level explanations, enabling us to tailor our interpretations to specific patient profiles. This personalized approach is critical in clinical practice, where understanding the unique contribution of features to a particular patient's prediction can guide targeted interventions and personalized treatment plans. By examining the SHAP values for individual patients, clinicians can gain insights into the primary drivers of risk for each patient, leading to more precise and effective medical decisions.

In addition to their interpretability, SHAP values offer robustness against common challenges in medical data analysis, such as missing data and varying feature scales. SHAP's consistency and adherence to properties like local accuracy ensure that the feature importance values remain reliable even in the presence of incomplete or noisy data. This robustness is crucial for maintaining the integrity of our analysis, especially when working with real-world medical datasets that often exhibit such characteristics. To illustrate the practical application of SHAP values in our study, we implement them in a two-step process. First, we train our machine learning models on the available clinical data, ensuring that the models are well-calibrated and generalize well to unseen data. After model training, we compute the SHAP values for each feature, providing us with a comprehensive view of feature contributions at both the global (model-level) and local (patient-level) scales. This dual perspective allows us to identify key predictors of mortality risk and understand their impact across different patient subgroups. Additionally, SHAP's approximation methods mitigate the issue of the high computational complexity of the Shapley values.

In summary, the use of SHAP-based FIV in our study not only enhances the interpretability and transparency of our machine learning models but also provides valuable insights into the interactions and contributions of clinical features where applicable. By addressing both the global and local aspects of feature importance, SHAP values empower us to make more informed and personalized clinical decisions, ultimately improving patient outcomes.

2.2 AUC Value

The area under the ROC curve (AUC) shown in Figure 1 below is a metric used to evaluate the performance of a binary classification model, and in our case, we are classifying the patient as either surviving 30 days after the MI diagnosis or not. AUC represents the overall quality of the model's predictions by measuring the trade-off between its true positive rate (sensitivity) and false positive rate (1 - specificity) across different classification thresholds [17].

The ROC curve as shown below is created by plotting the true positive rate (TPR) against the false positive rate (FPR) at various threshold settings. The AUC is then computed as the area under this curve, ranging from 0 to 1 [19]. An AUC value closer to 1 indicates a better classifier with higher discriminatory power, while a value of 0.5 represents a random or non-informative classifier. The image below shows a graphical representation of the ROC curve with values as obtained from our models. The AUC is a valuable metric for assessing a model's performance, capturing both its sensitivity and specificity, especially when the dataset size is sufficiently large. However, a cautionary note is warranted for imbalanced datasets. For instance, in a scenario where 95% of patients survive MI (the majority class), a model making nearly constant predictions can misleadingly achieve a high AUC value. In such contexts, an AUC value exceeding 0.95 should be viewed with suspicion, as it may indicate the model is not genuinely discriminating between classes as effectively as the score suggests.

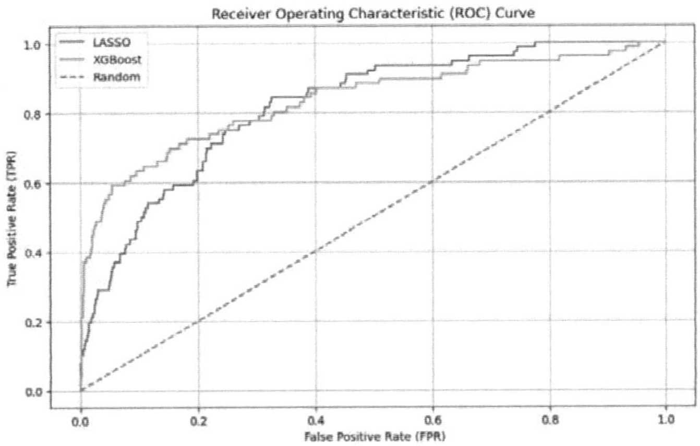

FIGURE 1. Receiver Operating Characteristic (ROC) Curves with an AUC (Area Under the Curve) value of 0.85 and 0.82, indicating strong predictive accuracy of MI 30 Day Survival.

2.3 Causal Discovery

Causal discovery is important in the analysis and modelling of complex datasets, particularly in medical research and understanding the underlying mechanisms. In our study, causal discovery is utilized to identify and map the relationships between variables within our dataset, which directly impacts the construction and refinement of our LASSO logistic regression model by adding appropriate interaction terms. By determining causal connections through expert opinion, we move beyond simple associations to uncover how variables interact and influence each other, providing deeper insights into the pathways leading to specific outcomes. The PC (Peter-Clark) algorithm [8, 14], a well-established method for causal discovery, has been used to generate a causal graph for our dataset. This algorithm may identify potential causal relationships by systematically testing conditional independencies between variables. The resulting graph offers a visual foundation that highlights direct and indirect influences among the variables. For instance, in our analysis, the causal graph reveals that the variable 'A65' (Age over 65) has a direct influence on both 'STE' (ST elevation on ECG) and 'ST4' (ST elevation on ECG with more than 4 leads). This kind of causal insight is important for model development, as it guides the selection of interaction terms to include in the LASSO model, in order to compare efficiency through AUC to the XGB model.

Incorporating interaction terms based on these identified causal relationships may enhance the model's capacity to capture the true complexity of the data. It allows us to account for the combined effects of multiple variables on the outcome, rather than evaluating each in isolation. This approach not only improves the accuracy of the model's predictions but also its interpretability. By understanding the causal pathways, we can explain why certain variable combinations are more predictive of MI survival, thereby making the model's outcomes more actionable in a clinical setting.

Moreover, causal discovery in conjunction with expert opinion provides a robust framework for ensuring that the variables and interactions included in the model are grounded in the real world. This avoids the pitfalls of overfitting and correlations of non-causal nature, which can lead to misleading predictions and interpretations. By aligning our modelling efforts with the discovered causal structure, we make our models more resilient to the complexities and nuances of real-world data, ultimately leading to more reliable and insightful predictions about MI outcomes. The integration of causal discovery into our modelling process may not only enhance the model's performance but also ensures that the insights derived from it are scientifically sound and clinically relevant. As we advance towards more personalized and precise predictive models, the role of causal discovery will become increasingly important in the development of robust and interpretable AI/ML-driven solutions in healthcare.

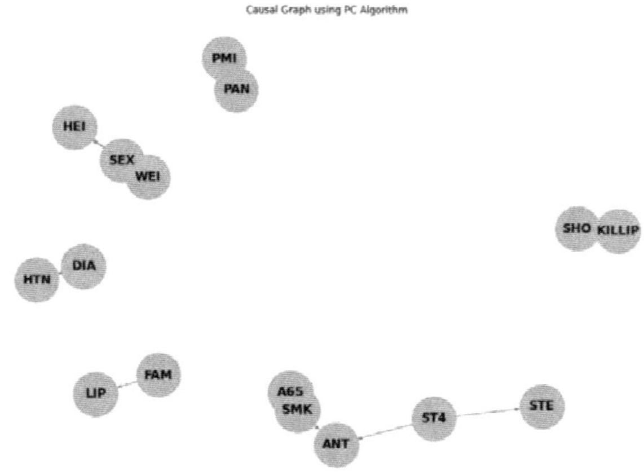

FIGURE 2. Visual causal representation of variables in the GUSTO-I dataset.

The causal graph generated in Figure 2 using the PC algorithm shows several potential causal relationships between variables in the GUSTO-I dataset, some of which align well with known medical knowledge, while others may be novel discoveries or, simply, spurious. For example, the relationship between HTN (Hypertension history) and DIA (Diabetes) is consistent with clinical understanding, as these conditions often co-occur due to shared risk factors like obesity and metabolic syndrome. The relationship between SEX (Gender) and WEI (Weight) is also plausible, given that gender can influence body composition. However, some suggested relationships, such as between ANT (Anterior Infarct Location) and A65 (Age >65), are less likely to represent true causal relationships and may instead be artifacts of the algorithm. Also, it is important to note that the GUSTO-I dataset is comprised exclusively of patients who have experienced an MI, which could contribute to the emergence of spurious causal relationships. The lack of a control group of healthy or non-MI patients complicates the ability to validate these relationships in a broader context.

Other relationships in the graph include the link between PMI (Previous Myocardial Infarction) and PAN (Previous Angina Pectoris), which is reasonable as both are indicators of prior cardiac events and might be causally related. The relationship between LIP (Lipids: hypercholesterolemia) and FAM (Family history of MI) could indicate a genetic predisposition to higher cholesterol levels leading to a family history of myocardial infarction, which is medically plausible. However, the exact nature of causality and the strength of these relationships should be carefully considered and further explored.

These observations underscore the importance of expert opinion in interpreting such graphs. By incorporating domain expertise, we can better distinguish true causal relationships from those that might be coincidental or artifacts of the data and algorithm, leading to more accurate and meaningful interpretations.

2.4 XGBoost Method

The Extreme Gradient Boosted trees (XGBoost) is a relatively recent ensemble method for constructing trees [10]. Trees in other ensemble algorithms are created in the conventional manner, i.e. either using Gini Impurity or Entropy. But XGBoost introduces a new metric, called *similarity score*, for node selection and splitting.

A very general (nonmathematical) description of the steps involved in creating a Decision Tree using the similarity score is shown below:

1) Create a single-leaf tree.
2) For the first tree, compute the average of the target variable as prediction and calculate the residuals using the chosen loss function. For any subsequent tree, the residuals come from prediction made by the previous tree.
3) Calculate the similarity score (for a pre-selected loss function) using the following formula:

$$Similarity\ Score = \frac{Gradient^2}{Hessian + \lambda}$$

   where the Hessian is equal to the number of residuals; Gradient^2 = squared sum of residuals; λ is a regularization hyperparameter.
4) Using the similarity score, we determine which part of the data (node) we should focus on. The goal is to find the spot in the tree where the data is most similar, meaning it's grouped together in a way that makes it easier to make accurate predictions. The higher the similarity score, the more alike the data points are within that node, which suggests that it's a good candidate for further splitting to improve the tree's predictions.
5) Using the similarity score, we calculate the *Information gain*. Information gain gives the difference between old similarity and new similarity and thus quantifies how much homogeneity is achieved by splitting the node at a given point. It is calculated using the following formula:

$$Information\ Gain = Left\ Similarity + Right\ Similarity - Similarity\ for\ Root$$

6) Create the tree of desired length using the above method. Pruning and regularization would be done by varying the regularization hyperparameter.
7) Predict the residual values using the Decision Tree you constructed.
8) The new set of residuals is calculated using the following formula:

$$New\ Residuals = Old\ Residuals + \rho\ \Sigma\ Predicted\_Residuals$$

   where ρ is the learning rate.
9) Go back to step 1 and repeat the process for all the trees.

The actual XGBoost algorithm is mathematically involved [9] and the details are omitted here.

2.5 Penalized LASSO Logistic Regression Model

Penalized Logistic Lasso Regression (*LASSO*) is a powerful statistical method that enhances the predictive ability of logistic regression by incorporating regularization techniques, making it particularly useful in high-dimensional data scenarios or when preventing overfitting. In the context of predicting the probability of death within 30 days following a myocardial infarction (MI), LASSO can effectively utilize various prognostic factors, such as age and blood pressure to improve prediction accuracy. The core of logistic regression is the logistic function, which models the probability of a binary outcome (e.g., death within 30 days) as a function of the input variables. This is mathematically represented as $P(Y = 1|\boldsymbol{X}) = \frac{1}{1+e^{-(\beta_0+\boldsymbol{\beta X})}}$, where $\boldsymbol{X}$ is the vector of input variables, β_0 is the intercept and $\boldsymbol{\beta}$ is the vector of coefficients.

LASSO incorporates a penalty term into the likelihood function to control model complexity and avoid overfitting. The most common types of regularization are L1 regularization (Lasso), which adds the absolute values of the coefficients to the penalty term, and L2 regularization (Ridge), which adds the squared values of the coefficients. Elastic Net is a combination of both L1 and L2 regularization. The penalized log-likelihood function, which LASSO aims to maximize, is given by $\mathcal{L}(\beta_0, \boldsymbol{\beta}) - \lambda \sum_{j=1}^{p} P(\beta_j)$, where $\mathcal{L}(\beta_0, \boldsymbol{\beta})$ is the log-likelihood of the logistic model and $P(\beta_j)$ is the penalty term [2].

The coefficients in LASSO provide insights into the relationship between each prognostic factor and the probability of death within 30 days post-MI, with larger absolute values indicating stronger associations. The model outputs a continuous probability, representing the likelihood of death within 30 days given the input variables. For binary classification, a threshold is applied to this predicted probability, often set at 0.5, to categorize outcomes as either death (1) or survival (0). Adjusting this threshold can help balance the trade-off between identifying high-risk patients and minimizing false positives. A specific feature of the LASSO as a regression approach is that the interpretable p-values and confidence intervals associated with the beta coefficients are generated, in difference with XGB.

2.6 Comparing Efficiency of XGB vs LASSO: Time Complexity

AI/ML models can take a significant amount of computational time, even with the help of a supercomputer, especially when working with very large datasets. Thus, it's crucial to review the time complexity of these two specific models. The comparison of these models is somewhat challenging since they are fundamentally different.

When analyzing the time complexities of different algorithms, we usually don't focus much on the constant multiplier before the 'n' term, where 'n' represents the problem size. Common time complexities include constant (O(1)), logarithmic (O(log n)), linear (O(n)), quasilinear (O(n log n)), polynomial ($O(n^a)$), exponential ($O(a^n)$), and factorial (O(n!)). Typically, algorithms with lower complexity run faster, while those with higher complexity run slower. Focusing on our comparison between XGBoost and LASSO, the explanation in the original paper by Tianqi Chen et al. clarifies that XGBoost's time complexity is essentially O(n log n) because it is greedy [9]. A greedy algorithm performs operations at each step in the most efficient way possible without revisiting or re-optimizing previous decisions. Regarding LASSO, its time complexity is influenced by the iterative nature of optimization algorithms used to minimize penalized loss functions, typically through techniques like gradient descent or coordinate descent. The time complexity of LASSO is polynomial, usually $O(n^a)$, where 'a' depends on the convergence criteria and the specific application of the regularization method, such as L1 or L2 regularization. This polynomial relationship indicates that as the dataset size (n) increases, the computational burden of LASSO also increases, though typically not as dramatically as in more complex AI/ML models. In conclusion, this analysis helps in understanding the time complexities of both XGB and LASSO. XGB's greedy nature leads to faster execution with a time complexity of O(n log n), while LASSO's polynomial complexity accounts for a relatively longer runtime, especially with larger datasets. These differences highlight the importance of selecting the appropriate algorithm based on the specific computational resources and time constraints of a given application.

2.7 Interpreting Model Prediction

Both *accuracy* and *interpretability* are important properties of models. The ability to correctly *interpret* a model:

- increases user's trust;
- provides insight into how a model may be improved; and
- supports understanding of the process being modeled.

Complex (nonlinear) models are, on average, more precise but less interpretable than simple models. Even for large data, despite having lower accuracy, simple models (e.g., *linear* or *logistic regression*) are sometimes preferred for their ease of interpretation. The area under the ROC curve (*AUC*) is often used to assess the predictive performance of the models and for model selection purposes.

For the LASSO model, it is possible to determine the probability p for a patient to survive 30 days after a myocardial infarction by the following equation:

$$\log\left(\frac{p}{1-p}\right) = \beta_0 + \beta_1 X_1 + \beta_2 X_2 + \cdots + \beta_k X_k \Rightarrow$$

$$\log\left(\frac{p}{1-p}\right) = -3.2268 + 0.0304 * \text{KILLIP} + 0.1596 * \text{TTR} + 0.2291 * \text{SEX} + 0.0462 * \text{HYP} + 0.1074 * \text{HRT} + 0.2808 * \text{HTN} + 0.2180 * \text{WEI} + 0.1951 * \text{SHO}$$

The *SHAP* (SHapley Additive exPlanations) values serve as importance metrics for predictor variables (also known as features) in generating specific predictions from a given AI/ML model like XGB. They provide essential interpretability in scenarios where deriving explicit equations, like the one shown above, is not feasible. [4-6]

2.8 Dataset and Software

We used the Python programming language and its environment for statistical computing in all calculations [24]. We analyzed a well-known dataset to determine the factors influencing mortality in myocardial infarction (MI) patients. This dataset was extracted from the GUSTO-I (Global Utilization of Streptokinase and Tissue Plasminogen Activator for Occluded Coronary Arteries) Study, which enrolled over 40,000 cardiac patients from 1,081 hospitals across 15 countries between December 1990 and June 1991. To the best of our knowledge, this dataset has not previously been analyzed using the specific methods presented in this paper. The GUSTO-I study was an observational clinical trial, and the main outcome variable in our analysis was 30-day survival following MI, with a focus on identifying the factors that most significantly influenced patient survival.

The variables included in the analyzed dataset are as follows:

Name	Description (N=40,830)	Name	Description
AGE	Age in years	HEI	Height in cm
A65	Age >65 years	WEI	Weight in kg
SEX	Gender(male = 0, female = 1)	SMK	Smoking (1 = never; 2 = ex; 3 = current)
KILLIP	Measure of left ventricular function	HTN	Hypertension history/ diagnosis
SHO	Shock: Killip class 3/4 versus 1/2	LIP	Lipids: hypercholesterolaemia
DIA	Diabetes	PAN	Previous angina pectoris
HYP	High Blood Pressure	FAM	Family history of MI
HRT	Heart rate: pulse >80	STE	ST elevation on ECG: number of leads
ANT	Anterior infarct location	ST4	ST elevation on ECG: >4 leads
PMI	Previous myocardial infarction	TTR	Time to relief of chest pain > 1 h
HIG	High risk: ANT or PMI		

TABLE 1. Variables in the GUSTO-I dataset.

3. RESULTS

Multiple datasets were merged before performing the actual data analysis. There were samples from different regions that had to be combined in order to analyze the data. The dataset was already clean and did not require imputation as there were no missing values. For the ML algorithms, the main dataset was randomly split into two: 80% for training and 20% for validation, with a separate dataset used for testing as illustrated in Figure 3.

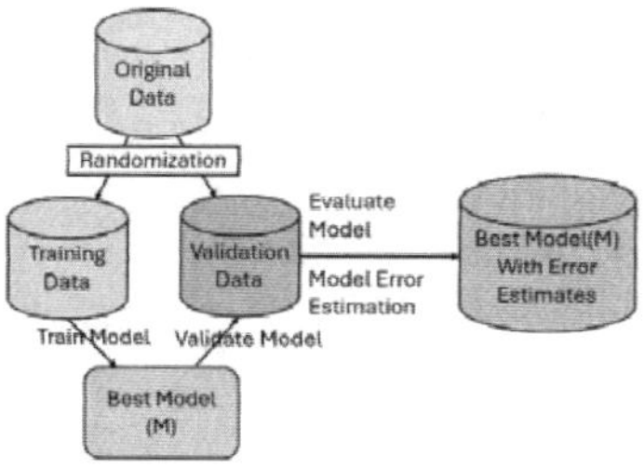

FIGURE 3. A diagram of tuning and testing of a machine learning algorithm in the general case.

For each of the two models (XGBoost and LASSO) used, the initial model was tuned based on the training data. This training dataset is used to create the best model, and the resulting model/algorithm was then validated on the testing data. Based on the predictor values in the testing data, the algorithm calculated the response variable (i.e., MI 30 Day survival) for every observation within the testing dataset. The true values of MI survival in the training dataset are known, so the predicted responses are compared to the actual responses and the AUC was calculated [10]. The real data results from the comparison of the two different approaches using the AUC and SHAP Values are shown numerically in Tables 2 and 3 as well as graphically in Figures 4 and 5 below. According to those tables, it is conclusive that both the LASSO and XGBoost models had AUC values of close to 0.85; hence, performed very well. XGB outperformed LASSO since its AUC value (0.8518) was higher than LASSO's AUC value (0.82). From both the XGB and LASSO model, the Shapley-based FIV of the features "KILLIP“, “Heart Rate” and “High Blood Pressure” have a relatively high impact on the model's predictions of MI survival compared to other features. This implies that changes in KILLIP, heart rate, and blood pressure have a strong influence on the predicted outcome.

Method	AUC
XGBoost	0.8518
LASSO	0.82

TABLE 2. AUC values for both models on the real dataset: XGBoost and LASSO.

SHAP-Based FIV	XGB	SHAP-Based FIV	LASSO
KILLIP	0.1	Hypertension Diagnosis	0.2808
Age	0.098	Gender	0.2291
High Blood Pressure	0.070	Weight	0.2180
Heart Rate	0.062	KILLIP CLASS	0.1951
>4 Leads on ECG	0.058	Time to Relieve Chest Pain	0.1596
Hypertension Diagnosis	0.054	Heart Rate	0.1074

TABLE 3. SHAP values for both models on the real dataset: XGBoost and LASSO.

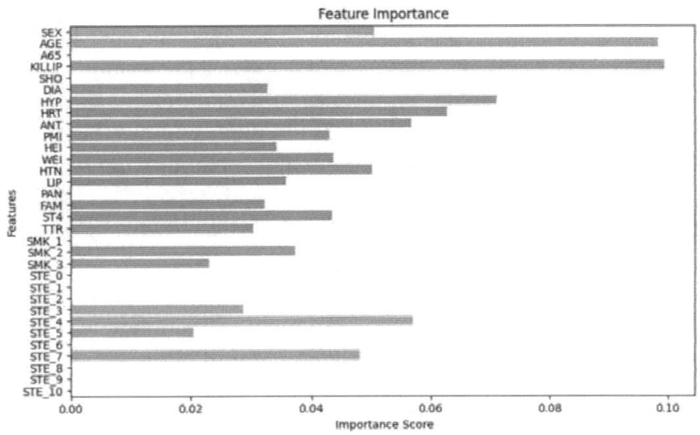

FIGURE 4. Diagrams of SHAP-based FIV for XGB

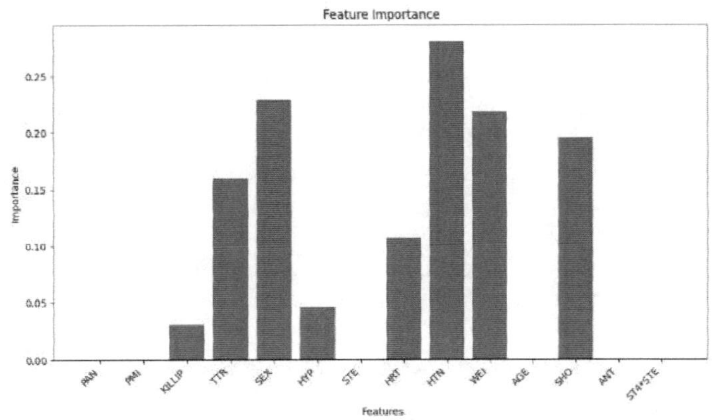

FIGURE 5. Diagram of SHAP-based FIV for LASSO

Journal of Physics: Conference Series **2910** (2024) 012016 doi:10.1088/1742-6596/2910/1/012016

In Section 2.7, we introduced interpretability in our predictive model using a LASSO penalized logistic regression equation; however, XGB does not provide a similar explicit equation for interpretation. To address this, we incorporated SHAP values, offering a method to enhance interpretability within the XGB framework. The Shapley-based feature importance analysis from our XGB model offers insights into the factors influencing 30-day survival after myocardial infarction within the GUSTO-I dataset. Notably, the variables SEX, AGE, and KILLIP were the most influential predictors, reflecting their well-established roles in cardiovascular outcomes. Interestingly, while A65, a binary variable indicating whether a patient is over 65 years old, was included in the model, its importance score was minimal. This suggests that the information it provides is already sufficiently captured by the continuous AGE variable, rendering A65 redundant in this context. The absence of significant importance for A65 highlights XGB's capability to handle potential multicollinearity effectively. This reinforces the robustness of our model, which effectively integrates age-related risk without relying on multiple, overlapping variables. KILLIP and SHO, both of which measure left ventricular function and shock, respectively, were also highly ranked, despite their conceptual overlap. This suggests that while both variables assess the severity of heart failure, they each contribute uniquely to survival predictions. SHO was particularly prominent, likely because it specifically represents the most severe cases (Killip class 3/4), which are critical in determining short-term mortality. Interestingly, cardiovascular risk factors such as DIA (Diabetes), HYP (High Blood Pressure), and HRT (Heart Rate >80 bpm) were also significant, reinforcing the multifactorial nature of post-myocardial infarction survival. Similarly, the importance of ANT (Anterior infarct location) and PMI (Previous myocardial infarction) aligns with clinical understanding that the location and history of infarction critically influence outcomes. Our analysis also included multiple variables related to ST-segment elevation (STE), which are known markers of myocardial injury. Despite potential concerns of multicollinearity, the model handled these variables effectively. XGB, with its tree-based approach, inherently mitigates multicollinearity by selecting the most informative splits and minimizing redundancy. For instance, although STE and ST4 both reflect ST-segment elevation, their individual contributions to the model indicate that each captures distinct and clinically relevant information about the extent of myocardial damage.

In summary, the feature importance analysis underscores the model's ability to integrate a wide range of clinical variables, including those that might traditionally be considered collinear, into a coherent and clinically interpretable prediction framework. These findings not only validate the model's predictive accuracy but also provide insights into the multifaceted nature of MI outcomes, highlighting key areas for future research and potential clinical application.

Incorporating advanced analytics and SHAP-based Feature Importance Values (FIV) into our model has allowed us to not only predict outcomes with greater accuracy but also to gain a deeper understanding of the underlying mechanisms. For instance, SHAP values have highlighted the significant roles that both age and left ventricular function play in determining MI outcomes. These insights reinforce the importance of personalized risk assessment and targeted treatment strategies, where interventions can be tailored based on an individual's specific risk profile, potentially leading to better management of the disease and improved survival rates. By focusing on both the predictive accuracy and interpretability of the model, we can provide actionable insights that are critical for improving patient outcomes in clinical settings [16, 18, 20, 21].

4. DISCUSSION AND CONCLUSIONS

We applied and compared the performance of two *Artificial Intelligence/Machine Learning* (AI/ML) approaches (XGB and LASSO) to a large-sized cardiac dataset. AI/ML approaches are popular for modeling large, complex data, including for the development of risk models in healthcare. For this dataset, the LASSO and XGB performed similarly well as risk models, with AUC value between 0.80 and 0.86.

Notably, by incorporating interaction effects derived from the causal model, the AUC for the LASSO model improved to 0.82, higher than the 0.787 AUC reported in [25] for a standard logistic regression model. However, this improvement still did not reach the AUC achieved by XGB. Our study investigates the performance disparities between these models, with a focus on different data types typically encountered in machine learning tasks. Clinical data is often characterized by high dimensionality, missing values, and complex, non-linear relationships between variables. XGB deals with these challenges effectively, outperforming LASSO regression in these key scenarios. XGB models high-dimensional datasets efficiently, which is a significant factor contributing to its superior performance. By leveraging an ensemble approach that combines multiple weak learners, XGB can effectively capture the effects of intricate interactions between clinical features on the target variable. LASSO, being inherently linear, often struggles to represent such non-linear dependencies and complex interactions, resulting in suboptimal performance on complex data structures. In the context of high-dimensional clinical datasets, XGB demonstrates superior capabilities. Though there are no missing values in the GUSTO-I dataset, the handling of missing data further distinguishes XGB from LASSO. XGB's ability to manage missing values internally during training—by determining the optimal direction for data splits when values are missing—often leads to enhanced model accuracy. In contrast, LASSO requires the explicit imputation of missing values before model fitting, which can introduce bias or lead to the loss of valuable information.

An additional advantage of XGB is its stability across different runs. During our analysis, we observed that LASSO can produce slightly different results when the Python kernel is refreshed, even when using the same training and testing data. This variability stems from the model's reliance on certain optimization algorithms, such as coordinate descent or stochastic gradient descent, which may converge differently depending on initial conditions or random seed states. This issue can lead to inconsistencies in model coefficients and predictions, potentially impacting the reproducibility of results. On the other hand, XGB consistently produces stable results across multiple runs due to its deterministic nature and careful handling of randomness, providing more reliable and reproducible outputs. This stability, despite similar AUC (Area Under the Curve) values between the two models, makes XGB a more dependable choice in practical applications.

However, these advantages come at the cost of interpretability. While XGB generally achieves higher accuracy, especially in complex tasks, its ensemble nature results in models that are less interpretable compared to the simpler, linear models produced by LASSO regression. LASSO's ability to shrink some coefficients to zero enhances its interpretability, offering a straightforward model that is easier to understand and communicate. Nevertheless, in contexts where accuracy, stability, and the ability to handle complex data structures are prioritized, XGB is the preferable choice. Lastly, XGB's robustness to outliers is noteworthy. LASSO, on the other hand, is more sensitive to outliers, which can adversely affect its performance.

In conclusion, XGB is significantly better suited than LASSO for datasets characterized by non-linear relationships, high dimensionality, missing values, sparsity, large scale, and scenarios requiring consistent, reproducible results. While LASSO remains a valuable tool for its interpretability and simplicity in linear settings, XGB is the superior choice for more complex data scenarios. Future work could explore hybrid approaches that combine the interpretability of LASSO with the predictive power and stability of XGB, potentially offering the best of both worlds.

5. REFERENCES

[1] Lundberg, S., & Lee, S.-I. (2017). A unified approach to interpreting model predictions. Neural Information Processing Systems, 30, 4768–4777.

[2] Hosmer, D. W., Lemeshow, S., & Sturdivant, R. X. (2013). Applied Logistic Regression (3rd ed.). Wiley.

[3] Mancia G, Grassi G. "Hypertension and diabetes mellitus as synergistic risk factors for cardiovascular disease: a global perspective" Journal of Hypertension. (2011)

[4] Shapley LS. A value for *n*-person games. In: Kuhn HW, Tucker AW (eds.) *Contribution to the Theory of Games* (AM-28) 2:307-318. (1953)

[5] Lipovetsky S, Conklin M. Analysis of regression in game theory approach. *Applied Stochastic Models in Business and Industry* 17(4):319-330. (2001)

[6] Hausken K, Mohr M. (2001) The value of a player in *n*-person games. *Soc Choice Welfare* 18: 465-483.

[7] Steyerberg, E. (2019). Clinical prediction models: A practical approach to development, validation and updating. Springer.

[8] Hyttinen, A., Eberhardt, F., & Hoyer, P. O. (2012). Learning linear cyclic causal models with latent variables. *Journal of Machine Learning Research, 13*(1), 3387-3439.

[9] Chen T, Guestrin C. XGBoost: A Scalable Tree Boosting System. KDD '16: Proceedings of the 22nd ACM SIGKDD International Conference on Knowledge Discovery and Data Mining, August 2016. pp 785-794.

[10] Haynatzki, R., Windle, T., & Windle, J. (2023). Building Machine Learning Models to Improve Hypertension Diagnosis. *Journal of Physics. Conference Series*, *2675*(1), 012001–012001. https://doi.org/10.1088/1742-6596/2675/1/012001

[11] Ichiishi, Tatsuro. *Game Theory for Economic Analysis*. New York: Academic Press. pp. 118–120. ISBN 0-12-370180-5. (1983)

[12] Lundberg, Scott M., and Su-In Lee. "A unified approach to interpreting model predictions." Advances in Neural Information Processing Systems (2017).

[13] Molnar, C. 5.9 Shapley Values | Interpretable Machine Learning. In *christophm.github.io*. Lulu.com. https://christophm.github.io/interpretable-ml-book/shapley.html. (2019)

[14] Ko, C. (2023, December 27). *Navigating the Inconsistencies of the PC Algorithm: Insights and Improvements*. Medium; The whispers of a data analyst. https://medium.com/the-whispers-of-a-data-analyst/navigating-the-inconsistencies-of-the-pc-algorithm-insights-and-improvements-ac55d774d0b7.

[15] Steyerberg, E. (1999). Stepwise Selection in Small Data Sets A Simulation Study of Bias in Logistic Regression Analysis. Journal of Clinical Epidemiology, 52(10), 935–942. https://doi.org/10.1016/s0895-4356(99)00103-1.

[16] Duband, B., Motreff, P., Marcollet, P., Gamet, A., Decomis, M., Bar, O., Saint Etienne, C., Hakim, R., Canville, A., Viallard, L., Beygui, F., Lesault, P., Bonnet, P., Durand, E., Boiffard, E., Collet, J., Benamer, H., Commeau, P., Cayla, G., Pereira, B., ... & Rangé, G. (2022). Early survival after acute myocardial infarction with ST-segment elevation: What could be improved? Insights from France PCI French registry. *Medicine, 101*(36), e30190. https://doi.org/10.1097/md.0000000000030190

[17] Google Developers. (2019). Classification: ROC Curve and AUC | Machine Learning Crash Course. Google Developers. https://developers.google.com/machine-learning/crash-course/classification/roc-and-auc.

[18] Del Buono, M. D., Montone, R., Rinaldi, R., Gurgoglione, F., Meucci, M., Camilli, M., Iannaccone, G., Sanna, T., Pedicino, D., Trani, C., Niccoli, G., & Crea, F. (2021). Clinical predictors and

prognostic role of high Killip class in patients with a first episode of anterior ST-segment elevation acute myocardial infarction. *Journal of Cardiovascular Medicine*, 22(1), 18–25. https://doi.org/10.2459/JCM.0000000000001168

[19] Shabani, F., Kumar, L., & Ahmadi, M. Assessing Accuracy Methods of Species Distribution Models: AUC, Specificity, Sensitivity and the True Skill Statistic. *Global Journal of Human-Social Science: B Geography, Geo-Sciences, Environmental Science & Disaster Management*, *18*(1). https://researchnow.flinders.edu.au/en/publications/assessing-accuracy-methods-of-species-distribution-models-auc-spe. (2018)

[20] Sádraque Gomes Amoras, T., Mendonça, T. S., Melo, G. S., Pereira, K. R. P., Zaninotto, C. V., de Oliveira, S. S., Silva, R. M., & Melo Neto, J. S. (2023). Sociodemographic and clinical variables as determinants of mortality and survival in patients with acute ST-elevation myocardial infarction in the Eastern Amazon. *Global Journal of Cardiovascular Diseases*, 12(1), 27-34. https://doi.org/10.1177/22799036221150062

[21] Scarparo, P., van Gameren, M., Wilschut, J., Daemen, J., den Dekker, W. D., Zijlstra, F., Van Mieghem, N., & Diletti, R. (2021). Impact of thrombus burden on long-term clinical outcomes in patients with either anterior or non-anterior ST-segment elevation myocardial infarction. *Thrombosis and Haemostasis*, 121(11), 1434-1441. https://doi.org/10.1007/s11239-021-02603-3

[22] Bonneville, E. (2023, January 20). *clinical-prediction-models*. GitHub. https://github.com/clinical-prediction-models

[23] American Heart Association. (2019). *Heart and Stroke Association Statistics*. Www.heart.org. https://www.heart.org/en/about-us/heart-and-stroke-association-statistics

[24] Van Rossum, G., & Drake, F. L. (2009). *Python 3 reference manual*. CreateSpace.

[25] Ennis, M., Hinton, G., Naylor, D., Revow, M., & Tibshirani, R. (1998). A comparison of statistical learning methods on the GUSTO database. *Statistics in Medicine*, *17*(21), 2501–2508. https://doi.org/10.1002/(sici)1097-0258(19981115)17:21%3C2501::aid-sim938%3E3.0.co;2-m

[26] An International Randomized Trial Comparing Four Thrombolytic Strategies for Acute Myocardial Infarction. (1993). *New England Journal of Medicine*, *329*(10), 673–682. https://doi.org/10.1056/nejm199309023291001

Exploring SO_2 air pollution in Plovdiv through multivariate adaptive regression splines: A case study

Atanas Ivanov[1*], Snezhana Gocheva-Ilieva[1] and Maya Stoimenova-Minova[1]

[1] Department of Mathematical Analysis, University of Plovdiv Paisii Hilendarski, Plovdiv, Bulgaria

*E-mail: aivanov@uni-plovdiv.bg

Abstract. This case study delves into the assessment of Sulfur dioxide (SO_2) air pollution in Plovdiv by employing Multivariate Adaptive Regression Splines (MARS) to model and understand the factors influencing daily SO_2 levels. By analyzing a dataset characterized by an average SO_2 pollution level of 0.43ppm, this study highlights the potency of MARS in capturing the non-linear relationships and complex interactions between SO_2 concentrations and measured meteorological and atmospheric time series in the form of quantitative and categorical variables. In particular, an increase in model performance is demonstrated by performing a modified Yeo-Johnson transformation on pollutant data and constructing additional predictors such as lag variables, date variable and dummies. A significant achievement of this investigation is the attainment of a coefficient of determination (R^2) exceeding 0.91. This high level of accuracy highlights the efficiency of MARS as a flexible and reliable machine learning tool thus emphasizing its potential in contributing to the urban air quality management toolbox.

1. Introduction

Sulphur dioxide (SO_2) is among the most dangerous pollutants, a major cause of serious illness and death. Air polluted with SO_2 and nitrogen oxides (NOx) have a negative impact on soils and waters, contributing to the formation of acid rain. The formation of SO_2 occurs during the burning of coal, oil and gas, which contain a large amount of Sulphur. A study [1] found that high SO_2 concentrations were associated with lower air temperatures, low wind speed and little precipitation. The pollutant is particularly dangerous for the human body, mainly affecting children, people with cardiovascular and chronic lung diseases. In the research paper [2] cases of allergic rhinitis, bronchitis and asthma in children who live near a petrochemical plant complex emitting SO_2 into the air were investigated. It was found that boys and younger children living near the petrochemical plant complex develop asthma and allergic rhinitis more often. Other studies report health hazards from even short-term exposure to SO_2 emissions [3, 4]. A systematic analysis of epidemiological publications in this area can be found in [3, 5,6].

There are a variety of methods for studying and predicting air pollutants. Machine learning (ML) methods are increasingly used in the scientific literature. A large number of papers apply ML methods such as Random Forests (RF), Artificial Neural Network (ANN), deep learning, Support Vector Machines (SVM), etc. The authors of [7] develop hybrid models for future

Content from this work may be used under the terms of the Creative Commons Attribution 4.0 licence. Any further distribution of this work must maintain attribution to the author(s) and the title of the work, journal citation and DOI.
Published under licence by IOP Publishing Ltd

predictions of several types of air pollutants, including SO_2. In the study [8], the ARIMA method was used to analyze PM_{10} concentrations in the city of Vidin, Bulgaria. Article [9] applies the XGBoost method for future forecasts of PM_{10} in the town of Ruse. ANN and deep-learning recurrent NN models to forecast SO_2 are developed in [10, 11]. A data-augmentation approach combined with different ML techniques to minimize bias is adopted in [12] for modeling long-term surface SO_2.

In recent years, the number of studies in which the MARS ML method is actively applied to model data of different air pollutants and forecast their future values has also increased. For example, in [13] the MARS method was applied to model PM_{10}, identifying the meteorological data that have an impact. The MARS method was implemented in [14] to analyze and predict the level of the air pollutant PM_{10} taking into account the influence of 7 factor variables. Paper [15] shows that MARS can be successfully adopted for air pollution forecasting. A hybrid model with MARS and Support Vector Regression (SVR) was developed to model SO_2 concentrations. The authors of [16] use three different methods for forecasting time series of SO_2. They argue that the MARS method is the secondary best method for monthly SO_2 forecasts. Other results with MARS air quality models are presented in [17, 18].

On a local scale, relatively very few scientific studies can be found on SO_2 air pollution in Bulgaria. Here we can mention the complex vector type model built in [19] with Temporal-causal modeling, based on Granger causality, investigating the relationships between 6 air pollutants, including SO_2. Analyses have been also done on fine particulate matter pollution, see [20, 21].

In the present study, the MARS method was applied to model and forecast SO_2 concentrations in the city of Plovdiv, Bulgaria. The purpose of the conducted research is to predict future pollution and determine the factors that influence high SO_2 concentrations. The results obtained on the basis of local independent measurements can serve to support the official authorities in implementing measures to control the quality of atmospheric air in the city.

2. Data and methods used

2.1 Study area and data

The harm of exposure to air with elevated concentrations of SO_2 is well known. The European Environment Agency, and more specifically the 2008 European Directive [22], has introduced limits on the permissible values of SO_2 in the air. The specified maximum permissible average day-night concentration of SO_2 is 125 $\mu g / m^3$ (or 0.047ppm), which must not be exceeded more than three times within a calendar year. The maximum allowable average annual limit of the pollutant is set at 20 $\mu g / m^3$ (or 0.006ppm) with 3 permitted exceedances each year [23].

The object of research in the present development is the air quality in the city of Plovdiv. Plovdiv is the second largest city in Bulgaria with a population of 360,000 inhabitants, located in the southern part of the country. The city is characterized by fogs, which are observed mainly during the cold winter periods of the year, relatively low altitude and weak winds.

This study is based on data on the pollutant sulfur dioxide SO_2, measured with the Alphasense automatic monitoring station with optical sensors, located on the roof of the new building of Plovdiv University at one of the exits of the city, near an intercity highway. The extracted data are for a period of 6 months from 26th September 2023 to 15th March 2024, i.e. a total of N=171 days. The missing data for the specified period are for 2 days or 1.2%. In the calculations, they are imputed by the method of linear interpolation. For the analysis, the initial measured data from 9 meteorological and atmospheric time series were used: minimum air temperature (*Min*), maximum air temperature (*Max*), relative humidity (*Humidity*), cloud coverage (*Cloud*), atmospheric pressure (*Pressure*), luminosity (*Luminosity*), rain (*Rain*) and the categorical variable for the wind direction *Winddirection$* with 16 different values. The atmospheric variable of categorical type *Weather$* is also included in the modeling process.

Descriptive statistics were performed on all quantitative initial variables included in the analysis. The results are presented in Table 1. It can be seen that the average daily value of the air pollutant SO_2 for the studied period is 0.39 ppm, which indicates a significant overshoot of the maximum permissible norms of 0.047 ppm compared to the European standards [23]. For the selected period, a maximum of 0.49 ppm was reached, which is 10 times above the specified safe health threshold. Figure 1 shows the time series plot of the pollutant SO_2 for the studied period. There is no linear trend, seasonality or cyclicality. Higher values are observed during cold winter days. From the initial analysis, it can be concluded that SO_2 in the city of Plovdiv has particularly high values and it is necessary to take measures to reduce it.

Table 1. Descriptive statistics of the initial measured data.

Variable /Statistics	Minimum	Maximum	Mean	Std. Dev.	Variance	Skewness	Kurtosis
SO_2 (ppm)	**0.39**	**0.49**	0.4316	0.017	0.0003	0.424	0.294
Max (°C)	1	30	14.39	7.176	51.49	0.341	-0.648
Min (°C)	-10	15	2.84	5.337	28.48	-0.077	-0.473
Humidity (%)	0.35	0.97	0.656	0.133	0.018	-0.182	-0.397
Cloud (%)	0	0.99	0.355	0.27	0.074	0.765	-0.443
Pressure_1 (mbar)	96828.81	101768.4	99552.7	765.6	586110.6	0.085	1.124
Luminosity (Luxe)	3	8553	2521.1	1618.9	2620981	0.6	0.276
Rain (mm)	0	25.5	1.53	4.205	17.68	3.686	14.69
Wind (m/s)	2	28	6.89	4.616	21.311	2.196	5.619

[a] Note: Std. Err. of Skewness for all variables is 0.186 and Std. Err. of Kurtosis is 0.369.

2.2 Brief description of MARS method

MARS is a flexible and efficient ML method for high dimensional data, created by the American statistician Friedman in 1991 [24]. MARS is a representative of the nonparametric regression methods and approximates the dependent variable with a formula similar in appearance to partial polynomial regression. Like most ML approaches, this method does not place strong restrictions on data distributions or is distribution free, i.e. does not depend significantly on the type of distribution of the variables, as in different subintervals different variables are included. However, an appropriate initial data transformation is recommended.

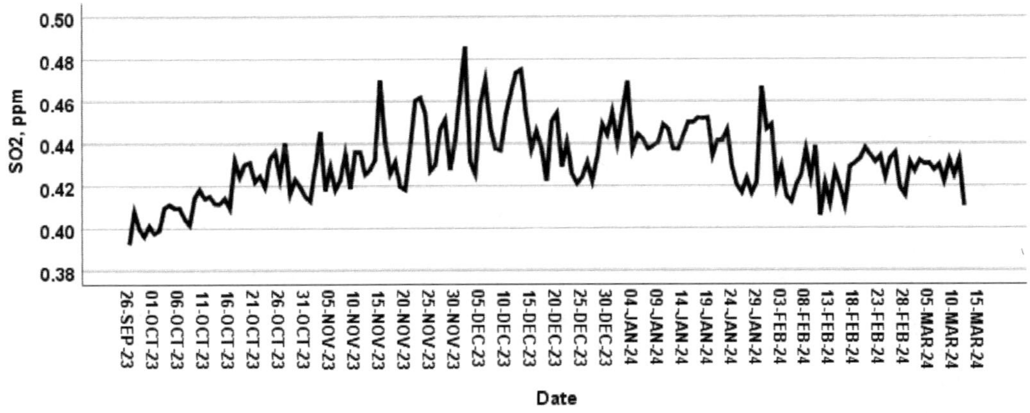

Figure 1. Measured SO_2 concentration in Plovdiv for the observed period of 171 days.

Let the dependent variable $Y = Y(X)$ (in this case SO_2) be modeled with N real values $Y = \{Y_1, Y_2, ...Y_t, ..., Y_N\}$, $Y_t \in \mathbb{R}$ depending on p predictor variables denoted by $X = (X_1, X_2, ..., X_p)$.

The task of predicting Y consists in finding approximations of its values (model) $\hat{Y}$ satisfying the equation

$$Y = \hat{Y} + \varepsilon, \quad \varepsilon = N(0, \sigma^2) \tag{1}$$

where ε is the residual (error) time series, σ^2 is its variance.

The general MARS model is represented by an expression of the form

$$\hat{Y}_{[M]} = b_0 + \sum_{j=1}^{M} b_j BF_j(X) \tag{2}$$

where b_0, b_j, $j = 1, 2, ..., M$ are constant coefficients, $BF_j(X)$ are the basis functions (BFs), M is their number. One-dimensional linear BFs have the form

$$BF_j(X) = \max_{X_k}(0, X_k - c_{k,j}) \text{ and } BF_j(X) == \max_{X_k}(c_{k,j} - X_k, 0), \tag{3}$$

where X_k is a predictor variable, $k = 1, 2, ..., p$, $c_{k,j} \in X_k$ are knots, determined by MARS to minimize a selected type of loss function, for example $MSE_{[L]} = \sum_{t=1}^{N} \left[Y_t - \hat{Y}_{[L],t}(\mathrm{X}) \right]^2$. Higher-order BFs are products of functions of the type (3).

To derive a specific model, the hyperparameters are first set, among which the main ones are the maximum number L_0 of BFs, maximum order of the interaction members in the BF, type of loss function and others. The algorithm is implemented in two consecutive steps - forward and backward. During the forward step, first some value b_0 in (2) is chosen and gradually BFs of the type (3) are added to the model one by one. The greedy algorithm seeks to determine the optimal values of the knots $c_{k,j} \in X_k$. At the backward step, the accuracy of the constructed MARS model is evaluated and the number of BFs is reduced by applying selected criteria for cross-validation or another type of training. BFs that turn out to be insignificant for the model and do not contribute to its accuracy are removed. The second step provides a solution for avoiding possible overfitting of the model [24].

Some of the most important advantages of the MARS method are: simple and easy-to-interpret models are created, it works with large and small datasets (with sample size N >50); the MARS models are highly efficient as they require little machine memory and run at very high speed, allowing the processing of huge datasets in real time (within 100 to 1000 times faster than ANNs).

We will note that the MARS method differs from the numerical approximation with standard splines in that the knots of the BFs, the choice of BFs and their number M are not known in advance and are controlled by hyperparameters, by the choice of training and test samples, by type of model training (cross-validation or independent holdout sample, with data not included in model building).

2.3 Model evaluation

MARS models are evaluated using the following statistical performance criteria: mean squared error (MSE), modified generalized cross-validation criterion (GCV), and coefficient of determination (R^2), calculated with the expressions

$$MSE = \frac{1}{N}\sum_{t=1}^{N}\left(Y_t - \hat{Y}_t\right)^2, \quad R^2 = \frac{\left\{\sum_{t=1}^{N}\left(\hat{Y}_t - \bar{\hat{Y}}\right)\left(Y_t - \bar{Y}\right)\right\}^2}{\sum_{t=1}^{N}\left(\hat{Y}_t - \bar{\hat{Y}}\right)^2 \cdot \sum_{t=1}^{N}\left(Y_t - \bar{Y}\right)^2}, \tag{4}$$

$$GCV_{[m]} = \frac{\sum_{t=1}^{N}\left(Y_t - \hat{Y}_{[m]}(X_t)\right)^2}{N\left(1 - C(m)/N\right)^2}, \quad m = 1, 2, \ldots, M, \tag{5}$$

where $\bar{Y}, \bar{\hat{Y}}$ are the mean values of the data and the predicted values from the MARS model, respectively, and the constants in (5) are $C(m) = m + \delta(m-1)/2,\ \ \delta \in [2,3]$, [24, p. 20]. In addition to these measures, some of their derivatives such as Adjusted R^2 and GCV R^2, embedded in the Salford Predictive Modeler (SPM) software we implement [25], can also be used. A "best model" of the type (1) with m BFs is selected, for which small MSE and GCV errors and large R^2 are reached.

3. Results and discussion

3.1 Data pre-processing and construction of additional predictors

In a first step, to improve the distribution and reduce the variation, a transformation of the dependent variable $Y = SO2$ was performed. For this purpose, the data are first scaled, then the Yeo-Johnson power transformation [26] was adopted in the case of non-negative values of the variable.

Thus, the modified Yeo-Johnson formula is written as

doi:10.1088/1742-6596/2910/1/012017

$$trY(k,\lambda,Y)=\frac{(kY+1)^{\lambda}-1}{\lambda},\quad Y\geq 0,\ \lambda\neq 0,\ \lambda\in[-2,2]\,, \tag{6}$$

In our case, the closest to the normal distribution was obtained with a scaling factor $k=1000$ and parameter $\lambda=-1.95$. The resulting dependent variable corresponding to Y hereafter is denoted by $trSO2\_195$.

We note that an appropriate initial data transformation typically increases the model performance by at least 5%. Attracting additional predictors also contributes to improving the statistical indicators and prognostic results of the models. In our case, such are the lagged variables one lag behind the dependent variable ($trSO2\_195<1>$), lag variables for some of predictors, a date variable to account for the data being time series, etc. The categorical variable Weather$ is coded with dummies to represent it numerically.

3.2 MARS model construction and model selection

Building and comparing MARS models was implemented using the MARS engine of SPM [25] with hyperparameter tuning. By specifying a certain set of hyperparameters, multiple models are generated simultaneously.

Different sets of hyperparameters were varied to run the modeling. For the initial forward step algorithm there were preset: for linear MARS models - a maximum L_0 = 30 BFs, for the second degree L_0 = 40 BFs, for the third degree $L_0=40$ BFs. The training of the models was carried out with 10-fold cross-validation and loss-function MSE and GSV. The models of the third degree turned out to have the best statistics.

According to (4) - (5), criteria GCV and MSE of the test sample at 10-fold cross-validation were used to account for the errors of the third degree models. Also, the GCV criterion is used as an indicator to determine the optimal number of BFs. Figure 2 illustrates a plot that visually shows how the errors change when increasing the number of BFs in the models. As a model with minimum GCV error, a model with 17 BFs was automatically selected. In our case, this choice coincides with the minimal MSE.

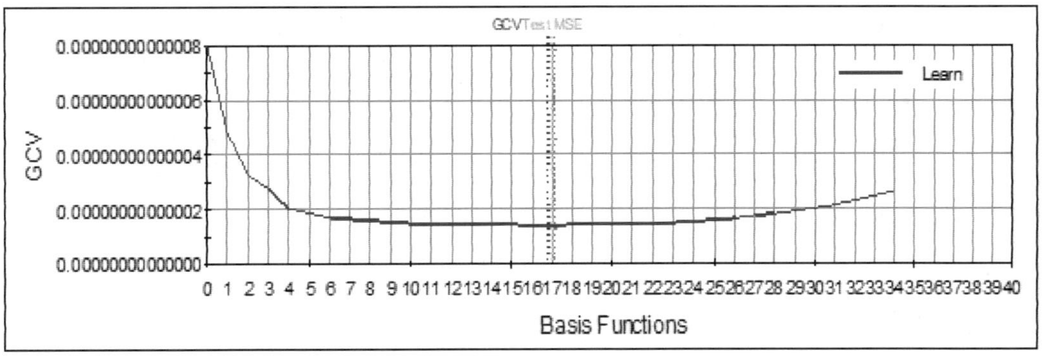

Figure 2. MARS selector error curve: MARS results for *trSO2_195*.

Another main indicator that we follow when choosing the best model is the coefficient of determination R^2. It shows how close the constructed model is fitted to the original data. In our case, R^2=0.91292 or approximately 91% is the fit of the model to the examined data. For a model with 17 BFs, the corresponding R^2 values for the training sample are given in Table 2.

Table 2. Model summary.

Statistics	Learn sample
R^2	0.91292
R^2 Normalized	0.90991
GCV R^2	0.82726
R^2 Adjusted	0.90305

The resulting best model for the transformed data $trY = trSO2\_195$ with 17 BFs has the following form

$$\begin{aligned} tr\hat{Y} &= 0.512816 - 5.86452.10^{-08} * BF1 + 5.58861.10^{-08} * BF2 + 3.14974.10^{-08} * BF3 \\ &-2.74104.10^{-08} * BF4 + 5.266.10^{-08} * BF5 + 4.84594.10^{-08} * BF10 \\ &-4.18029.10^{-10} * BF12 - 8.83487.10^{-11} * BF15 + 1.03634.10^{-08} * BF16 \\ &+8.54619.10^{-09} * BF18 - 1.1554.10^{-08} * BF20 + 0.0916942 * BF24 \\ &+2.57853.10^{-10} * BF26 - 1.5177.10^{-09} * BF31 - 6.02537.10^{-08} * BF32 \\ &+6.0391.10^{-09} * BF35 + 2.10597.10^{-08} * BF36 \end{aligned} \quad . \tag{7}$$

Equation (7) includes the following 20 BFs:

$$BF1 = max(0, TEMP\_1 - 15.2045);$$
$$BF2 = max(0,\ 15.2045 - TEMP\_1);$$
$$BF3 = max(0, MIN + 10); \quad BF4 = max(0, WIND - 5); \quad BF5 = max(0,\ 5 - WIND);$$
$$BF6 = \left(WEATHER\$\ in \left(\begin{array}{l} "Partly\ cloudy",\ "Cloudy",\ "Sunny",\ "Patchy\ light\ drizzle", \\ "Overcast",\ "Patchy\ rain\ possible", \\ "Patchy\ light\ rain\ with\ thunder",\ "Moderate\ rain", \\ "Patchy\ light\ snow" \end{array} \right) \right)$$
$$BF10 = max(0, MAX - 17) * BF2; \quad BF12 = max(0, DATE - 45308) * BF3;$$
$$BF15 = max(0, 910.5 - LUMINOSITY\_1) * BF4; \quad BF16 = max(0, RAIN - 1.3) * BF6;$$
$$BF17 = max(0, 1.3 - RAIN) * BF6; \quad BF18 = max(0, PMHUM\_1 - 49.9086) * BF17;$$

$$BF20 = (WEATHER\$ \ in \ ("Partly\ cloudy", "Cloudy", "Sunny", "Overcast", "Patchy\ rain\ possible", "Moderate\ or\ heavy\ rain\ with\ thunder", "Moderate\ rain\ at\ times", "Heavy\ snow", "Light\ freezing\ rain", "Moderate\ or\ heavy\ snow\ showers"))*BF2;$$

$$BF22 = (TR\_SO2\_\_195<1> ne\ .\ )*BF4;$$

$$BF24 = max(0, TR\_SO2\_\_195<1> -0.512817)*BF22;$$

$$BF26 = \left(WEATHER\$ \ in \left(\begin{array}{l} "Partly\ cloudy", "Cloudy", "Sunny", \\ "Patchy\ rain\ possible", \\ "Patchy\ light\ rain\ with\ thunder", \\ "Moderate\ or\ heavy\ rain\ with\ thunder", \\ "Light\ rain\ shower", "Patchy\ light\ snow" \end{array} \right) \right) *BF12;$$

$$BF31 = max(0, 45254 - DATE)*BF20;$$

$$BF32 = (WINDDIRECTION\$ \ in \ ("SSE", "WSW", "SSW", "S"))*BF5;$$

$$BF35 = max(0, 64.2965 - PMHUM\_1)*BF32;$$

$$BF36 = (WINDDIRECTION\$ \ in \ ("SE", "SSE", "SW", "WNW", "W"))*BF22; \quad (8)$$

Model (7) is of third order as it contains BFs that are products of several BFs. It contains 6 linear BFs, 11 second-order BFs, and 3 third-order BFs.

Figure 3 presents a comparison of the initial values of the pollutant SO_2 (in blue) with the model-predicted values, shown in red. It can be seen that the model closely approximates the actual measured data.

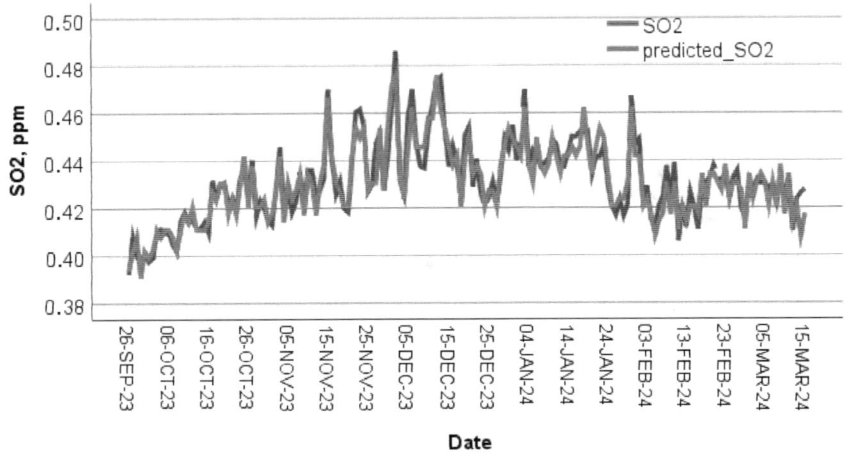

Figure 3. SO_2 concentration and fitment of selected MARS model.

3.3 Major factors determining SO_2 levels

Using the selected MARS model, the relative degree of influence of the investigated predictors on SO_2 levels was determined. Table 3 shows the score results of relative variable importance of the

Journal of Physics: Conference Series **2910** (2024) 012017 doi:10.1088/1742-6596/2910/1/012017

main ones, with an influence of more than 10 points compared to the first one, which is evaluated with the maximum 100 points.

Table 3 shows that the average daily ambient air temperature of the previous day has the greatest influence (100 points) followed by the minimum temperature (56), rain (54), weather conditions (53), relative humidity (52), wind direction (48), etc.

Table 3. Relative importance of the predictors in the model.

Variable name	Score
TEMP_1	100
MIN	56.31
RAIN	53.56
WEATHER$	52.77
PMHUM_1	52.13
WIND	47.69
MAX	32.32
LUMINOSITY_1	31.29
DATE	23.19
TRSO2__195<1>	17.45

3.4 Diagnostics of the model errors

An important stage in the analysis of an empirical time series model is the study of its residuals. Residuals are required to contain no serial correlation and to be white noise with zero mean and constant variance (see equation (1)).

We perform this check by building and studying the autocorrelation function (ACF) of the residuals of the built MARS model (see also [27]). The corresponding ACF is shown in Figure 4. Since there are no large values of the residuals outside the confidential intervals, we can conclude that the residuals do not contain serial correlation and are normally distributed. It follows that the resulting model is adequate and suitable for calculating forecasts and interpretation.

3.5 Application of the model for short-time forecasting of SO_2 concentrations

In a real situation, meteorological and atmospheric data can be predicted for a short future period of 5 to 10 days. In this regard, it is enough to find the predictions of the model within a few days ahead.

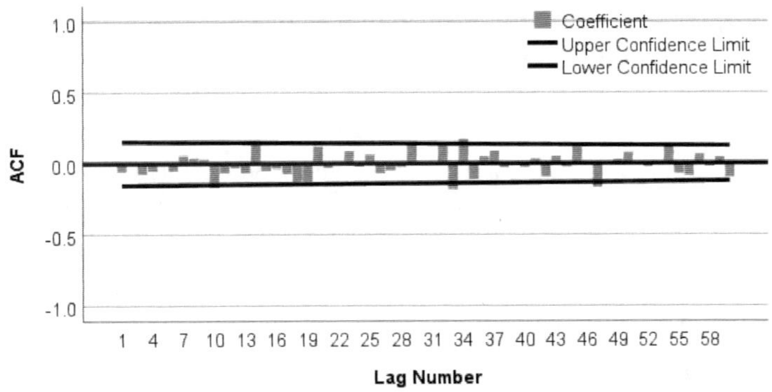

Figure 4. ACF of the model residuals for 60 lags.

Model (7) - (8) was applied for short-term forecasts of SO_2 concentrations within the last 6 days of the study period. For this purpose, the actual measured data of the air pollutant SO_2 these days are used, which are not involved in building the model, but are used for comparison with the forecasted values. Predictor data are saved. Figure 5 illustrates the comparison of the model-predicted data (dashed line) with the actual measured SO_2 data (solid line).

A very good agreement especially in the first 4 days of the period is observed. This gives us reason to consider that the method used and the built model are very suitable for short-term forecasts of SO_2 concentrations. The maximum relative error of the approximation (for 14th March 2024) is

$$\Delta_t = \left| \frac{Y_t - \hat{Y}_t}{Y_t} \right| = \left| \frac{0.426 - 0.408}{0.426} \right| = 0.0433, \tag{9}$$

which is less than 5%.

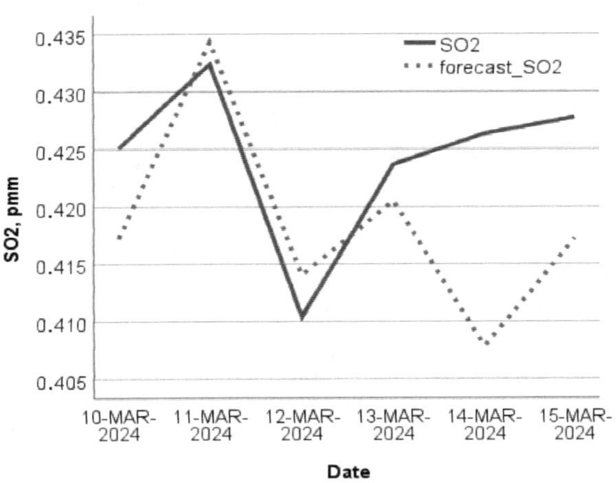

Figure 5. Short-term future forecasts of daily SO_2.

4. Conclusion

SO_2 is particularly harmful to the human body, causing severe diseases. It is of great importance to continuously monitor the quality of atmospheric air, as well as to determine the factors that contribute to its elevated values. From a physical point of view, the main causes of SO_2 formation in urban environments are emissions from vehicles, burnt fuels from production processes and from natural sources. However, on daily basis, these factors were considered as almost constant. These processes strongly interact with weather and atmospheric changes for each local geographic area.

In the present work, an approach is described and empirical ML models of urban air pollution with SO_2, depending on meteorological, atmospheric, and other conditions, are obtained and analyzed. A match of the best MARS model to the measured SO_2 values with up to over 90% was obtained. The models have been tested for short-term forecasts of SO_2 levels and show very good results. The built models enable adequate forecasting even in the absence of a part of the data.

The findings of this study are critical for environmental policymakers and urban planners in Plovdiv, offering a robust analytical framework for identifying key pollution drivers and formulating targeted interventions. Moreover, the methodology and insights gleaned from this research can be extrapolated to other urban settings grappling with similar air quality challenges, making a significant contribution to the global discourse on sustainable urban living and environmental conservation.

Acknowledgement

This study has emanated from a research conducted with the financial support of the Bulgarian National Science Fund (BNSF), Grant number KP-06-IP-CHINA/1.

References

[1] Luvsan ME, Shie RH, Purevdorj T, Badarch L, Baldorj B, Chan CC, 2012, The influence of emission sources and meteorological conditions on SO2 pollution in Mongolia, Atmospheric Environment, Vol. 61, pp. 542-549. https://doi.org/10.1016/j.atmosenv.2012.07.044

[2] Chiang TY, Yuan TH, Shie RH, Chen CF, Chan CC, 2016, Increased incidence of allergic rhinitis, bronchitis and asthma, in children living near a petrochemical complex with SO2 pollution, Environment International, Vol. 96, pp. 1-7. https://doi.org/10.1016/j.envint.2016.08.009

[3] Orellano P, Reynoso J, Quaranta N, 2021. Short-term exposure to sulphur dioxide (SO2) and all-cause and respiratory mortality: A systematic review and meta-analysis. Environment international, Vol. 150, Art. 106434. https://doi.org/10.1016/j.envint.2021.106434

[4] Adebayo-Ojo TC, Wichmann J, Arowosegbe OO, Probst-Hensch N, Schindler C, Künzli N, 2022. Short-term effects of PM10, NO2, SO2 and O3 on cardio-respiratory mortality in Cape Town, South Africa, 2006–2015, International Journal of Environmental Research and Public Health, Vol. 19, No. 13, Art. 8078. https://doi.org/10.3390/ijerph19138078

[5] Khreis H, Kelly C, Tate J, Parslow R, Lucas K, Nieuwenhuijsen M, 2017, Exposure to traffic-related air pollution and risk of development of childhood asthma: A systematic review and meta-analysis, Environment International, Vol. 100, pp. 1-31. https://doi.org/10.1016/j.envint.2016.11.012

[6] Barrera-Heredia AD, Zafra-Mejía CA, Cañas Arboleda A, Fernández Sánchez MJ, López-Kleine L, Rojas Moreno A, 2024, Trends in the use of air quality indexes in asthma studies, Atmosphere, Vol. 15, No. 7, Art. 847. https://doi.org/10.3390/atmos15070847

[7] Pakrooh P, Pishbahar E, 2016, Forecasting air pollution concentrations in Iran, using a hybrid model, Pollution, Vol. 5, No. 4, pp. 739-747. https://doi.org/10.22059/POLL.2019.274827.572

[8] Veleva E, Filipova IM, Zheleva I, 2022, Statistical study of particulate matter (PM10) air contamination in the city of Vidin, Bulgaria, AIP Conference Proceedings, Vol. 2522, pp. 050014-1-12. https://doi.org/10.1063/5.0101245

[9] Veleva E, Filipova M, Jordanova P, 2023, Machine learning approach to study and predict air pollution in Ruse, Bulgaria, AIP Conference Proceedings, Vol. 2953, pp. 030005 1-14. https://doi.org/10.1063/5.0178495

[10] Radojević D, Antanasijević D, Perić-Grujić A, Ristić M, Pocajt V, 2019, The significance of periodic parameters for ANN modeling of daily SO2 and NOx concentrations: A case study of Belgrade, Serbia, Atmospheric Pollution Research, Vol. 10, No. 2, pp.621-628. https://doi.org/10.1016/j.apr.2018.11.004

[11] Kurnaz G, Demir AS, 2022. Prediction of SO2 and PM10 air pollutants using a deep learning-based recurrent neural network: Case of industrial city Sakarya, Urban Climate, Vol. 41, Art. 101051. https://doi.org/10.1016/j.uclim.2021.101051

[12] Zhang S, Mi T, Wu Q, Luo Y, Grieneisen ML, Shi G, Yang F, Zhan Y, 2022, A data-augmentation approach to deriving long-term surface SO2 across Northern China: Implications for interpretable machine learning, Science of The Total Environment, Vol. 827, Art. 154278. https://doi.org/10.1016/j.scitotenv.2022.154278

[13] Tırınk S, Öztürk B, 2023, Evaluation of PM10 concentration by using Mars and XGBOOST algorithms in Iğdır Province of Türkiye, nternational Journal of Environmental Science and Technology, Vol. 20, pp. 5349–5358. https://doi.org/10.1007/s13762-022-04511-2

[14] Gocheva-Ilieva SG, Ivanov AV, Voynikova DS, Stoimenova MP, 2020, Modeling of PM10 air pollution in urban environment using MARS, Lecture Notes in Computer Science, Vol. 11958, Ch. 27, pp. 237-244. https://doi.org/10.1007/978-3-030-41032-2_27

[15] Alamoudi M, Taylan O, Keshtegar B, Abusurrah M, Balubaid M, 2022, Modeling Sulphur Dioxide (SO2) quality levels of Jeddah city using machine learning approaches with meteorological and chemical factors, Sustainability, Vol. 14, No. 23, Art. 16291. https://doi.org/10.3390/su142316291

[16] Kisi O, Parmar KS, Soni K, Demir V, 2017, Modeling of air pollutants using least square support vector regression, multivariate adaptive regression spline, and M5 model tree models, Air Quality, Atmosphere & Health, Vol. 10, pp. 873–883. https://doi.org/10.1007/s11869-017-0477-9

[17] Nieto PJG, Antón JCÁ, Vilán JAV, García-Gonzalo E, 2015, Air quality modeling in the Oviedo urban area (NW Spain) by using multivariate adaptive regression splines, Environmental Science and Pollution Research, Vol. 22, No. 9, pp. 6642-6659. https://doi.org/10.1007/s11356-014-3800-0

[18] Nieto PJG, J. Antón JCÁ, 2014, Nonlinear air quality modeling using multivariate adaptive regression splines in Gijón urban area (Northern Spain) at local scale, Applied Mathematics and Computation, Vol. 235, pp. 50-65. https://doi.org/10.1016/j.amc.2014.02.096

[19] Ivanov AV, Gocheva-Ilieva SG, Stoimenova-Minova MP, 2023, Temporal-causal modeling of air pollution in the city of Plovdiv, Bulgaria: a case study, Journal of Physics Conference Series, Vol. 2675, No. 1, pp. 012002. https://doi.org/10.1088/1742-6596/2675/1/012002

[20] Veleva E, Zheleva I, Georgiev I, 2020, Decomposition techniques for modelling the levels of particulate matter PM10 air pollutant in the city of Silistra, Bulgaria, AIP Conference Proceedings, Vol. 2302, pp. 060019-1-9. https://doi.org/10.1063/5.0033631

[21] Veleva E, Georgiev I, Zheleva I, Filipova M, 2020, Markov chains modelling of particulate matter (PM10) air contamination in the city of Ruse, Bulgaria, AIP Conference Proceedings, Vol. 2302, pp. 060018-1-9. https://doi.org/10.1063/5.0033630

[22] Directive 2008/50/EC of the European Parliament and of the council of 21 May 2008 on ambient air quality and cleaner air for Europe, 2008, Official Journal of the European Union, L 152/1, pp. 1-44. URL: https://eur-lex.europa.eu/eli/dir/2008/50/oj, Accessed on August 28, 2024.

[23] EU air quality standards, 2024. URL: https://environment.ec.europa.eu/topics/air/air-quality/eu-air-quality-standards_en, Accessed on August 28, 2024.

[24] Friedman JH, 1991, Multivariate adaptive regression splines (with discussion), Annals of Statistics, Vol. 19, No. 1, pp. 1-141. https://doi.org/10.1214/aos/1176347963.

[25] Salford Predictive Modeler (SPM): Machine Learning and Predictive Analytics Software, 2024, Minitab. URL: https://www.minitab.com/en-us/products/spm/

[26] Yeo IK, Johnson RA, 2000, A new family of power transformations to improve normality or symmetry, Biometrika, Vol. 87, No. 4, pp. 954–959. https://doi.org/10.1093/biomet/87.4.954. ISSN 0006-3444

[27] Gocheva-Ilieva S, Voynikova D, Stoimenova M, Ivanov A, Iliev I, 2019, Regression trees modeling of time series for air pollution analysis and forecasting, Neural Computing and Applications, Vol. 31, pp. 9023-9039. https://doi.org/10.1007/s00521-019-04432-1

Journal of Physics: Conference Series **2910** (2024) 012018 doi:10.1088/1742-6596/2910/1/012018

On the Estimation of Measurement Errors in Noise Studies of a Centrifugal Fan

I N Nikolaev[1], V A Mihova[2], N Y Kovachev [1] and V T Pavlov[2]

[1]Department of Heat, Hydraulics and Environmental Engineering, University of Ruse, 8 Studentska Str, 7017 Ruse, Bulgaria
[2]Department of Applied Mathematics and Statistics, University of Ruse, 8 Studentska Str, 7017 Ruse, Bulgaria

vmicheva@uni-ruse.bg

Abstract. The main objective of the present work is to describe an experimental setup to study the noise characteristics during centrifugal fan operation and the error of the measured quantities used to estimate the noise at different operating modes. The apparatus makes it possible to study the performance of the fan with impellers of different blade geometries. Determination of the error of the measured quantities is necessary in order to be able to take it into account later in the analysis of different experimental data.

1. Introduction

Accurate and reliable measurements are particularly important when analysing the noise, generated by machines such as centrifugal fans, pumps, etc. Noise not only degrades the quality of the working environment, but can also have significant consequences for human health. Accurate assessment of noise parameters and their error is critical for the design of effective noise reduction systems. The process of error assessment is complex and requires a detailed understanding of every part of the experimental set-up, from how data is generated to the methods of analysis and interpretation [1, 2].

For laboratory experimental studies, it is important to design an experimental setup that allows the parameters and operating modes of the setup to be varied. In these modes it must be possible to measure directly or indirectly the quantities that characterise the process and whose influence we want to study [3]. In order to know to what extend we can trust the results of the processing of the experimental data obtained with the instrument, it is important to determine the error of the measured quantities and then to take it into account when analysing the experimental data. This error is of two types: dynamic and steady-state. Dynamic uncertainties are observed in the measurement of dynamic processes [4, 5, 6] and are mainly due to the fastness of the measurement sensor. Established error [1, 2, 7, 8] is accounted for when measuring established physical processes in which the measurand remains constant over time [3].

In the present work, an experimental setup is described to investigate the noise characteristics of a centrifugal fan under different operating modes. The apparatus allows to study the noise during the operation of the fan with impellers of different blade geometry. A standardized nozzle with free inlet designed according to ISO 5801:2007 [9] is used to measure the flow rate Q. The analytical form of the flow rate as a function of the measured vacuum (transformation function) was obtained based on values calculated according to the standard over the entire measurement range.

Content from this work may be used under the terms of the Creative Commons Attribution 4.0 licence. Any further distribution of this work must maintain attribution to the author(s) and the title of the work, journal citation and DOI.
Published under licence by IOP Publishing Ltd

When studying fan noise, it is intended to take measurements only at established operating modes. It is therefore only necessary to take into account the established error (established error) of the measured sound pressure level and the total error of the fan flow calculated by the conversion function using the measured vacuum. Since the flow rate Q is an indirect measurand, the error of all quantities in the flow measurement and determination chain must be taken into account to estimate its total error. For this purpose, the error of the measured vacuum, the error of the flow rate calculated according to ISO 5801:2007 and the error coming from the approximation of the flow rate with the transformation function are defined and described.

2. Experimental setup

Figure 1 shows the experimental setup for noise studies. The fan (1) operates in a system with short suction (3) and discharge (12) ducting. Using wing nuts, it is possible to quickly to disassemble or install the front cover (11) of the fan and replace the impellers (6) under study. A throttle valve (5) is installed at the end of the discharge line to throttle the flow rate Q of the fan. By means of the frequency inverter (4), the flow rate Q can also be adjusted by changing the fan speed n. The frequency n can be established using the induction transducer (2) and the pulse counter (9). A standardized free inlet nozzle (10), designed according to ISO 5801:2007 and fabricated using 3D printing, is used to measure the flow rate Q at the inlet of the inlet duct. The vacuum pressure p_v acting in the narrow section of the nozzle is measured by the vacuum gauge (8) model test 521-1 and the sound pressure level is measured by the stand-mounted sound level meter (7).

Figure 1. Experimental setup for studying the noise characteristics at centrifugal fan operation.

To determine the flow rate, points from the nozzle characteristic Q vs p_v, were plotted, calculated according to ISO 5801:2007 (Figure 2) and then the data were approximated to obtain a transformation function of the type $Q = a \cdot p_{\text{в}}^b$. This is the relationship used to calculate the volumetric fan flow rate Q.

2.1. Plotting the characteristic according to ISO 5801:2007

Input calculation parameters

The following input calculation parameters are used in the centrifugal fan operation:

- Nozzle diameter $d = 82\ mm$;
- Absolute temperature before flow meter $T_u = 18$ °C;
- Absolute pressure before flow meter $p_u = 1012\ hPa$ (for nozzle with free inlet $p_u = p_{atm}$);
- Pressure drop between the nozzle inlet and the nozzle bottleneck Δp (for free inlet - the bottleneck vacuum p_v=Δp);
- Maximum vacuum $p_{v\,max} = 550\ Pa$

Main parameters and constants

The following coefficients are involved in the setup of the experimental setup:

- Adiabatic coefficient for air $k_u = 1.4$;
- Kinetic energy coefficient before the nozzle $\alpha_{Au} = 1$ (for nozzle with free inlet);
- Nozzle diameter coefficient $\beta = \frac{d}{D} = 0$ (for free inlet).

Sequence for calculating the characteristic

To calculate the tare characteristic, the following parameters are calculated sequentially:

1. Air density before nozzle ρ_u
   The air density ρ_u is read from reference data. For incompressible fluids it is determined by the temperature T_u and for compressible fluids by the absolute pressure $p_u = p_{\text{АТМ}}$.
2. Pressure coefficient – it is calculated by the following formula
$$r_d = 1 - \frac{p_v}{p_u}$$
3. Expansion coefficient – it is calculated by the following formula
$$\varepsilon = \left(\frac{k_u \cdot r_d^{\frac{2}{k_u}}}{k_u - 1} \cdot \frac{(1-\beta^4)}{1 - r_d^{\frac{2}{k_u}} \cdot \beta^4} \cdot \frac{1 - r_d^{\frac{(k_u-1)}{k_u}}}{1 - r_d} \right)^{0.5}$$
4. Reynolds number at nozzle exit - it is calculated by the following formula
$$Re_d = \frac{0.95\varepsilon d\sqrt{2\rho_u p_v}}{17.1 + 0.048T_u} \cdot 10^6$$
5. The nozzle flow coefficient - for $\frac{L}{d} = 0.6$ is calculated using the formula
$$\mu = \left(0.9986 - \frac{7.006}{\sqrt{Re_d}} + \frac{134.6}{Re_d}\right) \cdot \left(\frac{1}{\sqrt{1 - \alpha_{Au} * \beta^4}}\right)$$
6. Mass flow rate – it is calculated by the following formula
$$q = \varepsilon\mu\frac{\pi d^2}{4}\sqrt{2\rho_u p_v} \qquad (1)$$
7. Absolute nozzle orifice pressure is calculated by the following formula
$$p_d = p_u - p_v$$
8. Fluid density in the nozzle constriction - calculated by the following formula
$$\rho_d = \frac{p_d}{p_u}\rho_u$$
9. Volume flow rate in the nozzle narrow section - calculated by the following formula

$$Q_d = \frac{q}{\rho_d}$$

10. Characteristics limits

The lower limit of the characteristic is determined by the condition

$$Re_d > 12000,$$

and the upper limit is given by the condition

$$p_{v\,max} < 10000.$$

With the input parameters set according to the sequence described above, the points of the characteristic over the whole measurement range were calculated and are presented in Figure 2.

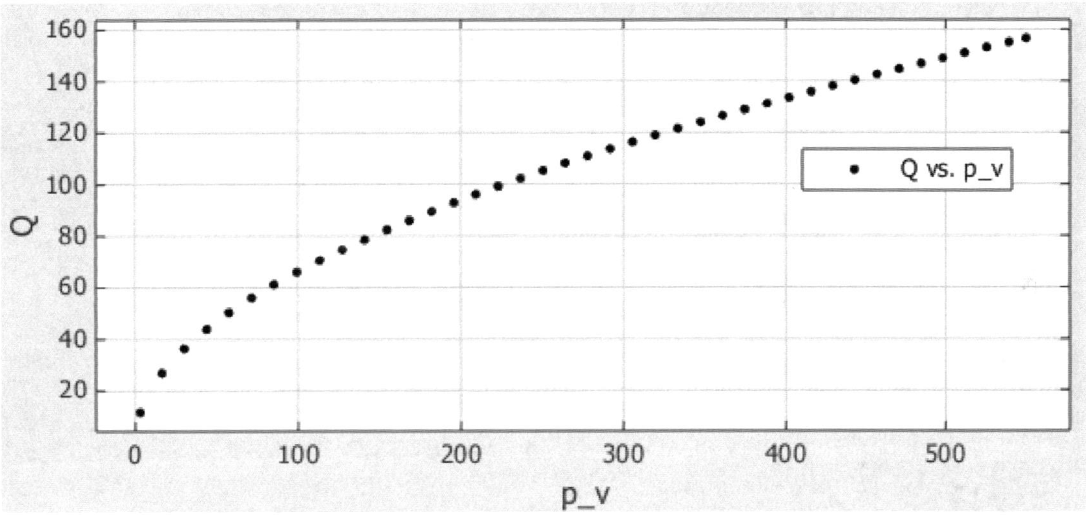

Figure 2. Nozzle characteristic points **Q** vs $\mathbf{p}_v$, calculated according to ISO 5801:2007.

2.2. Approximation of the characteristic

For the approximation of the characteristic, a model of the following form is used

$$Q = a \cdot p_v^{0,5}.$$

In the present work, a modification of this model is made, which gives more accurate results.

For this purpose a transformation function of type (2) is used

$$Q = a \cdot p_v^b. \quad (2)$$

The refinement of the coefficients of the mathematical model of type (2) was carried out after the approximation of the data Q vs p_v by the following methodology.

First the power model (2) was logarithmized and a model of the form (3) was obtained:

$$ln\, Q = ln\, a + b \cdot ln\, p_v. \quad (3)$$

Let us presume:

$$ln\, Q = y,$$
$$ln\, p_v = x,$$
$$ln\, a = a_1.$$

After substituting in (3), we obtain a linear model of the form (4):

$$y = a_1 + b \cdot x. \quad (4)$$

In stochastic form, model (4) has the form:

$$y = a_1 + b \cdot x + \varepsilon,$$

where ε is the stochastic error from the model and represents an undefined random variable.

The following point estimates for the parameters from the linear model were found using the least squares method:

$$\hat{a}_1 = 1.8553, \hat{b} = 0.5070.$$

With 95% confidence intervals for the parameters were also found as follows:

$$I_{a_1,95\%} = [\hat{a}_1 - t_{1-\frac{\alpha}{2},n-2}\hat{\sigma}_{\hat{a}_1}, \hat{a}_1 + t_{1-\frac{\alpha}{2},n-2}\hat{\sigma}_{\hat{a}_1}]=[1.8539, 1.8568],$$

$$I_{b,95\%} = [\hat{b} - t_{1-\frac{\alpha}{2},n-2}\hat{\sigma}_{\hat{b}}, \hat{b} + t_{1-\frac{\alpha}{2},n-2}\hat{\sigma}_{\hat{b}}]=[0.5068, 0.5073],$$

where

$t_{1-\frac{\alpha}{2},n-2}$ - t-quantile of the Student distribution, α - risk of error (α=0.05), n - sample size;

$\hat{\sigma}_{\hat{a}_1}$ - standard deviation for the parameter a_1;

$\hat{\sigma}_{\hat{b}}$ - standard deviation for parameter b.

The coefficients are significant and the model is adequate with a probabilistic guarantee bigger than 99.9%. The coefficient of determination is 0.999997.

After antilogarithmization of (4), the model to approximate the original data in stochastic form is obtained:

$$Q = \hat{\mathcal{E}}\hat{a} \cdot p_v^{\hat{b}} = \hat{a} \cdot p_v^{\hat{b}} + (\hat{\mathcal{E}} - 1)\hat{a} \cdot p_v^{\hat{b}}. \quad (5)$$

Let us denote by $\hat{\mathcal{E}}_0$ the estimate of the stochastic error in model (5). It has the following form:

$$\hat{\mathcal{E}}_0 = (\hat{\mathcal{E}} - 1)\hat{a} \cdot p_v^{\hat{b}}\ . \quad (6)$$

Substituting (6) into (5), we obtain:

$$Q = \hat{a} \cdot p_v^{\hat{b}} + \hat{\mathcal{E}}_0, \quad (7)$$

where the parameter estimates $\hat{a}$ and $\hat{b}$ are known and they have the following values:

$$\hat{a} = e^{\hat{a}_1} = e^{1.8553} = 6.3938,\ \hat{b} = 0.5070.$$

With a 95% confidence interval, the confidence intervals for the parameters were also found as follows:

$$I_{a,95\%} = [6.3846, 6.4030],$$
$$I_{b,95\%} = [0.5068, 0.5073].$$

Finally, the model (7) takes the form:

$$Q = 6.3938 \cdot p_v^{0.507} + \hat{\mathcal{E}}_0. \quad (8)$$

Graphically, model (8) is presented in Figure 3.

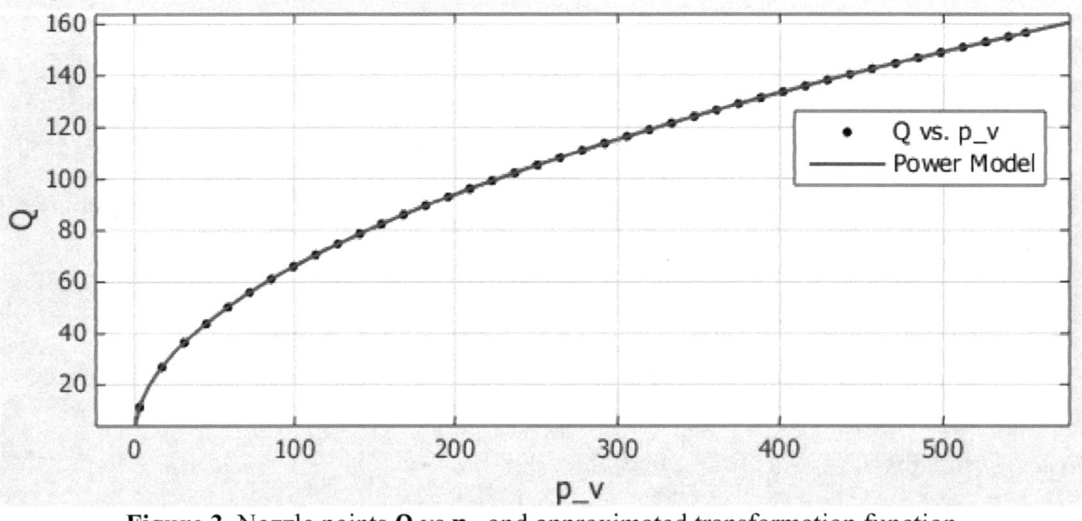

Figure 3. Nozzle points **Q** vs $\mathbf{p_V}$ and approximated transformation function **Q** vs $\mathbf{p_V}$.

For model (8), the values of the stochastic deviations $\hat{\mathcal{E}}_0$ were calculated using the relation (6) and confidence intervals for these deviations were constructed.

2.3. *Studying of the relationship between the length of the confidence intervals and* $\mathbf{p}_v$

It is interesting to note that the length of the confidence intervals for $\hat{\varepsilon}_0$ has a clear functional dependence on p_v. The following models were used to describe this dependence:

Lin-Log Model of the following form:

$$delta_{\varepsilon_0} = c_1 + c_2 \cdot ln(p_v), \quad (9)$$

where:

$delta_{\varepsilon_0}$- the length of the confidence intervals for $\hat{\varepsilon}_0$,

c_1 and c_2- coefficients of the model.

Using the least squares method, estimates for the unknown parameters are found and model (9) takes the form:

$$delta_{\varepsilon_0} = 0.002841 + 0.002309 \cdot ln(p_v).$$

With 95% confidence intervals for the parameters are also found as follows:

$$I_{c_1,95\%} = [0.001516, 0.004165],$$
$$I_{c_2,95\%} = [0.002064, 0.002554].$$

The adjusted coefficient of determination for this model is 0.9005 and the standard deviation is 0.0008155. This indicates a good fit of the model to the data. Graphically, this model is presented in Figure 4.

- Michaelis-Menten Model of the following form:

$$delta_{\varepsilon_0} = \frac{c_1^* \cdot p_v}{c_2^* + p_v}, \quad (10)$$

where:

$delta_{\varepsilon_0}$ - he length of the confidence intervals for $\hat{\varepsilon}_0$,

c_1^* and c_2^* – coefficients of the model.

Using the least squares method, estimates for the unknown parameters are found and model (10) takes the form:

$$delta_{\varepsilon_0} = \frac{0.01687 \cdot p_v}{15.5 + p_v}.$$

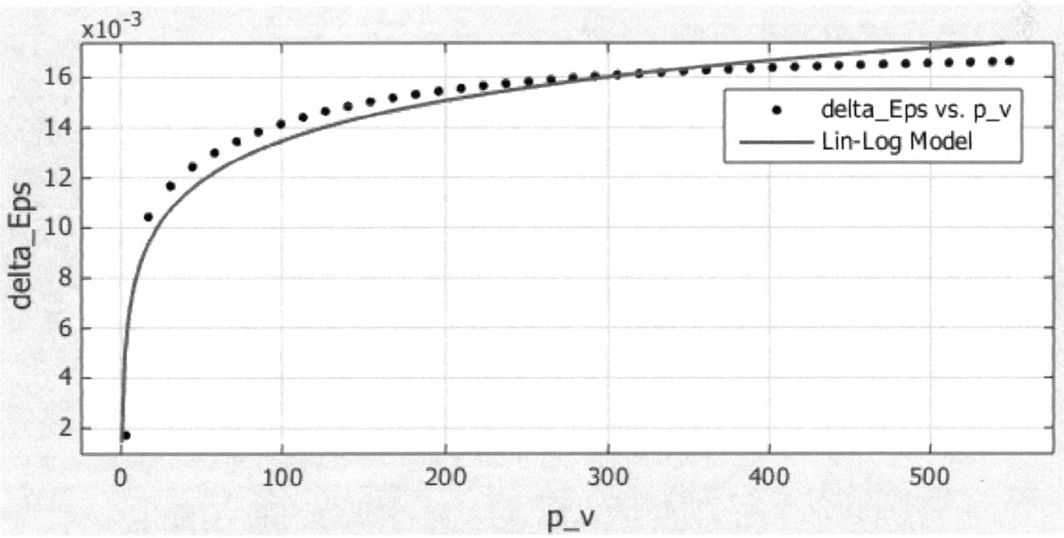

Figure 4. Lin-log Model to investigate the relationship between the length of confidence intervals and $\mathbf{p}_v$.

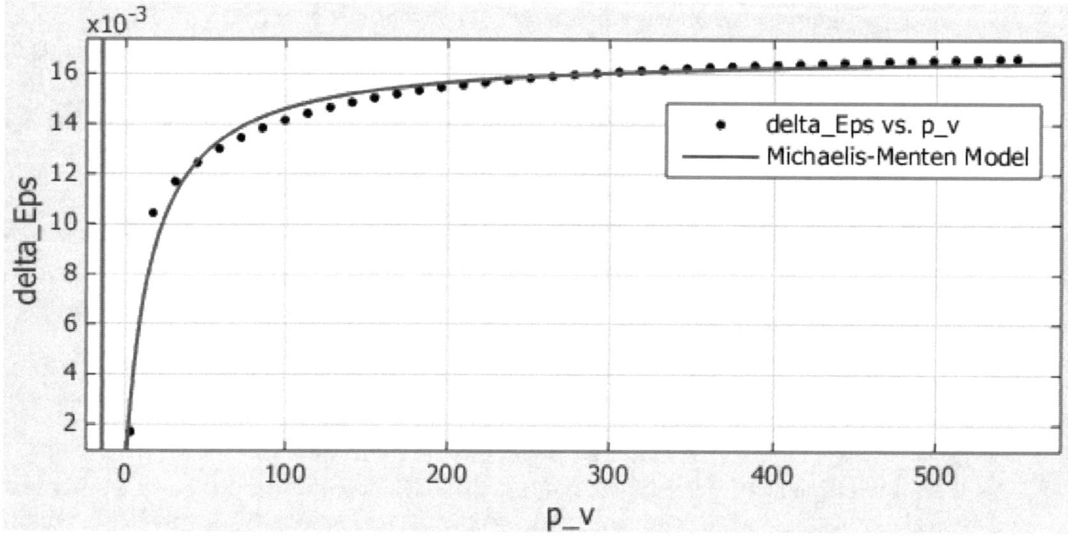

Figure 5. Michaelis-Menten Model to study the relationship between the length of confidence intervals and $\boldsymbol{p_v}$.

With 95% confidence intervals for the parameters are also found as follows:

$$I_{c_1^*,95\%} = [0.01667, 0.01706],$$

$I_{c_2^*,95\%} = [13.7, 17.3]$.The adjusted coefficient of determination for this model is 0.9765 and the root mean square deviation is 0.0003959. Here again we have a good fit of the model to the data. Graphically, this model is presented in Figure 5.

Table 1. Goodness of fit for the models of confidence interval length and p_V.

Model	SSE	R-square	Adjusted R-square	RMSE
Lin-Log Model	0.000025940	0.9030	0.9005	0.0008155
Michaelis-Menten Model	0.000006114	0.9771	0.9765	0.0003959

To compare the two models, let us examine Table 1. It can be seen that the Michaelis-Menten Model has a higher smoothed coefficient of determination and a lower root mean square deviation than the Lin-Log Model. This gives reason to believe that the Michaelis-Menten Model gives a more accurate approximation.

3. Error of the measured quantities

When studying the fan noise, it is necessary to consider the error of the measured sound pressure level δ_s and the total error of the fan flow δ_Q, calculated by the transformation function using the measured vacuum p_v.

1. Error of measured sound pressure level δ_s

This error is taken from the technical parameters of the sound level meter. In our case the noise was measured with a SVANTEK 971 Class 1 noise meter. The accuracy for a Type 1 instrument, according to IEC 61672:2002 Electroacoustics - Sound level meters - Part 1: Specifications [10], is:

- At a reference frequency of 1kHz, the tolerance limits for Class 1 are +/- 1.1 dB.
- At the lower and upper end of the frequency range the tolerances are wider. At 20 Hz the tolerances are +/- 2.5 dB. At 16 Hz the tolerances are +2.5 dB, -4.5 dB. At higher frequencies the same applies.
- The tolerances for a Class 1 instrument are tighter at frequencies above 8 kHz, with tolerances

at 10 kHz of +2.6 dB, -3.6 dB for Class 1. At frequencies higher than 16 kHz the tolerances are +3.5 dB, -17 dB for Class 1.

2. Total flow rate error δ_Q

Since the flow rate Q is an indirect measurand, the error of all quantities in the flow measurement and determination consequence must be considered to determine its total error.

Let us introduce the following notations:

Q_0- actual value of flow rate,

Q - flow rate calculated according to ISO 5801:2007,

Q_A - approximated value of the flow according to model (8).

- Maximum relative error of the measured vacuum $\delta_{p_{v_{max}}}$

This error is also taken from the technical parameters of the instrument. For vacuum measurement, a Differential pressure measuring instrument Testo 521-1 (accuracy ±0.2% of f.v.) is used, whose maximum relative error is:

$$\delta_{p_{v_{max}}} = 0.2\ \%.$$

- Maximum relative error of flow rate coefficient, calculated according to ISO 5801:2007

According to ISO 5801:2007 the flow rate is indirectly determined by relation (1). In this relation, also according to the standard, the error of the flow coefficient μ for $Re_d > 12000$ is

$$\delta_{\mu_{max}} = 1.2\ \%.$$

- Maximum relative flow rate error calculated according to ISO 5801:2007, including the measured vacuum error:

$$\delta_{Q_{0max}} = \frac{|Q_0 - Q|_{max}}{Q} \tag{11}$$

To determine $\delta_{Q_{0max}}$, the following formula for indirectly measured quantities was used (see [4]):

$$\delta_{Q_{0max}} = \sqrt{\delta^2_{\mu_{max}} + \frac{1}{4}\delta^2_{p_{v_{max}}}} = 1.204159458\%. \tag{12}$$

- Maximum relative error of the flow rate approximated by model (8):

$$\delta_{Q_{Amax}} = \frac{|Q - Q_A|_{max}}{Q_A} \tag{13}$$

To determine $\delta_{Q_{Amax}}$, the following formula was used:

$$\delta_{Q_{Amax}} = \frac{\frac{\sigma_{\hat{\varepsilon}_0}}{\sqrt{n}}}{Q_{max} - Q_{min}} t_{1-\frac{\alpha}{2},n-2} = 0.000597144\%, \tag{14}$$

where:

$\sigma_{\hat{\varepsilon}_0}$ – standard deviation of stochastic deviations $\hat{\varepsilon}_0$ from model (8),

$Q_{max} - Q_{min}$ – range of the flow rate, calculated by the ISO 5801:2007 (Q_{max} – maximum value of the flow rate, Q_{min} – minimum value of the flow rate),

$t_{1-\frac{\alpha}{2},n-2}$ - t-quantile of the Student distribution,

α - risk of error (α=0.05),

n - sample size.

- Total maximum relative flow rate error:

$$\delta_{Q_{max}} = \frac{|Q_0 - Q_A|_{max}}{Q_A} \tag{15}$$

The procedure to calculate the total maximum relative flow rate error is as follows.

From (11) we express the actual value of the flow rate:

$$Q_0 = \left(\delta_{Q_{0max}} + 1\right) Q. \tag{16}$$

From (13) we express the approximated value of the flow rate by model (8).

$$Q_A = \frac{Q}{1+\delta_{Q_{A max}}}. \tag{17}$$

By substituting (16) and (17) into (15) we get:

$$\delta_{Q_{max}} = \delta_{Q_{0 max}} + \delta_{Q_{A max}} + \delta_{Q_{0 max}} \delta_{Q_{A max}}. \tag{18}$$

Substituting in (18) with (12) and (14) for the total maximum relative flow rate error we get:

$$\delta_{Q_{max}} = 1.205475659\%.$$

4. Conclusions

The conclusions from the present work can be generalized as follows:

1) **Noise testing equipment is designed:** It allows to investigate the noise of a fan with different impellers and different operating modes. In future work, experiments will be carried out using the apparatus and the dependence of the fan noise on its flow rate and impeller blade profile will be sought.

2) **Design of a standardised smooth inlet:** A nozzle with a free inlet for volumetric flow measurement has been designed and built to the ISO 5801:2007 standard. Also according to the standard, points of the flowmeter's characteristic were calculated.

3) **Approximation of calibration characteristic:** To facilitate the "readout" of the fan flow rate, a simplified dependence of the flow rate on the measured vacuum was obtained by approximation. For this purpose, a power model of the form $Q = a \cdot p_v^b$ was used, which allows calculation of the volumetric flow rate by the measured vacuum value. After logarithmization, this model was reduced to a linear model and point estimates for the parameters were derived. With a 95% probabilistic guarantee, confidence intervals for the parameters were also calculated. The resulting model is adequate and its coefficients are significant with a probabilistic guarantee bigger than 99.9%, and its coefficient of determination is bigger than 0.9999.

4) **Studying of the relationship between confidence intervals and the vacuum:** It was empirically found that the length of the confidence intervals for the stochastic error in the Power model demonstrated a dependence on the vacuum. To describe this dependence, 2 models that describe the data well were used: the Lin-Log Model and the Michaelis-Menten Model. A comparative analysis of the two models was performed, showing that the Michaelis-Menten Model is better suited to describe the data.

5) **Error of the measurands:** the overall maximum relative error of flow as an indirect measurement variable was determined. The determination of the error takes into account the instrument errors in measuring the vacuum pressure, the error in calculating the flow coefficient according to ISO 5801:2007 and the approximation error.

The results obtained are essential for understanding the noise characteristics of a centrifugal fan and provide a methodology for the proper measurement and analysis of noise in similar systems.

Acknowledgments

This paper contains results of the work on project No 2024 - FNSE – 02, financed by „Scientific Research" Fund of Ruse University".

Credit roles: conceptualization, I.N. and V.P.; Experimental setup creation, I.N. and N.K., data curation, V.M. and N.K.; methodology - work on the standard, I.N. and N.K.; methodology - errors measurement, I.N. and V.M.; funding acquisition, V.M. and V.P.; project administration, V.M.; software, V.M. and I.N.; supervision, V.P.; validation, N.K. and V.P.; visualization, I.N. and V.M.; writing-original draft, I.N.; writing-review and editing, V.M. All authors have read and agreed to the published version of the manuscript.

References

[1] Kokalarov M 2019 *Predetermination of the Uncertainty Budget and Error Analysis of the Photogrammetric Method in Heavy Machinery* (*Proceedings of the 30th DAAAM International Symposium*) ed B Katalinic (Vienna: DAAAM International) pp 0914 - 0920

[2] Dichev D, Kogia F, Nikolova H and Diakov D 2018 *A Mathematical Model of the Error of*

Measuring Instruments for Investigating the Dynamic Characteristics (*Journal of Engineering Science and Technology Review* Volume 11 Issue 6) (Eastern Macedonia and Thrace Institute of Technology) pp 14-19
[3] Artiukh S, Kokalarov M, Kniazieva V, Georgiev I, Artiukh A and Milchev M 2023 *Metrological Aspects in Maintenance of Energy-Saving Automated Systems for Control of Pump Units in Pumping Stations (Proceedings of the 34th DAAAM International Symposium on Intelligent Manufacturing and Automation*) ed B Katalinic (Vienna: DAAAM International) pp 0268 - 0273
[4] Yordanov S, Ormandzhiev K, Mihalev G, Kostov K and Mitev V *2023 Studying the Influence of Working Fluid Temperature on the Performance of an Electrohydraulic Servo System in Dynamic Mode* (*International Conference Automatics and Informatics*) (IEEE) pp 211-216
[5] Ormandzhiev K, Yordanov S, Mihalev G and Kostov K 2011 *Parametric Optimization of Electrohydraulic Servo System* (PEPM'2020 *E3S Web Conf* Volume 207 04003) (E3S Web of Conferences)
[6] Jaremkiewicz M 2011 *Reduction of dynamic error in measurements of transient fluid temperature* (*Archives of Thermodynamics* 32(4)) pp 55-66
[7] Xiu D and Karniadakis G E 2002 *Modeling error in steady state diffusion problems via generalized polynomial chaos* (*Computer methods in applied mechanics and engineering* 191(43)) pp 4927-4948
[8] Zaharieva S, Borodzhieva A and Mutkov V 2019 *Metrological Assessment of an Electronic System for Monitoring the Geometrical Parameters of Cylindrical Profiles* (*IEEE Xplore Compliant Proceedings*) (IEEE) pp 404-407
[9] ISO 5801:2007 Industrial fans — Performance testing using standardized airways
[10] IEC 61672:2002 Electroacoustics - Sound level meters - Part 1: Specifications

Mathematical Modeling of the Operation of a Hydrophoric Installation

I N Nikolaev[1], E R Raeva[2] and V A Mihova[2]

[1]Department of Heat, Hydraulics and Environmental Engineering, University of Ruse, 8 Studentska Str, 7017 Ruse, Bulgaria
[2]Department of Applied Mathematics and Statistics, University of Ruse, 8 Studentska Str, 7017 Ruse, Bulgaria

eraeva@uni-ruse.bg

Abstract. In this paper, the performance of a pumping system is considered in its operation with a hydrophore vessel connected in parallel. The filling and emptying processes of the hydrophore and their influence on the flow rate to the system and the pump outlet pressure are experimentally investigated. An automated measuring system is used for the measurements of the flow rate and the pressure. That has cyclically recorded the values at equal time intervals. The measurement cycles are synchronized in time and the maximum errors of the measured quantities are known. From the experimental data, empirical relationships are obtained for flow rate and pressure as a function of time, as well as the relationships between the two quantities. The time periods of each of the two processes and the average cycle length are also empirically determined. Confidence intervals of the flow rate are constructed for vessel filling and emptying, respectively, and measurement errors were accounted for. Codes have been created to determine the filling time of a hydrophore vessel and to calculate confidence intervals for this type of pumping systems, thanks to which all calculations have been automated. The described dependencies can be used in planning future experiments.

1. Introduction

Water supply systems are an essential part of modern infrastructure, providing the necessary water resources for domestic, industrial, and agricultural needs. Key components of these systems are the pumping stations with hydrophore tanks (hydrophore). They play an important role in maintaining constant water pressure in the system and in ensuring guaranteed water supplies without delay in cases of significant interruptions. In such cases, hydrophore installations are particularly effective, delivering water across a wide range of operating modes (flow rates).

Hydrophore systems operate on the principle of periodic pump switching on and off, utilizing the compressed air and elasticity of the air bag in the hydrophore tank to store energy when the pump is running and use this energy when the pump is off. These systems enhance better pressure management and pump operation at optimal regimes. As a result, the pump system operates efficiently and economically, reducing pump wear and saving energy.

The aim of this paper is to model the processes in a hydrophore installation to find empirical dependencies for the water supply flow rate and the pressure in the hydrophore as a function of time during the filling mode of the hydrophore tank (pump on) $Q^+ = f(t)$ and $p_y^+ = f(t)$ and the emptying

Content from this work may be used under the terms of the Creative Commons Attribution 4.0 licence. Any further distribution of this work must maintain attribution to the author(s) and the title of the work, journal citation and DOI.
Published under licence by IOP Publishing Ltd

mode of the tank (pump off) $Q^- = f(t)$ and $p_y^- = f(t)$, based on experimental data. The relationship between the flow rate to the water supply system and the pressure in the hydrophore tank when the pump is on and off, respectively $Q^+ = f(p_y^+)$ and $Q^- = f(p_y^-)$, is also to be determined.

Using the obtained models, a methodology is to be developed to describe the cyclic operation mode of the hydrophore installation and to determine the parameters and values characterizing this mode.

2. Principle of Operation of the Hydrophore Installation

The system's diagram is shown in Figure 1 and operates as follows: the centrifugal pump (4), driven by the electric motor (6), draws water from the open reservoir (1) through the suction strainer (2) via the suction pipeline (3) and delivers it into the reservoir (18) via the discharge pipeline (13). At the pump's outlet, after the check valve (5), the hydrophore tank (12) is connected in parallel. Using valve K_2, the pressure and flow rate of the pump (4) and the flow rate to the hydrophore tank (12) are throttled.

The hydrophore tank is pre-charged with air at an initial pressure p_0. If the pressure at the hydrophore inlet p_y is less than p_0, no water enters the tank (it does not fill), and the system operates in a steady, non-cyclic mode. Conversely, part of the pump's flow rate enters the hydrophore. The air volume compresses, leading to an increase in p_y pressure in the tank, which in turn alters the pump's operating mode and the flow rate Q to the main discharge pipeline (13) and the hydrophore (12).

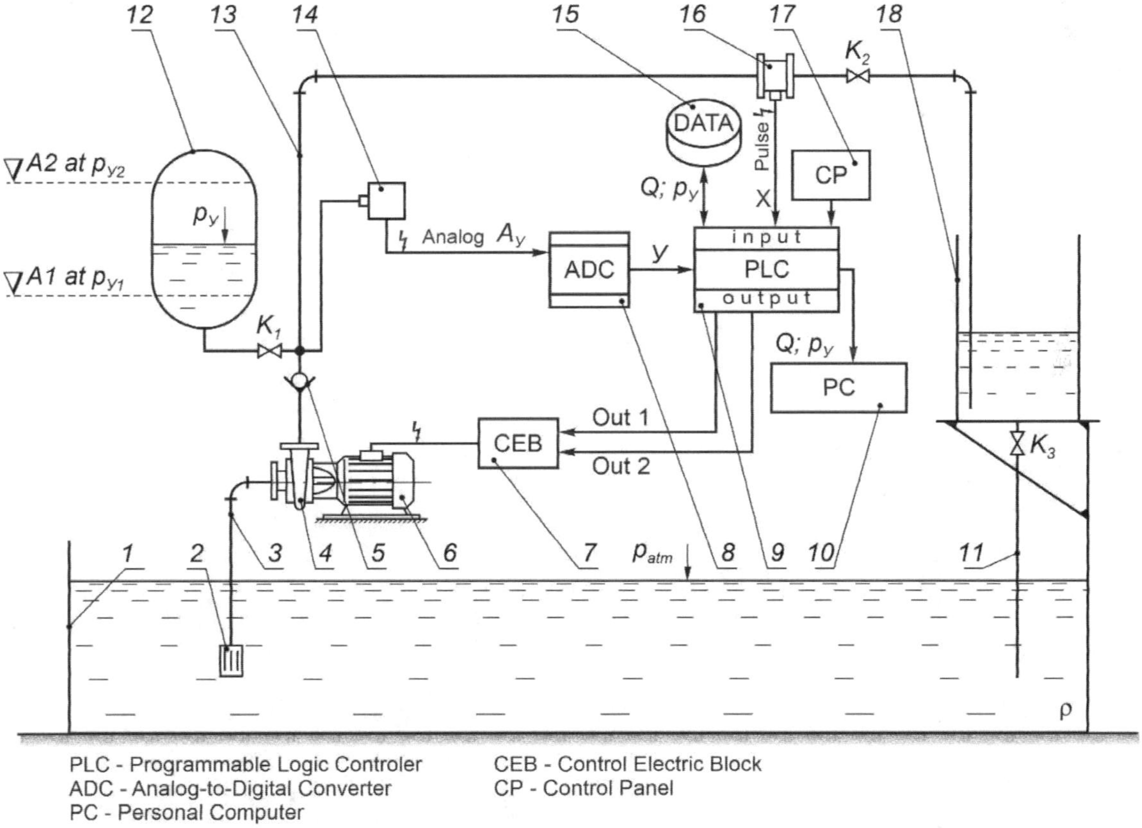

Figure 1. Diagram of a hydrophoric installation.

The most common operating mode for this type of installation is the cyclic mode, with periodic pump switching on and off, and continuous water supply through the discharge pipeline (13). The installation also includes an automated measurement system, which not only measures but also controls the cyclic operation of the hydrophore installation. The programmable logic controller (9, PLC) is preset with boundary control pressures p_{y1} and p_{y2}, greater than p_0, corresponding to the boundary levels A1 and A2 of the water in the hydrophore tank. The current pressure in the hydrophore p_y (current level A) is monitored by the PLC. When the water level is below the lower boundary level A1 ($p_y < p_{y1}$), the PLC sends an electrical signal Out 1 to the control electric block (7, CEB), and when the current level A is above the upper boundary level A2 ($p_y > p_{y2}$), it sends an electrical signal Out 2.

The electrical block (7) starts the pump when signal Out 1 is present, and the pump operates until signal Out 2 is received (indicating level A2), after which the control block stops the pump and does not restart it until signal Out 1 is sent again. During the pump's off-time, the stored energy from the compressed air in the hydrophore (12) is used to supply water to the main discharge pipeline (13), as the check valve (5) prevents water backflow through the pump. This achieves a cyclic operating mode and continuous water supply through pipeline (13).

If valve K_2 is not sufficiently closed, the pressure p_y in the hydrophore tank may not reach the upper pressure limit p_{y2}. In this case, a constant operating mode is also established, and the hydrophore remains partially filled. Valve K_1 is used to isolate the hydrophore tank from the pump system.

3. Automated Measurement System for Registering Flow Rate Q and Pressure p_y

The automated registering measurement system includes: the strain gauge pressure transducer (14) for measuring the pressure in the hydrophore tank p_y; the tachometric turbine transducer (16), which converts the flow rate Q into a pulse electrical signal X; the analog-to-digital converter (8, ADC) for converting the analog signal A_{py} received from the pressure transducer (14) into a digital signal Y; the programmable logic controller (9, PLC) with a recorded therein cyclic measurement program; the controller control panel (17); and the personal computer (10, PC) for extracting the obtained experimental data and monitoring current parameters.

The measurement system allows for the measurement and registration over time of the pressure p_y in the hydrophore tank and the flow rate Q in the discharge pipeline (13). The measurement and recording of values in the controller's memory (15) occur at equal time intervals, with the cycles of the measured values being synchronized in time.

The principle of operation of the measurement system, the cyclic measurement program, and the measurement methodology are described in detail in [1].

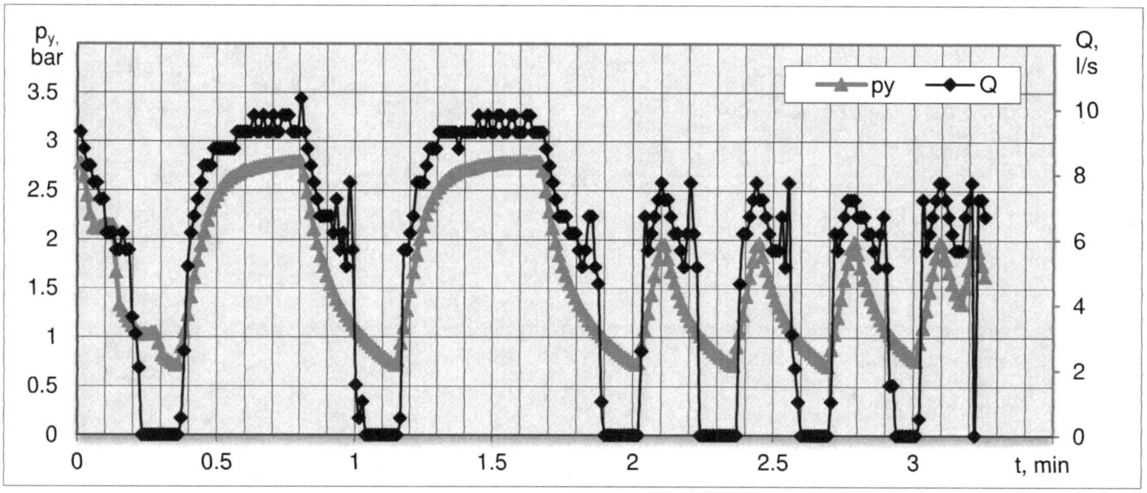

Figure 2. Experimental data in the cyclic mode of operation of a hydrophoric installation.

Experimental data from 279 measurements at equal time intervals (0.7 s) of pressure p_y and flow rate Q at different operating modes are recorded over time using the setup and are presented in Fig. 2.

Based on these data, the desired empirical dependencies describing the cyclic operation process are obtained using correlation and regression analysis.

4. Parameters Characterizing the Cyclic Operation Mode

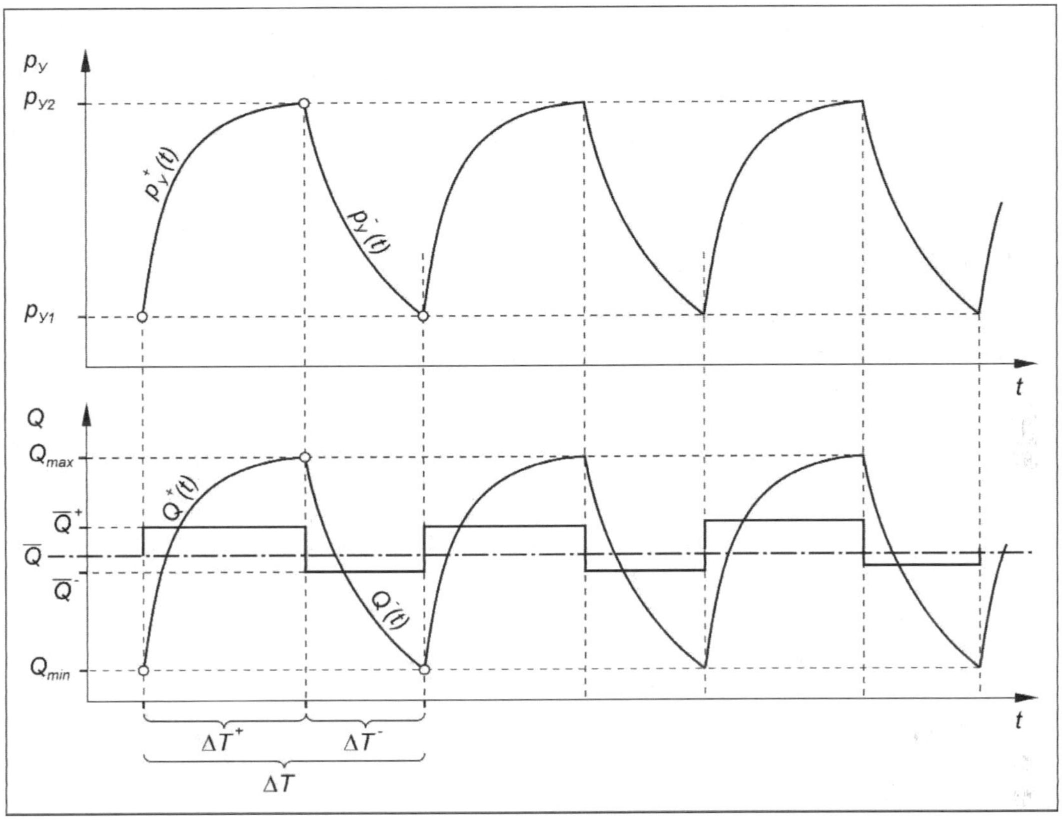

Figure 3. Parameters Characterizing the Cyclic Operation Mode.

The parameters characterizing the cyclic operation mode play a crucial role in the effective and reliable functioning of the system. Here are the main parameters:

- *Volume of the Hydrophore Tank* V_0 – This matches the volume of air in the tank when it is empty (no water).
- *Initial Pressure of the Hydrophore Tank* p_0 – The air pressure in the hydrophore tank when it is empty (no water).
- *Boundary Control Pressures* p_{y1} *and* p_{y2} – These are the boundary pressures at which the pump switches on and off. The lower boundary pressure p_{y1} is the minimum system pressure at which the pump activates, while the upper boundary pressure p_{y2} is the maximum pressure at which the pump deactivates.
- *Filling and Emptying Times of the Hydrophore Tank* ΔT^+ *and* ΔT^- – The filling time ΔT^+ is the period during which the pump operates and fills the hydrophore tank with water. The emptying time ΔT^- is the period during which the pump is off and water is supplied from the hydrophore tank.
- *Number of Pump Cycles* N_{cycles} – The number of cycles within a given period of time, most commonly expressed in cycles per hour (cycles/hour).
- *Cycle Period* ΔT – This is the duration of one complete operating cycle.

- *Minimum and Maximum Water Supply Flow Rates Q_{min} and Q_{max}* – These are the flow rate limits corresponding to pressures p_{y1} and p_{y2}, respectively.
- *Average Flow Rate During Filling and Emptying $\bar{Q}^{+}$ and $\bar{Q}^{-}$* – These are the time-averaged flow rates to the water supply system during the filling period $\bar{Q}^{+}$ and the emptying period $\bar{Q}^{-}$ of the hydrophore tank.
- *Average Flow Rate of the Hydrophore System $\bar{Q}$* – This is the time-averaged flow rate of the hydrophore system to the water supply system (overall average flow rate for the entire cycle period ΔT).
- *Coefficient of Variation (CV) of the Flow Rate* – This is a standard method for measuring the relative variability of the flow rate relative to the mean value and is expressed as a percentage.

The parameters characterizing the cyclic operation mode of the hydrophore installation are interrelated and play a critical role in the effective and reliable functioning of the system. Their proper adjustment and management can significantly improve the performance, energy efficiency, and longevity of the water supply system. Understanding these parameters and their importance is crucial for optimizing the operation of hydrophore pumping systems.

5. Empirical analysis

In order to obtain empirical dependences, we use the experimental data described in Item 3. They clearly show the cyclic mode of operation of the hydrophore at different boundary pressures. More fluctuating are the data for the flow rate Q. For this reason, first we find the empirical relationship between the flow rate and the pressure, and second – the empirical relationships of the flow rate and the pressure as a function of time. Dependencies are found by applying correlation analysis and a linear regression model [2]. The results are obtained using the statistical software SPSS [3].

5.1. Determining two components of p_y – "positive and negative"

The on and off pressure periods are determined by differentiating the data using the formula

$$p_{i+1} - p_i, i = 1, 2 \dots 279.$$

After differencing, a series of positive and negative values are obtained. The indices of the positive values define the component p_y^+. The indices with negative values define the component p_y^-. This way the flow rate is also divided into two parts, Q^- and Q^+.

5.2. Data modification of $Q^- = f(p_y^-)$

From Figure 4, a significant number of zero flow rate values can be observed when operating in off mode Q^-. A one-factor regression model between Q^- and p_y^- is considered to proxy the zeros in the Q^- flow rate data.

Table 1. Linear Model Coefficients for $Q^- = b_1 p_y^- + b_0$.

	Unstandardized Coefficients		Standardized Coefficients		
	B	Std. Error	Beta	t	Sig.
P_y	5.642	0.398	0.873	14.189	0.000
(Constant)	-3.617	0.553	–	-6.542	0.000

The coefficients of the resulting model are presented in Table 1. The low values in the last column are an indication that the coefficients in the model are significant with probabilistic guarantee close to 100%.

The resulting relationship takes the form

$$Q^- = 5.642 p_y^- - 3.617. \quad (1)$$

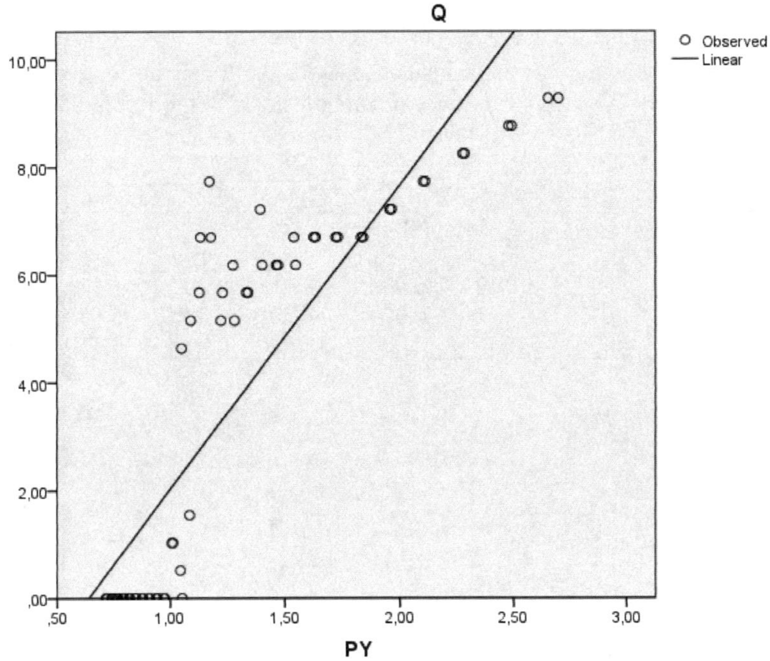

Figure 4. Linear model of flow rate Q^{-} versus pressure $p_{\bar{y}}$.

According to the ANOVA analysis conducted, the model is adequate with probabilistic guarantee close to 100%. The obtained coefficient of determination $R^2 = 0.762$ indicates that 76% of the variation in flow rate, is due to variation in pressure $p_{\bar{y}}$.

From Figure 4, it can also be noticed that a nonlinear model can be sought to better describe the relationship between the two variables. A good approach in this case is to approximate the null values to improve the homogeneity of the data. For this purpose, the above model described by equation (1) is used only to approximate the zero flow values. The modified data thus obtained is used in the following calculations.

5.3. Correlation between Q^{-} and $p_{\bar{y}}$

Table 2. Correlation matrix for flow rate Q^{-} and pressure p^{-}.

		$p_{\bar{y}}$	Q^{-}_{mod}
$p_{\bar{y}}$	Pearson Correlation	1	0.889**
	Sig. (2-tailed)		0.000
	N	65	65
Q^{-}_{mod}	Pearson Correlation	0.889**	1
	Sig. (2-tailed)	0.000	
	N	65	65

**. Correlation is significant at the 0.01 level (2-tailed).

Pearson's correlation coefficient with significance level $\alpha = 0.01$ is calculated to determine the degree of association between the variables (Table 2). A correlation coefficient of $\rho = 0.889$ is an indication of a strong correlational relationship, and the two stars following the value in the table indicate the presence of a statistically significant relationship between the two variables.

5.4. Comparing models for $Q^- = f(p_y^-)$

To determine the model of the flow rate as a function of pressure $Q^- = f(p_y^-)$ in the emptying mode of operation, different types of models are considered, with the logarithmic, quadratic and cubic models deserving attention. Table 3 presents summary results of the three models and Figure 5b shows the plots of the three models.

Table 3. Comparison of the models results.

Model	R^2	Regression F/Sig.	Constant / Sig		β_1 / Sig.		β_2 / Sig.		β_3 / Sig.	
Logarithmic	0.865	0.000	2.817	0.000	7.321	0.000				
Quadratic	0.879	0.000	-8.760	0.000	14.650	0.000	-3.084	0.000		
Cubic	0.892	0.000	-16.165	0.000	31.117	0.000	-14.150	0.001	2.269	0.010

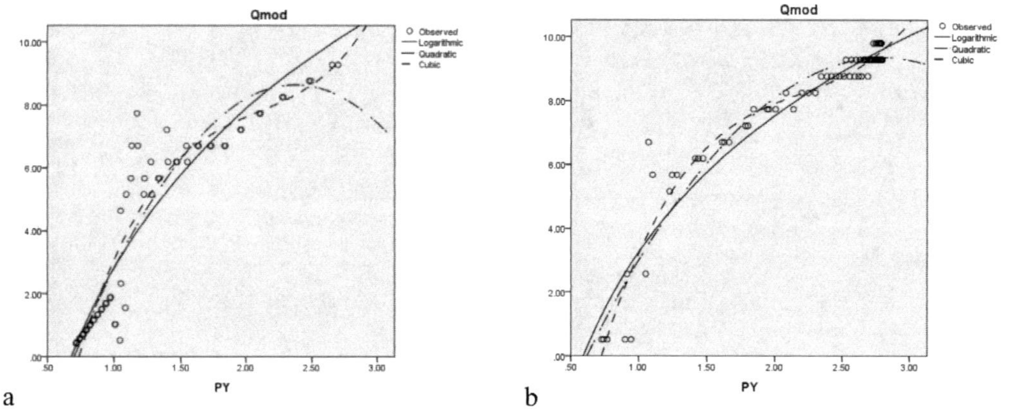

Figure 5. Comparison of the results from logarithmic, quadratic and cubic models: (a) Q^+ and (b) Q^-.

After a comparison between the three models, the cubic model is selected because of the largest coefficient of determination and the smallest range and standard deviation of the error (Table 4). As a result of a Kolmogorov-Smirnov test, the errors of the cubic model are normally distributed at the 5% significance level. Finally, the selected model takes on the form

$$Q_{mod}^- = 2.269 * (p_y^-)^3 - 14.15 * (p_y^-)^2 + 31.117 * p_y^- - 16.165. \quad (2)$$

Table 4. Estimates of the errors from the three models for Q_{mod}^-.

Error for Qmod	N	Range	Minimum	Maximum	Mean	Std. Deviation
LOGARITHMIC model	65	6.36918	-2.61842	3.75077	0E-7	1.10993932
QUADRATIC model	65	6.21439	-2.66130	3.55309	0E-7	1.04949734
CUBIC model	65	6.17560	-2.96889	3.20671	0E-7	0.99287870

5.5. Comparing models for $Q^+ = f(p_y^+)$

For the Q^+ values, there are only three zero values. In this case, they are replaced by the smallest value of Q^+ after excluding the zeros.

The analysis for the periods of increase in flow Q^+, relative to the periods of increase in pressure p_y^+ is carried out in a similar way as for $Q^- = f(p_y^-)$.

Journal of Physics: Conference Series **2910** (2024) 012019 doi:10.1088/1742-6596/2910/1/012019

The correlation coefficient with a value of 0.975 is tested for significance and the result shows a strong and significant correlation with significance level $\alpha = 0.01$.

Again, logarithmic, quadratic and cubic models are considered. Based on the same criteria, the third degree model is selected as the most appropriate approximation. The results are shown in the following three tables: Table 5, Table 6 and Table 7, and Figure 5a shows the plots of the three models.

Table 5. Coefficient of determination for cubic model for $Q^+ = f(p_y^+)$.

R	R Square	Adjusted R Square	Std. Error of the Estimate
0.975	0.951	0.949	0.530

Table 5 shows the coefficient of determination of the selected model. It can be seen from the table that the Adjusted R-Square is 0.949, which indicates that 94.9% of the variation in flow rate, is due to variation in pressure p_y^+.

Table 6. ANOVA table for cubic model for $Q^+ = f(p_y^+)$.

	Sum of Squares	df	Mean Square	F	Sig.
Regression	468.699	3	156.233	556.950	0.000
Residual	24.405	87	0.281	–	–
Total	493.104	90	–	–	–

The ANOVA table (Table 6) shows that the cubic model is adequate with probabilistic guarantee close to 100%.

Table 7. Coefficients for cubic model for $Q^+ = f(p_y^+)$.

	Unstandardized Coefficients		Standardized Coefficients	t	Sig.
	B	Std. Error	Beta		
p_y^+	28.894	3.329	7.828	8.679	0.000
$(p_y^+)^2$	-12.497	1.924	-13.129	-6.495	0.000
$(p_y^+)^3$	1.903	0.346	6.284	5.496	0.000
(Constant)	-15.110	1.758	–	-8.597	0.000

The following cubic model is obtained with the coefficients from Table 7

$$Q^+ = 1.903 * (p_y^+)^3 - 12.497 * (p_y^+)^2 + 28.894 * p_y^+ - 15.110. \tag{3}$$

5.6. Time depending model for p_y

Different models are considered to approximate the variable corresponding to the pressure p_y. The pump on and off periods are analyzed separately for the values of p_y^+ and p_y^-, respectively.

a) *Time-dependent model in pump-off mode* $p_y^- = f(t)$

Of the models investigated, a comparison is made in particular between quadratic, power and cubic models. The different forms of prediction errors estimation MAD, MSE, MAPE, MPE and R^2 are calculated. These forms are used as the criterion for model selection.

Mean Absolute Deviation is calculated as follows

$$MAD = \frac{\left|p_{y_i} - \widehat{p_y}_i\right|}{n},$$

where p_{y_i} are the observed values of the dependent variable, $\widehat{p_y}_i$ are the predicted values from the model, n is the number of observations.

Mean Squared Error is calculated as shown

$$MSE = \frac{(p_{y_i} - \widehat{p_y}_i)^2}{n}.$$

Mean Absolute Percentage Error is calculated by the formula

$$MAPE = \frac{1}{n}\left|\frac{p_{y_i} - \widehat{p_y}_i}{p_{y_i}}\right|.$$

Mean Percentage Error is calculated as follows

$$MPE = \frac{1}{n}\frac{(p_{y_i} - \widehat{p_y}_i)}{p_{y_i}}.$$

Adjusted R^2 is calculated by the formula

$$Adjusted\ R^2 = \frac{(p_{y_i} - \widehat{p_y}_i)^2/(n-k)}{(p_{y_i} - \overline{p_y})^2/(n-1)},$$

where k is the number of parameters, $\overline{p_y}$ is the mean value of the observed values of the dependent variable.

All the three models are adequate according to the ANOVA analysis done for model adequacy. Table 8 shows the results for the first period when the hydrophore is operated in the emptying mode and Figure 6 shows their plots.

Table 8. Errors estimation for two periods of pump-off mode.

$\bar{p_y}_{(1)}$	Quadratic	Power	Cubic
MAD	0.0497	0.1206	0.0155
MSE	0.0034	0.0435	0.0003
MAPE	3.95%	8.10%	1.26%
MPE	-0.023%	-0.498%	0.000%
R^2	0.988	0.933	0.999

$\bar{p_y}_{(2)}$	Quadratic	Power	Cubic
MAD	0.0518	0.1196	0.0178
MSE	0.0038	0.0424	0.0004
MAPE	4.07%	8.00%	1.41%
MPE	-0.027%	-0.482%	0.000%
R^2	0.987	0.935	0.999

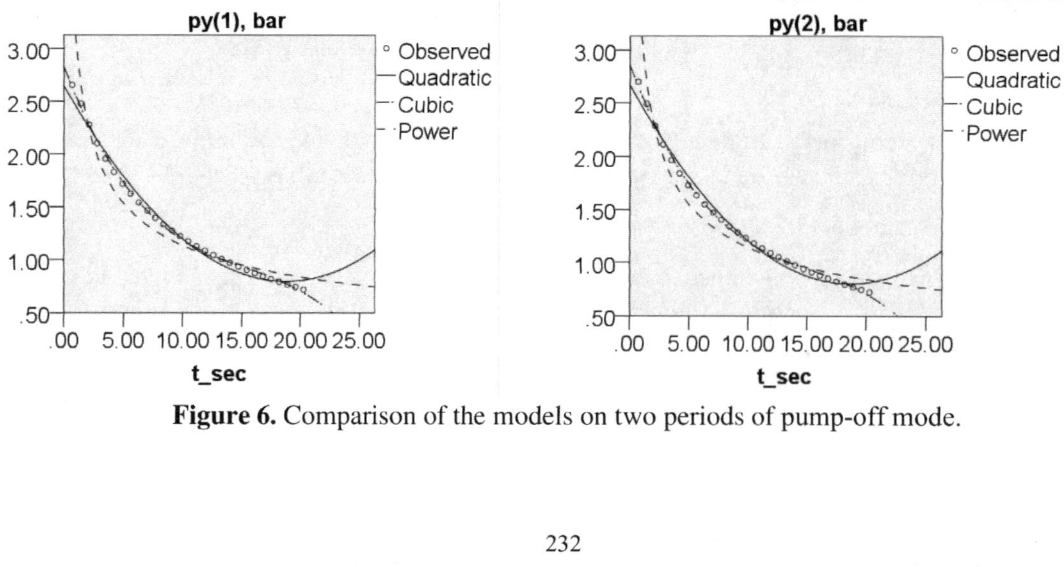

Figure 6. Comparison of the models on two periods of pump-off mode.

From Table 8 it can be seen that the cubic model is preferable in all the criteria considered. Therefore, we dwell on this model with coefficients presented in Table 9. The last column of Table 9 shows that all coefficients from the model are significant and we obtain the following relationships:

$$p_{y(1)}^{-} = -0.000354 * t^3 + 0.016411 * t^2 - 0.292477 * t + 2.825615,$$

$$p_{y(2)}^{-} = -0.000373 * t^3 + 0.017089 * t^2 - 0.300164 * t + 2.857535.$$

Table 9. Coefficients of the cubic models for the pressure for the two periods of pump-off mode.

$p_{(1)}^{-}$	Unstandardized Coefficients B	Unstandardized Coefficients Std. Error	t	Sig.
t	-0.292477	0.007	-44.244	0.000
t^2	0.016411	0.016	22.638	0.000
t^3	-0.000354	0.000	-15.585	0.000
Const	2.825615	0.016	173.297	0.000

$p_{(2)}^{-}$	Unstandardized Coefficients B	Unstandardized Coefficients Std. Error	t	Sig.
t	-0.300164	0.008	-39.478	0.000
t^2	0.017089	0.001	20.496	0.000
t^3	-0.000373	0.000	-14.260	0.000
Const	2.857535	0.019	152.373	0.000

b) Time-dependent model in pump-on mode $p_y^+ = f(t)$

Similar considerations are made for the filling mode pressure values.

Table 10. Errors estimation for two periods of pump-on mode.

$p_{y(1)}^{+}$	Quadratic	Power	Cubic
MAD	0.1056	0.1934	0.0269
MSE	0.0140	0.0493	0.0010
MAPE	5.37%	8.50%	1.46%
MPE	-0.939%	-0.46%	0.019%
R^2	0.96	0.921	0.997

$p_{y(2)}^{+}$	Quadratic	Power	Cubic
MAD	0.1306	0.2084	0.0417
MSE	0.0228	0.0562	0.0023
MAPE	6.62%	9.05%	1.84%
MPE	-1.279%	-0.54%	-0.12%
R^2	0.923	0.886	0.992

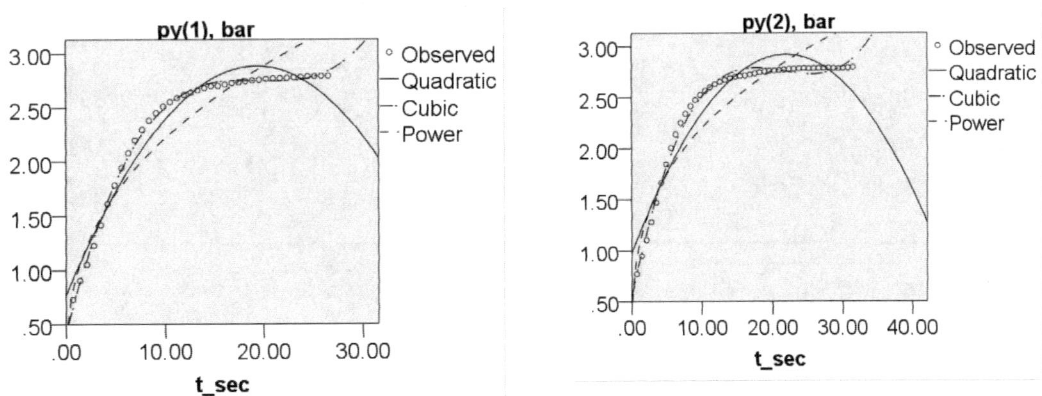

Figure 7. Comparison of the models on two periods of pump-on mode.

Table 10 gives the error values for the first and second periods with the pump on, and Figure 7 shows their plots. Both confirm the cubic model as the best approximation (see Table 11).

$$p_{y(1)}^{+} = 0.000322 * t^3 - 0.018883 * t^2 + 0.365497 * t + 0.413686,$$

$$p_{y(2)}^{+} = 0.000243 * t^3 - 0.015783 * t^2 + 0.330269 * t + 0.543737.$$

Table 11. Coefficients of the cubic models for the pressure for the two periods of pump-on mode.

$p_{y(1)}^{+}$	Unstandardized Coefficients B	Unstandardized Coefficients Std. Error	t	Sig.
t	0.365497	0.005	48.516	0.000
t^2	-0.018883	0.000	-29.653	0.000
t^3	0.000322	0.000	21.005	0.000
Const	0.413686	0.024	17.194	0.000

$p_{y(2)}^{+}$	Unstandardized Coefficients B	Unstandardized Coefficients Std. Error	t	Sig.
t	0.330269	0.009	38.060	0.000
t^2	-0.015783	0.001	-25.353	0.000
t^3	0.000243	0.000	19.110	0.000
Const	0.543737	0.033	16.668	0.000

For the model selection of the two periods at the two modes with pump on and pump off, respectively, the MSE and MAPE errors are calculated. It turns out that the prediction of $p_{(1)}^{-}$ is closer to the actual values of $p_{(2)}^{-}$. Similarly, the prediction of $p_{(1)}^{+}$ deviates less from the actual values of $p_{(2)}^{+}$. Taking these results into account, the following models are finally selected

$$p_y^{-} = p_{y(1)}^{-} = -0.000354 * t^3 + 0.016411 * t^2 - 0.292477 * t + 2.825615, \quad (4)$$

$$p_y^{+} = p_{y(1)}^{+} = 0.000322 * t^3 - 0.018883 * t^2 + 0.365497 * t + 0.413686\,. \quad (5)$$

In order to be able to determine the time periods of the operating modes when setting specific limit pressures, the inverse functions $t = f(p_y)$ are also found.

$$t^{-} = 15.4529 + 0.00094162\left(919.719\sqrt{3.38353\times10^{18}(p_y^{-})^2 - 6.21584\times10^{18}p_y^{-} + 2.93798\times10^{18}} - 1.69177\times10^{12}p_y^{-} + 1.55396\times10^{12}\right)^{\frac{1}{3}} - \frac{38879.1}{\left(919.719\sqrt{3.38353\times10^{18}(p_y^{-})^2-6.21584\times10^{18}p_y^{-}+2.93798\times10^{18}}-1.69177\times10^{12}p_y^{-}+1.55396\times10^{12}\right)^{\frac{1}{3}}}, \quad (6)$$

$$t^{+} = 19.5476 - 0.00294396\left(174.439\sqrt{1.21716\times10^{17}(p_y^{+})^2 - 6.68959\times10^{17}p_y^{+} + 9.19159\times10^{17}} - 6.0858\times10^{10}p_y^{+} + 1.6724\times10^{11}\right)^{\frac{1}{3}} - \frac{1273.16}{\left(174.439\sqrt{1.21716\times10^{17}(p_y^{+})^2-6.68959\times10^{17}p_y^{+}+9.19159\times10^{17}}-6.0858\times10^{10}p_y^{+}+1.6724\times10^{11}\right)^{\frac{1}{3}}}. \quad (7)$$

From formulas (2) and (4), the flow rate versus time relationship in emptying mode is obtained

$$Q^{-} = -1.0065710^{-10}\cdot t^9 + 1.3999\cdot10^{-8}\cdot t^8 - 8.98467\cdot10^{-7}\cdot t^7 + 0.0000337978\cdot t^6 - 0.00080139\cdot t^5 + 0.0119779\cdot t^4 - 0.10752\cdot t^3 + 0.525155\cdot t^2 - 1.60858\cdot t + 9.97327. \quad (8)$$

And from formulas (3) and (5) the flow rate versus time relationship in filling mode is obtained

$$Q^{+} = 6.3534 \cdot 10^{-11} \cdot t^{9} - 1.11775 \cdot 10^{-8} \cdot t^{8} + 8.71827 \cdot 10^{-7} \cdot t^{7} - 0.0000392386 \cdot t^{6} + 0.00111285 \cdot t^{5} - 0.0204007 \cdot t^{4} + 0.239106 \cdot t^{3} - 1.72276 \cdot t^{2} + 7.13865 \cdot t - 5.16092. \tag{9}$$

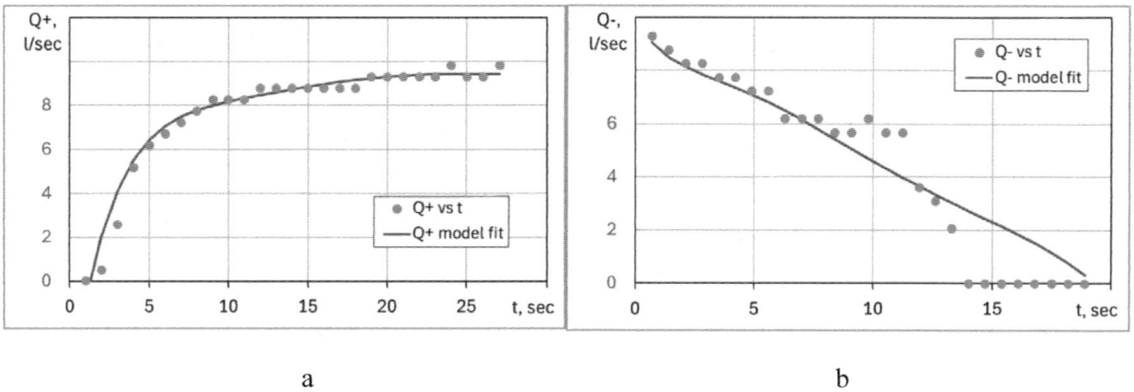

Figure 8. Graphs on the empirical dependency of flow rate versus time: (a) in pump on mode, (b) in pump off mode.

Figure 8 shows the plots of the above two equations of flow rate versus time, with the plots of the two filling and emptying regimes shown respectively on Figure 8a and Figure 8b.

6. Methodology for Constructing the Cyclic Mode and Determining the Operating Parameters

The models created in section 5 describe the processes occurring during the filling and emptying modes of the hydrophore when the installation is operating without changing the position of valve K2 and the characteristics of the pipelines (Figure 1). Under these conditions, the processes remain the same for other boundary control pressures p_{y1} and p_{y2}. Changing p_{y1} and p_{y2} leads to different cyclic operating modes and other parameters characterizing these cycles.

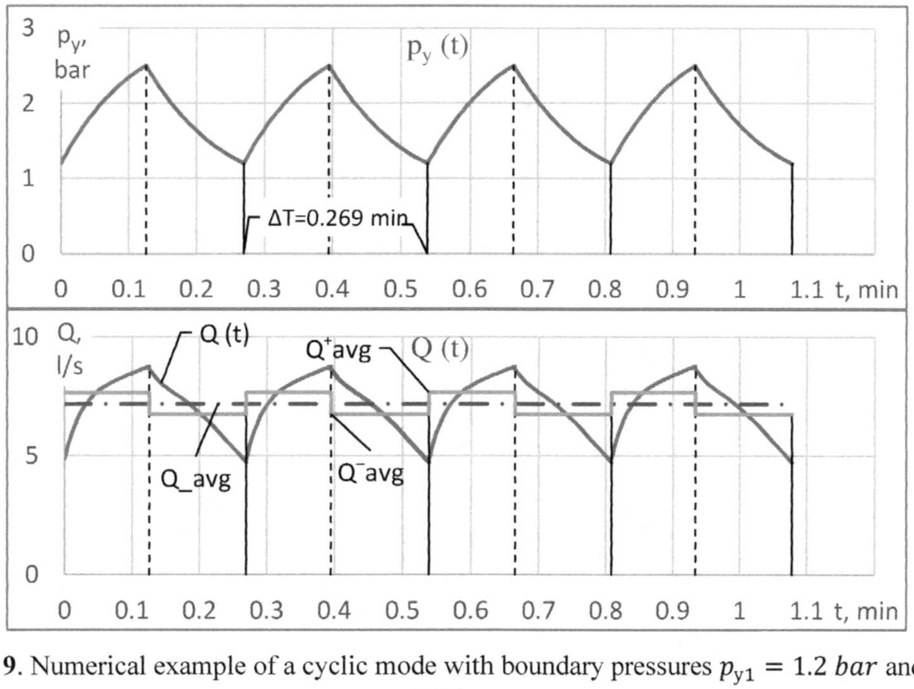

Figure 9. Numerical example of a cyclic mode with boundary pressures $p_{y1} = 1.2\ bar$ and $p_{y2} = 2.5\ bar$.

For constructing these modes and determining their parameters, the following methodology is developed:

- *Set the boundary control pressures* p_{y1} *and* p_{y2}.
- *Determine the times for the filling mode:* Using equation (7) for the inverse function of the model $p_y^+ = f(t)$ with p_{y1} and p_{y2}, determine the respective times for the start t_{y1}^+ and the end t_{y2}^+ of the filling mode of the hydrophore.
- *Determine the times for the emptying mode*: Similarly, using equation (6) for the inverse function $p_y^- = f(t)$, determine the respective times t_{y1}^- and t_{y2}^- for the emptying mode.
- *Describe the cyclic mode for the pressure variation* p_y*:* Using equation (5), construct the model for filling the hydrophore $p_y^+ = f(t)$ in the interval $t_{y1}^+ \le t_y \le t_{y2}^+$, and using equation (4), construct the model for emptying $p_y^- = f(t)$ in the interval $t_{y1}^- \le t_y \le t_{y2}^-$. In this cyclic mode, the two processes alternate sequentially, starting with the filling model and initial time $t_0 = 0$.
- *Describe the cyclic mode for the flow rate variation* Q: Use equation (9) to construct the model $Q^+ = f(t)$ in the interval $t_{y1}^+ \le t_y \le t_{y2}^+$ and equation (8) for $Q^- = f(t)$ in the interval $t_{y1}^- \le t_y \le t_{y2}^-$.
- *Determine the periods for filling and emptying:* The filling period is calculated as

$$\Delta T^+ = t_{y2}^+ - t_{y1}^+,$$

and the emptying period as

$$\Delta T^- = t_{y2}^- - t_{y1}^-.$$

- *Calculate the period of the cyclic mode:* The period of the cyclic mode is

$$\Delta T = \Delta T^+ + \Delta T^-.$$

- *Calculate the number of cycles per hour:* Using the equation [4]

$$N_{cycles} = \frac{3600}{\Delta T}, \frac{\text{cycles}}{\text{hour}}.$$

- *Determine the minimum and maximum flow rates* Q_{min} *and* Q_{max}*:* Using equation (9) with the times at the start t_{y1}^+ and end t_{y2}^+ of the filling process, calculate

$$Q_{min} = Q^+(t_{y1}^+)$$

and

$$Q_{min} = Q^+(t_{y1}^+).$$

- *Calculate the average flow rate during filling:* The average flow rate during filling $\bar{Q}^+$ is calculated by numerical integration of the model $Q^+ = f(t)$ (equation 9) in the interval $t_{y1}^+ \le t_y \le t_{y2}^+$ [5, 6, 7]. For higher accuracy, the trapezoidal method is used [8]:

$$\bar{Q}^+ = \int_{t_{y1}^+}^{t_{y2}^+} Q^+(t)\,dt \approx \frac{1}{\Delta T^+}\sum_{i=1}^{n}\left(\frac{Q^+(t_i)+Q^+(t_{i-1})}{2}\right)\Delta t,$$

where $\Delta t = \frac{\Delta T^+}{n}$.

- *Calculate the average flow rate during emptying:* The average flow rate during emptying $\bar{Q}^-$ is calculated similarly using equation (8) for the model $Q^- = f(t)$ in the interval $t_{y1}^- \le t_y \le t_{y2}^-$:

$$\bar{Q}^- = \int_{t_{y1}^-}^{t_{y2}^-} Q^-(t)\,dt \approx \frac{1}{\Delta T^-}\sum_{i=1}^{n}\left(\frac{Q^-(t_i)+Q^-(t_{i-1})}{2}\right)\Delta t.$$

– *Calculate the overall average flow rate* $\bar{Q}$*:* The overall average flow rate of the hydrophore installation $\bar{Q}$ in cyclic mode is calculated as the weighted average [4] of the two modes (filling and emptying):

$$\bar{Q} = \frac{\bar{Q}^{+}\Delta T^{+} + \bar{Q}^{-}\Delta T^{-}}{\Delta T}.$$

– *Calculate the coefficient of variation (CV) of the flow rate:* The coefficient of variation is calculated using the formula [4]:

$$CV = \frac{Q_{max} - Q_{min}}{\bar{Q}} 100\%.$$

Using this methodology, a numerical example is solved with different boundary pressures $p_{y1} = 1.2\ bar$ and $p_{y2} = 2.5\ bar$. The cyclic mode of operation is constructed and presented in Figure 9. The parameters characterizing the mode are determined as follows: $\Delta T^{+} = 7.52\ s$; $\Delta T^{-} = 8.65\ s$; $\Delta T = 16.17\ s$; $N_{cycles} = 222.6\ cycles/hour$; $Q_{min} = 4.86\ l/s$; $Q_{max} = 8.75\ l/s$; $\bar{Q}^{+} = 7.65\ l/s$; $\bar{Q}^{-} = 6.75\ l/s$; $\bar{Q} = 7.17\ l/s$; $CV = 54.4\%$.

7. Conclusions

As a result of this work, a mathematical model has been developed to describe the cyclic mode of operation of the hydrophore plant. The model includes empirical relationships for the water supply flow rate and the pressure in the hydrophore as functions of time, for both filling mode (pump on) and emptying mode (pump off) operation. The empirical relationships were obtained based on experimental data and using regression analysis. With the help of these models, a methodology for the determination of the main parameters characterizing the cyclic mode of operation has been compiled.

The work provides the following advantages and opportunities for future research:

- By selecting different governing boundary pressures and using the developed models and methodology, other cyclic regimes of the system can be constructed and their parameters can be determined. This can help to identify the optimum parameters under different operating conditions and in different water supply systems.
- Specifically, optimization may include:
  - Pressure limit optimization: reducing pump wear and increasing energy efficiency by optimizing pressure limits.
  - Cycle period extension: Reducing the number of on-off cycles to extend pump and motor life.
  - Flow control: Ensuring stability in flow rate and reducing variation to achieve better system efficiency.
  - Selection of optimum parameters in terms of system energy efficiency.
- The development of such models can be integrated into control systems to dynamically change the control limit pressures in order to obtain optimal cycling modes according to the current demand and required water supply flow.
- With the obtained models and methodology, programs for energy-saving consumption can be compiled. It is an algorithm for finding the necessary values of the regulated quantities, at which the efficiency of the control object will be maximum while maintaining the necessary parameters of the process [9].

Acknowledgments

This paper contains results of the work on project No 2024 - FNSE – 02, financed by „Scientific Research" Fund of Ruse University".

Credit roles: conceptualization, I.N.; data curation, E.R. and V.M.; modelling physical processes, I.N., mathematical modelling V.M. and E.R., funding acquisition, V.M.; methodology, I.N., E.R. and V.M.; project administration, E.R.; software, E.R. and V.M.; supervision, I.N.; validation, V.M.;

Journal of Physics: Conference Series **2910** (2024) 012019 doi:10.1088/1742-6596/2910/1/012019

visualization, I.N. and E.R.; writing-original draft, E.R. and I.N.; writing-review and editing, V.M. All authors have read and agreed to the published version of the manuscript.

References

[1] Husein J, Yankov Y, Nikolaev I 2019 *Automated Measuring System for Recording Hydraulic Magnitudes* (*Proceedings of the 58-th Annual Scientific Conference of University of Ruse „Angel Kanchev“ and Union of Scientists – Ruse*) (Ruse: PUBLISHING HOUSE University of Ruse "Angel Kanchev") pp. 40-47
[2] Pavlov V, Mihova V, Raeva E 2023 *Statistics with SPSS* (Ruse: Avangard Print, in Bulgarian)
[3] SPSS IBM Statistics <http://www-01.ibm.com/software/analytics/spss/>
[4] Montgomery D C and Runger G C 2018 *Applied Statistics and Probability for Engineers* (John wiley & sons)
[5] White F M 2016 *Fluid Mechanics* (New York: McGraw-Hill Education) Chapter. 6
[6] Fischer H B, List E J, Koh R C Y, Imberger J & Brooks N H 1979 *Mixing in Inland and Coastal Waters* (New York: Academic Press)
[7] Chaudhry M H 2013 *Applied Hydraulic Engineering* (New York: Springer)
[8] Chapra S C and Canale R P 2015 *Numerical Methods for Engineers* (New York: McGraw-Hill Education)
[9] Artiukh S, Kokalarov M, Kniazieva V, Georgiev I, Artiukh A, Milchev M 2023 *Metrological Aspects in Maintenance of Energy-Saving Automated Systems for Control of Pump Units in Pumping Stations* (*Proceedings of the 34th DAAAM International Symposium on Intelligent Manufacturing and Automation*) ed B Katalinic (Vienna: DAAAM International) pp. 0268 - 0273

On the Power Series Method for Solitons and Peakons of Nonlinear PDE

Erika Olivares[1] and Vesselin Vatchev[1]

[1]The University of Texas Rio Grande Valley, One West University Blvd. Brownsville, Texas 78520, U.S.

E-mail: vesselin.vatchev@utrgv.edu, erika.olivares01@utrgv.edu

Abstract. In this paper we discuss the power series method to obtain traveling wave solutions of nonlinear PDE in one space variable. The method, including a recursive procedure for the coefficients and error estimate for the partial sums, is presented for the Camassa-Holm equation. Numerical examples of approximate single soliton and single peakon solutions are included.

1 Introduction

In this paper we consider the classical power series method for finding traveling wave solutions of certain class of nonlinear partial differential equations(PDE) in one space variable. The equations are evolution type and can be described by considering a polynomial in two variables

$$Q(y,z) = \sum_{i=0}^{N} c_j y^i + \sum_{i,j=0}^{N} a_{i,j} y^i z^j$$

with real coefficients $c_j, a_{i,j}$ and then substituting both powers y^i and z^j by derivatives of the the type $\frac{\partial^m}{\partial x^m \partial t^{m-i}} u$ of a function $u(x,t)$, and setting the resulting expression to 0. Our main example will be the Camassa-Holm equation(CH), [1],

$$u_t + 2ku_x - u_{xxt} + 3uu_x - 2u_x u_{xx} - uu_{xxx} = 0.$$

It is a well known model for propagation of water waves in shallow water traveling in a narrow channel with a flat bottom. The equation can be derived from the Euler equations in fluid mechanics and it is also shown to model gas dynamics in certain situations.

In the case when we require the relation $u(x,t) = u(x-ct)$ between the time and space variables we consider traveling wave solutions. Traveling waves are important solutions of nonlinear PDE not only for their physical significance but also for establishing theoretical properties of the underlying PDE. Solitonic type traveling solutions are defined as a lump traveling without changing its shape in time and space. The solitonic waves are expected to be smooth functions. We also consider peakon solutions which have discontinues at their peaks.

For many classical nonlinear PDE the existence of solitonic type waves is known and in many cases they can be found explicitly. Unfortunately increasing the order of the equations it becomes quite complicated even impossible to find explicit solitonic waves.

The purpose of this note is to study the possibilities of finding those type of solutions in the form of power series, where the coefficients are recursively defined.

In section 2 we introduce and discuss the method on CH equation, providing the recurrence relations and the error analysis. We demonstrate the method on an explicit soliton solution to CH and discuss a composite approximation in section 3. In section 4 we discuss applications in approximating exact traveling waves by compactly supported traveling type functions and provide several examples for solitons and peakons. Section 5 contains conclusions and further research directions.

Content from this work may be used under the terms of the Creative Commons Attribution 4.0 licence. Any further distribution of this work must maintain attribution to the author(s) and the title of the work, journal citation and DOI.
Published under licence by IOP Publishing Ltd

2 The Power Series Method for Camassa-Holm Equation

The Camassa-Holm equation(CH) was introduced in [1] and reads,

$$u_t + 2ku_x - u_{xxt} + 3uu_x - 2u_x u_{xx} - uu_{xxx} = 0, \tag{1}$$

where k is an environmental real constant. It is a well known model for propagation of water waves in shallow water traveling in a narrow channel with a flat bottom. Integrability of the equation, Lax pair, was presented in [1], [7], and an implicit single and multi-soliton solutions were studied in [5], [3],[4], [7]. CH has a very interesting peakon type solution, [5], [8], for $k = 0$, which suggests the possibility of breaking of water waves.

We will look for traveling wave type solutions in the form

$$u(x,t) = \sum_{n=0}^{\infty} a_n (x - ct - \alpha)^n,$$

where c is the speed of translation, α is a real displacement and $a_n, n = 0, 1, \ldots$ are unknown coefficients. Setting $y = x - ct$, the above series is a Taylor series of $u(y)$ in $(y-\alpha)^n, n = 0, 1, \ldots$, centered at α. The series is expected to converge in an interval $y \in [\alpha - \beta, \alpha + \beta]$, for some real β. We will show that $\beta > 0$.

Straightforward computations lead to

$$\begin{aligned}
u(y) &= \sum_{n=0}^{\infty} a_n y^n, \\
u_t(y) &= \sum_{n=0}^{\infty} (-c)(n+1)a_{n+1} y^n, \\
u_x(y) &= \sum_{n=0}^{\infty} (n+1)a_{n+1} y^n, \\
u_{xx}(y) &= \sum_{n=2}^{\infty} (n+2)(n+1)a_{n+2} y^n, \\
u_{xxx}(y) &= \sum_{n=3}^{\infty} (n+3)(n+2)(n+1)a_{n+3} y^n, \\
u_{xxt}(y) &= \sum_{n=3}^{\infty} (-c)(n+3)(n+2)(n+1)a_{n+3} y^n.
\end{aligned}$$

The product terms in (1) become

$$\begin{aligned}
u(y)u_x(y) &= \sum_{n=0}^{\infty} B_n y^n, B_n = \sum_{i=0}^{n} (i+1)a_{i+1}a_{n-i} \\
u_x(y)u_{xx}(y) &= \sum_{n=0}^{\infty} C_n y^n, C_n = \sum_{i=0}^{n} (i+2)(i+1)(n-i+1)a_{i+2}a_{n-i+1} \\
u(y)u_{xxx}(y) &= \sum_{n=0}^{\infty} D_n y^n, D_n = \sum_{i=0}^{n} (i+3)(i+2)(i+1)a_{i+3}a_{n-i}.
\end{aligned}$$

By substituting in (1) and collecting the coefficients of $y^n, n = 0, 1, \ldots$ we get the equation

$$(2k-c)(n+1)a_{n+1} + c(n+3)(n+2)(n+1)a_{n+3} + 3B_n - 2C_n - D_n = 0$$

for $a_n, n = 0, 1, \ldots$. Let a_0, a_1, a_2 be free parameters and c and k given, then solving the above equation for the coefficients with the highest index, $n+3$, we get

$$\begin{aligned}
a_{n+3} = \frac{1}{(n+3)(n+2)(n+1)(c-a_0)} \Bigg(& (c-2k)(n+1)a_{n+1} + -3\sum_{i=0}^{n} (i+1)a_{i+1}a_{n-i} \\
& + 2\sum_{i=0}^{n} (i+2)(i+1)(n-i+1)a_{i+2}a_{n-i+1} + \sum_{i=0}^{n-1} (i+3)(i+2)(i+1)a_{i+3}a_{n-i} \Bigg).
\end{aligned} \tag{2}$$

The above relation is a quadratic recurrence for a_n with initial values a_0, a_1, a_2. In general, a quadratic recurrence cannot be solved in a closed form but it is a convenient and fast way to compute large number of terms. The first two terms, a_3, for $n=0$ and a_4, for $n=1$ are

$$a_3 = \frac{1}{6(c-a_0)}\Big((c-2k)a_1 - 3a_0a_1 + 4a_1a_2\Big), \quad (3)$$

$$a_4 = \frac{1}{24(c-a_0)}\Big(2(c-2k)a_2 - 3(a_1^2+2a_0a_2) + 2(4a_2^2+6a_1a_3) + 6a_1a_3\Big). \quad (4)$$

For appropriate initial values we can establish the rate of convergence of the sequence a_n and thus the interval of existence of the exact solution u_α. In the next theorem we provide that and the approximation rate of the partial sum

$$u_{\alpha,N}(y) = \sum_{i=0}^{N} a_i(y-\alpha)^i,$$

to the exact solution u_α as a function of y.

Theorem 2.1 *Let c and k be any real numbers and $0<b<1$ be such that $|a_0| \le b, |a_1| \le b, |a_2| \le \frac{b}{4}$, and*

$$\max(16|b|, |c-2k|) \le \frac{1}{3}|c-a_0|,$$

then for any natural n,

$$|a_n| \le \frac{b}{n^2}$$

and

$$|u_\alpha(y) - u_{\alpha,N}(y)| \le \frac{b}{1-\gamma}\frac{\gamma^{N+1}}{(N+1)^2},$$

for $|y-\alpha| \le \gamma < 1$.

Proof: The proof for the rate of convergence of a_n is done by induction. The initial coefficients a_0, a_1, a_2, set as above, satisfy the inductive assumption. Assuming $|a_{n+2}| \le \frac{b}{(n+2)^2}$ for $n \ge 0$ then from (2), we have the inequality

$$|a_{n+3}| \le \frac{b^2}{(n+3)(n+2)(n+1)|c-a_0|}\Bigg(\frac{|c-2k|}{b(n+1)} + 3\frac{|a_0|}{n+1}$$
$$+3\sum_{i=0}^{n-1}\frac{1}{(i+1)(n-i)^2} + 2\sum_{i=0}^{n}\frac{i+1}{(i+2)(n-i+1)} + \sum_{i=0}^{n-1}\frac{(i+1)(i+2)}{(i+3)(n-i)^2}\Bigg).$$

From the assumptions on b and a_0, it follows that the sum of the first two terms is less than $\frac{b}{2(n+3)^2}$. For the three sums, after changing the summation index, we have the following estimates

$$S_1 = \sum_{i=0}^{n-1}\frac{1}{(i+1)(n-i)^2} \le \sum_{i=1}^{n}\frac{1}{n^2} \le 1$$

$$S_2 = \sum_{i=0}^{n}\frac{i+1}{(i+2)(n-i+1)} \le \sum_{i=1}^{n}\frac{1}{n} \le 2\ln n < 2n$$

$$S_3 = \sum_{i=0}^{n-1}\frac{(i+1)(i+2)}{(i+3)(n-i)^2} \le \sum_{i=1}^{n}\frac{n-m+1}{m^2} \le 2(n+1) - \ln n \le 2(n+1),$$

and it follows, that

$$\frac{b^2(3S_1+2S_2+S_3)}{(n+3)(n+2)(n+1)|c-a_0|} \le \frac{b}{2(n+3)^2}.$$

Combining the two estimates we get that $a_{n+3} \le \frac{b}{(n+3)^2}$ for any $n = 0, 1, \ldots$

For the approximation with the partial Taylor polynomial $u_{\alpha,N}$ the standard arguments provide the estimate. The proof is completed.

Remark: From Theorem 2.1, it follows that $u_{\alpha,N}$ convergence to the exact solution u_α for y at least on the interval $[\alpha-0.5,\alpha+0.5]$and thus the solution can to be extended by constructing a Taylor series centered at $\alpha \pm 0.5$. An example is considered in section 3.two

The method described in this section and Theorem 2.1 allow constructing power series traveling waves solutions for the CH. In the next section we consider application to special type of solutions, solitons and peakons.

3 Composite C^2 Approximation of an Explicit Soliton for CH

Explicit soliton and peakon solutions in some special cases for CH with $k=0$ were shown in [9]. For example, if $y = x - ct$ then the function

$$U(y) = \frac{c}{11}\frac{11\cos\left(\frac{2}{3}\arccos(\operatorname{sech}\frac{|y|}{2})\right)-1}{\cos\left(\frac{2}{3}\arccos(\operatorname{sech}\frac{|y|}{2})\right)+1},$$

is a smooth soliton for CH defined for any real y with $\lim_{y\to\pm\infty} U(y) = A$, for some real A.

The method considered in section 2 constructs a power series $u_{\alpha,N}$ that converges to the exact solution u on a finite interval containg α. We construct Taylor polynomial $u_{0,N}$ on the interval $[-1.5,1.5]$ with the initial parameters a_0, a_1, a_2 defined from the explicit solution U, at $y=0$, and afterwards we construct Taylor polynomial on the adjacent interval $[1.5,4.5]$ with the initial parameters defined from $u_{0,N}$ at $y = 1.5$.

Let $N = 12$ then by setting the IC at $y = 0$ as

$$a_0 = U(0) = 0.4545, a_1 = U_x(0) = 0, a_2 = U_{xx}(0) = -0.0152,$$

and using the recurrence relation (2) we get that

$$\begin{aligned} u_{0,12}(y) =& 0.4545 - 0.0152y^2 + 9.8524\times 10^{-4}y^4 - 6.3193\times 10^{-5}y^6 \\ &+ 4.2417\times 10^{-6}y^8 - 3.0064\times 10^{-7}y^{10} + 2.2371\times 10^{-8}y^{12} \end{aligned}$$

for $y \in [-1.5,1,5]$.The numerical error of approximation is $\|su - u_{0,12}\|_{[-1.5,1.5]} = |su(1.5) - u_{0,12}(1.5)| \leq 1.4\times 10^{-4}$. Next we consider a Taylor series centered at 1.5. For the initial conditions

$$a_0 = u_{0,12}(1.5) = 0.4247, a_1 = u_{0,12,x}(1.5) = -0.0347, a_2 = u_{0,12,xx}(1.5) = -0.0056,$$

we obtain the following Taylor polynomial,

$$\begin{aligned} u_{1.5,12}(y) =& 0.4247 - 0.0347(y-1.5) - 0.0056(y-1.5)^2 + 0.0030(y-1.5)^3 - 1.5595\times 10^{-4}(y-1.5)^4 \\ &- 1.3446\times 10^{-4}(y-1.5)^5 + 3.1006\times 10^{-5}(y-1.5)^5 + 1.7429\times 10^{-6}(y-1.5)^6 \\ &- 2.0173\times 10^{-6}(y-1.5)^7 + 2.8681\times 10^{-7}(y-1.5)^8 + 5.9443\times 10^{-8}(y-1.5)^9 \\ &- 2.8861\times 10^{-8}(y-1.5)^{10} + 2.3704\times 10^{-9}(y-1.5)^{11} + 1.2777\times 10^{-9}(y-1.5)^{12}, \end{aligned}$$

for $y \in [1.5,4,5]$ with numerical error of approximation $\|su - u_{1.5,12}\|_{[1.5,4.5]} = |su(4.5) - u_{0,12}(4.5)| \leq 0.0011$.

Figure 1 is a visualization of the process. It is clear that $u_{0,12}$ approximates U on a symmetric interval centered at zero. The error of approximation can be estimated form Theorem 2.1 but in this case we empiracaly pick the left end of the interval as 1.5. The initial conditions for $u_{1.5,12}$ are the approximated values to U but still the numerical error is small on the large interval $[-1.5,4.5]$. By either increasing the degree of u_N, taking smaller interval, or both the approximation rate can be improved. We can also continue the piece-wise approximation along the horizontal axes and obtain a C^2 approximation to U on larger interval.

The composed approximation presented above suggests that one can construct piece-wise smooth approximation on large intervals but further investigation is needed. Another interesting direction is to consider Christov functions,[2] in lieu of the power functions. The Christov functions are smooth complete orthogonal basis in $L_2(-\infty,\infty)$ and their derivatives have a recursive expression with a lower order derivatives which could provide the corresponding u_N to approximate U on the entire real line.

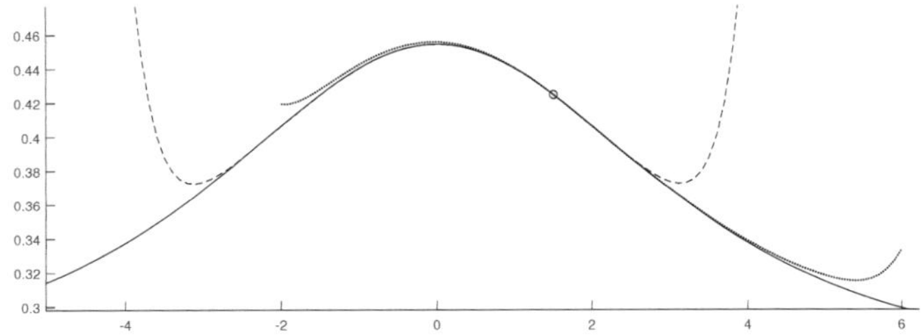

Figure 1: Two-piece approximation to the actual soliton solution U. The continuous line is the graph of U, the dashed line is the graph of $u_{0,12}$ and the dotted is $u_{1.5,12}$. The circle is the point with coordinates $(1.5, u_{0,12}(1.5))$.

4 Traveling Waves Approximation for CH

The method described in section 2 constructs recursively an exact solution (on the interval of convergence) to (1). Since there is no explicit formula for a_n we cannot expect to have explicit solutions but rather to adopt $u_{\alpha,N}$. Since all a_n are functions of c,k and the initial conditions a_0,a_1,a_2 we have 2 plus 3 degrees of freedom to model $u_{\alpha,N}$. Recalling the CH equation,

$$u_{xxx} = \frac{2c+3u-2u_{xx}}{u-c}u_x,$$

and adding the IC at a real point b,

$$u(b) = u_0, u_x(b) = u_1, u_{xx}(b) = u_2,$$

we see that the IVP has a unique solution if $u_0 < |c|$.

For the initial conditions $u_0 = u_1 = u_2 = 0$ the CH has only the trivial solution $u(x,t) = 0$. In this section we will consider u_N such that $u_N(b) = u'_N(b) = 0$ and extend beyond $y > b$ by the trivial solution $u = 0$. The resulting functions will be approximated solutions to CH with discontinuity in the second derivative at b. We also consider choices of the initial conditions as well as c and k that provide $u_N^{(k)}(b) = 0$ for $k = 0,1,2$. The resulting functions are C^2smooth on the whole real line and are approximation to an exact solution of CH. We will set $\alpha = 0$ in all of the following examples.

4.1 *Approximate Compactly Supported C^1 Soliton*

For solitons, we consider only even powers y^{2n}, and hence u_N will be even function in $y = x - ct$. Let c and k be fixed then for

$$u_4(y) = a_0 + a_2y^2 + a_4y^4,$$

by setting $a_0 = sb^2, a_2 = -2b^2$ and $a_4 = s$ we see that for those choice of parameters $u_4(y) = s(y^2-b^2)^2$. By using the recursive expressions (3) and (4), for a_4 we get the equation

$$\frac{32b^4s^2 + 12b^6s^2 - 2b^2s(2c-4k)}{24(sb^4-c)} = s$$

for s and b. Setting $b = 3/4$ we get $s = 1.3221c - 0.2266k$. From Theorem 2.1. it follows that the error of approximation by u_4 is

$$|u(y) - u_4(y)| \leq \frac{|1.3221c - 0.2266k|}{6.25}\left(\frac{3}{4}\right)^9.$$

For $b = 3/4, s = 1.0955, c = 1, k = 1$ and $a_0 = 0.3466, a_2 = -1.2324, a_4 = 1.0955$, the C^1 approximate soliton is visualized in Figure 2, left part. From Theorem 2.1, in this case the error of approximatation can be estimated by $|u - u_4| \leq 0.0132$.

4.2 *Approximate Compactly Supported C^2 Soliton*

An interesting case is when the polynomial u_6 satisfies the initial conditions $u^{(k)}(\pm b) = 0$ for $k = 0,1,2$ i.e. the solution has discontinuities in the third derivative (jolt, jerk) at the end points $\pm b$.

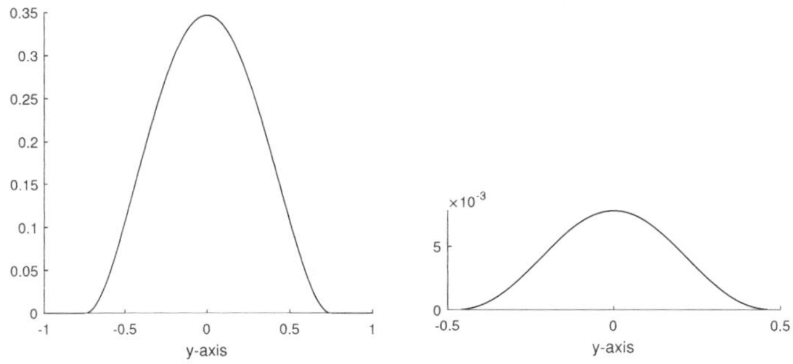

Figure 2: Profile of a C^1, left, and C^2, right, smooth compactly supported approximate solitons

We can pick the parameters c, k, a_0, a_2 in such a way that u_6 has a zero of multiplicity 3 at a real point b, i.e $u_6^{(k)}(b) = 0$ for $k = 0, 1, 2$. Adding the higher terms the zero will be removed but from Theorem 2.1 for $b < 1$ we have that u_6 approximates u with error of approximation $|u(x,t) - u_6(x,t)| \leq \frac{\beta\gamma^7}{7^2(1-\gamma)}$ when $|x - ct| < \gamma$ and $0 < \beta < 1$.

In this case b and s are free parameters and

$$u_6(y) = a_0 + a_2 y^2 + a_4 y^4 + a_6 y^6 = s(y^2 - b^2)^3$$

when

$$\begin{aligned}
c &= 2.25sb^8 + 26sb^6, \\
k &= \frac{3sb^8 + 24sb^6 + cb^2 + 12c}{2b^2}, \\
a_0 &= -b^6 s, \\
a_2 &= 3b^4 s, \\
a_4 &= \frac{12b^8 s^2 + 3b^{10} s^2 + b^4 cs - 2b^4 ks}{4(sb^6 + c)}, \\
a_6 &= -0.15s \frac{3sb^8 + 28sb^6 - 8c}{sb^6 + c}.
\end{aligned}$$

For $b = 0.5, s = -0.5$ then for the equation with $k = -5.4709$ has a soliton with speed $c = -0.2075$ and

$$a_0 = 0.0078, a_2 = -0.0938, a_4 = 0.3750, a_6 = -0.5000.$$

The graph is visualized in Figure 2, right part.

4.3 *Approximate Compactly Supported C^3 Peakon*

The IVP has unique solution for $u(b) \neq c$ and thus $u^{(k)}(\beta) = 0$ for $k = 0, 1, 2$ results in the trivial solution. We can construct a compactly supported peakon in the form

$$u_4(y) = a_0 + a_1 y + a_2 y^2 + a_3 y^3 + a_4 y^4 = s(y - b)^4,$$

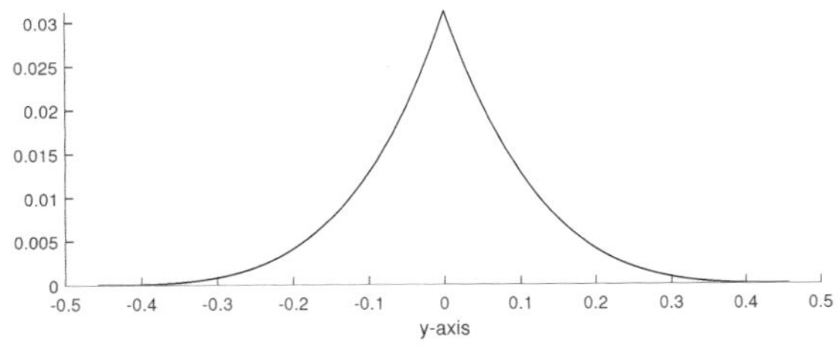

Figure 3: Profile of a C^3 piece-wise smooth compactly supported approximate peakon.

In terms of b ansd s, the formulas for the coefficients satisfying the above relation for any real y are

$$\begin{aligned} c &= sb^6 - 5sb^4, \\ k &= -\frac{3sb^6 - 30sb^4 - cb^2 + 6c}{2b^2}, \\ a_0 &= sb^4, \\ a_1 &= -4sb^3, \\ a_2 &= 6sb^2, \\ a_3 &= -\frac{2b^3 s(-3sb^4 + 24sb^2 + c - 2k)}{-3sb^4 + 3c}, \\ a_4 &= s. \end{aligned}$$

The profile of the peakon type wave for $y \geq 0$ and $b = 0.5, s = 0.5$ and for $y < 0$ and $b = -0.5, s = 0.5$ is visualized in Figure 3. The function is compactly supported on the interval $[-0.5, 0, 5]$ and is a C^3 smooth except at the origin, or the peak of the wave.

5 Discussion and Conclusion

The method described above can be readily extended to the CH Hierarchy, [7]

$$u_t + 2ku_x - u_{xxt} + (q+1)uu_x - qu_x u_{xx} - uu_{xxx} = 0$$

for integer q. For $q = 2$ we get Camassa-Holm and for $q = 3$ we get Degasperis–Procesi equations.

In the case when the nonlinear PDE is of order higher than 5 then we can construct C^2 approximate compactly supported solitons just by choosing the initial parameters which in the case will be a_0, a_2, a_4. Few examples for equations for which the power series method can be applied are the Kaup–Kupershmidt equation, [6],

$$u_t - (u_{xxxxx} + 10u_{xxx}u + 25u_{xx}u_x + 20u^2 u_x) = 0,$$

the KdV of order five, [6]

$$u_t + 6uu_x + u_{xxx} + u_{xxxxx} = 0,$$

and the Sawada-Kotera equation, [10]

$$u_t + 45u^2 u_x + 15u_x u_{xx} + 15uu_{xxx} + u_{xxxxx} = 0.$$

It is our intention to study the power series solutions in Christov complete orthogonal basis, [2].

6 Acknowledgments

Erika Olivares thanks for the doctoral GTA support from the President of the University of Texas Rio Grande Valley.

References

[1] R. Camassa and D. Holm, An integrable shallow water equation with peaked solitons, 1993, *Phys. Rev. Lett.*, 71, 1661.

[2] C. Christov, A Complete Orthonormal System of Functions in $L_2(-\infty,\infty)$ Space, 1982, *SIAM Journal on Applied Mathematics*, 42, 6, 1337-1344

[3] D. Holm and R. Ivanov, Smooth and peaked solitons of the CH equation, 2010, *J. Phys. A: Math. Theor.* , 43, 434003.

[4] R. Ivanov, T. Lyons and N. Orr, A dressing method for soliton solutions of the Camassa-Holm equation, 2017, *AIP Conf. Proc.*, 1895, 040003.

[5] A. Parker, On the Camassa-Holm equation and a direct method of solution I. Bilinear form and solitary waves, 2004, *Proc. R. Soc. London, Ser. A*, 460, 2929.

[6] A. Polyanin and V. Zaitsev, Handbook of Nonlinear Partial Differential Equations, 2004, 2-nd Ed. p 1034, CRC PRESS

[7] Z. Qiao, The Camassa-Holm hierarchy, N-dimensional integrable systems, and algebro-geometric solution on a symplectic submanifold, 2003, *Communications in Mathematical Physics*, 239, 309-341.

[8] Z. Qiao, A new integrable equation with cuspons and W/M-shape-peaks solitons, 2006, *Journal of Mathematical Physics*, 47, 112701-09.

[9] Z. Qiao and G. Zhang, On peaked and smooth solitons for the Camassa-Holm equation, 2006 *EPL*, 73, 657.

[10] K. Sawada and T. Kotera, A Method for Finding N-Soliton Solutions of the K.d.V. Equation and K.d.V.-Like Equation, 1974, *Prog. Theor. Phys.*, 51, 1355

A Review of the Influence of Global Anthropogenic Activity on the State of Biosphere

Mihai Petrov[1], Velichka Traneva[1*], Stoyan Tranev[2*] and Venelin Todorov[3,4]

[1] Department of Mathematics, Informatics and Physics, Prof. Asen Zlatarov University, 1 Prof. Yakimov Blvd., Bourgas 8010, Bulgaria
[2] Department of Industrial technology and management, Prof. Asen Zlatarov University, 1 Prof. Yakimov Blvd., Bourgas 8000, Bulgaria
[3] Institute of Mathematics and Informatics, Bulgarian Academy of Sciences, 8 Akad. G. Bonchev str., Sofia 1113, Bulgaria
[4] Institute of Information and Communication Technologies, Bulgarian Academy of Sciences, 25A Acad. Georgi Bonchev str., 1113 Sofia, Bulgaria

*E-mail: veleka13@gmail.com (V.N.T), tranev@abv.bg (S.T.), venelinltodorov@gmail.com(V.T.), mihpetrov@abv.bg (M.P.)

Abstract. The current stage of the World is characterized by a very intense anthropogenic activity with the aim of providing energy for whole of humanity. The pollutants emitted as a result of anthropogenic activity accumulate in the atmosphere and some of them have a greenhouse effect. Greenhouse gases lead to an increase in atmospheric temperature and as a result we have negative consequences on flora and fauna with frequent natural cataclysms (floods, drought, torrential rains, strong winds, spontaneous natural wildfires, etc.). The main purpose of the work is the systematization of the very recent current literature with the elaboration of the qualitative and quantitative description of the atmospheric state influenced by pollutants. It is found the following moment from this generalization that the increase in the concentration of pollutants of the atmosphere leads to the increase of the Albedo values of the unique complex Earth-Atmosphere system. The proportionality coefficient between the Albedo and concentration values contains the time interval required for combustion as well as the power of the solar flux and the specific heat of combustion of the fuels. The following fact can be found by the quantitative expression that when the fuels with a high specific heat of combustion are used, then the level of pollutants in the atmosphere is as minimal as possible with minimal negative effects on the flora and fauna.

Key words: anthropogenic activity, pollution, Albedo, specific heat of burning, carbon dioxide, green house gases

1. Introduction

The Biosphere, which is a complex global ecosystem for which the life exists, depends on a series of complex factors that ensure the stability and evolution. Atmospheric oxygen, which is the main element for ensuring the life and the complex biochemical processes of living organisms throughout the long stage of evolution, is subject to the changes in its content depending on the evolution of flora and fauna. [1] The human being, namely the modern stage, is characterized by the large-scale production of modern technologies that provide the necessities of humanity. Natural energy resources ensure the necessities of human existence. Most of the

Content from this work may be used under the terms of the Creative Commons Attribution 4.0 licence. Any further distribution of this work must maintain attribution to the author(s) and the title of the work, journal citation and DOI.
Published under licence by IOP Publishing Ltd

components of the Biosphere in the form of natural and energy deposits ensure both the energy needs of humanity and the needs for the ensuring of production. [2] Many manufactured products are followed by waste disposal both in the form of chemical waste in solid and liquid forms as well as in the form of gases. Polluting gases such as sulphur oxides, nitrogen oxides, as well as large amounts of carbon dioxide accumulate in the atmosphere, which have a greenhouse effect. These greenhouse gases lead to the increase of the global temperature of the atmosphere. The last 40-50 years are characterized with the increase of atmospheric temperature by approximately 1^{0}C. [3], [4] It is considered that this rise of the atmospheric temperature has negative secondary effects on the state of the Biosphere in general and the health of humanity. [5],[6] The vast modern anthropogenic activity leads to the result of the decipherment of the forest surfaces. [7] As a result of this process, a change of the general physico-chemical properties of the respective geographical area is observed, such both the increase of the Albedo value of the ground's soil and of the clouds due to of greenhouse gases and aerosol pollutants. [7], [8] The high Albedo values of the soil of the respective geographical area correspond to such states of the Biosphere which contains the reduced flora and fauna, extreme temperature variations, drought and floods, frequently strong winds, etc. [9], [10]
Recently, beside the mentioned natural cataclysms that are frequently repeated, the natural spontaneous wildfires are also observed that is a consequence of atmospheric pollution. The relationship of the dependence of the Albedo variation values on the variation of the concentration of polluting gases was deduced [11]. Respectively, the Albedo values of the Atmosphere-Earth system increase with the increasing of pollutant gas concentrations, and the respective increasing of fuel pollutant gases in the atmosphere leads to more probable initiation of spontaneous wildfires [11].
The fact that the temperature of the atmosphere increases with the increase of the concentration of greenhouse gases is proven by the physical relationship of the thermodynamic phenomenology that contains the adiabatic constant of the atmospheric air and that is a method of the validation of the suggested method of the study of the paper [12].
The aim of the recent work is the complex combined study of the methods described in [11] and [12] and the generalization of the quantitative description of the elaborated expression that contains the Albedo variation as the dependence of the atmospheric temperature variation.

2. Material and methods

In this study, an important parameter that is being studied and that has an enormous influence on the climate state of the various geographical regions is Albedo.
It is necessary to specify the definition of the notion of Albedo. In nature we have different substances that have very varied shades of colours and backgrounds (from the white to the black backgrounds) in the presence of sunlight. The possibility of substances in nature to reflect the light that falls on them is related to the notion of Albedo.

Albedo (from Latin albedo 'whiteness') is the fraction of sunlight that is diffusely reflected by a body. It is measured on a scale from 0 (corresponding to a black body that absorbs all incident radiation) to 1 (corresponding to a body that reflects all incident radiation). The whiter is the body then the higher is its Albedo and the value almost equals to one. For example, black soils have smallest values of Albedo (almost equals to zero).

Any Albedo in visible light falls within a range of about 0.9 for fresh snow to about 0.04 for charcoal, one of the darkest substances. Deeply shadowed cavities can achieve an effective Albedo

Journal of Physics: Conference Series **2910** (2024) 012021 doi:10.1088/1742-6596/2910/1/012021

approaching the zero of a black body. When seen from a distance, the ocean surface has a low Albedo, as do most forests, whereas desert areas have some of the highest Albedos among landforms. Most land areas are in an Albedo range of 0.1 to 0.4 [21]. The Albedo of desert sands is of the order 0.4 and just higher. [24] The average Albedo of Earth is about 0.3 [22]. This is far higher than for the ocean primarily because of the contribution of clouds.

Forests generally have a low Albedo because the majority of the ultraviolet and visible spectrum is absorbed through photosynthesis. For this reason, the greater heat absorption by trees could offset some of the carbon benefits of afforestation (or offset the negative climate impacts of deforestation). In other words: The climate change mitigation effect of carbon sequestration by forests is partially counterbalanced in that reforestation can decrease the reflection of sunlight (Albedo) [23].

Studying the origin of spontaneous wildfires with the elaboration of quantitative calculation expressions, an important analytical expression of the dependence of the Albedo variation ΔA on the variation of the concentrations of fuel polluting gases ΔC in the atmosphere was obtained in [11]. This expression is:

$$\Delta A = \frac{k.\Delta C}{t.P} \qquad (1)$$

where k is the specific heat of burning of green house fuel gases or solid fuels as dry grass, forests, etc. ; P - Sun flux; t is the time of ignition. The time of ignition is of the order 100 ms for the masses of dry solid materials (dry grass, dry leaves, woods, etc.) of the order 200 mg that is contained within the area of 1 m^2 of the soil. The values of Sun fluxes are of the order 500 – 1000 W/m^2 [11].

This expression (1) can be generalized widely and it can be considered as that when the concentration of polluting gases increases with the value ΔC, then the Albedo value of whole system Atmosphere-Earth increases with the value ΔA.

The expression (1) can be simply generalized to a linear analytical one with a proportionality coefficient Z and then this expression is represented as follows:

$$\Delta A = Z \cdot \Delta C; \;\; Z = \frac{k}{t \cdot P} \qquad (2)$$

The dependence of the variation of Albedo as the function of the concentration variation is represented on the Fig. 1. The slope tgα of the graphic from the Fig. 1 is the proportional coefficient *Z* *(Z=tgα).*
According to the known value of *Z*, one of the parameters can be calculated from the totality of all the parameters that are included inside of the propionality coefficient Z. We can calculate, for example, the value of *P* if the fixed values of *t=100 ms=0.1 s* are known, γ=1.4; T=288 K. In this way, the following fact is established, that if the value of *P* of Sun fluxes is obtained within the order of 500 - 1000 W/m^2, then the suggested developed analytical method is valid and it can be widely applied with the aim of the quantitative description of the state of the Biosphere.

Taking into consideration the fact that the current anthropogenic activity is characterized by the final result of high level of pollutant emissions in the atmosphere with the result of the increasing of the Albedo values of this complex system Earth-Atmosphere [14], then empirically it can be found that the expression (1) can really be generalized to a very general case. The use of fuels for the purpose of providing of energy for World population has resulted in the burning of fuels and the accumulation of excess heat in the atmosphere. The effect of greenhouse gases really leads to an increase of the temperature of the atmosphere with the recent final result of the increasing of natural cataclysms (droughts, floods, extreme winds, etc.) [15]. Qualitatively, the expression (1) can be explained by a general modality. The parameter *k* can generally be considered as the specific heat of combustion of fuels with the purpose of anthropogenic activity. As a result of this process of the using fuels, we have an increase in the concentration of the atmosphere by the value ΔC, due to of the increase of pollutants. As a result of the deforestation of massive geographical areas, we have the result of the increasing the Albedo values for those large geographical regions. There is a direct proportionality between the variations of Albedo values and the variations of concentrations described by the expression (1). The proportionality coefficient Z from the expression (2) contains these very important parameters such as *k, P, t.* It is possible to explain at the empirical-qualitative level the phenomenon of the emerging of natural spontaneous wildfires described in [11] by this expression (1).

Taking into consideration the expression (2), then the following moment can be observed, that when we have high values of P, then we will have lower values for the ignition time. From a

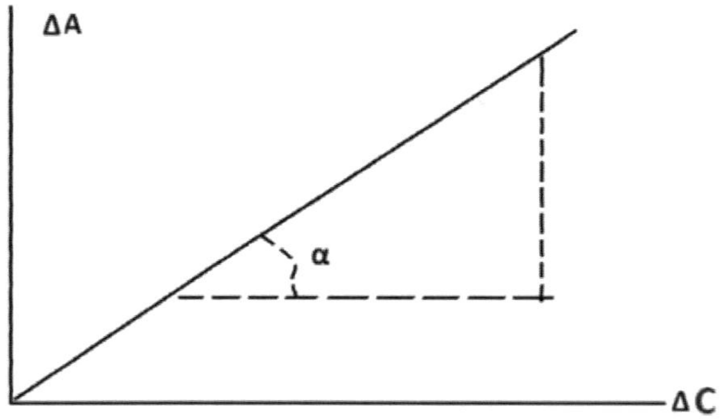

Figure 1. General dependence of the variation of Albedo as the function of the variation of the concentration of atmosphere

mathematical point of view, in order to have one and the same value of the unchanged

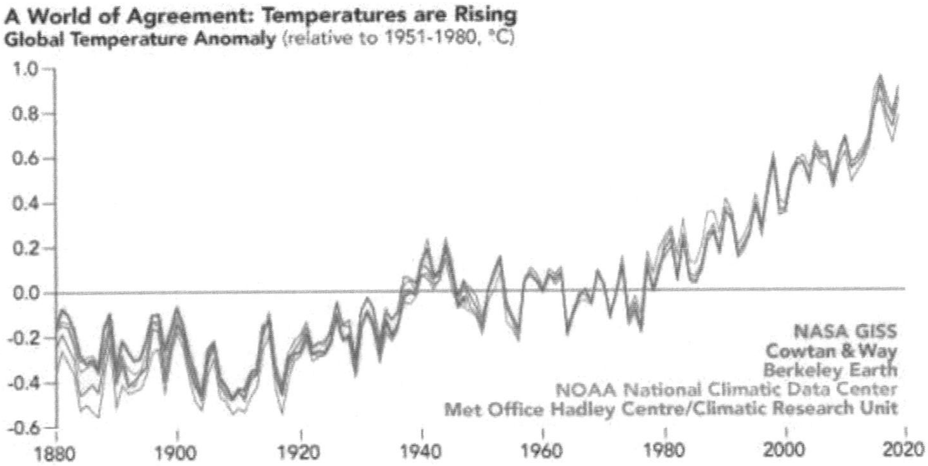

Figure 2. The history of global change of the temperature [16]

proportionality coefficient Z, it is necessary that the small values of the ignition time to correspond to higher values of the Sun fluxes P. In reality, it is clearly observed that when the solar flux power P is high, then the initiation of a fire occurs during shorter time intervals.
The paper [13] studies this case that small ignition time intervals correspond to small values of specific heat of burning *k* and to bigger values of energy fluxes P.

3. RESULTS AND DISCUSSION

The registration of the increasing of the temperature of atmosphere with time is represented on the Fig. 2 [16].
From all types of pollutant gases with a greenhouse effect, the largest contribution is given by carbon dioxide of around 75-80%. [4]. Carbon dioxide in the atmosphere is the result of major processes such as oxidation processes which is predominant with respect to the processes of photosynthesis. Oxidation processes are mostly related to the anthropogenic activity of the using of fuels. The last 40-50 years, the processes of photosynthesis from Earth's flora are small in comparison to the processes of accumulation of carbon dioxide in the atmosphere that comes from anthropogenic activity and has the result of negative impact on the climate and atmosphere with all consequences of natural cataclysms [4], [18]. Therefore, it can be considered by some approximation for the expression (2) that the variation of the concentration ΔC can be considered in general as the variation of the concentration of carbon dioxide.
The registration of the concentration of carbon dioxide over time is represented on the Fig. 3.
As a result of the accumulation of pollutants in the atmosphere and Biosphere in general, we have changes of the physical-chemical properties of the components of the Biosphere. One of the parameters that can be changed is the Albedo. The calculated values of Albedo [7] of the unique complex system Atmosphere-Biosphere as a result of the accumulation of carbon dioxide in the atmosphere show an increasing trend over time. These values are presented in Table 1.

Journal of Physics: Conference Series **2910** (2024) 012021 doi:10.1088/1742-6596/2910/1/012021

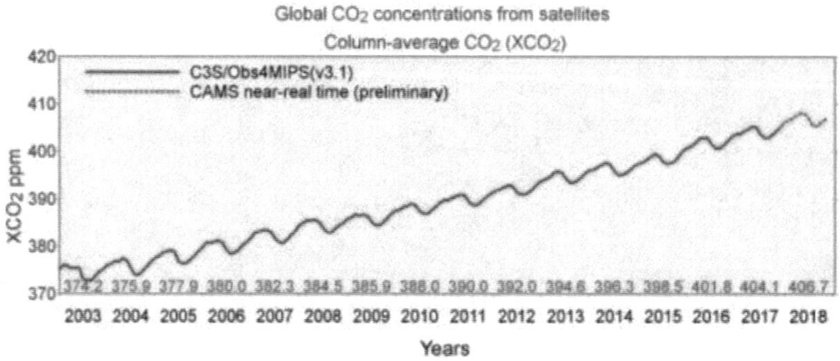

Figure 3. The dynamics of the change of the concentration of CO_2 [17]

The values of the temperatures and of the concentrations from the figures 2 and 3 are presented on the Table 1.

The conversion of the concentrations from *ppm* into *mol/l* is performed by the expression described by [19]:

$$\Delta C[mol/l] = \frac{\Delta C_{ppm}}{1000 \cdot M[g/mol]} \quad (3)$$

In general the value of Albedo of the unique complex system Earth-Atmosphere is A=0,3 with the tendency of gradually increasing over time due to of excess of pollutants.

Table 1. The dynamics of the accumulation of carbon dioxide in atmosphere

Year	ΔT; °C	C (CO_2); ppm	ΔC (CO_2); ppm	ΔC (CO_2); mol/l	ΔC (CO_2); mol/m³	ΔA	A
1980	0	338	0	0	0	0	0.3
1981	0.03	340	2	4.54545E-05	0.0454545	2.08E-07	0.300000208
1982	0.05	341	3	6.81818E-05	0.0681818	6.37E-08	0.300000064
1983	0.07	343	5	0.000113636	0.1136364	1.953E-07	0.300000195
1984	0.09	345	7	0.000159091	0.1590909	4.4123E-07	0.300000441
1985	0.11	347	9	0.000204545	0.2045455	7.159E-07	0.300000716
1986	0.13	348	10	0.000227273	0.2272727	8.6265E-07	0.300000863
1987	0.15	349	11	0.00025	0.25	9.9867E-07	0.300000999
1988	0.18	350	12	0.000272727	0.2727273	1.1407E-06	0.300001141
1989	0.19	350.5	12.5	0.000284091	0.2840909	1.2073E-06	0.300001207

1990	0.22	351	13	0.000295455	0.2954545	1.2741E-06	0.300001274
1991	0.26	352	14	0.000318182	0.3181818	1.4173E-06	0.300001417
1992	0.32	353	15	0.000340909	0.3409091	1.5609E-06	0.300001561
1993	0.34	355	17	0.000386364	0.3863636	1.8551E-06	0.300001855
1994	0.36	357	19	0.000431818	0.4318182	2.1502E-06	0.30000215
1995	0.37	359	21	0.000477273	0.4772727	2.4547E-06	0.300002455
1996	0.39	360	22	0.0005	0.5	2.6038E-06	0.300002604
1997	0.41	361	23	0.000522727	0.5227273	2.7362E-06	0.300002736
1998	0.42	362	24	0.000545455	0.5454545	2.8604E-06	0.30000286
1999	0,43	363	25	0,000568182	0.5681818	3.0079E-06	0.300003008
2000	0.45	365	27	0.000613636	0.6136364	3.3162E-06	0.300003316
2001	0.46	369	31	0.000704545	0.7045455	3.9123E-06	0.300003912
2002	0.47	373	35	0.000795455	0.7954545	4.5102E-06	0.30000451
2003	0.5	375	37	0.000840909	0.8409091	4.7907E-06	0.300004791
2004	0.54	376	38	0.000863636	0.8636364	4.9452E-06	0.300004945
2005	0.58	378	40	0.000909091	0.9090909	5.2532E-06	0.300005253
2006	0.59	379	41	0.000931818	0.9318182	5.4002E-06	0.3000054
2007	0.6	380	42	0.000954545	0.9545455	5.5477E-06	0.300005548
2008	0.61	381	43	0.000977273	0.9772727	5.6898E-06	0.30000569
2009	0.62	382	44	0.001	1	5.8178E-06	0.300005818
2010	0.63	383	45	0.001022727	1.0227273	5.9636E-06	0.300005964
2011	0.65	385	47	0.001068182	1.0681818	6.2759E-06	0.300006276
2012	0.68	393	55	0.00125	1.25	7.5403E-06	0.30000754
2013	0.7	398	60	0,001363636	1.3636364	8.3351E-06	0.300008335
2014	0,74	401	63	0.001431818	1.4318182	8.8164E-06	0.300008816
2015	0.78	402	64	0.001454545	1.4545455	8.9825E-06	0.300008982

2016	0.8	406	68	0.001545455	1.5454545	9.6188E-06	0.300009619
2017	0.81	410	72	0.001636364	1.6363636	1.0239E-05	0.300010239
2018	0.82	412	74	0.001681818	1.6818182	1.0561E-05	0.300010561
2019	0.83	413	75	0.001704545	1.7045455	1.0719E-05	0.300010719
2020	0.85	415	77	0.00175	1.75	1.1043E-05	0.300011043

The graphic of the dependence of $\Delta A=f(\Delta C)$ is represented on Fig. 4.

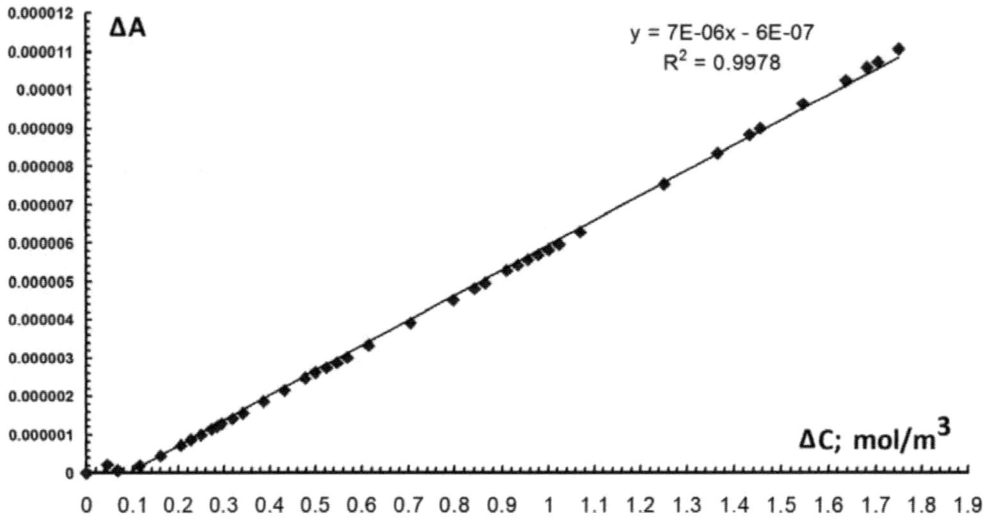

Figure 4. The dynamics of the change of the Albedo as the function of the change of the concentration of CO_2

The proportional coefficient of the graphic of the Fig. 4 is: $Z=7\cdot 10^{-6}$. Then, it is possible to calculate the interval of time t if the value of Sun flux could be of the order $P=500$ W/m^2 and the specific heat of burning $k=10^7$ (J/kg). [20]. Then:

$$t=\frac{k}{P.Z}=\frac{10^7}{500\cdot 7\cdot 10^{-6}}=\frac{10^{13}}{3500}=0{,}029\cdot 10^{11}=2{,}9\cdot 10^{9}(s)\approx 80(years)\rightarrow 100(years) \qquad (4)$$

It means that the last 80-100 years the value of Albedo of this complex system Earth-Atmosphere is increased from $A\approx 0.3$ till $A\approx 0.300011043$.
Regarding the expression (4), it could be possible to calculate the total mass m of fuels that are burnt last 100 years. The value $k=\frac{\Delta Q_t}{m}$, where ΔQ_t is the total excess quantity of heat accumulated in the atmosphere due to of anthropogenical activity. This quantity is $\Delta Q_t \approx 4.88\cdot 10^{21}$ (J). Then, the mass m is calculated from the expression:

$$t = \frac{k}{P.Z} = \frac{10^7}{500 \cdot 7 \cdot 10^{-6}} = \frac{10^{13}}{3500} = 0{,}029 \cdot 10^{11}(s) \approx 80(years) \rightarrow 100(years) \tag{5}$$

The simplified chemical reaction of the burning of fuels is:

$$C + O_2 \rightarrow CO_2 + \Delta Q_t \tag{6}$$

The quantity of moles:
$$\frac{m}{M(C)} = \frac{m(CO_2)}{M(CO_2)} \tag{7}$$

Then, the quantity of carbon dioxide accumulated in Atmosphere is:

$$m(CO_2) = \frac{M(CO_2) \cdot m}{M(C)} = \frac{44 \cdot 3 \cdot 10^{15}}{12} \cong 10^{16}(kg) = 10^9 \cdot 10^4(tones) = 10000(Gt)$$

Considering for one year, then it is of the order 100 (Gt/year).

In such a way we have the estimative interpretation of the state of Biosphere and Atmosphere in general by the application of the values of Albedo, Sun flux and the variation of the concentration of atmospheric air.

The expression (2) could be written also as:

$$\Delta C = \frac{t.P.\Delta A}{k} \tag{8}$$

Here of course, logically is that for more accumulation of pollutants is necessary more time. When k is smaller, but wood has the value of k the order 10-12 MJ/kg, then more pollutants accumulate in atmosphere than in comparison with bigger k = 141.8 MJ/kg [25] for hydrogen which give almost zero pollutants.

Conclusions

This empirical study suggested in this paper has a very important significant result. The broad and in-depth qualitative description that generalizes the current data of the literature about atmospheric pollution as a result of anthropogenic activity gives clear results that allow to estimate the prognostic situations for the future both about the quantities of fuel that are used to satisfy the anthropogenic activity as well as the actual state of atmospheric air and what can be the negative consequences in general on the flora and fauna of Earth and humanity in general.

ACKNOWLEDGMENTS

The work is supported by a project of the university Prof. Dr. Asen Zlatarov Burgas NIX-482/2023 "Application of Intuitionistic Fuzzy Logic in the Decision Making Process".

CRediT roles: Conceptualization, M.P.; data curation, M.P.,S.T.. and V.N.T.; formal analysis, M.P.; funding acquisition, V.N.T.. and S.T.; investigation, M.P. and S.T.; methodology, M.P. and S.T.; project administration, V.N.T.; resources, M.P.,V.N.T.,S.T. and V.T.; supervision, V.N.T.; validation, M.P.,S.T.,V.N.T. and V.T.; visualization, M.P., S.T. and V.N.T.; writing–original draft, M.P.; writing–review and editing, M.P. and V.N.T.

References

[1] Bartsev, S.; Degermendzhi, A. The Evolutionary Mechanism of Formation of Biosphere Closure. *Mathematics* **2023**, *11*, 3218. https://doi.org/10.3390/math11143218

[2] Kalak, T. Potential Use of Industrial Biomass Waste as a Sustainable Energy Source in the Future. *Energies* **2023**, *16*, 1783. https://doi.org/10.3390/en16041783

[3] NOAA National Centers for Environmental Information (2024). Annual 2023 Global Climate Report. Accessed January 17, 2024, from https://www.ncei.noaa.gov/access/monitoring/monthly-report/global/202313.

[4] Mihai Petrov, Major contribution of carbon dioxide excess in atmosphere to the green house effect, Oxidation Communications 44, No 4, 870–907 (2021), https://scibulcom.net/en/journal/0209-4541/issue/2021-44-4/

[5] Shivanna K. R. Climate change and its impact on biodiversity and human welfare. Proc.Indian Natl. Sci. Acad. 2022;88(2):160–71. doi: 10.1007/s43538-022-00073-6. Epub 2022 May 2. PMCID: PMC9058818.

[6] Mihai Petrov, Zdravka Nikolaeva, Aleksandar Dimitrov, Cyclical aspect of the complex factors of anthropogenic activity on the global ecosystem, Oxidation Communications 46(2), (2023) Pages: 509-518

[7] Mihai Petrov, Relational empirical study of outgoing longwave radiation-Albedo-temperature of the atmosphere-earth system under the influence of carbon dioxide pollutants. Oxidation Communications 45, No 2, 350-382 (2022)

[8] S. TWOMEY, POLLUTION AND THE PLANETARY ALBEDO, Atmospheric Environment, Vol 8, pp. 1251-1256, Pergamon Press 1974., Printed in Great Britain. PII: 0004-6981(74)90004-3 (climate-dynamics.org)

[9] S. E. Nicholson, C. J. Tucker and M. B. Ba*, *Desertification, Drought, and Surface Vegetation: An Example from the West African Sahel.*, Bulletin of the American Meteorological Society. https://www.researchgate.net/publication/249615095_Desertification_Drought_and_Surface_Vegetation_An Example_from_the_West_African_Sahel

[10] Mihai Petrov, Zdravka Nikolaeva, Aleksandar Dimitrov, Cyclical aspect of the complex factors of anthropogenic activity on the global ecosystem, Oxidation Communications 46(2), (2023) Pages: 509-518

[11] Mihai Petrov, Zdravka Nikolaeva, Aleksandar Dimitrov, Spontaneous global wildfires as the consequence of climate variability, Oxidation Communications 46(2), (2023) Pages: 519-546

[12] Mihai Petrov, Zdravka Nikolaeva, Aleksandar Dimitrov, Velichka Traneva, Thermodynamic empirism for describing atmospheric pollutants, Environment. Technology. Resources. Rezekne, Latvia, Proceedings of the 15th International Scientific and Practical Conference. Volume I, 300-305 Print ISSN 1691-5402, Online ISSN 2256-070X https://doi.org/10.17770/etr2024 vol1.

[13] Shi L, & Chew MYL. (2013). Experimental study of woods under external heat flux by autoignition: Ignition time and mass loss rate. Journal of Thermal Analysis and Calorimetry, 111(2), 1399-1407

[14] Mihai Petrov, The evolution of Albedo values of the Earth-atmosphere system under the influence of carbon dioxide pollutant concentrations, INTERNATIONAL SCIENTIFIC JOURNAL "INDUSTRY 4.0", WEB ISSN 2534-997X; PRINT ISSN 2534-8582, YEAR VII, ISSUE 1, P.P. 36-41 (2022) 36.full.pdf (stumejournals.com)

[15] Cassia R, Nocioni M, Correa-Aragunde N, Lamattina L. Climate Change and the Impact of Greenhouse Gasses: CO_2 and NO, Friends and Foes of Plant Oxidative Stress. Front Plant Sci. 2018 Mar 1;9:273. doi: 10.3389/fpls.2018.00273. PMID: 29545820; PMCID: PMC5837998.

[16] Rebecca Lindsey and LuAnn Dahlman, "Climate Change: Global Temperature", 2024, Climate Change: Global Temperature | NOAA Climate.gov

[17] "Last four years have been the warmest on record – and CO2 continues to rise", 7th January 2019, https://climate.copernicus.eu/

[18] Mihai Petrov, Zdravka Nikolaeva, Aleksandar Dimitrov, The impact of anthropogenic activity on the global environment, INTERNATIONAL SCIENTIFIC JOURNAL "SCIENCE. BUSINESS. SOCIETY", WEB ISSN 2534-8485; PRINT ISSN 2367-8380, YEAR VIII, ISSUE 2, P.P. 59-64 (2023), 59.full.pdf (stumejournals.com)

[19] Bhola R. Gurjar, Luisa T. Molina and Chandra S. P. Ojha, Air pollution, Health and environmental impacts, Taylor & Francis Group, 2010, pp. 109-249

[20] Ralph S. Jessup and Edward Prosen, Heats of Combustion and Formation of Cellulose and Nitrocellulose (Cellulose Nitrate), Research Paper RP2086 Volume 44, April 1950, htpps://nvlpubs.nist.gov/nistpubs/jres/44/jresv44n4p387_A1b.pdf

[21] "Albedo – from Eric Weisstein's World of Physics". (2011) Scienceworld.wolfram.com.

[22] Goode, P. R.; et al. (2001). "Earthshine Observations of the Earth's Reflectance". Geophysical Research Letters. 28 (9): 1671-1674. Bibcode:2001GeoRL..28.1671G. doi:10.1029/2000GL012580. S2CID 34790317.

[23] Zhao, Kaiguang; Jackson, Robert B (2014). "Biophysical forcings of land-use changes from potential forestry activities in North America" (PDF). Ecological Monographs. 84 (2): 329–353. Bibcode:2014EcoM...84..329Z. doi:10.1890/12-1705.1. S2CID 56059160.

[24] Tetzlaff, G. (1983). Albedo of the Sahara. Cologne University Satellite Measurement of Radiation Budget Parameters. pp. 60–63.

[25] Wenting Sun, Subith Vasu, Matthew S. Blais, Chapter 2 - Fundamentals, Editor(s): Klaus Brun, Timothy Allison, Machinery and Energy Systems for the Hydrogen Economy, Elsevier, (2022), Pages 11-30, ISBN 9780323903943, https://doi.org/10.1016/B978-0-323-90394-3.00015-1.

Journal of Physics: Conference Series **2910** (2024) 012022 doi:10.1088/1742-6596/2910/1/012022

On solving and optimizing contact problems with doubly connected domains taking the nonlinear law of roughness deformation into account

G Shyshkanova * and A Walther

Department of Mathematics, Humboldt-Universität zu Berlin, Berlin, Germany

*E-mail: mathzntu@meta.ua

Abstract. A detailed study of the real properties of contacting bodies stimulated the development of the theory of contact problems in the direction of considering these properties. As a result, contact problems for rough surfaces were formulated. In this paper, an indentation of a doubly connected punch into an elastic rough half-space is investigated taking into account a nonlinear law of change in the deformation of the surface roughness. With a power dependence of the displacement due to the deformations of microasperity on the pressure, the main integral equation is the Hammerstein equation. Two-dimensional integral equations are transformed into one-dimensional ones using the small parameter method and the obtained expansion of the potential of the simple layer at an internal point. The potential expansion is applied to reduce the problem of indenting a non-circular annular punch into an elastic rough half-space into the similar problems for the contact domain with the circular ring form. Successive approximations are used for the solution. The role of the minimizing functional is played by the root-mean-square deviation of the normal pressure distribution arising under the punch from a certain optimal distribution. The result of the solution is shown by examples for ring contact domain.

1. Introduction

The advancement of technology introduces new challenges in the study of machine performance and their components. Enhancing reliability and durability is crucial to boosting competitiveness in today's markets. Moreover, even a slight extension of the service life of machines and equipment, particularly in highly technology-saturated industries, can translate into substantial gains in production capacity and efficiency.

An analysis of failures of mechanical engineering machines and constructions has revealed that, under operational conditions, the primary cause of failure is not breakage but rather wear and damage to working surfaces [1].

Currently, researches are investing considerable efforts into developing methodologies for quantifying contact stresses in interacting mechanical components. This is a crucial area of research, as understanding contact problems in deformable solids is essential for advancing from the calculation of material wear to the prediction of structural wear resistance [2].

One method for reducing wear is to optimize the contact characteristics of deformable bodies [3, 4]. The optimization of contacting surfaces subject to wear is receiving increasing attention due to the dual concerns of safety and reliability [5]. In particular, the contact geometry has been

Content from this work may be used under the terms of the Creative Commons Attribution 4.0 licence. Any further distribution of this work must maintain attribution to the author(s) and the title of the work, journal citation and DOI.
Published under licence by IOP Publishing Ltd

demonstrated to exert a significant influence on the characteristics and behavior of fretting wear in contact between nuclear fuel rods [6]. One of the most significant factors affecting wear intensity is contact pressure, which is largely determined by contact geometry [3]. A comprehensive investigation into the optimal limiting contact geometry in fretting wear in the stick-slip mode can be found in [7, 8]. The necessity for minimizing wear during the initial wearing-in period can be addressed through the design of an initial contact geometry that reduces the time required for transient contact geometry adaptation [2]. The optimal contact shape design can facilitate the implementation of a uniform pressure distribution, which in turn enables the avoidance of stress concentrations. In order to mitigate the singularities of the contact pressure under a punch with sharp edges, it is recommended that the edges undergo a rounding-off procedure [9,10]. A topology optimization approach in frictionless contact interaction was developed in [11] to achieve the uniformity of pressure distribution. The general framework for contact-constrained topology optimization [12] was extended to robust topology optimization in [13]. The minimized integral functional describing the discrepancy between the actual contact pressure and the required one was utilized and the problem decomposition was proposed in [4, 14]. An effective analytical approach to structural optimization of contact geometry was developed in [5]. A combined shape and material stiffness optimization method is developed based on evolutionary optimization techniques [15].

Real-world surfaces are never perfectly smooth; instead, they exhibit microscopic irregularities that affect their behavior when in contact with other surfaces. These irregularities, known as roughness, significantly influence the contact mechanics, frictional properties, wear rates, and even the thermal and electrical conductivity of materials. Understanding the relevance of roughness is vital for optimizing the design, manufacturing, and maintenance of engineering systems, whether in aerospace technologies or industrial equipment.

The investigation of rough-surface interactions dates back many years, with contemporary understanding of these contacts developing around the 1950s. First, it was suggested to include an additional term corresponding to the displacement of the surface due to deformation of asperities, proportional to normal pressure in the work [17]. However, experimental results in [18, 19] show that the displacement of rough contacting surfaces due to deformation of asperities has a nonlinear dependence. In most cases it is a power function of pressure with an exponent less than one. Based on such a dependence, plane and axisymmetric contact problems with a simply connected contact domain were studied in [3].

A useful summary of the statistics of rough surfaces was given in [1], in which it was shown that measurements of surface roughness confirm that a nominally flat surface will always have a distribution of height variations, possibly superimposed on a longer wavelength "form" or regular waviness often related to the way in which the surface was manufactured. Contact problem for wavy surfaces were researched in many works such as [19-21].

A comprehensive and detailed analysis of the effects of rough-surface contact was conducted in [22], providing a valuable reference for further investigation. An article [23] reviewed recent advances and findings on the emergence of roughness on surfaces, focusing on a common feature: self-affine fractal roughness. The impacts of key factors, such as fractal dimension, fractal roughness, material properties, friction coefficient, and load ratio, are thoroughly analyzed and examined in [24]. A combined effect of the surface roughness and coatings was considered in [25]. The use of analytical and numerical research methods makes it possible not only to evaluate the influence of microgeometry on the characteristics of contact interaction, but also to control these processes by selecting the optimal parameters of the surface relief.

Nevertheless, despite the aforementioned successes, it remains challenging to achieve optimal results when considering local design features and the microgeometry of the interacting bodies. Given the inherent complexity of analyzing contact problems with a non-simply connected contact domain, it is notable that only a limited number of researchers have addressed this area of inquiry. In the present paper, we consider an indentation of a doubly connected punch into an elastic rough half-space taking into account a nonlinear law of change in the deformation of the surface roughness and revisit the corresponding optimization problem evolving approaches developed in [5, 26, 27].

2. Statement of the problem

An application of a vertical force Q results in indentation of a rigid punch into a rough elastic half-space $x_3 \geq 0$. The punch initially touches the surface of the half-space by closed line cognate to a circumference. We choose the origin of the coordinate system in the center of this line. The contact area Ω occupies an annular domain in the plane Ox_1x_2, bounded by rings Γ_1 and Γ_2 containing points on the punch surface located after deformation of the half-space. The remaining part of the half-space is free from loading, $\Omega_0 \subset Ox_1x_2$. A desired pressure distribution satisfies the condition $p(x) \geq 0$, $x \in \Omega$, where $x(x_1, x_2, 0)$ is a point on the plane Ox_1x_2.

To study the contact of elastic bodies with a rough surface, an approach was first introduced in [16] where the total displacements of the points of the half-space surface can be represented as the sum of two terms, one due to the deformations in the elastic body and the second due to the deformations of the roughness. The latter deformations are local in nature and therefore depend only on the pressure applied at that point. In the simplest case, this dependence can be linear; in this case, a method to reduce the solution of the problem of indenting a doubly connected punch with a flat base into a rough elastic half-space to one-dimensional integral equations using the expansion of the potential of a simple layer is developed in the work [28]. Then the method of successive approximations is applied, and to eliminate the singularities of the kernels the method of smoothing or composition is used. The given methods of successive approximation can also be applied to the solution of nonlinear integral equations, which are obtained by taking into account the displacements due to the crumpling of microprotrusions, which are related to the contact pressure $p(\rho,\theta)$ by various types of nonlinear dependence, where (ρ,θ) are polar coordinates; $x_1 = \rho\cos\theta$, $x_2 = \rho\sin\theta$.

The problem of the punch indentation into a rough elastic half-space consists of solving a system of equilibrium equations

$$Q = \iint_{\Omega} p(\rho,\theta)\,\mathrm{d}\Omega\,,\ M_{x1} = \iint_{\Omega} \rho\sin\theta \cdot p(\rho,\theta)\,\mathrm{d}\Omega,\ M_{x2} = -\iint_{\Omega} \rho\cos\theta \cdot p(\rho,\theta)\,\mathrm{d}\Omega, \tag{1}$$

where Q, M_{x1}, M_{x2} are the main vector and the main moments of the forces applied to the punch, $d\Omega = \rho d\rho d\theta$, and an equation for the displacement of points of the elastic half-space in the direction of the axis Ox_3 in the contact domain [3] is given by

$$\tilde{\varphi}(B, p(\rho_0,\theta_0)) + \kappa \iint_{\Omega} p(\rho,\theta)/r\ d\Omega = g(\rho_0,\theta_0)\,, \tag{2}$$

where $\tilde{\varphi}(B, p(\rho,\theta))$ is a function that describes the law of roughness deformation due to pressure; B is the coefficient that characterizes the deformation properties of the roughness of the elastic half-space surface; $\kappa=(1-\nu^2)\cdot(\pi E)^{-1}$; E is the modulus of elasticity of the half-space; ν is Poisson's ratio; $g(\rho,\theta)=\delta-\beta_2\rho_0\cos\theta_0+\beta_1\rho_0\sin\theta_0-f(\rho_0,\theta_0)$ is the function of the punch indentation; $r^2=\rho^2+\rho_0^2-2\rho\rho_0\cos(\theta-\theta_0)$; $(\rho_0,\theta_0)\in\Omega$; δ is the vertical displacement; β_1, β_2 are the vector projections of the small rotation of the punch, $f(\rho,\theta)$ is a function describing the punch shape.

It is also necessary to take the next boundary conditions

$$\sigma_{31}=\sigma_{23}=0,\ \sigma_{33}=-p(x),\text{ when } x(\rho,\theta)\in\Omega;\text{ and } \sigma_{33}=\sigma_{31}=\sigma_{23}=0,\text{ when } x\in\Omega_0, \quad (3)$$

into account, where σ_{ij} ($i,j=1,2,3$) are the stress tensor components.

The solution of this problem exists, and is unique in the case of linear dependence of micro asperity deformation from pressure and it could be find using simple layer potential [28]. The existence and uniqueness theorems can be found in [29] for the case of the power dependence.

3. Solving Methods

Assume that the contact area can be described as follows

$$\Omega=\{(\rho,\theta): a(1+\tilde{f}(\varepsilon,\theta))\le\rho\le b(1+\tilde{f}(\varepsilon,\theta)),\quad 0\le\theta\le2\pi\}, \quad (4)$$

where a and b are scalars which characterize the punch shape, and the function $\tilde{f}(\varepsilon,\theta)$ can be represented as $\tilde{f}(\varepsilon,\theta)=\sum_{i=1}^{\infty}f_i(\theta)\varepsilon^i$, $0<\varepsilon<1$.

A ray emanating from the origin intersects each of the boundaries at no more than one point. The method for reducing a problem with a nearly circular doubly connected contact region to a circular region is fully presented in [30]. We will apply it here. Considering that the function characterizing the distribution of normal pressure $p(\rho,\theta)$ and the right-hand side $g(\rho,\theta)$ of equation (2) depend on a small parameter, on which the equations of the boundaries of the contact area depend, we assume that they can be represented as series in powers of ε. That is

$$p(\rho,\theta)=\sum_{k=0}^{\infty}p_k(\rho,\theta)\varepsilon^k;\quad g(\rho,\theta)=\sum_{k=0}^{\infty}g_k(\rho,\theta)\varepsilon^k, \quad (5)$$

Then we represent the integral in equation (2) as a series in powers of ε. To do this, we use the mapping of the region Ω onto a circular ring S

$$\rho=R(1+f(\varepsilon,\varphi)),\quad \theta=\varphi;\ S=\{(R,\varphi): a\le R\le b, 0\le\varphi\le2\pi\};\ (\rho_0,\theta_0)\in\Omega\to(R_0,\varphi_0)\in S. \quad (6)$$

In the new variables, the potential density and the right-hand side of equation (2) are transformed into series

$$p(\rho(R,\varphi,\varepsilon),\varphi)=\sum_{i=0}^{\infty}P_i(R,\varphi)\varepsilon^i;\quad g(\rho(R,\varphi,\varepsilon),\varphi)=\sum_{i=0}^{\infty}G_i(R,\varphi)\varepsilon^i. \quad (7)$$

Assuming that the function $\tilde{\varphi}(B,p(\rho,\theta))$ is continuously differentiated the necessary number of times, using the representation of the potential of a simple layer in the form of a power series ε [26], we obtain a recurrent system of equations for determining $P_i(R,\varphi)$

$$\tilde{\varphi}(B,P_0)+\kappa\iint_{\Omega}P_0/r\,dS=\tilde{G}_0(P_0),$$

$$\widetilde{\Phi}_i(B,P_0)P_i(R_0,\varphi_0)+\kappa\left(\iint_{\Omega}P_i/r\,dS+\Phi_i(P_0,P_1,...,P_{i-1})\right)=\widetilde{G}_i(P_0,P_1,...,P_{i-1}), \tag{8}$$

$$i=1,2...;\ \ S=\{(R,\varphi):a\le R\le b,\ 0\le\varphi\le 2\pi\},$$

where $\widetilde{\Phi}_i=\widetilde{\varphi}'_p(B,P_0)$, $i\neq 0$; Φ_i are the same operators as in [26], $\widetilde{G}_i$ are known functions that are expressed in terms of already defined functions. As one can see from equations (8), to determine the coefficients $P_i(R,\varphi)$ of the expansion of the unknown function $P(R,\varphi)$, starting from $i=1$, we obtain Fredholm's linear integral equations of the second kind.

Let us consider the case of the power dependence of the displacement due to the deformation of the microprotrusions from the contact pressure. In this case equation (2) takes the form

$$Bp^k(\rho_0,\theta_0)+\kappa\iint_{\Omega}\frac{p(\rho,\theta)}{r}d\Omega=g(\rho_0,\theta_0), \tag{9}$$

and can be reduced to the Hammerstein equation, as done in the paper [3]:

$$\psi(\rho_0)+\lambda\int_a^b K(\rho_0,\rho)[\psi(\rho)+g_1]^{\frac{1}{k}}d\rho=0, \tag{10}$$

$$\lambda=\frac{2\pi}{B_1^{1/k}};\ p_1(\rho)=[\psi(\rho)+g_1]^{1/k}B_1^{-1/k};\ B_1=\frac{B}{a}\kappa^{-k};\ Q_1(\rho)=\kappa\cdot Q(\rho);\ \ a=0.5;\ \ b=1.$$

In order to eliminate the singularities of the kernel, we apply the kernel smoothing method, similar to [28], and transform the resulting equation into the following

$$\psi(\rho_0)+\lambda\int_a^b K(\rho_0,\rho)\left[[\psi(\rho)+g_1]^{1/k}-[\psi(\rho_0)+g_1]^{1/k}\right]d\rho+$$

$$+\ \lambda\cdot[\psi(\rho_0)+g_1]^{1/k}\int_a^b K(\rho_0,\rho)\,d\rho=0, \tag{11}$$

Successive approximations to the solution of this equation are determined by the dependence

$$\psi_n(\rho_0)=-\lambda\int_a^b K(\rho_0,\rho)\left[[\psi_{n-1}(\rho)+g_1]^{1/k}-[\psi_{n-1}(\rho_0)+g_1]^{1/k}\right]d\rho- \tag{12}$$

$$-\ \lambda\cdot[\psi_{n-1}(\rho_0)+g_1]^{1/k}\int_a^b K(\rho_0,\rho)\,d\rho,$$

Using the classical calculus of variations, the problem of optimizing pressure under a rigid punch with a doubly connected base is studied here, as it was done in [5] for the simply connected base. The shape of the punch is taken as the desired function, and the role of the minimizing functional is played by the root-mean-square deviation of the pressure distribution under the punch. The values of the total forces and moments applied to the punch are given, which leads to restrictions imposed on the pressure distribution by the equilibrium conditions.

It is assumed that the desired punch shape $f_*(\rho,\theta)$ is a continuous and smooth function of coordinates. The normal pressure should be nonnegative $p(\rho,\theta)\ge 0$, and the stresses outside the contact domain and at its boundaries are equal to zero. As an optimizing functional, we consider an integral of the following form [5,14]

$$J_f=J(p(f_*))=\iint_{\Omega}(p_{tag}-p(f_*))^2d\Omega, \tag{13}$$

which characterizes the mismatch between the pressure distribution $p(\rho,\theta)$ corresponding to a certain shape of the punch and a given target pressure $p_{tag}(\rho,\theta)$. For example, it is possible to specify the function $p_{tag}(\rho,\theta)$ in the form of a constant (average) [5]. The optimization problem is to find the function $f_*(\rho,\theta)$ delivering a minimum to the mismatch functional, satisfying the equilibrium conditions and non-negativity of normal pressure values.

An approach was proposed in [5, 14], where the feasible set and the following auxiliary problem are

$$\Lambda_p = \left\{p:(p)_\Omega \geq 0,\ \iint_\Omega p d\Omega = Q^*,\ \iint_\Omega \rho\sin\theta\cdot p d\Omega = M^*_{x1},\ \iint_\Omega \rho\cos\theta\cdot p d\Omega = M^*_{x2}\right\}, \quad (14)$$

$$J^*_p = J(p_*) \to \min_{p\in\Lambda_p} J(p) = \min_{p\in\Lambda_p} \iint_\Omega \left[p_{tag} - p\right]^2 d\Omega. \quad (15)$$

It was shown in [5] that solving problem (15) and finding the pressure $p_* = p_*(\rho,\theta)$ allows us to determine the corresponding shape function as a solution to the second boundary problem and $\min_{f\in\Lambda_f} J(p(f_*)) = \min_{p\in\Lambda_p} J(p)$. Thus, the formulated optimization problem, as it was shown in [5], admits decomposition and reduces to two effectively solvable problems. When searching for the optimal pressure distribution $p_*(\rho,\theta)$ based on the solution of problem (14), (15), an auxiliary variable $\phi(\rho,\theta)$ is introduced, satisfying the relation $p(\rho,\theta) - \phi^2(\rho,\theta) = 0$. The solution method that we use here for contact pressure optimization is developed in [5]. An extended Lagrange functional is given by

$$J^L = \iint_\Omega \left(\left(p - p_{tag}\right)^2 - \lambda_1 p - \lambda_2 x_1 p - \lambda_3 x_2 p - \chi(p - \phi^2)\right) d\Omega, \quad (16)$$

where the constants λ_i $(i = 1,2,3)$ are the Lagrange multipliers, and $\chi(\rho,\theta)$ is the Lagrange function. With respect to the variables p and ϕ the optimality conditions are given by

$$p = p_{tag} + 0.5\cdot(\lambda_1 + \lambda_2 x_1 + \lambda_3 x_2 + \chi),\quad \phi\chi = 0,\ x \in \Omega\ . \quad (17)$$

The expression for p gives the desired pressure distribution. To determine λ_i, the integral equalities included in (14) are used, and the function χ is found using the second equality in (17). These conditions generally allow the domain Ω_f to be divided into two subdomains [5] Ω_f^+, Ω_f^- $(\Omega_f^+ + \Omega_f^- = \Omega_f)$, where $p(\rho,\theta) > 0$, $(\rho,\theta) \in \Omega_f^+$, $p(\rho,\theta) = 0$, $(\rho,\theta) \in \Omega^-$. Then we have $\phi(\rho,\theta) = 0$ when $(\rho,\theta) \in \Omega^+ \subset \Omega$, if $\phi(\rho,\theta) \neq 0$; and $p(\rho,\theta) = 0$ when $(\rho,\theta) \in \Omega^- \subset \Omega$, if $\phi(\rho,\theta) = 0$. The optimal pressure distribution $p_*(\rho,\theta)$ is given in [5] by

$$p_*(\rho,\theta) = \max\left\{0, p^+(\rho,\theta)\right\},\quad p^+(\rho,\theta) = p_{tag}(\rho,\theta) + 0.5(\lambda_1 + \lambda_2\rho\cos\theta + \lambda_3\rho\sin\theta). \quad (18)$$

The maximum before the curly brackets in (18) means sampling at each fixed $(\rho,\theta) \in \Omega$ the maximum of the two numbers written in brackets. In this case, the optimization problem can be reformulated in abs-smooth form and solved by the methods developed in [31]. For values of the problem parameters, when the given force Q^* significantly exceeds the values of the applied moments M^*_{x1} and M^*_{x2}, on the domain $\Omega = \Omega^+$ one can receive next optimal pressure

$$p_* = p_{tag} + 0.5(\lambda_1 + \lambda_2 \rho \cos\theta + \lambda_3 \rho \sin\theta). \quad (19)$$

In this case, if we consider the contact domain Ω, which has symmetry with respect to the axes Ox_1 and Ox_2 the unknown Lagrange multipliers are determined as follows

$$\lambda_1 = \frac{2(Q^* - \iint_\Omega p_{tag} d\Omega)}{\iint_\Omega d\Omega}, \ \lambda_2 = \frac{2(M^*_{x2} - \iint_\Omega p_{tag} \rho^2 \cos^2\theta d\Omega)}{\iint_\Omega \rho^2 \cos^2\theta d\Omega}, \ \lambda_3 = \frac{2(M^*_{x1} - \iint_\Omega p_{tag} \rho^2 \sin^2\theta d\Omega)}{\iint_\Omega \rho^2 \sin^2\theta d\Omega}. \quad (20)$$

If, in addition, the function $p_{tag}(\rho,\theta)$ has symmetry with respect to the axes Ox_1 and Ox_2, then the main moments will be equal to zero. In the case where the average pressure is taken as the given pressure distribution, the following relations are satisfied $p_{tag} = Q^*/S^*$, where S^* is aria of the contact domain Ω, and $Q(p_{tag}) = Q^*$. Moreover, therefore, we have

$$p_*(\rho,\theta) = Q^*/S^* + M^*_{x2}\left(\iint_\Omega \rho^2 \cos^2\theta d\Omega\right)^{-1} \rho\cos\theta + M^*_{x1}\left(\iint_\Omega \rho^2 \sin^2\theta d\Omega\right)^{-1} \rho\sin\theta. \quad (21)$$

The considered case of full contact is realized under certain restrictions on the parameters of the problem. The condition ensuring contact in the domain Ω is given in [5]. Note also that the construction of the optimal distribution carried out here is based on the necessary conditions for optimality. Checking the sufficient conditions of the minimum of the optimized functional requires additional assessments, which are generally not given here.

The algorithm for solving this optimization problem is described in detail in [5, 14]. The second problem proposed in [5] is to find normal displacements $\delta_*(\rho,\theta)$ corresponding to the optimal pressure distribution $p_*(\rho,\theta)$ and determining the optimal shape of the punch $f_*(\rho,\theta)$. To determine the normal displacements, a spatial boundary value problem of the elasticity theory for a half-space is necessary to solve, the solution method for which is developed in the previous paragraph using the method of simple layer potential expansion.

4. Results and numerical examples

A numerical study of the effect of roughness, which varies according to a power law, on the distribution of normal pressure over the contact area was carried out. In this section, for simplicity, we consider the cases of a centrally applied vertical force and no moments. The proposed method is applicable for simply connected domains also. The numerical results for the circular punch coincides with the ones given in [3], with the same parameter values, which confirms the reliability of the obtained results.

Let us consider circular ring punch with flat base. Figure 1 shows the dependences of the dimensionless normal pressure distribution $\widetilde{p}(\rho) = 2\pi\kappa b^2 p(\rho)/Q$ on the annular contact area with the power-law deformation of the microprotrusions. Curve 1 is a graph of the dimensionless pressure distribution under the ring punch (fig. 1), constructed with the following values of the dimensionless parameters: $a/b = 1/5$, $\delta' = 0.1$, $B_1 = 0.6$, $k = 1/3$. The following lines differ from line 1 in only one parameter: 2 – $a/b = 2/3$, 3 – $B_1 = 0.4$, 4 – $k = 0.4$, 5 – $\delta' = 0.14$.

Figure 2 represents the relationship between the indentation depth of the cylindrical circular ring flat punch and the dimensionless load applied to it. Lines 1, 2, 3, and 4 correspond to roughness coefficient B_1 values of 0.7, 0.6, 0.4, and 0.3, respectively. The other dimensionless

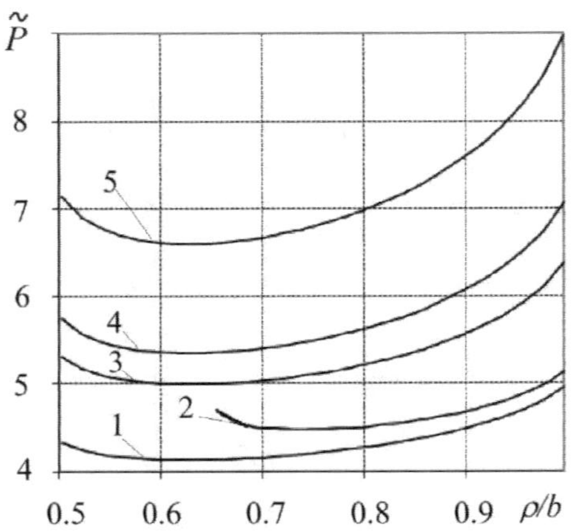

Figure 1. Normal pressure distribution from the roughness coefficient.

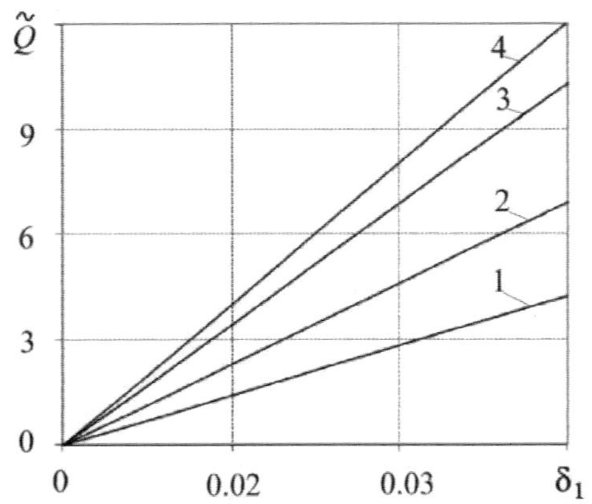

Figure 2. Dependence between the pressing force and the indentation depth.

parameters are set to $a/b=1/2$; $k=0.8$. When the same load is applied, the values of the indentation of the punch and the pressure under it depend on the roughness of the surface: less indentation corresponds to lower roughness (increasing contact stiffness $p(\rho,\theta)/\delta'$).

For values B_1 close to or equal to zero, we obtain an approximate solution of the integral equation of the first kind. In this case, the successive approximations quickly converge, since the solution is a continuous function in the domain of integration.

Let us consider indentation of three non-flat punches, annular in plan, into the elastic rough half-space (fig. 3). Suppose that the first punch is limited by parabola which vertex moves along a circle. (fig. 3 *a*), that is, the right-hand side of equation (9) has the form $g(\rho,\theta)=\delta-D\cdot(\rho-\rho_0)^2$.

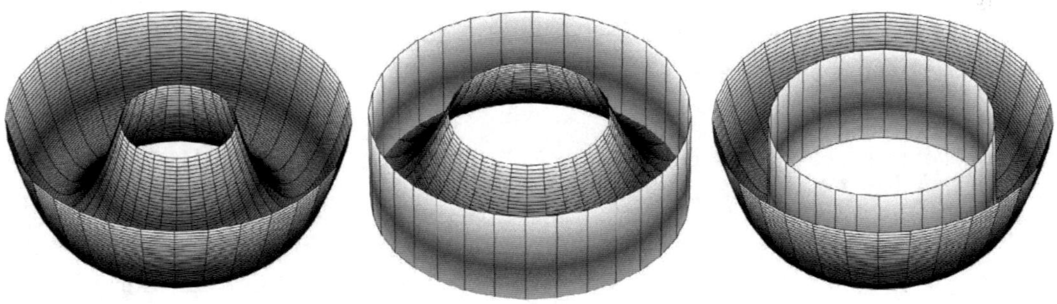

(*a*) (*b*) (*c*)

Figure 3. Scheme of annular ring in plan punches.

In the following examples we take the next values of dimensionless parameters $k=0.8$ and the parameter $D'=Db^3(\kappa Q)^{-1}$ is chosen to keep $a/b=0.5$, b=1. Figure 4*a* shows graphs of the dependence of the dimensionless normal pressure distribution $\tilde{p}$ on the contact area for a circular ring at $D'=0.5$ and B_1=0.7; 0.6; 0.2; 0, correspondingly to lines 1,2,3,4. The line 4 is the case for an absolutely smooth contact, the exact solution for which was received in [27]. The maximum pressure is shifted relative to the central line of the section. It follows from the given

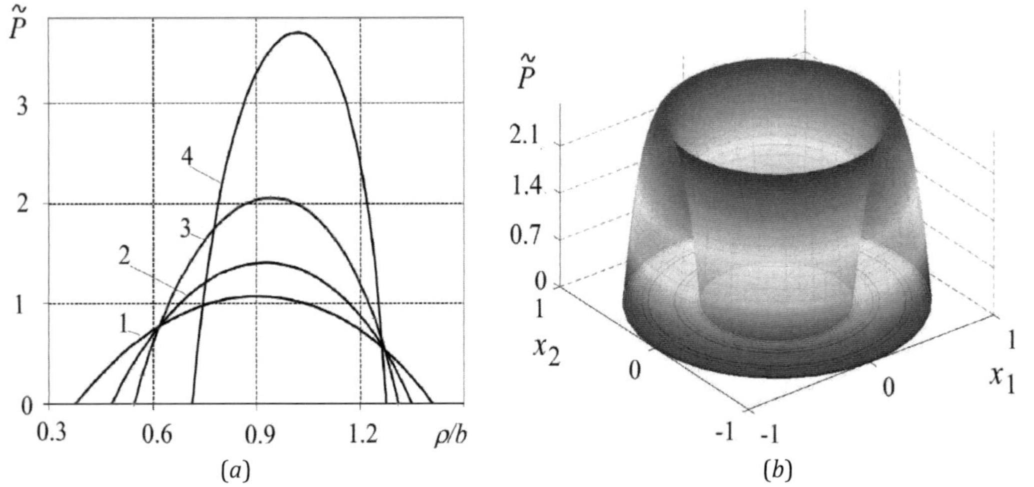

Figure 4. Normal pressure distribution under the annular ring punch limited by parabola which vertex moves along a circle.

dependences that with an increase in roughness, the size of the area of the contact area increases and the pressure distribution over it is evened out.

The surface of the dimensionless normal pressure distribution $\widetilde{p}$, over the annular contact domain for the dimensionless ρ/b, is shown in fig. 4*b* at B_1 =0.7 and $D'=0.5$.

Let us also consider the problem of pressing into the elastic rough half-space of a ring punch, which is limited by cylindrical and parabolic surfaces with initial contact around the circle. There are two cases (fig. 3). In the case (fig. 3*b*), when a cylindrical surface with the equation $\rho-b=0$ limits the punch from the outside, and from the inside – a parabolic surface with the equation $z=D(\rho-b)^2$, which is created by the movement of the parabola along the direction of the cylinder. Figure 5*a* shows the pressure distribution surface over the contact area at $D'=2.9671$ and $B_1=0.7$. The normal pressure at the points of the outer boundary increases with decreasing roughness. There are lines of equal pressure on the plane Ox_1x_2, which are concentric circles, since the contact area is a circular ring.

When the ring punch is bounded from the inside by a cylindrical surface $\rho-a=0$, and from the outside by a parabolic surface with the equation $z=D(\rho-a)^2$ (fig.3*c*), the pressure distribution surface is shown in Figure 5*b* at $B_1=0.7$ and $D'=4.4436$. The normal pressure at the points of the inner boundary increases with decreasing roughness. The contact area is a circular ring, the size of which decreases with decreasing roughness.

The considered methods of solving contact problems taking into account the roughness of the elastic half-space can be considered as regularization methods, when the regularization parameter characterizes the roughness of the elastic half-space. The solution of an integral equation of the first kind using regularization methods is reduced to the solution of an equation of the second kind. That is, we obtain an approximate solution to the problem of pressing a punch into a smooth half-space, when $B_1=0$. The solution of the problem of indentation of the punch taking into account the roughness is the solution of the equation of the second kind and it is continuous under the given conditions in the contact region, including the boundary.

Journal of Physics: Conference Series **2910** (2024) 012022 doi:10.1088/1742-6596/2910/1/012022

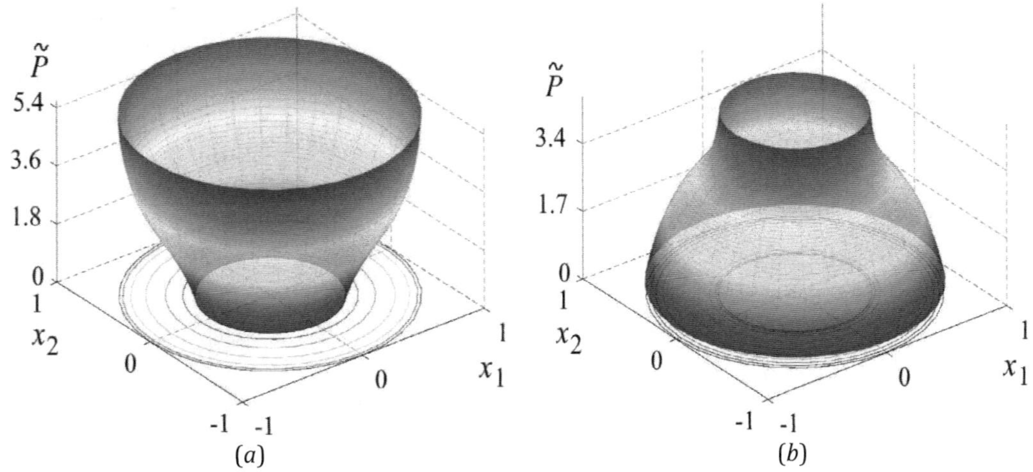

Figure 5. Normal pressure distribution under the annular ring punches limited by cylindrical surface from one side and half of parabola which vertex moves along a circle.

For $B_1 \approx 0$, in the small neighborhood of the sharp edge boundary points, the pressure increases, approaching the solution of the problem that ignores roughness. However, the pressure remains finite even at the sharp edge boundary. This differs from the "smooth" case when the pressure approaches the infinity.

5. Conclusions

The analytic-numerical method is developed for solving contact problems with a doubly connected domain, taking into account a power law of roughness deformation from pressure.

The main integral equation is the Hammerstein equation characterising the displacement of the punch into the rough elastic half-space. Nonlinear integral equations are reduced to a sequence of linear equations in the case where the integration boundaries depend on a parameter. Two-dimensional integral equations are transformed into one-dimensional ones using the small parameter method and the obtained expansion of the simple layer potential at an internal point. The potential expansion is applied to reduce the problem of indenting a non-circular annular punch into an elastic rough half-space into the similar problems for the contact domain with the circular ring form. Successive approximations are used for the solution. In order to eliminate the singularities, the algebraic transformations of the desired function was proposed and the subsequent interpolation of the terms. A punch shape was taken as a desired function, and the role of a minimizing functional is played by deviation of the pressure distribution from the optimal one. The optimization problem was transformed into two successively solved problems, according to [5]. Optimal shapes were found using the simple layer potential expansion.

Numerical examples are given, which confirm the effectiveness of the proposed method. It is shown that with an increase in roughness coefficient makes the pressure distribution more even. The maximum pressure decreases and the minimum pressure increases. The pressure at the sharp boundaries of the contact area increases abruptly, but remains finite, in contrast to the idealized smooth contact, when it goes to infinity at the boundaries. Maximal pressure value has a displacement relatively to the center of the cut of the ring.

The geometrical dimensions of the punch also have an influence on the pressure distribution. A decrease in the contact area leads to an increase in the maximum pressure value and a more uneven distribution. To receive the more even pressure distribution, the punch shape should be mostly flat with rounded edges [2, 32].

Examples were considered for the punch with the doubly connected base bounded by circular ring.

Acknowledgments

The authors thank the Berliner Programm zur Förderung der Chancengleichheit für Frauen in Forschung und Lehre (BCP) and Humboldt-Universität zu Berlin for their support.

References

[1] Greenwood J A 2013 Contact of Rough Surfaces: The Greenwood and Williamson/Tripp, Fuller and Tabor Theories *Encyclopedia of Tribology* ed Q J Wang and Y-W Chung (Boston: Springer) p 517-522 https://doi.org/10.1007/978-0-387-92897-5_524

[2] Argatov I I and Chai Y S 2020 A note on optimal design of contact geometry in fretting wear *Int. J. Mech. Mater. Des.* **16** 415–422 https://doi.org/10.1007/s10999-019-09467-9

[3] Goryacheva G 2013 *Contact Mechanics in Tribology* (Dordrecht: Springer)

[4] Banichuk N, Ragnedda F and Serra M 2010 Some Optimization Problems for Bodies in Quasi-Steady State Wear *Mechanics Based Design of Structures and Machines* **38(4)** 430–439 https://doi.org/10.1080/15397734.2010.483574

[5] Banichuk N V and Ivanova S Yu 2017 *Optimal Structural Design: Contact Problems and High-Speed Penetration* (Berlin, Boston: De Gruyter)

[6] Kim H-K, Lee Y-H and Heo S-P 2006 Mechanical and experimental investigation on nuclear fuel fretting *Tribol. Int.* **39(10)** 1305–1319 https://doi.org/10.1016/j.triboint.2006.02.027

[7] Popov V L 2014 Analytic solution for the limiting shape of profiles due to fretting wear *Sci. Rep.* **4** 3749 https://doi.org/10.1038/srep03749

[8] Chai Y S and Argatov I I 2019 Fretting wear accumulation in partial-slip circular Hertzian contact *Mech. Res. Commun.* **96**, 45–48 https://doi.org/10.1016/j.mechrescom.2019.02.005

[9] Argatov I 2002 The pressure of a punch with a rounded edge on an elastic half-space *J. Appl. Math. Mech.* **66 (4)** 637–643 https://doi.org/10.1016/S0021-8928(02)00082-5

[10] Shishkanova S F 1987 Indentation of an elliptical die with a rounded edge into an elastic half-space *Mechanics of solids* **22(3)** 74–77

[11] Niu C, Zhang W and Gao T 2019 Topology optimization of continuum structures for the uniformity of contact pressures *Struct. Multidisc. Optim.* **60** 185–210 https://doi.org/10.1007/s00158-019-02208-8

[12] Strömberg N and Klarbring A 2010 Topology optimization of structures in unilateral contact *Struct. Multidisc. Optim.* **41** 57–64 https://doi.org/10.1007/s00158-009-0407-z

[13] Schmidt T, Kriegesmann B and Seifried R 2024 Robust contact-constrained topology optimization considering uncertainty at the contact support *Struct. Multidisc. Optim.* **67** 46 https://doi.org/10.1007/s00158-024-03750-w

[14] Banichuk N V, Ivanova S Yu and Makeev E V 2010 Finding of Rigid Punch Shape and Optimal Contact Pressure Distribution *Mechanics Based Design of Structures and Machines* **38(4)** 417-429 https://doi.org/10.1080/15397734.2010.483573

[15] Zhou Y, Lin Q, Hong J and Yang N 2020 Combined interface shape and material stiffness optimization for uniform distribution of contact stress *Mechanics Based Design of Structures and Machines* **51(2)** 1001–1015 https://doi.org/10.1080/15397734.2020.1860086

[16] Shtaerman I Ya 1970 *Contact problem of the theory of elasticity* (Moscow: Gostekhizdat)

[17] Kragelsky I V, Dobychin M N and Kombalov V S 2013 *Friction and wear: calculation methods* (Elsevier)

[18] Kragelsky, I V & Demkin, N B. (1960). Contact area of rough surfaces. Wear, 3(3), 170-187 https://doi.org/10.1016/0043-1648(60)90136-8

[19] Wang X, Xu Y and Jackson R L 2017 Elastic sinusoidal wavy surface contact under full stick conditions *Tribol. Lett.* **65(4)** 156 https://doi.org/10.1007/s11249-017-0937-2

[20] Tsukanov I Y 2019 An extended asymptotic analysis for elastic contact of threedimensional wavy surfaces *Tribol. Lett.* **67(4)** 107 https://doi.org/10.1007/s11249-019-1220-5

[21] Kozachok O P and Martynyak R M 2018 Contact problem for wavy surfaces in the presence of an incompressible liquid and a gas in interface gaps *Math. Mech. Solids.* **24(11)** 3381-3393 https://doi.org/10.1177/1081286518781679

[22] Taylor R I 2022 Rough surface contact modelling - a review *Lubricants* **10** 98 https://doi.org/10.3390/lubricants10050098

[23] Aghababaei R, Brodsky E E, Molinari J F and Chandrasekar S 2022 How roughness emerges on natural and engineered surfaces *MRS Bulletin* **47**, 1229–1236 https://doi.org/10.1557/s43577-022-00469-1

[24] Huang G, Liu C, Xie W and Jiang D 2024 Tangential contact stiffness modeling between fractal rough surfaces with experimental validation *Arch. Appl. Mech.* **94** 719–736 https://doi.org/10.1007/s00419-024-02547-z

[25] Goryacheva I and Yakovenko A 2024 Modeling of the combined effect of the surface roughness and coatings in contact interaction *Lubricants* **12(3)** 68 https://doi.org/10.3390/lubricants12030068

[26] Shyshkanova G and Walther A 2023 Optimization of a punch shape with a doubly connected domain *J. Phys.: Conf. Ser.* **2609** 012009 https://doi.org/10.1088/1742-6596/2609/1/012009

[27] Roitman A B and Shishkanova S F 1980 Pressing of a bevel annular punch into an elastic half-space *Soviet Applied Mechanics* **16(4)** 293-298 https://doi.org/10.1007/BF00884245

[28] Shyshkanova G 2018 About Stability of First Kind Equation Solving *IEEE First International Conference on System Analysis & Intelligent Computing (SAIC)* 8516805 https://doi.org/10.1109/SAIC.2018.8516805

[29] Galanov B A 1985 The method of boundary equations of the Hammerstein-type for contact problems of the theory of elasticity when the regions of contact are not known *Journal of Applied Mathematics and Mechanics*, **49(5)** 827-835. https://doi.org/10.1016/0021-8928(85)90084-X

[30] Shyshkanova G 2011 Solution of the integral equations in the three-dimensional nonsymmetrical contact problems with the friction taken into account *TWMS J. of Pure and Appl. Math.* **2(1)** 134-145.

[31] Griewank A and Walther A 2020 Polyhedral DC decomposition and DCA optimization of piecewise linear functions *Algorithms* **13(7)** 166 https://doi.org/10.3390/a13070166

[32] Liu B, Li T, Han Y, He Y and Fu C 2022 The problem of wedge indenter with flat-rounded bottom indenting half-plane elastic body *International Journal of Applied Mechanics* **14(03)** 2250010 https://doi.org/10.1142/S1758825122500107

An Intercriteria Approach to Assessing the Impact of Air Pollutants on Premature Deaths in the European Union

Velichka Traneva[1], **Mihai Petrov**[1], **Stoyan Tranev**[2] **and Venelin Todorov**[3,4]

[1] Department of Mathematics, Informatics and Physics, "Prof. Asen Zlatarov" University, 1 "Prof. Yakimov" Blvd., Bourgas 8010, Bulgaria
[2] Department of Industrial technology and management, "Prof. Asen Zlatarov" University, 1 "Prof. Yakimov" Blvd., Bourgas 8000, Bulgaria
[3] Institute of Mathematics and Informatics, Bulgarian Academy of Sciences, Acad. G. Bonchev Str., Bl. 8, Sofia 1113, Bulgaria
[4] Institute of Information and Communication Technologies, Bulgarian Academy of Sciences, 25A Acad. Georgi Bonchev Str., 1113 Sofia, Bulgaria

E-mail: veleka13@gmail.com (V.N.T.), mihpetrov@abv.bg (M.P.), tranev@abv.bg (S.T.), venelin@iict.bas.bg (V.T.)

Abstract. Air pollution is the largest environmental health risk in Europe, with heart disease and stroke being the most common causes of premature death due to air pollution. This study estimates the health effects of exposure to fine particulate matter ($PM_{2.5}$), nitrogen dioxide (NO_2), and ozone (O_3) in terms of years of life lost (YLL) and premature mortality within the European Union over a five-year period from 2018 to 2022. The InterCriteria Approach (ICrA) is employed, utilizing data provided by the European Environment Agency. ICrA, which is based on the theories of intuitionistic fuzzy sets (IFSs) and index matrices (IMs), offers a modern interpretation of classical correlation analysis under parameter uncertainty. ICrA refrains from using the term "correlation" between criteria and instead employs the terms "positive consonance", "negative consonance", and "dissonance". IFS theory, an extension of Zadeh's fuzzy sets, handles uncertain data, while IMs digitize intuitionistic fuzzy information for ICrA application. The results highlight the primary air pollutants that significantly influence years of life lost and premature mortality, providing valuable insights for public health policies in the European Union.

Key words: Air pollutant, InterCriteria analysis, Intuitionistic fuzzy logic, Premature deaths

1. Introduction

Increased mortality, early mortality, and morbidity have all been linked to exposure to high levels of ambient air pollution [9]. According to a World Health Organization (WHO) report [8], the combined effects of ambient and household air pollution cause 7 million premature deaths each year. Numerous negative effects on health have been linked to ambient air pollutants such as particulate matter (PM), nitrogen dioxide (NO2), and ozone (O3) [25]. Lung disease and lung cancer are the next most common causes of premature mortality linked to air pollution, behind heart disease and stroke [23]. The largest quantities of particulate matter were found in Italy and Central-eastern Europe [24], mostly as a result of solid fuel combustion for home heating and industrial use. Every EU nation reported ozone and nitrogen dioxide concentrations higher than the WHO's health-based recommended thresholds. Central Europe and the Mediterranean region have the greatest ozone levels [24]. One of the air pollutants most detrimental to human health is particulate matter (PM) having a diameter of 2.5 μm or less; this type of PM is commonly referred to as $PM_{2.5}$ [24]. 96% of urban EU residents are exposed to dangerously

Content from this work may be used under the terms of the Creative Commons Attribution 4.0 licence. Any further distribution of this work must maintain attribution to the author(s) and the title of the work, journal citation and DOI.
Published under licence by IOP Publishing Ltd

high levels of fine particulate matter $PM_{2.5}$ [25]. When inhaled, these particles enter the respiratory tract deeply, causing oxidative stress and systemic inflammation. They then directly travel into the systemic circulation, disrupting the autonomic nerve system [7]. It was shown that $PM_{2.5}$ exposure was linked to fatalities from stroke and Ischemic Heart Disease (IHD) [10]. According to the study [14], the health effects of $PM_{2.5}$, NO_2, and O_3 pollutants in the cities under study resulted in a high number of adult fatalities from chronic bronchitis, premature infant deaths, and respiratory disorders.

In the 27 EU Member States (EU-27), a considerable number of early fatalities were caused by air pollution. Over a five-year period from 2018 to 2022 in the EU-27, exposure to fine particle matter concentrations above the World Health Organization guideline level caused 1711984 premature deaths; exposure to nitrogen dioxide above the corresponding guideline level caused 410166 premature deaths. In the EU-27, acute ozone exposure resulted in 2024960 premature deaths between 2018 and 2022. In human decision systems, uncertainties are categorized as statistical or non-statistical [17]. A statistical uncertainty can be used to express the probability that an event will occur in the future [1, 12, 13, 15, 18]. When there are unclear or missing data, typical statistical analysis is not an option. Classical statistical analysis cannot be used to identify the correlations between premature mortality, years of life lost, and air pollution when there are fuzzy parameters.

In the studies [4, 5], Atanassov et al. presented the InterCritera analysis (ICrA) based on the ideas of index matrices (IMs) and intuitionistic fuzzy sets (IFSs). It can use both precise and imprecise data [11] to uncover hidden correlations between criteria. In this work, we use the Intuitionistic Fuzzy Pairs (IFPs) dataset provided by the European Environment Agency [24, 25] to estimate the health effects of exposure to fine particulate matter ($PM_{2.5}$), nitrogen dioxide (NO_2), and ozone (O_3) in terms of years of life lost (YLL) and premature mortality within the European Union over a five-year period from 2018 to 2022. The work's scientific contribution comes from applying the ICrA to ascertain how years of life lost (YLL) and premature mortality within the European Union during the study period relate to $PM_{2.5}$, NO_2 and O_3. The sections of the remaining article are as follows: The ideas of IFPs and IM are introduced, and the ICrA is briefly explained, in Section 2. To identify the impact of specific air pollutants on years of life lost and premature mortality, an ICrA application is given in the Section 3. Possible directions for future research and conclusions are suggested in Section 4.

2. Several Essential Definitions

2.1. Intuitionistic Fuzzy Pair (IFP) Definition

The **IFP** is an object used to evaluate a proposition p [6], of the form $\langle a, b\rangle = \langle \mu(p), \nu(p)\rangle$, where $a, b \in [0, 1]$ and $a + b \leq 1$. The variables $mu(p)$ and $nu(p)$, respectively, define the degrees of membership and non-membership. Let $x = \langle a, b\rangle$ and $y = \langle c, d\rangle$ be two IFPs. We can compare x and y using the following relations [3]:

$$x \geq y \quad \text{iff} \quad a \geq c \text{ and } b \leq d; \quad x \geq_{\Box} y \quad \text{iff} \quad a \geq c; \tag{1}$$

Let $R_{\langle a,b\rangle} = 0.5(2 - a - b)(1 - a)$ [16]. Next, according to [22], we can rank IFPs as follows:

$$x \geq_R y \quad \text{iff} \quad R_{\langle a,b\rangle} \leq R_{\langle c,d\rangle}. \tag{2}$$

2.2. Three-Dimensional Intuitionistic Fuzzy IM (also known as 3-D IFIM) definition

Let $\mathcal{I}$ be a fixed set of indexes and $\mathcal{X}$ be a finite set of objects. Then, the definition of 3-D IFIM with index sets K, L and H ($K, L, H \subset \mathcal{I}$) is [4]:

$$[K, L, H, \{a_{k_i,l_j,h_g}\}] \equiv \left\{ \begin{array}{c|ccccc} h_g & l_1 & \ldots & l_j & \ldots & l_n \\ \hline k_1 & a_{k_1,l_1,h_g} & \ldots & a_{k_1,l_j,h_g} & \ldots & a_{k_1,l_n,h_g} \\ \vdots & \vdots & \ddots & \vdots & \ddots & \vdots \\ k_m & a_{k_m,l_1,h_g} & \ldots & a_{k_m,l_j,h_g} & \ldots & a_{k_m,l_n,h_g} \end{array} \middle| h_g \in H \right\}.$$

Specific operations with 3-D IFIMs, which are defined in [4, 19, 20, 21] in a manner akin to that of classical matrices, include addition, transposition, multiplication, projection, substitution, aggregation operations, internal subtraction of IMs' components, termwise multiplication, and subtraction.

2.3. The approach known as InterCriteria (ICrA)

The ICrA is a hybrid approach to correlation analysis that blends intuitionistic fuzzy logic with traditional statistical techniques. The following is the algorithm:

Let us consider an IM

$$A = \begin{array}{c|ccccccc} & O_1 & \dots & O_k & \dots & O_l & \dots & O_n \\ \hline C_1 & a_{C_1,O_1} & \dots & a_{C_1,O_k} & \dots & a_{C_1,O_l} & \dots & a_{C_1,O_n} \\ \vdots & \vdots & \vdots & \vdots & \vdots & \vdots & \vdots & \vdots \\ C_i & a_{C_i,O_1} & \dots & a_{C_i,O_k} & \dots & a_{C_i,O_l} & \dots & a_{C_i,O_n} \\ \vdots & \vdots & \vdots & \vdots & \vdots & \vdots & \vdots & \vdots \\ C_j & a_{C_j,O_1} & \dots & a_{C_j,O_k} & \dots & a_{C_j,O_l} & \dots & a_{C_j,O_n} \\ \vdots & \vdots & \vdots & \vdots & \vdots & \vdots & \vdots & \vdots \\ C_m & a_{C_m,O_1} & \dots & a_{C_m,O_k} & \dots & a_{C_m,O_l} & \dots & a_{C_m,O_n} \end{array}, \tag{3}$$

in which for every $p, q,\ (1 \le p \le m, 1 \le q \le n)$:

- C_p is a criterion, participating in the evaluation,
- an object, O_q, is being evaluated.
- a_{C_p,O_q} is a real number or another object, that is comparable about relation R with the other a-objects, so that for each i, j, k: $R(a_{C_k,O_i}, a_{C_k,O_j})$ is defined for each i, j, k. Let $\overline{R}$ be the dual relation of R in the sense that if R is satisfied, then $\overline{R}$ is not satisfied and vice versa. For example, if "R" is the relation "<", then $\overline{R}$ is the relation ">", and vice versa.

We can transform the data in IFPs such that, in the event of unclear or missing data, they can be utilized as the starting IM for the ICrA by using [3, 22] and the methodology from [3, 22]. The exact values of the measured parameters can likewise be used with ICrA.

Let us consider the set of intervals $\left[i_1, i_I\right]$ and let $A_{min,i} = \min\limits_{i_1 \le j \le i_I} x_{i,j} < \max\limits_{i_1 \le j \le i_I} x_{i,j} = A_{max,i}$ for $1 \le i \le m$. We build IFPs [2] for the interval $\left[i_1, i_I\right]$ in the following manner:

$$\nu_{i,j} = \frac{A_{max,i} - x_{i,j}}{A_{max,i} - A_{min,i}}, \mu_{i,j} = \frac{x_{i,j} - A_{min,i}}{A_{max,i} - A_{min,i}}.$$

The following are fulfilled: $0 \le \mu_{i,j} + \nu_{i,j} \le 1$. In the event that the data is unclear, we can use the expert approach, which is fully described in [2], to convert it to IFPs. The number of occurrences in which $R(a_{C_k,O_i}, a_{C_k,O_j})$ and $R(a_{C_l,O_i}, a_{C_l,O_j})$ are satisfied concurrently is denoted by $S^{\mu}_{k,l}$. Let $S^{\nu}_{k,l}$ represent the number of cases when $\overline{R}(a_{C_l,O_i}, a_{C_l,O_j})$ and $R(a_{C_k,O_i}, a_{C_k,O_j})$ are satisfied concurrently.

It is evident that

$$S^{\mu}_{k,l} + S^{\nu}_{k,l} \le \frac{n(n-1)}{2}.$$

We define now, for any k, l, such that $1 \le k < l \le m$ and for $n \ge 2$,

$$\mu_{C_k,C_l} = 2\frac{S^{\mu}_{k,l}}{n(n-1)}, \quad \nu_{C_k,C_l} = 2\frac{S^{\nu}_{k,l}}{n(n-1)}.$$

$\langle \mu_{C_k,C_l}, \nu_{C_k,C_l} \rangle$ is an IFP as a result. The following IM A is then created.

$$\begin{array}{c|ccc} & C_1 & \dots & C_m \\ \hline C_1 & \langle \mu_{C_1,C_1}, \nu_{C_1,C_1} \rangle & \dots & \langle \mu_{C_1,C_m}, \nu_{C_1,C_m} \rangle \\ \vdots & \vdots & \vdots & \\ C_m & \langle \mu_{C_m,C_1}, \nu_{C_m,C_1} \rangle & \dots & \langle \mu_{C_m,C_m}, \nu_{C_m,C_m} \rangle \end{array},$$

that establish the degrees to which criteria $C_1, ..., C_m$ correspond.

Assume that $\alpha + \beta \leq 1$ and that $\alpha, \beta \in [0, 1]$. The criteria C_k and C_l are in

- (α, β)-positive consonance, if and only if $\mu_{C_k,C_l} > \alpha$ and $\nu_{C_k,C_l} < \beta$;
- (α, β)-negative consonance, if and only if $\mu_{C_k,C_l} < \beta$ and $\mu_{C_k,C_l} > \alpha$;
- (α, β)-dissonance, otherwise.

The term "intercriteria" refers to the fact that the $n \times n$ IMs A^μ, A^ν, and A^* from the ICrA application contain the membership degrees of the IFPs of A^*, the non-membership degrees of the IFPs of A^*, and the IFPs A^* determining the grades of the relationship between each pair of criteria. To ensure terminology clarity, the ICrA use the phrases "positive consonance," "negative consonance," and "dissonance" in place of the term "correlation" between criteria. Regarding the requirements that the pertinent IFPs A^* components must fulfill, the following details In the case of $\mu \in [0.75; 1]$ and $\nu \in [0; 0.25]$, statistically significant positive consonance is found, whereas an area of statistically significant negative consonance is derived for $\mu \in [0; 0.25]$ and $\nu \in [0.75; 1]$. The remaining cases are regarded as dissonance. The program that is detailed in [5] is used to calculate the intercriteria pairings.

3. Intercriteria Evaluation of the Health Effects of Exposure to O_3, NO2, and PM2.5 in the European Union (2018–2022)

96% of urban EU residents are exposed to dangerously high levels of fine particulate matter, or PM_2.5. The majority of people living in metropolitan areas throughout the EU are still exposed to harmful levels of important air pollutants. Specifically, nearly all urban residents were exposed to PM2.5 concentrations that exceeded the WHO annual guideline level of 5 μg/m^3, ozone O_3 concentrations that exceeded the short-term guideline level of 100 μg/m^3, and NO2 concentrations that exceeded the 10 μg/m^3 [25]. The primary sources of particulate matter PM_2.5 emissions are industrial processes, road transportation, and the burning of solid fuels for home heating. Ammonia emissions from agriculture play a major role in the formation of fine particulate matter in the atmosphere. Methane, another potent greenhouse gas, and nitrogen oxides undergo chemical reactions in the presence of heat and light, which results in the formation of ozone O_3, a pollutant in the atmosphere. Road transportation is the main source of nitrogen dioxide NO_2. It releases NO_2 near the ground, primarily in places with high people density, increasing exposure to the gas. There are various questions regarding whether the high premature mortality rate and years of life lost inside the EU (2018–2022) are dependent on the levels of air pollution. Using the dataset provided by the European Environment Agency [24, 25] in the form of Intuitionistic Fuzzy Pairs (IFPs), let us apply ICrA [5] to estimate the health effects of exposure to fine particulate matter ($PM_{2.5}$), nitrogen dioxide (NO_2), and ozone (O_3) in terms of years of life lost (YLL) and premature mortality within the European Union over a five-year period from 2018 to 2022. The ICrA's primary components are:

- C_1 (NO_2 Air Pollution Population Weighted Average [ug/m3])
- C_2 (Early deaths due to NO_2)
- C_3 (Life Years Lost from NO_2)
- C_4 (O_3 Air Pollution Population Weighted Average [ug/m3])
- C_5 (Premature deaths owing to O_3)
- C_6 (Years of life lost due to O_3)

- C_7 ($PM_{2.5}$ Air Pollution Population Weighted Average [ug/m3])
- C_8 (Premature deaths due to $PM_{2.5}$)
- C_9 (Years of life lost as a result of $PM_{2.5}$)

With the structure shown in (3), the starting data are provided in a 3-D IM $A[C, O, H]$. The values of the European Union's member states $O = \{O_1, O_2, \ldots, O_{27}\}$, as determined by the evaluation criteria from the set $C = \{C_1, C_2, \ldots, C_9\}$, at a time-moment $h_g \in H$ (for $g = 1, ..., 5$ - from 2018 to 2022). In the μ - ν view, the ICrA software [5] findings are displayed as IM. The intuitionistic fuzzy correlations' membership and non-membership degrees are represented by μ-data and ν-data, respectively (see Table 1). The matrix's data are colored so that the highest values are represented by the color green and the lowest by the color red.

Table 1. The Intercriteria IFPs' levels of membership and non-membership.

	C_1	C_2	C_3	C_4	C_5	C_6	C_7	C_8	C_9
C_1	--	$\langle 0.9;0.0\rangle$	$\langle 0.9;0.0\rangle$	$\langle 0.2;0.7\rangle$	$\langle 0.0;0.9\rangle$	$\langle 0.2;0.7\rangle$	$\langle 0.7;0.0\rangle$	$\langle 0.7;0.2\rangle$	$\langle 0.8;0.1\rangle$
C_2	$\langle 0.9;0.0\rangle$	--	$\langle 1.0;0.0\rangle$	$\langle 0.2;0.8\rangle$	$\langle 0.0;1.0\rangle$	$\langle 0.2;0.8\rangle$	$\langle 0.8;0.0\rangle$	$\langle 0.8;0.2\rangle$	$\langle 0.9;0.1\rangle$
C_3	$\langle 0.9;0.0\rangle$	$\langle 1.0;0.0\rangle$	--	$\langle 0.2;0.8\rangle$	$\langle 0.0;1.0\rangle$	$\langle 0.2;0.8\rangle$	$\langle 0.8;0.0\rangle$	$\langle 0.8;0.2\rangle$	$\langle 0.9;0.1\rangle$
C_4	$\langle 0.2;0.7\rangle$	$\langle 0.2;0.8\rangle$	$\langle 0.2;0.8\rangle$	--	$\langle 0.8;0.2\rangle$	$\langle 1.0;0.0\rangle$	$\langle 0.1;0.7\rangle$	$\langle 0.0;1.0\rangle$	$\langle 0.1;0.9\rangle$
C_5	$\langle 0.0;0.9\rangle$	$\langle 0.0;1.0\rangle$	$\langle 0.0;1.0\rangle$	$\langle 0.8;0.2\rangle$	--	$\langle 0.8;0.2\rangle$	$\langle 0.0;0.8\rangle$	$\langle 0.2;0.8\rangle$	$\langle 0.1;0.9\rangle$
C_6	$\langle 0.2;0.7\rangle$	$\langle 0.2;0.8\rangle$	$\langle 0.2;0.8\rangle$	$\langle 1.0;0.0\rangle$	$\langle 0.8;0.2\rangle$	--	$\langle 0.1;0.7\rangle$	$\langle 0.0;1.0\rangle$	$\langle 0.1;0.9\rangle$
C_7	$\langle 0.7;0.0\rangle$	$\langle 0.8;0.0\rangle$	$\langle 0.8;0.0\rangle$	$\langle 0.1;0.7\rangle$	$\langle 0.0;0.8\rangle$	$\langle 0.1;0.7\rangle$	--	$\langle 0.7;0.1\rangle$	$\langle 0.8;0.\rangle$
C_8	$\langle 0.7;0.2\rangle$	$\langle 0.8;0.2\rangle$	$\langle 0.8;0.2\rangle$	$\langle 0.0;1.0\rangle$	$\langle 0.2;0.8\rangle$	$\langle 0.0;1.0\rangle$	$\langle 0.7;0.1\rangle$	--	$\langle 0.9;0.1\rangle$
C_9	$\langle 0.8;0.1\rangle$	$\langle 0.9;0.1\rangle$	$\langle 0.9;0.1\rangle$	$\langle 0.1;0.9\rangle$	$\langle 0.1;0.9\rangle$	$\langle 0.1;0.9\rangle$	$\langle 0.8;0.0\rangle$	$\langle 0.9;0.1\rangle$	--

Next, as shown in Fig. 1, the intuitionistic fuzzy (IF) interpretational triangle displays the results produced by the ICrA software. The premature deaths in the European Union from 2018 to 2022 were shown in Fig. 2. The following Table 2 and Fig. 3 present the premature deaths in EU from 2018 to 2022 by air pollutant levels.

Table 2. The premature deaths in EU27 from 2018 to 2022 by air pollutant levels.

Air pollutant	Value of pollutant	Value of premature deaths
NO_2	16.32 ug/m^3	82033
O_3	7409.44 ug/m^3	37103
$PM_{2.5}$	12.6 ug/m^3	322396

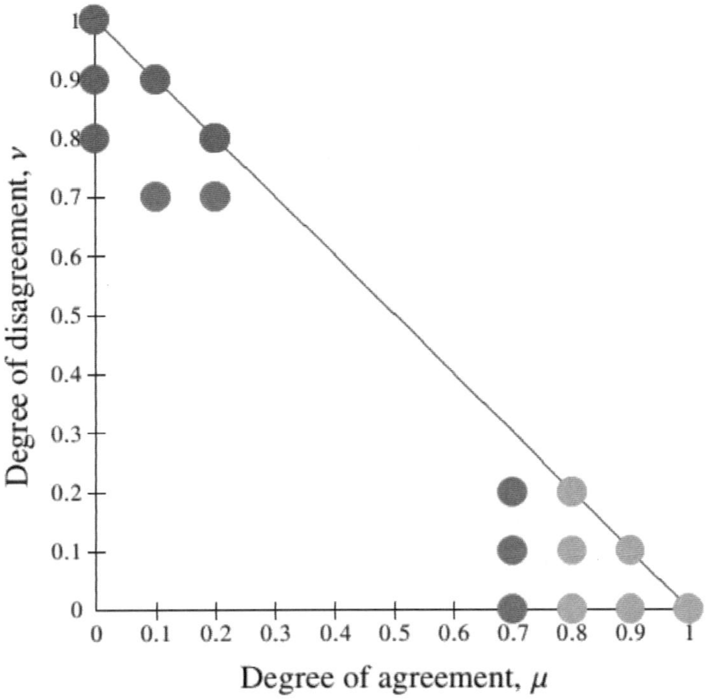

Figure 1. The triangle of IF interpretation with ICrA IFPs.

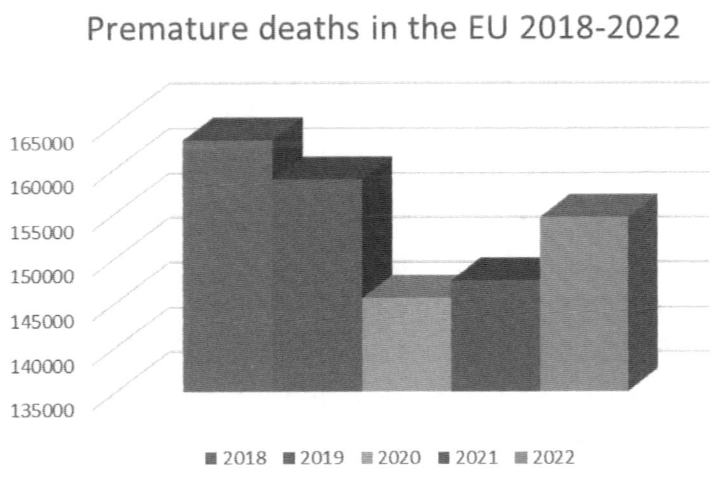

Figure 2. Distribution of Premature Deaths in the EU27 by Air Pollutant Levels $PM_{2.5}$, O_3, and NO_2 from 2018 to 2022.

Each set of criteria's dependencies are calculated in Table 3:

The pairings with the strongest correlations are identified by measuring each pair's distance from the pair $\langle 1, 0\rangle$ [16] $d_{\langle a,b\rangle} = \sqrt{(1-a)^2 + b^2}$, which are provided in Table 4 along with the pairs' order.

The following is demonstrated by the ICrA results:

- The population-weighted average of O_3 (ozone) air pollution and the years of life lost as a result of O_3 have a significant positive consonance (the ICrA pair is $\mu = 1, \nu = 0$). This highlights the severe

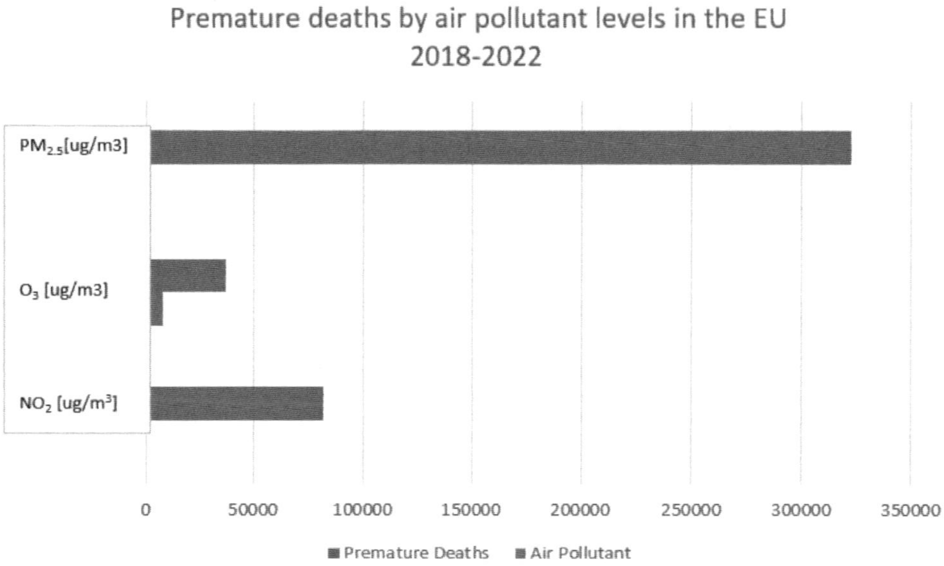

Figure 3. Bar chart illustrating the number of premature deaths associated with different levels of air pollutants $PM_{2.5}$, O_3, and NO_2 across the EU27 countries from 2018 to 2022.

Table 3. Intercriteria pairs

Strength of the correlation	Pairs of criteria
strong positive consonance $[0,95;1]$	2-3, 4-6
positive consonance$[0,85;0,95)$	1-2, 1-3, 2-9, 8-9, 9-3
weak positive consonance $[0,75;0,85)$	1-9, 2-7, 2-8, 3-7, 3-8, 4-5, 5-6, 7-8, 7-9
weak dissonance $[0,67;0,75)$	1-7, 1-8, 8-7
dissonance $[0,57;0,67)$, strong dissonance $[0,43;0,57)$	–
dissonance $[0,33;0,43)$, weak dissonance $[0,25;0,33)$	–
weak negative consonance $[0,15;0,25)$	1-4, 1-6, 2-4, 2-6, 3-4, 3-6, 4-1, 4-2, 4-3, 5-8
negative consonance $[0,05;0,15)$	4-7, 4-9, 5-9, 6-7, 6-9
strong negative consonance $[0;0,05]$	1-5, 2-5, 3-5, 4-8, 5-7, 6-8

Table 4. The couples with the highest correlation

No	d_{C_i,C_j}	Criteria	μ_{C_i,C_j}	ν_{C_i,C_j}
1.	0	$C_2 - C_3$	1	0
1.	0	$C_4 - C_6$	1	0
2.	0.1	$C_1 - C_2$	0.9	0
2.	0.1	$C_1 - C_3$	0.9	0
3.	0.22	$C_2 - C_9$	0.9	0.1
3.	0.22	$C_8 - C_9$	0.9	0.1
3.	0.22	$C_9 - C_3$	0.9	0.1

negative health impacts of O_3 exposure by showing that greater levels of O_3 air pollution are closely related with an increase in years of life lost.

- The analysis demonstrates a positive consonance between the population-weighted average of NO_2 air pollution and the premature deaths resulting from NO_2 ($\mu = 0.9, \nu = 0$). Furthermore, the population-weighted average of NO_2 air pollution and the number of years of life lost as a result of NO_2 ($\mu = 0.9, \nu = 0$) have a similar positive correlation. These findings highlight the harmful health impacts of NO_2 exposure by indicating that greater levels of NO_2 are linked to an increase in premature death and years of life lost.
- The population-weighted average of $PM_{2.5}$ air pollution and the number of premature deaths attributed to $PM_{2.5}$ ($\mu = 0.8, \nu = 0.2$) as well as the number of years of life lost owing to $PM_{2.5}$ ($\mu = 0.8, \nu = 0$) show a mild positive consonance. These dependencies imply that higher $PM_{2.5}$ levels are still associated with a higher number of early deaths and years of life lost, even though the link is not as strong.

These results highlight the substantial harm that air pollution poses to public health in EU member states. The correlations shown among different pollutants (O_3, NO_2, and $PM_{2.5}$) and detrimental health consequences underscore the necessity of strict rules governing air quality and efficient methods for mitigating pollution. By addressing these contaminants, public health might be significantly improved, and the number of years lost to diseases linked to air pollution and premature death could be decreased.

4. Conclusion

The health effects of ozone (O_3), nitrogen dioxide (NO_2), and fine particulate matter ($PM_{2.5}$) on years of life lost (YLL) and premature mortality in the European Union from 2018 to 2022 were assessed using an ICrA technique. The findings demonstrate how seriously air pollution affects public health and how much it contributes to preventable deaths throughout the EU. Based on our analysis, $PM_{2.5}$ is the most dangerous pollutant and is consistently linked to the greatest number of premature deaths and YLL. A smaller degree than $PM_{2.5}$, NO_2 and O_3 also exhibit notable health effects. These results are consistent with previous research and support the necessity for strict air quality laws. Adverse health effects underscore the need for strict air quality laws and efficient pollution control methods. The intercriteria approach was useful in determining the relative impacts of different pollutants (O_3, NO_2, and $PM_{2.5}$). In summary, the primary goal of public health initiatives in the EU should be to reduce exposure to $PM_{2.5}$, NO_2, and O_3. It is imperative to implement policy measures that target the reduction of emissions from major sources, including industry, transportation, and domestic heating. To measure success and modify methods in response to new trends and discoveries, ongoing study and monitoring are essential. In the end, lowering air pollution will enhance public health results as well as the general wellbeing and standard of living of Europeans. Research on air pollution and its impact on the growth of different economic sectors will be conducted in the future using the ICrA.

4.1. Acknowledgments

Work on Sect. 1 and 3 is supported by Assen Zlatarov University through project Ref. No. NIX-486/2023, titled "Modeling Management Decisions with New Analysis Tools in a Fuzzy Environment." Work on Sect.2 and 4 is supported by Assen Zlatarov University through project Ref. No. NIX-482/2023, titled "Application of Intuitionistic Fuzzy Logic in the Decision Making Process."

CRediT roles: Conceptualization, V.N.T.; data curation, V.N.T. and S.T.; formal analysis, V.N.T.,S.T. and M.P.; funding acquisition, V.N.T.. and S.T.; investigation, V.T.. and S.T.; methodology, V.N.T. and S.T.; project administration, V.N.T.; resources, M.P.,V.N.T.,S.T. and V.T.; supervision, V.N.T.; validation, M.P.,S.T.,V.N.T. and V.T.; visualization, V.N.T., S.T. and V.T.; writing–original draft, V.T.; writing–review and editing, V.N.T. and S.T.

References

[1] Atanasov A.Z., Georgiev S.G., Vulkov L.G.: Reconstruction analysis of honeybee colony collapse disorder modelling. Optimization and Engineering vol. 22 (4), pp. 2481 – 2503 (2021). DOI: 10.1007/s11081-021-09678-0

[2] Atanassov, K.: Intuitionistic fuzzy sets. In Proceedings of VII ITKR's Session, Sofia, (1983) (in Bulgarian)

[3] Atanassov, K.: On Intuitionistic Fuzzy Sets Theory. STUDFUZZ. **283**. Springer, Heidelberg (2012)

[4] Atanassov, K.: Index Matrices: Towards an Augmented Matrix Calculus. Studies in Computational Intelligence **573**, Springer, Cham (2014)

[5] Atanassov, K., Mavrov, D., Atanassova, V.: Intercriteria decision making: a new approach for multicriteria decision making, based on index matrices and intuitionistic fuzzy sets. In: Issues in Intuitionistic Fuzzy Sets and Generalized Nets **11**, pp. 1-8 (2014)

[6] Atanassov, K., Szmidt, E. , Kacprzyk, J.: On intuitionistic fuzzy pairs. Notes on Intuitionistic Fuzzy Sets **19**(3), 1-13 (2013)

[7] Basith, S., Manavalan, B., Shin, TH., Park, CB., Lee, WS., Kim, J., et al.: The impact of fne particulate matter 2.5 on the cardiovascular system: a review of the invisible killer. Nanomaterials **12** (1), 2656 (2022)

[8] Brauer, M., Casadei, B., Harrington, RA., Kovacs, R., Sliwa, K.: Taking a stand against air pollution: the impact on cardiovascular disease. Glob Heart **16** (1), 1-6 (2021)

[9] Cohen, AJ., Brauer, M., Burnett, R., Anderson, HR., Frostad, J., Estep, K. et al.: Estimates and 25-year trends of the global burden of disease attributable to ambient air pollution: an analysis of data from the Global Burden of diseases Study 2015. Lancet **389** (10082):, 1907–18 (2017)

[10] Corda, M.O., Charalampous, P., Haagsma, J.A. et al.: Mortality burden of cardiovascular disease attributable to ambient PM2.5 exposure in Portugal, 2011 to 2021. BMC Public Health **24**, 1188 (2024). doi10.1186/s12889-024-18572-0

[11] Fidanova, S., Zhivkov, P., Roeva, O., InterCriteria Analysis Applied on Air Pollution Influence on Morbidity. Mathematics **10** (7), 1195 (2022).

[12] Georgiev S.G., Vulkov L.G., Computation of the unknown volatility from integral option price observations in jump–diffusion models. Mathematics and Computers in Simulation, 188, pp. 591 – 608 (2021). DOI: 10.1016/j.matcom.2021.05.008

[13] Grozev D., Milchev M., Georgiev I.: Analysis of the load on the taxi system in a medium-sized city. IOP Conference Series: Materials Science and Engineering **664** (1) (2019).

[14] Kazemi, Z., Kazemi, Z., Jafari, A., Farzadkia, M., Hosseini, J., Amini, P., Shahsavani, A., Kermani, M.: Estimating the health impacts of exposure to Air pollutants and the evaluation of changes in their concentration using a linear model in Iran. Toxicology Reports **12**, 56-64 (2024)

[15] Pencheva V., Asenov A., Sladkowski A., Ivanov B., Georgiev I.:Current Issues of Multimodal and Intermodal Cargo Transportation. Studies in Systems, Decision and Control **400**, pp. 51 – 124. DOI: 10.1007/978-3-030-87120-8_2

[16] Szmidt, E., Kacprzyk, J.: Amount of information and its reliability in the ranking of Atanassov intuitionistic fuzzy alternatives, In: Rakus-Andersson (eds.) Recent Advances in Decision Making, vol. 222, pp. 7-19. SCI, Springer, Heidelberg (2009)

[17] Todorov, V., Dimov, I. T.: Monte Carlo methods for multidimensional integration for European option pricing. AIP Conf. Proc. **1773**, 100009 (2016). DOI:10.1063/1.4965003

[18] Todorov, V., Dimov, I.: Efficient Stochastic Approaches for Multidimensional Integrals in Bayesian Statistics. In: Lirkov, I., Margenov, S. (eds) Large-Scale Scientific Computing. LSSC 2019. Lecture Notes in Computer Science, vol. 11958. Springer, Cham, pp. 454-462. (2019). DOI:10.1007/978-3-030-41032-2_52

[19] Traneva, V.: Internal operations over 3-dimensional extended index matrices. Proceedings of the Jangjeon Mathematical Society **18**(4), 547-569 (2015)

[20] Traneva, V., Tranev, S.: Index Matrices as a Tool for Managerial Decision Making. Publ. House of the Union of Scientists, Bulgaria (2017) (in Bulgarian)

[21] Traneva, V., Tranev, S., Stoenchev, M., Atanassov, K. Scaled aggregation operations over 2- and 3-dimensional IMs. Soft computing **22**(15), 5115-5120 (2018)

[22] Traneva, V., Tranev, S., Intuitionistic Fuzzy Analysis of Variance of Ticket Sales. In: Kahraman, C. (eds.) INFUS 2020, Advances in Intelligent Systems and Computing, vol. 1197, pp. 363-340. Springer, Cham (2021)

[23] WHO. Burden of disease from the joint efects of household and ambient air pollution for 2016: summary of results. Geneva: World Health Organization (2018)

[24] `https://www.eea.europa.eu/publications/europes-air-quality-status-2023`. Last accessed 20 July2024

[25] `https://www.eea.europa.eu/publications/air-quality-in-europe-2022/health-impacts-of-air-pollution`. Last accessed 20 July 2024

Journal of Physics: Conference Series **2910** (2024) 012024 doi:10.1088/1742-6596/2910/1/012024

Design and realization of self-tuned absorber control system

Qian Wang[2], Yunan Zhu[1],Xiangying Guo[1*]

[1] Beijing Key Laboratory of Nonlinear Vibrations and Strength of Mechanical Structures, School of Mathematics, Statistics and Mechanics, Beijing University of Technology, Beijing,100124, China
[2] Beijing Key Laboratory of Nonlinear Vibrations and Strength of Mechanical Structures, College of Mechanical & Energy Engineering, Beijing University of Technology, Beijing,100124, China

*Email: guoxy@bjut.edu.cn

Abstract. A self-tuning vibration absorber is capable of adjusting its intrinsic frequency in real time to track the excitation frequency and achieve effective vibration control over a wide frequency range. In this study, a self-tuning vibration absorber control system based on STM32 is proposed. The system acquires vibration signals using MPU6050 sensors, calculates the vibration frequency in real time, and controls the motor rotation to achieve self-tuning vibration absorber control. The system employs the STM32F407ZGT6 as the core control chip and is designed in terms of the overall scheme, hardware circuit, and software. For hardware, the system uses the STM32F407 as the control core and the MPU6050 sensor to obtain vibration signals while controlling motor rotation using the TMC2209 motor driver chip. Keil μVision software is applied for program writing and debugging. The system mainly includes modules for data acquisition, signal processing, control algorithm, and drive output. The signal processing module employs the fast Fourier transform (FFT) algorithm, which can achieve fast frequency calculation and vibration frequency tracking. Finally, experiments verify the system's feasibility and effectiveness. The experimental results demonstrate that the system can accurately control the self-tuning vibration absorber, with excellent control performance and practicality.

1. Introduction

Vibration-induced fatigue and structural damage pose significant challenges in long-term engineering systems. Over time, structural fatigue undermines mechanical stability, ultimately compromising structural integrity. Consequently, controlling harmful vibrations has become essential. The concept of dynamic vibration absorbers as a vibration damping technique was first introduced in 1911. However, traditional passive vibration absorbers are limited by their fixed parameters, resulting in a narrow frequency range. When the frequency of external excitation changes, the damping efficacy of these passive absorbers diminishes markedly, restricting their utility in complex vibration environments.In response to these limitations, self-tuning vibration absorber technology has been developed. Self-tuning vibration absorbers can dynamically adjust their intrinsic frequency in real-time to match the characteristics of the external excitation frequency. [1-5]This capability enables them to achieve superior damping effects over a broader

Content from this work may be used under the terms of the Creative Commons Attribution 4.0 licence. Any further distribution of this work must maintain attribution to the author(s) and the title of the work, journal citation and DOI.
Published under licence by IOP Publishing Ltd

frequency range. As a result, self-tuning absorbers exhibit enhanced adaptability and performance under varying vibration conditions.[6-9]

The performance of self-tuning vibration absorbers in practical applications depends not only on their internal structural design but also, more critically, on external control mechanisms that enable dynamic frequency adjustment. Consequently, the study of control systems has become a pivotal aspect of self-tuning vibration absorber technology.[10-12] To comprehensively understand the interaction between the vibration absorber and the external system, and to enhance their synergy, it is essential to analyze the mechanisms and control methods of various control systems. This analysis facilitates the optimization of performance and the development of more effective vibration damping solutions.[13-16]

In 1996, Franchek et al. designed a feedback control algorithm using hardware circuits to control a mechanical self-tuning vibration absorber with a coil spring. The feedback gain of this system was manually adjusted through analog circuits. They utilized the 8MHz, 8-bit microcontroller 8052AH from Intel as the core component of their hardware control system. Despite the ease of development and cost-effectiveness of this core chip, it fell short of meeting the real-time control requirements necessary for optimal performance.[17]

In 2001, Davis et al. used the floating-point control card DS1102 as the hardware foundation for their project. They employed Matlab's Real-Time Workshop to convert the simulation file of the fuzzy control algorithm, designed in Simulink, into C code, which was then downloaded to the control card. This setup enabled the vibration damping control of their solid-state piezoelectric self-tuning power absorber, achieving a damping effect of up to 20dB. However, the control card could not operate independently and required insertion into a PC, leading to high design costs. In 2001, Kidner et al. utilized a Data Acquisition Card (DAQ) to acquire and output signals. They implemented a control algorithm that combined fuzzy control and simple proportional control on a PC using the C++ programming environment. This setup enabled the joint control of stiffness and damping for their mechanical self-tuning vibration absorber, achieving effective vibration damping results. However, this design lacked independent control capability, as the control program had to be run on a PC, making it dependent on external hardware for operation.[18, 19] In 2007, to implement vibration damping control for the magnetorheological elastomer self-tuning absorber designed by Deng Huaxia et al., Lianhua Wang et al. employed a USB7325B data acquisition module for signal input and a 4NIC-CK-HL programmable control power supply for control signal output. These components communicated with the PC via USB and RS232 interfaces, respectively. They developed a genetic control algorithm in a C/C++ programming environment to regulate the excitation coil current, achieving an optimization time of nearly 2 minutes. While this approach significantly improved control time and accuracy, the control program still ran on a PC, preventing the system from achieving embedded operation.[20, 21]

Duan Kebo et al. designed a digital vibration control system with the TMS320F2407 as the core, complemented by an FPGA. This design successfully meets the real-time requirements of the system, provided that it does not need to execute complex algorithms. In 2017, Jinghao Shen et al. developed a hardware control system based on NI Compact RIO for their variable-section beam power absorber. They integrated a frequency-following particle swarm optimization (PSO) control algorithm to effectively manage the cantilever beam power absorber, achieving resonance state control. This approach significantly improved control time compared to using a simple PSO control algorithm, enhancing the system's real-time performance considerably.

In this paper, a cantilever beam-type self-tuning absorber is examined as the research object. The frequency shift characteristics are analyzed in detail using simulation methods, focusing on

Journal of Physics: Conference Series **2910** (2024) 012024 doi:10.1088/1742-6596/2910/1/012024

practical application demands and evaluating the controlled quantities. The hardware circuit design features the STM32F407 as the core processor. Through hardware selection and design, the system meets various control requirements while adapting to space-constrained environments by minimizing the size of the embedded control system, thereby enhancing flexibility in installation and deployment.Additionally, the software control system for the cantilever beam self-tuned vibration absorber is developed by integrating the control algorithm proposed in [20] with its implementation in computer language. A corresponding human-computer interface is constructed to improve operational convenience and enhance the transparency of the self-tuning process.[22-25] The control system developed in this study operates independently, with a simple hardware composition that is easy to extend through open-source resources. The human-computer interface further enhances operational transparency and convenience, providing a solid foundation for subsequent applications.[26]

2. Cantilever Beam Self-Tuning Absorber Structure and Performance

2.1 Structure of cantilever beam self-tuning absorber

The structure of the cantilever beam self-tuning vibration absorber is depicted in the accompanying Figure 2. 1. The vibration absorber has overall dimensions of 300 mm x 35 mm x 50 mm and a total mass of 4 kg. It comprises several components: a motor equipped with a screw, a slider, a slide, a slide rail, a slide rail base, a solid support frame, and a metal plate.The motor, slider, slide rail, slide rail base, and support frame form a relatively fixed motor module. The variable stiffness element of the absorber is a metal plate. One end of the metal plate is fixed by a bracket, while the other end is attached to a slider that can move along the screw rod. The stepping motor, following the control system's instructions, rotates the screw to drive the slider's movement. This movement alters the effective length of the metal plate by shifting the motor module relative to the fixed end of the metal plate, thereby changing the absorber's stiffness. Specifically, the cantilever beam extends from the fixed end to the distance of the fixed frame, adjusting the vibration absorber's stiffness accordingly.The slider is connected to the slide rail to ensure a fixed direction of motion. The slide rail base provides enhanced support for the slide rail and increases the dynamic mass of the vibration absorber. This increase in both static and dynamic mass ratios improves the vibration damping effect of the absorber.

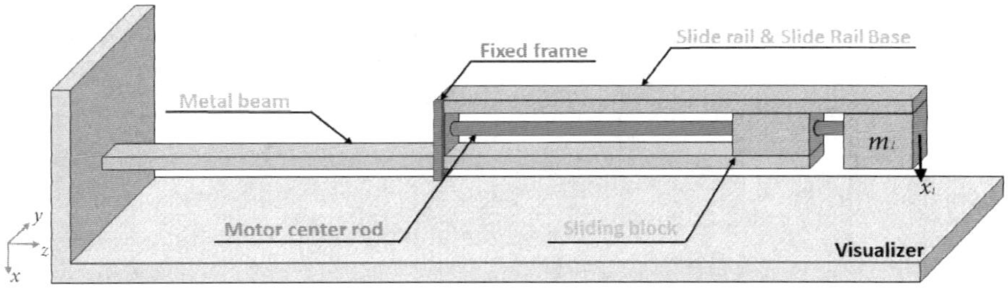

Figure 2. 1 Cantilever Beam Type Adaptive Tuned Absorber Structure

2.2 Analysis of controlled quantities in cantilever beam self-tuning absorbers

In order to understand the dynamics of this absorber, the variable stiffness structure was simulated and analyzed, and finally the correspondence between the cantilever beam self-tuned absorber intrinsic frequency and the effective spreading length of the metal plate was obtained as shown in the following Figure 2. 2.

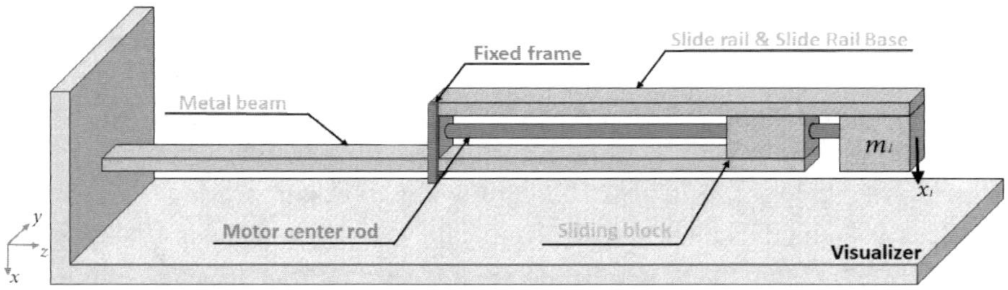

Figure 2. 2 Correspondence between the intrinsic frequency of a self-tuned absorber and the effective spreading length of a metal plate

There is a good linear relationship between the intrinsic frequency of the cantilever beam self-tuned vibration absorber and the effective spreading length of the metal beam. The relationship between the intrinsic frequency of the vibration absorber, f, and the effective spreading length of the metal beam, l, is obtained by numerical fitting as shown in Eq. (1), which serves as the basis for the vibration absorber to undergo the self-tuning process.

$$f = -0.2842l + 78.765 \tag{1}$$

From the results of the structural and dynamic analysis of the cantilever beam self-tuning power absorber, it can be seen that the absorber can be adjusted to adjust the intrinsic frequency of the absorber by changing the spreading length of the cantilever beam, while the adjustment of the spreading length is driven by the rotation of the stepping motors, with the smallest unit of a stepping angle, so the control system outputs directional stepping impulses that are the controllable variables of the cantilever beam self-tuning power absorber.

3. Control system hardware design

The complete vibration damping hardware system workflow is: the main system is subject to external excitation, the acceleration sensor to collect the vibration of the acceleration signal, after charge amplification is converted to a voltage signal, by AD conversion to complete the conversion of analog signals to digital signals. The core circuit reads the signal at this time, through the on-chip resources for DSP commands to identify the vibration signal frequency at this time. According to this frequency to perform the control process, get the target distance signal, the distance sensor to collect the actual distance signal from the vibration absorber system at this time, the two distance signal processing to get the required direction of the pulse value at this time output to the motor drive circuit to regulate the span, so as to adjust the frequency of the vibration absorber, to achieve the purpose of self-tuning vibration damping. At the same time through the serial port and the host computer communication, in order to achieve real-time process data visualization.

The synergy of each hardware system during the working process is illustrated in the Figure3. 1. The acceleration sensors and distance sensors collect data, which is then processed by the master control system. The system utilizes a frequency identification algorithm and a control algorithm to analyze the signals and output control signals to the motor drive, thereby controlling motor operation.The hardware modules of the control system designed in this paper primarily include the main control circuit, signal acquisition circuit, and motor drive circuit. The motor drive circuit is conventionally connected.

Journal of Physics: Conference Series **2910** (2024) 012024 doi:10.1088/1742-6596/2910/1/012024

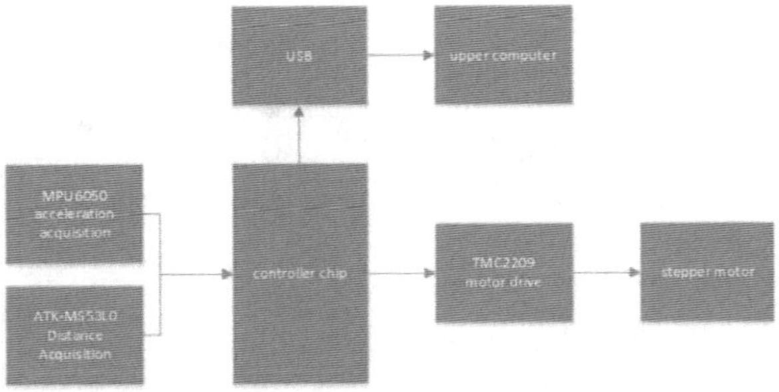

Figure3. 1 Hardware System Connections

3.1 main control circuit

Considering the design requirements and cost considerations of the control system, the CPU chip used in this study is the STM32F407. This 32-bit ARM microcontroller, based on the Cortex-M4 core and produced by STMicroelectronics, serves as the main processor. It integrates a wealth of I/O ports and on-chip peripheral resources, providing an excellent platform for signal processing, motor control, and other industrial control applications. The functional block diagram is shown in the Figure3. 2.

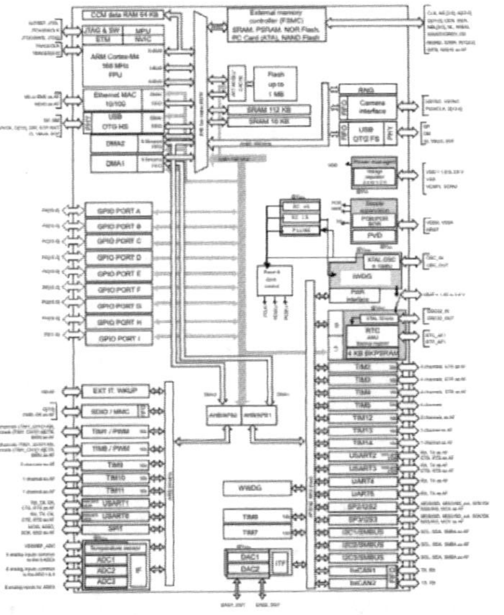

Figure3. 2 On-Camera Peripheral Resources

From the figure, we can understand the chip's extensive on-chip peripheral resources. The table compares the control system's required functions with the chip's related resources, serving as a basis for selecting the main processor.

Table 1. Comparison table of functions required by the control system and related resources of the chip

Control System Functional Requirements	On-Camera Peripheral Resources
Guaranteed high-speed real-time	Operating frequency up to 168 MHz
Real-time online debugging of code	Serial Wire Debug (SWD), JTAG interface
Multiple analog signal sampling	Three 12-bit ADCs

Journal of Physics: Conference Series **2910** (2024) 012024 doi:10.1088/1742-6596/2910/1/012024

High-speed signal processing	Implements a complete set of DSP instructions
Precise control of multiple motors	12 general-purpose 16-bit timers (including 2 PWM timers for motor control), 2 general-purpose 32-bit timers
Multi-environment peripheral circuit interface	I2C, SPI, USART, USB

With the STM32F407 as the main processor, the complete core circuit schematic is shown in Figure3. 3. The minimum system includes the chip, along with externally connected power supply circuits, reset circuits, clock circuits, and program download port circuits.The reset circuit uses a CMOS technique for the reset pin input driver. By default, the pin is pulled up, and a low-level input is accessed through an external key. Pressing the key initiates the reset function with a low-level input. The clock circuit consists of an external high-speed crystal and an external low-speed crystal to enhance clock accuracy. The power supply circuit is integrated with the program download circuit.

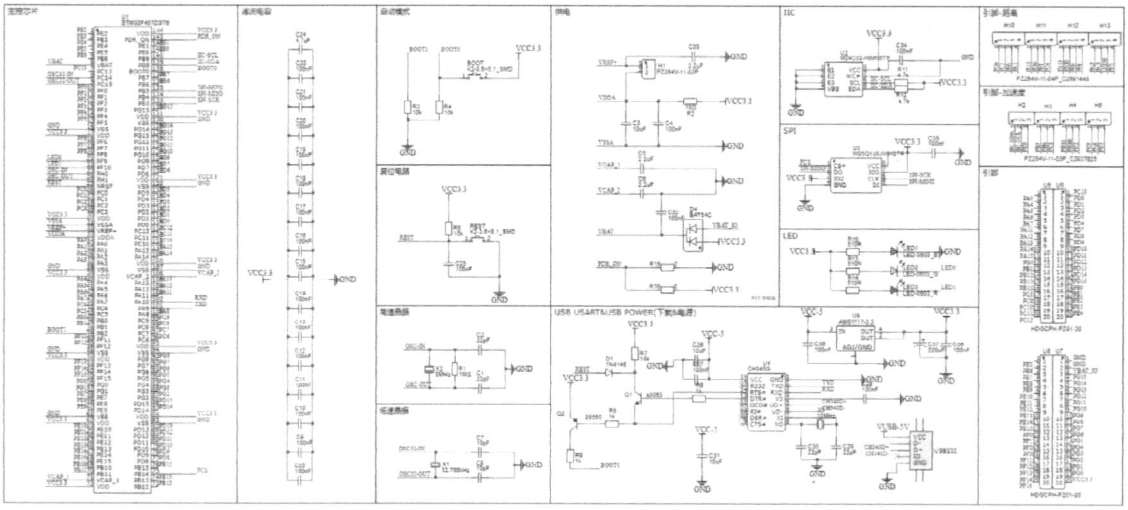

Figure3. 3 Core Circuit Schematic

As shown in the Figure3. 4, the power supply interface for the core circuit is USB232, which operates at 5 V. To enable the CPU chip, it is essential to align the core functions, on-chip functions, and I/O pins with the power supply, which requires a voltage of 3.3 V. This design incorporates the AMS1117 chip circuit to convert the 5 V USB power supply to the required 3.3 V for the chip. Additionally, the CH340 chip circuit is integrated to ensure that the USB interface not only supplies power but also facilitates program downloading and communication with the host computer. This integration reduces the number of components required.

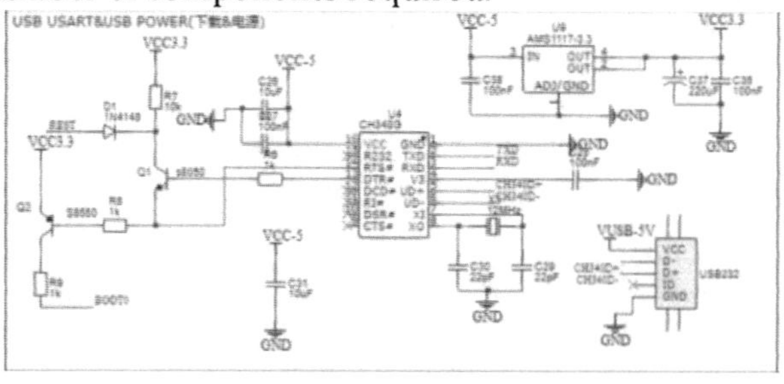

Figure3. 4 power circuit

In the working process of the control system, due to the large number of collected data and the requirement of fast processing speed. Therefore, the Flash circuit with W25Q as the core is integrated in the core circuit, as shown in the Figure3. 5. Its large storage capacity, fast reading and writing speed, low power consumption and other features to meet the needs of this paper on the control system. And compared with other memories, Flash will not lose data due to power loss, so it is used to store important system data.

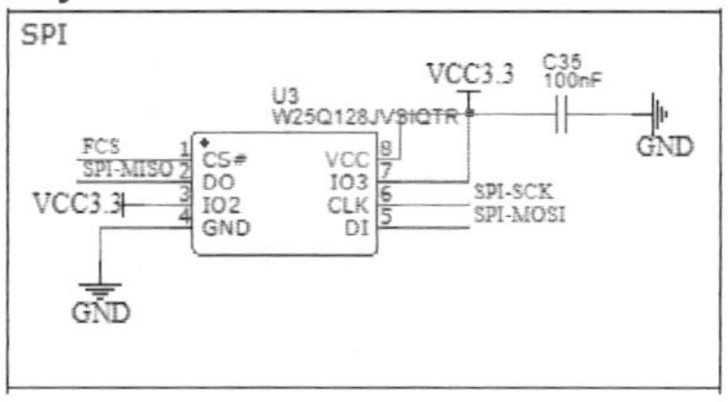

Figure3. 5 Flash Circuit

In addition, the LED circuit is added to the circuit to test the circuit connection, the EEPROM circuit extends the signal acquisition function to realize multi-sensor signal acquisition, and the main control chip pin out for function expansion. The resulting final master control circuit diagram as shown in Figure3. 6.

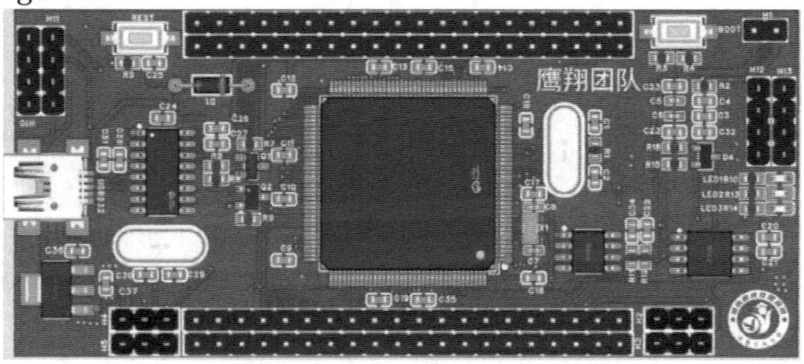

Figure3. 6 Main Control Circuit Diagram

3.2 Acceleration acquisition circuit

MPU6050 contains 3-axis gyroscope and 3-axis accelerometer, which the main role of the gyroscope is to measure the angular velocity of the object around the chip's three axes, the principle is that high-speed rotation of the rotor pointing in the direction of the direction will remain unchanged, that is, the so-called gyroscopic effect, the accelerometer is to measure the acceleration of the three axes, the same also refers to the chip's three coordinate axes. As shown in the Figure3. 7, the principle can be imagined as a cubic box inside a weightless suspended ball, the ball will remain in the middle of the box, the pressure on all six sides is 0.

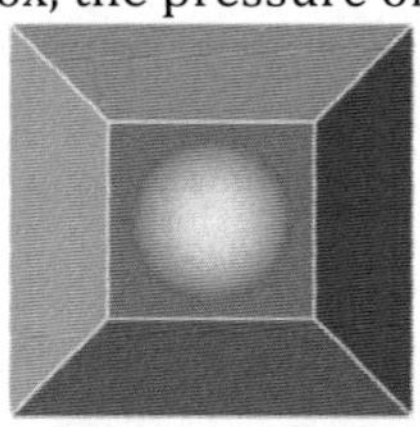

Figure3. 7 Balls of weightlessness

When the accelerometer starts to move, the ball will be displaced relative to the cubic box and touch the inner wall of the box, at which time the ball will exert a pressure of corresponding magnitude on the box. As shown in the Figure3. 8 if the accelerometer suddenly moves to the left, due to inertia the ball will move to the right relative to the box, and the right side of the box will be subjected to a corresponding force.

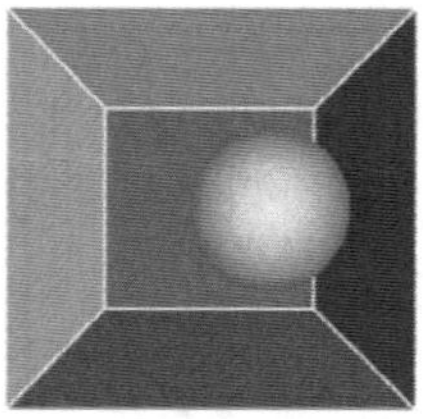

Figure3. 8 Sampler moves to the left

As shown in the Figure3. 9 replacing the cubic box with the new model created, X, Y, and Z are the three coordinate vectors representing the horizontal, vertical, and vertical directions, respectively.

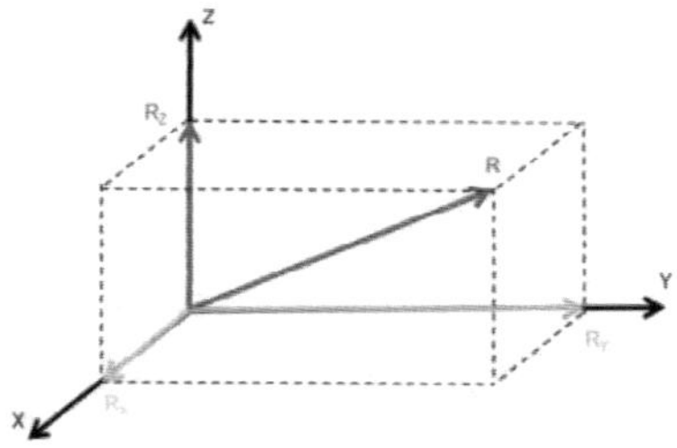

Figure3. 9 Simulated Coordinate Plot

R is the direction vector that replaces the movement of the ball, and the projection on the X, Y, and Z axes are RX, RY, and RZ, respectively, i.e., the real realistic acceleration, and there exists the relation Eq:

$$R^2 = {R_X}^2 + {R_Y}^2 + {R_Z}^2$$

The analog values are converted to digital values by function calculations, the gravity acceleration constant in carrier coordinates is unitized, and the unit gravity vectorization of the coordinate system is then converted to the on-board coordinate system using a rotation matrix. The digitized acceleration value at this time is stored in the function *ax, *ay, *az in order to facilitate the subsequent function call.

The acceleration acquisition circuit is connected to the main control circuit. MPU6050 as the core of the signal acquisition circuit, and the circuit is connected as shown in Figure3. 10.

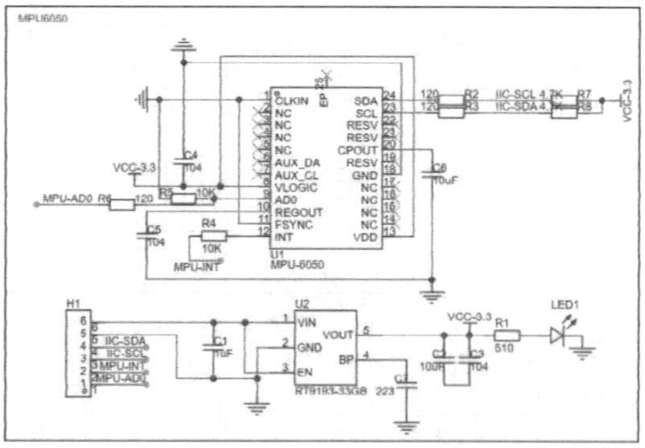

Figure3. 10 Acquisition Circuit Connection Diagram

4. Control system software design

The control process of the cantilever beam type adaptive tuned vibration absorber is as follows: the vibration signals are collected using acceleration sensors, and the Fast Fourier Transform (FFT) is performed to determine the main frequency of the excitation signals, which is preliminarily adjusted by the correspondence between the frequency and the span. Then the intrinsic frequency of the vibration absorber continues to be adjusted through the optimization algorithm to optimize the vibration absorption effect. At the same time, the working process of the vibration absorber is monitored and recorded in real time through the human-machine interface developed according to the control system. The resulting flowchart of the entire control algorithm is shown in Figure4. 1 Control Algorithm Flowchart. The software of the control system mainly consists of the optimization result program, the distance correspondence program, and the driving motor program. All the above programs are written in C language and call the hardware firmware library design.

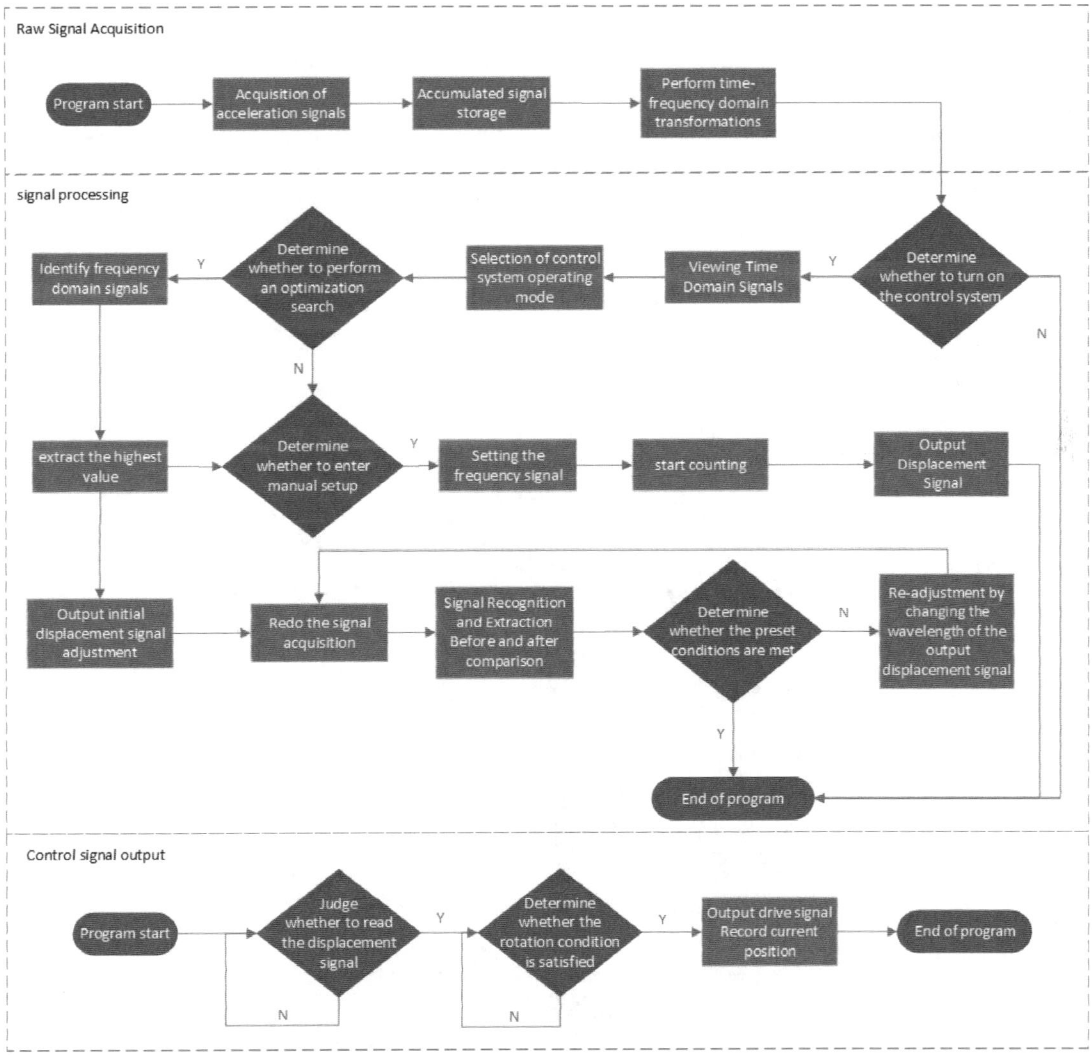

Figure4. 1 Control Algorithm Flowchart

4.1 Optimizing the results program

The control of an adaptively tuned shaker aims to make the intrinsic frequency and excitation frequency of the shaker always consistent. However, it is difficult to achieve the desired results in practice due to certain factors. Therefore, an optimization process is added to the control method to better optimize the vibration damping effect of the vibration absorber. Based on this purpose, the optimization control strategy is investigated in this paper. Firstly, a coarse tuning process is carried out, aiming at adjusting the intrinsic frequency of the vibration absorber to the vicinity of the excitation frequency as soon as possible. Then, the optimization process is carried out, and the magnitude of the acceleration rms is continuously compared during the step size halving process for optimization, and the optimal position is considered to be the smallest value of the acceleration rms. The expression for the root mean square of acceleration rms is:

$$\mathrm{rms} = \sqrt{\sum_{i=1}^{N} a_i^2 \Big/ N}$$

The flowchart of the optimization program is shown in Figure4. 2 Flowchart of the search for excellence. At the beginning, the initial acceleration rms0 is considered to be the minimum value and is assigned to rmsk-1, then the first step of optimization is started with the step size K=K/2, and then the acceleration rmsk at this position is calculated, and the next step of optimization is carried out by comparing the values of the acceleration rms of the two acceleration rmss. By analogy, the subsequent optimization process can be carried out by comparing the sizes of rmsk and rmsk-1. rmsk>rmsk-1 indicates that the vibration is suppressed, and the optimization is continued by halving the step size in the original direction. rmsk<rmsk-1 indicates that the vibration process has worsened, and the optimization is carried out by halving the step size in the opposite direction. This method ensures the rapidity and accuracy of the control process and better optimizes the vibration damping effect of the absorber.

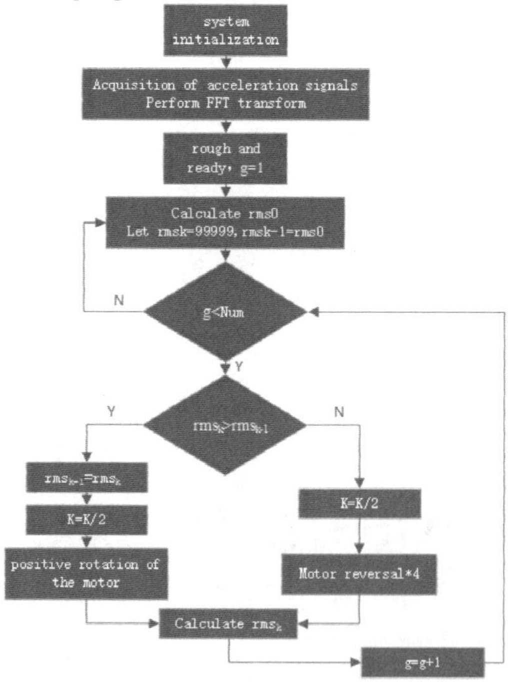

Figure4. 2 Flowchart of the search for excellence

Journal of Physics: Conference Series **2910** (2024) 012024 doi:10.1088/1742-6596/2910/1/012024

4.2 human-computer interface

In practical applications, a well-designed human-computer interface can significantly enhance the user's ability to operate the system and monitor and record the vibration absorption process in real-time. This paper presents the design of a human-computer interface panel for adaptive tuning of a vibration absorber using LabVIEW, as illustrated in Figure4. 3 Human-computer interactive panels. This interface allows users to control the start/stop function of the vibration absorber, select and manually set its operating mode, and observe and record vibration signals. When the control system is deactivated, the vibration absorber functions as a passive device. However, when the control system is active, users can choose between manual and automatic adjustment modes. In the manual setting mode, users can input specific parameters to fix the shaker at a desired frequency. In the auto-tuning mode, the shaker automatically adjusts its settings. Additionally, the observation panel provides a visual representation of the vibration signal changes, facilitating real-time monitoring and operation.

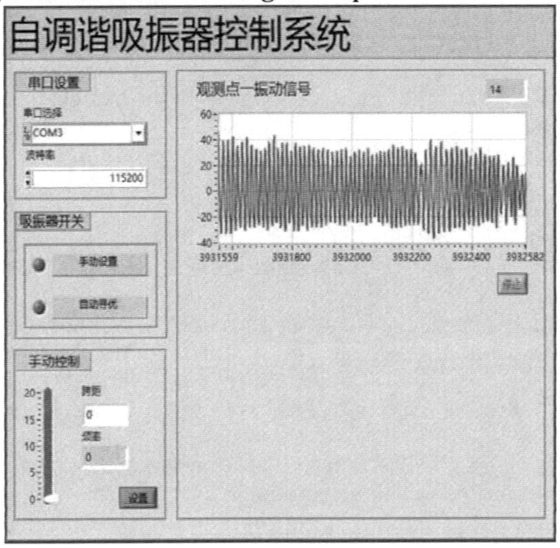

Figure4. 3 Human-computer interactive panels

5. Conclusions

In this paper, a set of self-tuning vibration absorber control system based on STM32 is designed through the analysis of hardware and software, which realizes the embedded operation of the code and improves the precision and stability of the control, and at the same time, the human-machine interface is developed, which realizes the visualization of the damping process under the premise of guaranteeing the basic work of the vibration absorber.

References

[1] S. Bakre, R.J.I.J.o.S.S. Jangid, Dynamics, Optimum multiple tuned mass dampers for base-excited damped main system, 4(04) (2004) 527-542.

[2] R. Lewandowski, J.J.J.o.C.E. Grzymisławska, Management, Dynamic analysis of structures with multiple tuned mass dampers, 15(1) (2009) 77-86.

[3] C.R. Fuller, R.L.J.N.C.E.J. Harne, Passive distributed vibration absorbers for low frequency noise control, 58(6) (2010) 627-635.

[4] T. Bandivadekar, R.J.J.o.V. Jangid, Control, Optimization of multiple tuned mass dampers for vibration control of system under external excitation, 19(12) (2013) 1854-1871.

[5] S. Elias, V. Matsagar, T.J.E.S. Datta, Effectiveness of distributed tuned mass dampers for multi-mode control of chimney under earthquakes, 124 (2016) 1-16.
[6] Y.-Y. Zhao, J.J.J.o.S. Xu, Vibration, Effects of delayed feedback control on nonlinear vibration absorber system, 308(1-2) (2007) 212-230.
[7] S. Megahed, A.K.J.J.o.S. Abd El-Razik, Vibration, Vibration control of two degrees of freedom system using variable inertia vibration absorbers: Modeling and simulation, 329(23) (2010) 4841-4865.
[8] S.-T. Wu, Y.-J.J.J.o.S. Shao, Vibration, Adaptive vibration control using a virtual-vibration-absorber controller, 305(4-5) (2007) 891-903.
[9] F. Beltrán-Carbajal, G.J.J.o.S. Silva-Navarro, Vibration, Active vibration control in Duffing mechanical systems using dynamic vibration absorbers, 333(14) (2014) 3019-3030.
[10] L. Suresh, K.J.P.P.o.S.D. Mini, Construction, Effect of multiple tuned mass dampers for vibration control in high-rise buildings, 24(4) (2019) 04019031.
[11] S. Nagarajaiah, N.J.E.S. Varadarajan, Short time Fourier transform algorithm for wind response control of buildings with variable stiffness TMD, 27(3) (2005) 431-441.
[12] M.J.J.S. Brennan, Vibration, Some recent developments in adaptive tuned vibration absorbers/neutralisers, 13(4-5) (2006) 531-543.
[13] M. Khazaei, R. Vahdani, A.J.S. Kheyroddin, Vibration, Optimal location of multiple tuned mass dampers in regular and irregular tall steel buildings plan, 2020 (2020) 1-20.
[14] A. Matin, S. Elias, V.J.P.o.t.I.o.C.E.-S. Matsagar, Buildings, Distributed multiple tuned mass dampers for seismic response control in bridges, 173(3) (2020) 217-234.
[15] L.F. Fadel Miguel, G.J.S. Piva dos Santos, Vibration, Optimization of multiple tuned mass dampers for road bridges taking into account bridge-vehicle interaction, random pavement roughness, and uncertainties, 2021 (2021) 1-17.
[16] S.G. Hill, S.D.J.J.V.A. Snyder, Design of an adaptive vibration absorber to reduce electrical transformer structural vibration, 124(4) (2002) 606-611.
[17] M. Franchek, M. Ryan, R.J.J.o.S. Bernhard, Vibration, Adaptive passive vibration control, 189(5) (1996) 565-585.
[18] M. Kidner, M.J.S.M. Brennan, Structures, Real-time control of both stiffness and damping in an active vibration neutralizer, 10(4) (2001) 758.
[19] M. Kidner, M.J.J.V.A. Brennan, Varying the stiffness of a beam-like neutralizer under fuzzy logic control, 124(1) (2002) 90-99.
[20] Lianhua Wang, Xinglong Gong, Huaxia Deng, Zhengchao Ni, Qinghe Kong, Magnetorheological elastomer self-tuned absorber and its optimal control, Experimental mechanics, 2007.
[21] KANG Cunjun, GONG Xinglong, CHEN Xianmin, LIAO.J. Vibration and Shock, A study of active self-tuned vibration absorber control system for magnetorheological elastomers, 31(6) (2012) 27-31.
[22] K. Nagaya, A. Kurusu, S. Ikai, Y.J.J.o.s. Shitani, vibration, Vibration control of a structure by using a tunable absorber and an optimal vibration absorber under auto-tuning control, 228(4) (1999) 773-792.
[23] S.M.R. Rasid, T. Mizuno, Y. Ishino, M. Takasaki, M. Hara, D.J.J.o.S. Yamaguchi, Vibration, Design and control of active vibration isolation system with an active dynamic vibration absorber operating as accelerometer, 438 (2019) 175-190.
[24] C.M. Pappalardo, D.J.E.L. Guida, Development of a New Inertial-based Vibration Absorber for the Active Vibration Control of Flexible Structures, 26(3) (2018).
[25] M. Tso, J. Yuan, W.O.J.E.s. Wong, Design and experimental study of a hybrid vibration absorber for global vibration control, 56 (2013) 1058-1069.
[26] K. Williams, G.-C. Chiu, R.J.J.o.s. Bernhard, vibration, Dynamic modelling of a shape memory alloy adaptive tuned vibration absorber, 280(1-2) (2005) 211-234.

Numerical Study of the Good Boussinesq Equation

Krassimir Angelow[1] **and Veselina Vucheva**[2]

[1 2]Institute of Mathematics and Informatics, Bulgarian Academy of Sciences, Acad., G. Bonchev Bl.8, 1113, Sofia, Bulgaria

E-mail: angelow@math.bas.bg

Abstract. In this paper we evaluate propagating wave solutions to the one-dimensional good Boussinesq Equation (BE). Two numerical methods are used to obtain a solution. The first is a conservative finite difference scheme, and the second exploits Taylor series expansions around the time variable t. The solutions are computed over nested meshes to examine the convergence of both approaches. The main tool for testing the convergence rate of all examined finite difference schemes and Taylor expansions is the Runge's method. It is shown that for a fixed time interval the numerical methods preserve the shape of the waves after interaction and mass of the solution. The waves obtained by the two methods are compared with the exact solution. Both methods produce similar results in case of $O(h^2+\tau^2)$ approximation order, with errors measured in L_2 and L_∞ norms which are 100 times or less than the solution maximum.

1 Introduction

The Boussinesq type equations are famous with the approximation of shallow water waves or also weakly non–linear long waves. It is often used for simulation of various physical processes e.g. turbulence in fluid mechanics, vibrations in acoustics, etc. For the numerical interaction of Boussinesq traveling waves (TW) one needs the shape of a stationary wave in order to build the Initial Condition (IC) u_0, u_1.

The shallow water model developed by Boussinesq and the equaitons he derived in this conjuction were first described in [1]. The equation is known to have an explicit solution in the case of one moving solitary wave. It is also known that it has a solution with two solitary waves that results in their temporary combination and subsequent reemergence in their original forms after the interaction [7, 8]. An improvement of the Boussinesq model was done by Christov's "energy-consistent approximation" [3] a.k.a the Boussinesq Paradigm Equation (BPE). Christov states that the original model with the initial value problem defined by Boussinesq is incorect in the sense of Hadamard (see [3]), i.e. even small perturbations in the initial or boundary conditions could lead to significant differences in the behavior of the solution over time. Unfortunately the BPE is not known to posses an explicit solution in the case of two solitary waves. We would like to compare the numerical and the exact solutions in the case of two solitary waves. This could be done in the case of the original (a.k.a good) Boussinesq model:

$$\begin{aligned}&\frac{\partial^2}{\partial t^2}u(x,t) = \frac{\partial^2}{\partial x^2}u(x,t) - \frac{\partial^4}{\partial x^4}u(x,t) - 3\frac{\partial^2}{\partial x^2}f(u(x,t)) \\ &u(x,0) = u_0(x), \quad \frac{\partial}{\partial t}u(x,0) = u_1(x), \\ &u(x,t) \to 0, \quad \frac{\partial}{\partial t}u(x,t) \to 0 \quad \text{for} \quad |x| \to \infty,\end{aligned} \tag{1}$$

where $f(u) = u^2$. The domain of the unknown function u is defined by:

$$u : \Omega \times [0,T] \to \mathbb{R}, \tag{2}$$

Content from this work may be used under the terms of the Creative Commons Attribution 4.0 licence. Any further distribution of this work must maintain attribution to the author(s) and the title of the work, journal citation and DOI.
Published under licence by IOP Publishing Ltd

Journal of Physics: Conference Series **2910** (2024) 012025 doi:10.1088/1742-6596/2910/1/012025

where $\Omega \equiv (-\infty,\infty)$ and $T > 0$. The continuous energy of the solution is defined by

$$E\left(u(x,t)\right) = \frac{1}{2}\left\|(-\partial_x^2)^{-1/2}\frac{\partial u}{\partial t}(x,t)\right\|^2 + \frac{1}{2}\left\|u(x,t)\right\|^2 + \frac{1}{2}\left\|\nabla u(x,t)\right\|^2 + \int_R u(x,t)^3 dx. \tag{3}$$

The mass of the solution is defined by

$$M(u(x,t)) = \int_{\mathbb{R}} u(x,t)dx. \tag{4}$$

Equation (1) is equivalent to the BPE if the mixed time and space derivative is omited. E.g. if we set $\beta_1 = 0$, $\beta_2 = 1$ and $\alpha = 3$ in equation (2) from article [2] (Chertock, Christov, Kurganov) then it is equivalent to (1) defined above. Equation (1) obtains two wave soliton solution (see formula (15) defined in section Numerical Results). The soliton is localized wave, which moves at constant velocity (ω_i, $i = 1,2$ in case of (15)) and maintains its shape. When a soliton interacts with another soliton, it emerges from the "collision" unchanged, i.e. its amplitude, shape, and velocity are conserved.

The goal of the article is to show that one could obtain a numerical solution for the 1D problem (1) in the frame of two solitary waves moving towards each other. Furthermore, the numerical analysis shows that increasing the approximation order and decreasing the space step, produces numerical solution with smaller error norms. The mass and the energy (in case of the Conservative Finite Difference Scheme) norms are preserved with the same accuracy as the numerical solution. The results and observations for the good Boussinesq equation (1) could be further used in order to find a two wave solitary solution for the 1D and 2D BPE.

The paper is organized as follows. Section 2 introduces the numerical methods that are used for equation (1): a Conservative Finite Difference Scheme (FDS) and Taylor method, which uses Taylor Series (TS) expansions around the time variable t. Section 3 concerns the numerical results. It states the initial condition, boundary condition, space and time domains. The numerical calculations are done over nested meshes to examine the convergence of both methods. The goal is to justify the TS approach by showing that both methods exhibit similar results. Furthermore the TS method could be used with higher approximation order which produces a finer solution. Subsection 3.1 shows the convergence speed of the Conservative FDS and energy measured with Rugne's rule. The next section shows the convergence speed of the Taylor method. Furthermore, for each method, the mass of the solution is calculated which is described in Subsection 3.3. In Subsection 3.4, the shape, mass and energy of the solution are investigated one more time, over a larger time span, where the solitary waves pass through each other. The last section gives an overview of the achieved goals and summarize the results.

2 Numerical Methods

The space domain $(-\infty,\infty)$ is replaced with a finite interval $[L_1,L_2]$, where $|L_i|$, $i = 1,2$ are sufficiently large numbers, so that the exact solution to (1) and its derivatives are negligible outside this interval. A uniform grid x_i, $i = 0,1,\cdots N$ is introduced with step $h = |L_2 - L_1|/(N-1)$ in the interval $[L_1,L_2]$. The space domain is one dimensional, and the Δ_h operator below denotes the approximation of the second derivative $\partial^2/\partial x^2$. Let τ be the time step in $[T_0,T]$ and $T = T_0 + K\tau$ be the final time.

2.1 *Conservative Finite Difference Scheme with weight σ and approximation error $O(h^2+\tau^2)$*

We use the following finite differences

$$y^k_{\bar{t}t,i} = \frac{y_i^{k+1} - 2y_i^k + y_i^{k-1}}{\tau^2}, \quad \Delta_h y_i^k = \frac{y_{i+1}^k - 2y_i^k + y_{i-1}^k}{h^2} \tag{5}$$

$$\Delta_h^2 y_i^k = \frac{y_{i+2}^k - 4y_{i+1}^k + 6y_i^k - 4y_{i-1}^k + y_{i-2}^k}{h^4}. \tag{6}$$

Here y_i^k denotes the approximation of the unknown function u, on node x_i, at moment $t_k = T_0 + k\tau$, i.e. $y_i^k \approx u(x_i,t_k)$. Replacing the derivatives (5) and (6) in (1) leads to the following equation:

$$y^k_{\bar{t}t,i} = \Delta_h y_i^{\sigma,k} - \Delta_h^2 y_i^{\sigma,k} - 3\Delta_h g(y_i^k). \tag{7}$$

Here $y_i^{\sigma,k} = y_i^k + \sigma\tau^2 y^k_{\bar{t}t,i}$ and the nonlinear term is approximated as in [6] with:

$$g(y_i^k) = \frac{1}{3}\frac{(y_i^{k+1})^3 - (y_i^{k-1})^3}{y_i^{k+1} - y_i^{k-1}}. \tag{8}$$

This leads to the following Conservative finite difference scheme:

$$\frac{\sigma\tau^2}{h^4}y_{i-2}^{k+1}+\left(-4\frac{\sigma\tau^2}{h^4}-\frac{\sigma\tau^2}{h^2}\right)y_{i-1}^{k+1}+\left(1+6\frac{\sigma\tau^2}{h^4}+2\frac{\sigma\tau^2}{h^2}\right)y_i^{k+1}+\left(-4\frac{\sigma\tau^2}{h^4}-\frac{\sigma\tau^2}{h^2}\right)y_{i+1}^{k+1}+\frac{\sigma\tau^2}{h^4}y_{i+2}^{k+1}=$$
$$=2y_i^k-y_i^{k-1}-2\sigma\tau^2\Delta_h y_i^k+\sigma\tau^2\Delta_h y_i^{k-1}+2\sigma\tau^2\Delta_h^2 y_i^k-\sigma\tau^2\Delta_h^2 y_i^{k-1}+\tau^2\left(\Delta_h y_i^k-\Delta_h^2 y_i^k-3\Delta_h g(y_i^k)\right). \quad (9)$$

The last equation needs to be resolved for the upper time layer y^{k+1} and the nonlinear term also depends on y^{k+1}, which makes (9) implicit. For this, we do Piccard iterations at each time step.

The complexity of the Conservative FDS applied for the Good Boussinesq equation is

$$O(N_x N_t)$$

where N_x is the number of points in Ω_h and $N_t = T/\tau$. The most time consumable operation is inverting the following five band matrix

$$\left[\frac{\sigma\tau^2}{h^4},\left(-4\frac{\sigma\tau^2}{h^4}-\frac{\sigma\tau^2}{h^2}\right),\left(1+6\frac{\sigma\tau^2}{h^4}+2\frac{\sigma\tau^2}{h^2}\right),\left(-4\frac{\sigma\tau^2}{h^4}-\frac{\sigma\tau^2}{h^2}\right),\frac{\sigma\tau^2}{h^4}\right]$$

which results from equation (9). This is an operation with $O(N_x)$ algorithmic complexity. The use of Picard iterations (which are not more than six per time step) affects complexity by a constant.

2.2 *Discrete energy*

Let $E_h(y^k)$ be the discrete energy, which is defined in the following way:

$$E_h(y^k) := E_{h,L}(y^k)+\frac{1}{2}((y^{k+1})^3+(y^k)^3,1). \quad (10)$$

The linear part $E_{h,L}$ is given by

$$E_{h,L}\left(y^k\right) := \frac{1}{2}\left(A^{-1/2}y_t^k, A^{-1/2}y_t^k\right)+\left(\frac{\sigma\tau^2}{2}-\frac{\tau^2}{8}\right)\left(y_t^k,y_t^k\right)+$$
$$+\left(\frac{\sigma\tau^2}{2}-\frac{\tau^2}{8}\right)\left(Ay_t^k,y_t^k\right)+\frac{1}{8}\left((I_d+A)\left(y^{k+1}+y^k\right),y^{k+1}+y^k\right)$$

where $A=-\Delta_h$ and I_d is the Identity matrix. The discrete energy $E_h(y^k)$ approximates the energy (3) with error $O(h^2+\tau^2)$.

Discrete conservation law

$$E_h(y^k)=E_h(y^0),\ k=1,2,...,K.$$

The proof is a consequence of the stability results from the book of Samarskii [4] and it is derived in [6].

2.3 *Taylor Series Approach with Method of Lines*

Here we proceed as in [5] and use similar approach as the one defined for the BPE. The finite differences along space and time discretization require Taylor Series (TS) expansions of $u(x,t)$. Therefore it is assumed that the solution is $p+1$ times infinitely differentiable with respect to x and t, i.e. $u\in C^{p+1,p+1}(\Omega\times T)$. For the Taylor method two different approximations of the spatial differential operator Δ_h are used. The following central finite differences along the x asis are applied:

$$u_{\hat{x}\hat{x},p}(x,y) := \Delta_h u = \frac{1}{h^2}\sum_{i=-p/2}^{p/2} d_i u(x+ih,t). \quad (11)$$

The weights d_i are defined below in Table (1). E.g. if $p=2$, then (11) is equivalent to the spatial finite difference in (5). The approximation error of formulas (11) is $O(h^p)$. Let $u_i(t)$ and $u_{i,\hat{x}\hat{x},p}(t)$ be the approximations of the unknown function u and its second derivative u_{xx} at arbitrary mesh point (x_i) for arbitrary time t. Then one obtains a system of ODEs:

$$\frac{\partial^2}{\partial t^2}u(x_i,t)=(u_i-u_{i,\hat{x}\hat{x},p}-3u_i^2)_{\hat{x}\hat{x},p}(t) \quad (12)$$

$p=2$		1	-2	1	
$p=4$	$-\frac{1}{12}$	$\frac{4}{3}$	$-\frac{5}{2}$	$\frac{4}{3}$	$-\frac{1}{12}$

Table 1: Finite differences used for the approximation of the Laplace operator.

for all mesh points $i = 0..N_x$. For each ODE in the system we do TS expansion along the time variable:

$$u(x_i, t+\tau) = u(x_i, t) + \tau\frac{\partial u}{\partial t}(x_i, t) + \ldots \frac{\tau^p}{p!}\frac{\partial^p u}{\partial t^p}(x_i, t) + O(\tau^{p+1}) \tag{13}$$

for some natural number $p \geq 2$. The approximation order of the time discretization depends on p, i.e. the number of terms included in the TS expansion. Each point on the mesh represents a starting point of a line and the line itself is described by the TS expansion (13). Evaluation of formula (13) is done by evaluating each term separately. E.g. for $t = 0$ the first two terms are known from the IC (u_0, u_1). The third term is evaluated from the discrete equation (12). With subsequent differentiaton of equation (12) one could obtain higher time derivatives $\frac{\partial^3 u}{\partial t^3}$, $\frac{\partial^4 u}{\partial t^4}$, etc. This is an iterative procedure where e.g. the fourth time derivative requires 2nd, 1st and 0th derivatives, 3rd time derivative requires 1st and 0th derivatives. After all necessary time derivatives are calculated one could substitute those in (13) and gets the following approximations: for $p = 2$ it is $O(|h|^2 + \tau^2)$ and for $p = 4$ it is $O(|h|^4 + \tau^4)$. Note that another TS expansion must be calculated for the first time derivative $u_t(x_i, t+\tau)$, which is analogous to (13) with the same approximation order. The last is required in order to calculate the solution on the next time layer because the pair (u, u_t) serve as basis and all higher time derivatives could be expressed only by that pair.

The complexity of the algorithm is

$$O(N_x N_t)$$

where N_x is the number of points in Ω_h and $N_t = T/\tau$. Current choices of p affect complexity by a constant which resutls in a linear time graph depending on the number of points in the discrete domain.

3 Numerical Results

The convergence rates for the Conservative FDS and Taylor method are calculated. At the end, the properties of the solution that are obtained numerically by the two different mechanisms are compared. The mass is a vector of size N_t and is calculated for each iteration step. The tool for testing the convergence rate ξ of all examined finite difference schemes and TS expansions is the Runge's method

$$\xi = ln\frac{\|u_h - (u_{ex})_h\|_\kappa}{\|u_{h/2} - (u_{ex})_{h/2}\|_\kappa}|/\ln(2), \tag{14}$$

where u_{ex} is the exact solution (15), which is defined below, over the corresponding grid with step size h or $h/2$. Here, κ denotes the used norm, which is either L_2 or L_∞. The time step τ is kept constant.

$$\begin{aligned} u(x,t) =& -2\frac{\partial lnF(x,t)}{\partial x^2}, \\ F(x,t) =& a_0 + a_1 e^{k_1 x + \omega_1 t + b_1} + a_2 e^{k_2 x + \omega_2 t + b_2} + a_{12} e^{(k_1+k_2)x + (\omega_1+\omega_2)t + b_1 + b_2}, \\ |k_i| <& 1,\ \omega_i = \sqrt{k_i^2(1-k_i^2)},\ k_i \neq 0, \quad i = 1,2, \\ a_0 =& \frac{a_1 a_2}{a_{12}}\frac{4k_1^4 + 4k_2^4 - 3k_1^2 - 3k_2^2 - 2k_1^2k_2^2 + 6\sqrt{k_1^2 - k_1^4}\sqrt{k_2^2 - k_2^4}}{(k_1+k_2)^2(4(k_1^2 + k_1k_2 + k_2^2) - 3)}. \end{aligned} \tag{15}$$

Most of the work for the derivation of the exact solution, with two solitary waves, could be found in [7, 8]. Some additional calculations were done by Vasil Vasilev where he provided formulas (15) (private communication). The parameters in (15) for the numerical tests are defined as:

$$k_1 = 1/3,\ k_2 = -1/2,\ b_1 = -20\sqrt{(k_1^2(1-k_1^2))},\ b_2 = -20\sqrt{(k_2^2(1-k_2^2))},\ a_1 = a_2 = a_{12} = 1. \tag{16}$$

The exact solution, with the above parameters (see Figure (1)), is tested against the differential equation (1), using free software mathematical applications, which justified the validity of (15).

The numerical tests for both, the Conservative FDS and Taylor method ($p = 2$), show that the time step needs to be smaller than the space step squared $\tau < h^2 < 1$. For the Taylor method with $p = 4$, the restriction is even

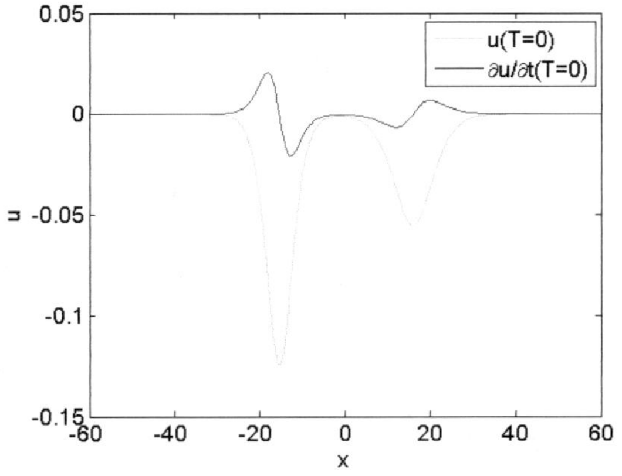

Figure 1: The exact solution (15) and its time derivative with parameters defined in (16). These are the two waves which move against each other.

stronger. Otherwise, the solution diverges and becomes non-smooth (e.g. the graph is jagged) and there is no convergence (see Figure (2)b)). The last explains the choice of h, τ when measuring the convergence speed in Table (3), Table (4) and Table (5). A short numerical analysis on the this statement could be found in Subsection 3.2 and Figure (2), where we measure the effect of decreasing the time step on the convergence speed. It is also inapplicable to measure the time convergence as the space errors are much higher ($\tau < h^2 < 1$). Thus the time step τ is kept constant when measuring convergence speed with Runge's rule (14).

	Variant a	Variant b	
Grid 1	$h = 0.4$	$h = 0.2$	$\tau = 2.5e-04$
Grid 2	$h = 0.4$	$h = 0.2$	$\tau = 1.0e-06$
Grid 3	$h = 0.8$	$h = 0.4$	$\tau = 6.25e-05$

Table 2: Space and time steps used to benchmark the Conservative FDS and the Taylor method.

The numerical methods described in the article are tested (at least) on six different grids defined in Table (2). The Conservative FDS uses second approximation order on all grids while the Taylor method uses second (on Grid 1 and Grid 2) and fourth approximation order (on Grid 2 and Grid 3). Later, other combinations of h and τ are also used and are explicitly defined.

The discrete time interval for the numerical simulations is $[0, 10]$ unless stated otherwise. For that time, the two waves move against each other and at the moment $T = 10$ are seen just before their collision. There is one additional experiment, with a time interval $[0, 35]$, where the waves pass through each other (i.e. collide) and the distance between them at the final moment $T = 35$ is approximately the same as at the initial moment.

All computations are done with $-L_1 = L_2 = 60$. Zero boundary condition is used on the computational edges of the space domain $[L_1, L_2]$. In case of the second derivative, with fourth approximation order, used inside the Taylor method, the finite difference stencil extends an extra point outside the L_1 and L_2 boundaries and the numerical solution is approximated again with 0. The same is also valid for the fourth spatial derivative, with second approximation order, used in the Conservative FDS. This decision is taken because, the unknown function near the computational boundaries is calculated to be lower than $1e-10$ (using (15)). Furthermore, it is confirmed that this approximation does not affect the convergence speed caclulated with Runge's rule (see Table (5) and (3)).

3.1 Convergence Rate for the Conservative FDS

Table (3) measures the convergence speed based on the results of the numerical solution u_h. The σ parameter in (9) is set to $\sigma = 0.75$. The Conservative FDS applies only second approximation order, thus $p = 2$ for all completed tests in this subsection. Three nested meshes are used with different step sizes which are present in the second column. The next two columns show the errors $\|u_h - (u_{ex})_h\|_\kappa$, $\|u_{(h/2)} - (u_{ex})_{h/2}\|_\kappa$ and convergence speed ξ from (14), which are measured in L_2 and infinity norms. The time step $\tau = 0.001$ is kept constant.

FDS	h, τ	errors $E_i$inL_2	Conv. Rate	errors $E_i$inL_∞	Conv. Rate
	0.4, 0.001	0.0007711747		0.0005850814	
$O(h^2+\tau^2)$	0.2, 0.001	0.0001231543	2.65	0.0001279704	2.19
	0.1, 0.001	0.0000193704	2.67	0.0000236392	2.44

Table 3: Space convergence speed for the solution obtained by the Conservative FDS with zero boundary conditions and approximation errors $O(h^2)$. Errors E_i are measured in L_2 and L_∞ norms

FDS	h, τ	errors $E_i$inL_2	Conv. Rate	errors $E_i$inL_∞	Conv. Rate	$max_k\|E(y^0)-E(y^k)\|$
	0.4, 0.001					8.6644e-07
	0.2, 0.001	0.0000006687		0.0000149559		8.6649e-07
$O(h^2+\tau^2)$	0.1, 0.001	0.0000001757	1.93	0.0000039284	1.93	8.6490e-07
	0.05, 0.001	0.0000000482	1.87	0.0000010792	1.86	8.6685e-07

Table 4: Convergence speed for the discrete energy using the Conservative FDS and approximation errors $O(h^2)$. Errors E_i are measured in L_2 and L_∞ norms. The last column measures the deviation of the energy.

Table (4) is analogous to Table (3) and measures the convergence speed based on the results of the discrete energy (10). The only difference is that in the last Table (4), the energy is calculated over four nested meshes and no exact formula is used in the Runge's rule. The convergence rate of both solution and energy is found to be in correspondence with the second spatial approximation order that is used. The last column in table measures the deviations of the energy with respect to its initial value (equal to 0.0889) and is computed using formula (10). There, the measured differences are lower than the errors $\|u_h-(u_{ex})_h\|_K$ of the numerical solution, measured in Table (3).

In conclusion, we see that the energy is preserved with even higher accuracy than the numerical solution and both of those are calculated with a convergence rate of about 1.86 or higher. Decreasing the step size h decreases the errors of the numerical solution.

3.2 Convergence Rate for the TS Approach with Method of Lines

Table (5) measures the convergence rate based on the results of the numerical solution u_h. The Taylor method applies second and fourth approximation order, i.e. $p=2$ on Grid 1 and Grid 2 and $p=4$ on Grid 2 and Grid 3, which is defined in the first column. Three nested meshes are used with different step sizes which are present in the second column. The next two columns show the errors $\|u_h-(u_{ex})_h\|_K$, $\|u_{h/2}-(u_{ex})_{h/2}\|_K$ and convergence speed ξ from (14), which are measured in L_2 and infinity norms. The convergence results for the Taylor solution correspond to the second and fourth approximation order of the space derivatives that are used.

TS	h, τ	errors $E_i$inL_2	Conv. Rate	errors $E_i$inL_∞	Conv. Rate
	0.4, 2.5e-04	0.0017022652		0.0012462726	
$O(h^2+\tau^2)$	0.2, 2.5e-04	0.0002766992	2.62	0.0002926296	2.09
	0.4, 1.0e-06	0.0007969647		0.0006080163	
$O(h^2+\tau^2)$	0.2, 1.0e-06	0.0001398547	2.51	0.0001508256	2.01
	0.4, 1.0e-06	0.0000091188		0.0000070642	
$O(h^4+\tau^4)$	0.2, 1.0e-06	0.0000003715	4.61	0.0000003563	4.31
	0.8, 6.25e-05	0.0001962184		0.0001058293	
$O(h^4+\tau^4)$	0.4, 6.25e-05	0.0000104438	4.23	0.0000068085	3.96

Table 5: Space convergence rate for the solution obtained by the Taylor method with zero boundary conditions and approximation errors E_i of second and fourth order - $O(h^2)$ and $O(h^4)$ - measured in L_2 and L_∞ norms.

Figure (2) shows the effect of modifing the time step τ on the solution convergence. The nested space grids obtain the following step sizes: $h=0.8, 0.4$, with end time $T=10$ and fourth approximation order $p=4$. For

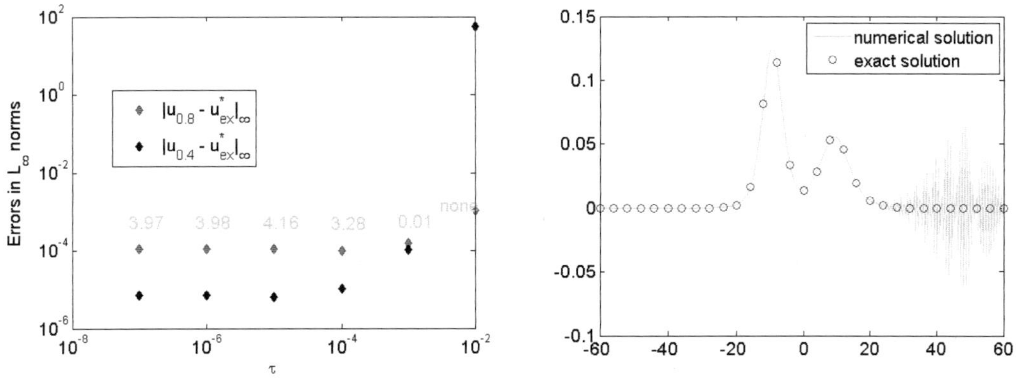

2a) 2b)

Figure 2: 2a) On the left panel we see the effect of decreasing the time step τ. Errors E_i are measured in L_∞ norms and fourth approximation order ($p = 4$). The green values mark the convergence speed, measured with Runge's rule (14), based on the measured errors E_i. 2b) On the right panel, the solution diverges as the time step is not small enough. This is numerical example, using the TS approach, with $h = 0.4$, $\tau = 0.01$ and end time $T = 7$.

$\tau > 1e-4$, the stability of the Taylor method breaks as the L_2 and L_∞ errors become much higher compared to the maximum of the solution wave $|u_h|_\infty$. The solution diverges and becomes non-smooth (e.g. the graph is jagged) and there is no convergence. On the other hand, if the time step decreases and $\tau \leq 1e-4$, convergence speed is near 4, which reflects the fourth approximation order that is used. The respective convergence speeds and results when using the L_2 error norms are similar and are omitted. From this analysis, we could see that $\tau < h^4$ (see Figure (2b)) is a necessary condition for the stability of the Taylor method with $p = 4$.

In conclusion, we see that the numerical solution is calculated with a convergence rate of about 2.01 or higher when $p = 2$ and 3.96 or higher when $p = 4$. Decreasing the step sizes h or τ, or increasing the approximation order p, decreases the errors of the numerical solution.

3.3 *Numerical Results for the Mass*

The mass is compared on Grid 1 and Grid 3 from Table (2) and the results are shown in Table (6). Similar results are obtained for Grid 2, when using the Conservative FDS or Taylor method with $p = 2$ and $p = 4$, and thus are omitted. The first column in Table (6) shows the method which is used - Conservative FDS or Taylor method. The space and time steps are present in the second column. For the cases with $p = 2$, we use Trapezoidal rule for integration. In case of $p = 4$ and Taylor method, we use Simpson (1/3) rule. The results, with the initial mass and its maximum deviation, are displayed in the third and fourth columns respectively. One could see that all deviations

method	(h, τ)	$M(y^0)$	$max_k\|M(y^0) - M(y^k)\|$
Cons FDS	(0.4, 2.5e-4)	1.666666e+00	1.959443e-07
Taylor, $p = 2$	(0.4, 2.5e-4)	1.666666e+00	4.659617e-08
Cons FDS	(0.2, 2.5e-4)	1.666666e+00	3.424187e-07
Taylor, $p = 2$	(0.2, 2.5e-4)	1.666666e+00	3.499779e-08
Cons FDS	(0.8, 6.25e-5)	1.666667e+00	8.198307e-07
Taylor, $p = 4$	(0.8, 6.25e-5)	1.666666e+00	3.720412e-08
Cons FDS	(0.4, 6.25e-5)	1.666667e+00	2.714807e-06
Taylor, $p = 4$	(0.4, 6.25e-5)	1.666666e+00	3.002585e-08

Table 6: The mass and its deviations from initial value measured on different grids.

are lower (at least 10 times) than the errors measured when comparing the numerical solution with the exact one (see Table (3) and Table (5)). On Figure (3) one could see the mass plotted for Grid 1. The values of the mass over the other grids are similar and thus are omitted. Table (6) and Figure (3) confirm that the mass is preserved with better accuracy than the numerical solution, over a fixed time interval $[0, 10]$.

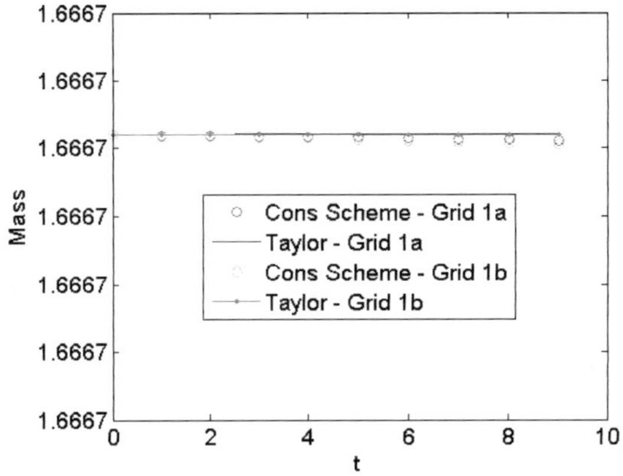

Figure 3: The mass of the solution on Grid 1, using the solutions from both Conservative FDS and Taylor method.

3.4 Numerical Results for the shape of the solution and its mass and energy on a longer time interval $T = 35$
Here, we present results for a longer time interval $[0, 35]$, obtained using the Conservative FDS and Taylor method. On Figure (4) one could see the evolution of the shape for that time, using the Conservative FDS. An analogous picture is obtained for the Taylor method and is omitted. The results discussed below are summarized in Table (7).

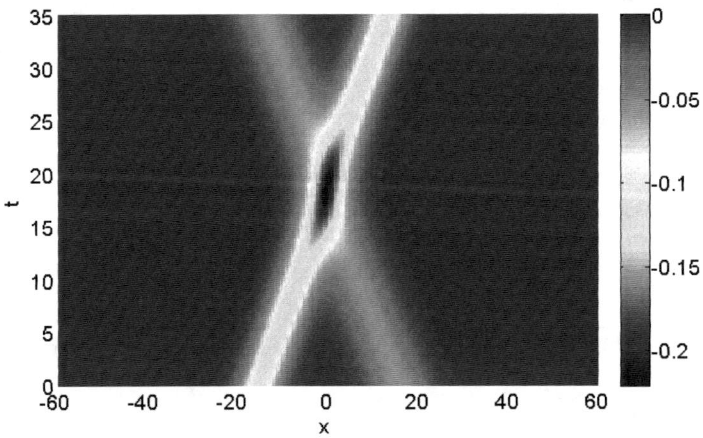

Figure 4: Evolution of the shape over a larger time interval $[0, 35]$. The soluiton is produced using the Conservative FDS.

The errors measured in L_2 and L_∞ norms for the Conservative FDS are 0.001257 and 0.000640. The space and time steps are as follow: $h = 0.4$ and $\tau = 0.05$, with $\sigma = 0.75$. The initial energy equals 0.0889 and its maximum deviation from the initial value equals $1.4076e-04$. The initial mass equals 1.66 and its maximum deviation from the initial value equals $2.3608e-04$.

The errors measured in L_2 and L_∞ norms for the Taylor method are 0.000629 and 0.000385. The space and time steps are as follow: $h = 0.4$ and $\tau = 0.001$. The initial mass equals 1.66 and its maximum deviation from the initial value equals $1.1196e-05$.

At the final moment, the difference between the exact and numerical solutions, measured in L_2 and L_∞ norms, is at least 100 times less than the absolute value of the maximum of the wave $|u_h|_\infty = 0.1248$. For that longer time period, the energy and mass norms are also preserved with high accuracy.

Method	(h,τ)	$\|\|u_h-(u_{ex})_h\|\|_{L_2}$	$\|\|u_h-(u_{ex})_h\|\|_\infty$	$max_k\|E_h-E_0\|_\infty$	$max_k\|M_h-M_0\|_\infty$
Cons FDS	(0.4, 0.05)	0.001207	0.000640	$1.4076e-04$	$2.3608e-04$
Taylor, $p=4$	(0.4, 0.001)	0.000629	0.000385	—	$1.1196e-05$

Table 7: Error norms of the solution, energy (in case of Conservative FDS) and mass. Here u_{ex} is the exact solution (15) and u_h is the numerical solution obtained either with the Conservative FDS or with the Taylor method, which is defined in the first column.

4 Conclusion

The good BE is solved using the Conservative FDS and TS approach with method of lines. The first utilizes second order approximations and preserves the discrete energy with accuracy higher than the numerical solution itself. The second exploits TS expansions around the time variable t, with up to the fourth time derivative and fourth order finite difference approximation over the space derivative Δu. The numerical solution is compared with the exact one, using L_2 and L_∞ norms. From the numerical tests conducted, we observe that decreasing the space step h twice, results in four times lower errors norms $\|u_h-(u_{ex})_h\|_K$, in case of second approximation order, and sixteen times lower, in case of fourth approximation order. Increasing the approximation order (from second to fourth) and keeping h,τ fixed, when using the Taylor method, leads to (accuracy boost with) 85 times or much lower error norms (see Table (5)). The deviations of the discrete mass are lower than the error norms of the numerical solution. In a conducted numerical example for a larger time interval $T=35$, we observe that the error norms are at least 100 times smaller than the absolute value of the solution maximum, at the final time. In this example, the two waves collide with each other and interchange positions on the x-axis. The collision of the two waves is elastic. For that longer time period, the discrete energy and mass norms are also preserved with high accuracy.

References

[1] Boussinesq, J.V., Theorie des ondes et des remous qui se propagent le long d'un canal rectangulaire horizontal, en communiquant au liquide contenu dans ce canal des vitesses sensiblement pareilles de la surface au fond. *Journal de Mathematiques Pures et Appliquees*, **17** (1872), 55-108.

[2] Chertok, A., Christov, C.I., Kurganov, A., Central-Upwind Schemes for the Boussinesq Paradigm Equations, *Computational Science and High Performance Computing IV, Notes Numer. Fluid Mech.*, **113** (2011), 267-281.

[3] Christov, C.I., An energy-consistent dispersive shallow-water model, *Wave Motion*, **34** (2001), 161-174.

[4] Samarskii, A., The Theory of Difference Schemes, Marcel Dekker Inc., New York, 2001.

[5] Angelow K., Comparison between two numerical methods for solution of 2D Boussinesq paradigm equation, *AIP Conference Proceedings*, **2522**, (2022), 090001

[6] Kolkovska N., Dimova M., A new conservative finite difference scheme for Boussinesq Paradigm equation, *Central European Journal of Mathematics*, **10** (2012), 3

[7] A.-M. Wazwaz. Multiple-soliton solutions for the Boussinesq equation. *Applied Mathematics and Computation*, **192(2)** (2007), 479–486

[8] N. Fenyvesi and G. Bene. Collision of water wave solitons., *Cent. Eur. J. Phys.*, textbf11(11) (2013), 1605–1615

Numerical investigation of the Klein-Gordon and sine-Gordon equations using the Christov expansion

M A Christou[1]

[1] Department of Computer Science, University of Nicosia, 46 Makedonitissas Avenue, CY-2417 Nicosia, Cyprus

E-mail: `christou.ma@unic.ac.cy`

Abstract. The numerical interactions of solitary waves given by the Klein-Gordon and sine-Gordon equations are demonstrated. To do that we apply the "Christov expansion method" , which is a Galerking type expansion. The Christov expansion is based on the Christov functions which are functions that form a complete orthonormal set of functions on $L^2(-\infty, \infty)$ that allow us to expand derivatives, nonlinear products back into the same basis. Here we show how can implement this method to solitary wave equations that form kink or anti-kink type of solutions. We also show how can speed up the algorithms by rewriting the expressions for the nonlinear terms more efficiently. To this end we need to mention that the numerical integration is performed using a Crank–Nicolson type scheme. Our results are in a very good agreement with other published works.

1. Introduction

In this work we would like to simulate the Klein-Gordon (KG) and the sine-Gordon (sG) type of equations

$$u_{tt} - u_{xx} + G(u) = 0 \tag{1}$$

with the potential terms

$$G_1(u) = \sin u + \mu_1 \sin(2u) + \mu_2 \sin(3u) \quad \text{where} \quad \mu_1, \mu_2 \quad \text{are constants} \tag{2}$$

$$G_2(u) = -u + u^3/\pi^2 \tag{3}$$

$$G_3(u) = \lambda(u - u^3) \tag{4}$$

We note that potentials (2) and (3) are used in [1] and (4) is used in [2]. Moreover, (2) is used in the triple sine-Gordon equation in [3]. Here we should note that (4) can be obtained if we rescale (3) as $u = \pi u$ and set $\lambda = 1$.

Equation (1) along with the associated potentials is a very useful model for a plethora nonlinear wave phenomena: for instance, torsion waves in strings, phase differences in Josephson junctions, dislocations propagation and the motion of suspended cables.

Content from this work may be used under the terms of the Creative Commons Attribution 4.0 licence. Any further distribution of this work must maintain attribution to the author(s) and the title of the work, journal citation and DOI.
Published under licence by IOP Publishing Ltd

For the various types of potentials (2),(3),(4) we have the permanent wave solutions

$$u(\xi) = 2\tan^{-1}\Big(\mathrm{M}\,\mathrm{cosech}\,(\mathrm{M}\,\xi)\Big),\ \mathrm{M} = \sqrt{2\mu_1+1}, \xi = \frac{x-ct}{\sqrt{1-c^2}}, \text{for } \mu_2 = 0,\ (\text{see [1]}) \tag{5}$$

$$u(\xi) = 4\tan^{-1}\Big(\mathrm{e}^{(\pm\xi)}\Big),\quad \xi = \frac{x-ct}{\sqrt{1-c^2}},\ \text{for}\quad \mu_1 = \mu_2 = 0,\ (\text{see [1]}) \tag{6}$$

$$u(\xi) = \pi\tanh(\xi),\quad \xi = \frac{x-ct}{\sqrt{1-c^2}},\ (\text{see [1]}) \tag{7}$$

$$u(\xi) = \frac{\sqrt{2}}{\cosh(b\xi)},\quad b = \sqrt{-\frac{\lambda}{c^2-1}}, \xi = x - ct,\ (\text{see [2]}) \tag{8}$$

$$u(\xi) = \pm 2\tan^{-1}\Big(\sqrt{\frac{5\mu_2}{(1-m^2)c}}\mathrm{sn}\,\xi\Big),\ \xi = Z - cT\ (\text{see [3]}) \tag{9}$$

where in (9) sn and $m, 0 \leq m \leq 1$ are the Jacobi elliptic function and its modulus. Note that:

- (5) is obtained after applying (2)
- (6) is obtained after applying (2)
- (7) is obtained after applying (3)
- (8) is obtained after applying (4)
- (9) is obtained after applying (2)

To simulate these equations we will apply the Christov spectral expansion method. This method was applied in different works of the author, see [4], [5] and in previous works [6]. The Christov method is based on the Christov functions which form a complete orthonormal set of functions belonging in the $L^2(-\infty,\infty)$ space and allow us to expand derivatives and nonlinear products back into the same basis, see [7]. In the current work we will show how we can apply this method on problems that obtain kink and anti-kink solutions. We will also show how we can rewrite in a more efficient way the nonlinear term, when we have products of the unknown profile function, and speed up the convergence of our algorithms. The obtained results are in a very good agreement with other published works, see [1] and [2].

2. The Christov expansion method

The real valued functions (S_n, C_n) where defined by Christov in [7]

$$S_n(x) = \frac{\rho_n(x) + \rho_{-n-1}(x)}{\mathrm{i}\sqrt{2}},\quad C_n(x) = \frac{\rho_n(x) - \rho_{-n-1}(x)}{\sqrt{2}},\qquad n = 0, 1, 2, \ldots, \tag{10}$$

in terms of Wiener's orthonormal complex-valued functions $\{\rho_n\}$ [8, p. 35],

$$\rho_n(x) = \frac{(\mathrm{i}x-1)^n}{\sqrt{\pi}(\mathrm{i}x+1)^{n+1}} \tag{11}$$

where the functions (S_n, C_n) are odd end even respectively.

In a paper by J. P. Boyd [9] these functions, are named as "Christov functions." Various properties of the Christov functions where demonstrated in [7, 10, 6, 4]. These properties arise by the basic properties of the Wiener functions.

Here, we summarize the key results needed to tackle the construction of a Galerkin spectral method for the KG equation (1). First, because of 3 and 4 we need the triple product formulæ for the Christov even functions (10):

$$\begin{aligned} C_l C_n C_k = \frac{1}{8\pi}\,[&C_{l+n+k+2} - 2C_{l+n+k+2} + C_{l+n+k} - 2C_{l-n-k-1} + C_{l-n-k} - C_{l+n-k+1} \\ &- C_{l-n+k+1} + 2C_{l+n-k} + 2C_{l-n+k} - C_{l-n+k-1} - C_{l+n-k-1}C_{l-n-k-2}] \end{aligned} \tag{12}$$

If we use the relations from [7]: $C_{-n} = -C_{n-1}$ and after some manipulations one gets:

$$C_l C_n C_k \stackrel{\text{def}}{=} \sum_{m=0}^{\infty} \beta_{lnk,m} C_m(x)\,, \quad \text{where} \tag{13}$$

$$\begin{aligned}
&\beta_{lnk,m} = \delta_{m,n+k+l+2} - 2\delta_{m,n+k+l+1} + \delta_{m,n+k+l}\\
&-2\mathrm{sgn}(l-n-k-0.5)\,\delta_{m,[|l-n-k-0.5|]} - \mathrm{sgn}(l-n+k-0.5)\,\delta_{m,[|l-n+k-0.5|]}\\
&-\mathrm{sgn}(l+n-k-0.5)\,\delta_{m,[|l+n-k-0.5|]} + \mathrm{sgn}(l-n-k-1.5)\,\delta_{m,[|l-n-k-1.5|]}\\
&+\mathrm{sgn}(l-n-k+0.5)\,\delta_{m,[|l-n-k+0.5|]} + 2\mathrm{sgn}(l+n-k+0.5)\,\delta_{m,[|l+n-k+0.5|]}\\
&+2\mathrm{sgn}(l-n+k+0.5)\,\delta_{m,[|l-n+k+0.5|]} - \mathrm{sgn}(l+n-k+1.5)\,\delta_{m,[|l+n-k+1.5|]}\\
&-\mathrm{sgn}(l-n+k+1.5)\,\delta_{m,[|l-n+k+1.5|]}\,,
\end{aligned}$$

where $[a]$ is the largest integer less than or equal to a, and δ is Kronecker's delta-function. Similar formulas were found in [4], for all the triple product combinations of S_n and C_n, i.e $S_l S_n S_k$, $S_l C_n C_k$, $S_l S_n C_k$.

We also need the second derivative expansion formulæ for the Christov functions (10):

$$C_n'' = \sum_{m=0}^{\infty} \chi_{nm} C_m, \qquad S_n'' = \sum_{m=0}^{\infty} \chi_{nm} S_m, \tag{14a}$$

where χ_{mn}, is a symmetric pentadiagonal matrix. Its components can be explicitly given by:

$$\begin{aligned}
\chi_{mn} = &-\frac{1}{4}n(n-1)\delta_{m,n-2} + n^2\delta_{m,n-1} - \frac{1}{4}(n+1)(n+2)\delta_{m,n+2}\\
&-\frac{1}{4}n^2 + (2n+1)^2 + (n+1)^2\delta_{m,n} + (n+1)^2\delta_{m,n+1}.
\end{aligned} \tag{15}$$

Taking though a closer look on the expressions for the triple product, one might wonder about their computational effectiveness. For that reason we find it useful to discuss the way to bypass this difficulty. We first recall the product formula of two even functions:

$$C_n C_k = \frac{1}{2\sqrt{2\pi}}\left[\ C_{n+k+1} - C_{n+k} - C_{n-k} + C_{n-k-1}\right]\,, \tag{16}$$

Using again the relation $C_{-n} = -C_{n-1}$ the product of two even functions is rendered to:

$$C_n C_k = \sum_{m=1}^{\infty} \beta_{nk,m} C_m, \quad \text{where} \tag{17a}$$

$$\alpha_{nk,m} = \frac{1}{2\sqrt{2\pi}}\left\{\delta_{m,n+k+1} + \delta_{m,|n-k|} - \delta_{m,n+k} - \mathrm{sgn}\left[|n-k| - \tfrac{1}{2}\right]\delta_{m,[|n-k|-1/2]}\right\} \tag{17b}$$

It is clear that the direct treatment of the square nonlinear term would require N^2 operations (multiplications) per node, or N^3 for the whole system. So, we derive here "convolution" sums that acknowledge the fact that the expressions from Eq. (17) involve extremely sparse matrices. In fact each of them has only four nontrivial elements. Consider a function from the $L^2(-\infty,\infty)$ space, say $U(x)$, then

$$U(x) = \sum_{n=0}^{\infty}\left(u_n C_n + v_n S_n\right)$$

Then for U^2 we have the following Theorem.

Theorem 1. *For the nonlinear term*

$$U^2(x) = \sum_{n=0}^{N}\sum_{m=0}^{M}\Big(u_n u_m C_n C_m + 2u_n u_m S_n C_m + v_n v_m S_n S_m\Big) = W_1 + W_2 + W_3 \tag{18}$$

where

$$W_1 = \sum_{n=0}^{N}\sum_{m=0}^{M} u_n u_m C_n C_m, \quad W_2 = 2\sum_{n=0}^{N}\sum_{m=0}^{M} u_n v_m S_n C_m, \quad W_3 = \sum_{n=0}^{N}\sum_{m=0}^{M} v_n v_m S_n S_m \tag{19}$$

the following expansions holds

$$W_1 = \sum_{n=0}^{N}\sum_{m=0}^{N} u_n u_m C_n C_m = \sum_{l=0}^{N}\left[\sum_{n=0}^{N}\sum_{m=0}^{N}\beta_{nm,l} u_n u_m\right] C_l \stackrel{\text{def}}{=} \frac{1}{2\sqrt{2\pi}}\sum_{l=0}^{N} b_l C_l,$$
$$b_l = \sum_{n=0}^{l-1} u_n u_{l-1-n} - \sum_{n=0}^{l} u_n u_{l-n} - 2\sum_{n=l}^{N} u_n u_{n-l} + 2\sum_{n=l+1}^{N} u_n u_{n-l-1}. \tag{20}$$

$$W_3 = \sum_{n=0}^{N}\sum_{m=0}^{N} v_n v_m S_n S_m = \sum_{l=0}^{N}\left[\sum_{n=0}^{N}\sum_{m=0}^{N}\alpha_{nm,l} u_n u_m\right] C_l \stackrel{\text{def}}{=} \frac{1}{2\sqrt{2\pi}}\sum_{l=0}^{N} a_l C_l,$$
$$a_l = -\sum_{n=0}^{l-1} u_n u_{l-1-n} + \sum_{n=0}^{l} u_n u_{l-n} - 2\sum_{n=l}^{N} u_n u_{n-l} + 2\sum_{n=l+1}^{N} u_n u_{n-l-1}. \tag{21}$$

$$W_2 = \sum_{n=0}^{N}\sum_{m=0}^{N} u_n v_m S_n C_m = \sum_{l=0}^{N}\left[\sum_{n=0}^{N}\sum_{m=0}^{N}\gamma_{nm,l} u_n v_m\right] S_l \stackrel{\text{def}}{=} \frac{1}{2\sqrt{2\pi}}\sum_{l=0}^{N} g_l C_l,$$
$$g_l = -\sum_{n=0}^{l-1} v_n u_{l-n} + \sum_{n=0}^{l} v_n u_{l-n+1} + \sum_{n=l}^{N} v_n u_{n-l+1}$$
$$-\sum_{n=l+1}^{N} v_n u_{n-l} - \sum_{n=l+1}^{N} u_n v_{n-l+1} + \sum_{n=l+1}^{N} u_n v_{n-l}. \tag{22}$$

Proof. Here we will show (20). To prove (20) we rearrange the summation in (17) making use of (16), namely

$$\sum_{n=0}^{N}\sum_{m=0}^{N} u_n u_m C_n C_m = \frac{1}{2\sqrt{2\pi}}\left[\sum_{n=0}^{N}\sum_{m=0}^{N} u_n u_m C_{n+m+1} - \sum_{n=0}^{N}\sum_{m=0}^{N} u_n u_m C_{n+m}\right.$$
$$\left. - \sum_{n=0}^{N}\sum_{m=0}^{N} u_n u_m C_{n-m} + \sum_{n=0}^{N}\sum_{m=0}^{N} u_n u_m C_{n-m-1}\right]$$

Let us take a closer look of the first term in the r.h.s. of last equation. Denote $l = m + n + 1$ or which is the same $m = l - n - 1$. One can use the last formula for $n \leq l - 1$ since $m \geq 0$. Hence

$$\sum_{n=0}^{n=N}\sum_{m=0}^{m=N} u_n u_m C_{n+m+1} = \sum_{l=1}^{N}\sum_{n=0}^{l-1} u_n u_{l-1-n} C_l. \tag{23}$$

We augment the definition for the sum for $l = 0$ as $\sum_{n=0}^{-1} \equiv 0$ and then (23) is valid for any l which immediately gives

$$\sum_{n=0}^{n=N}\sum_{m=0}^{m=N} u_n u_m C_{n+m+1} = \sum_{l=0}^{N} b_l^{(1)} C_l, \qquad \text{where} \qquad b_l^{(1)} = \sum_{n=0}^{l-1} u_n u_{l-1-n} C_l.$$

Likewise, we get

$$-\sum_{n=0}^{n=N}\sum_{m=0}^{m=N} u_n u_m C_{n+m} = \sum_{l=0}^{N} b_l^{(2)} C_l, \qquad \text{where} \qquad b_l^{(2)} = -\sum_{n=0}^{l} u_n u_{l-n} C_l.$$

The situation with the third and fourth term in (20) is more complicated because of the fact that $(n - m)$ can take either positive or negative values. Now, using again the identities from [7] $C_{-i-1} = -C_i$ and $S_{-i-1} = S_i$, which we render to the form $C_{-i} = -C_{i-1}$ and $S_{-i} = S_{i-1}$. Then for the third term we get

$$\sum_{n=0}^{N}\sum_{m=0}^{N} u_n u_m C_{n-m} = \sum_{n\geq m}^{N}\sum_{m=0}^{N} u_n u_m C_{n-m} - \sum_{m>n}^{N}\sum_{n=0}^{N} u_n u_m C_{m-n-1}$$

Upon changing the indexes $m = i + 1$, $(i \leq n)$ in the last term one obtains

$$\sum_{m>n}^{N}\sum_{n=0}^{N} u_n u_m C_{m-n-1} = \sum_{i>n}^{N}\sum_{n=0}^{N} u_n u_m C_{i-n}.$$

In the last term we can rename the dummy indexes $(i \to n, n \to m)$ to get

$$-\sum_{m>n}^{N}\sum_{n=0}^{N} u_n u_m C_{m-n-1} = \sum_{n\geq m}^{N}\sum_{m=0}^{N} u_n u_m C_{n-m},$$

which gives

$$\sum_{n=0}^{N}\sum_{m=0}^{N} u_n u_m C_{n-m} = 2\sum_{n\geq m}^{N}\sum_{m=0}^{N} u_n u_m C_{n-m}.$$

Upon substitution $l = n - m \geq 0$ (or $m = n - l$) one obtains

$$-\sum_{n=0}^{N}\sum_{m=0}^{N} u_n u_m C_{n-m} = -2\sum_{l=0}^{N}\sum_{n=l}^{N} u_n u_{n-l} C_l = \sum_{l=0}^{N} b_l^{(3)} C_l, \quad b_l^{(3)} = -2\sum_{n=l}^{N} u_n u_{n-l}.$$

Repeating the same manipulations for the last term in (20) we obtain

$$\sum_{n=0}^{N}\sum_{m=0}^{N} u_n u_m C_{n-m-1} = 2\sum_{l=0}^{N}\sum_{n=l+1}^{N} u_n u_{n-l-1} C_l = \sum_{l=0}^{N} b_l^{(4)} C_l, \quad b_l^{(4)} = 2\sum_{n=l+1}^{N} u_n u_{n-l-1}.$$

In the last formula the definition of the sum is once again tacitly augmented for the case $l = N$, namely $\sum_{n=N+1}^{N} = 0$. Noting that $b_l = b_l^{(1)} + b_l^{(2)} + b_l^{(3)} + b_l^{(4)}$, proves the theorem. □

We can find a similar expression for the cubic nonlinearity but actually there is no need for that. What we can do instead is the following:

$$U(x) = \sum_{n=0}^{\infty} \Big(u_n C_n + v_n S_n \Big), \quad \text{spectral expansion for } U \tag{24}$$

$$U^2(x) = \frac{1}{2\sqrt{2\pi}} \sum_{l=0}^{N} \Big((a_l + b_l) C_l + 2 g_l S_l \Big), \quad \text{from (20), (21), (22)} \tag{25}$$

$$\text{then, } U^3(x) = U \cdot U^2 = \frac{1}{2\sqrt{2\pi}} \sum_{n=0}^{\infty} \sum_{l=0}^{N} \Big(u_n (a_l + b_l) C_n C_l + 2 u_n g_l C_n S_l + v_n (a_l + b_l) S_n C_l + 2 v_n g_l S_n S_l \Big) \tag{26}$$

Then the terms in (26) are evaluated using (20),(21),(22). This idea can be applied for all non-linearities of the form

$$U^2, U^3, U^4, \cdots, U_x U, U_x^2, U_{xx} U^3, U_x U_{xx}^2, \cdots \tag{27}$$

Now, if we take a closer look on the exact solutions (5), (6), (7), (8) and (9) we can see that most of them are not in the $L^2(-\infty, +\infty)$ space. That means we can not develop them using the Christov functions.

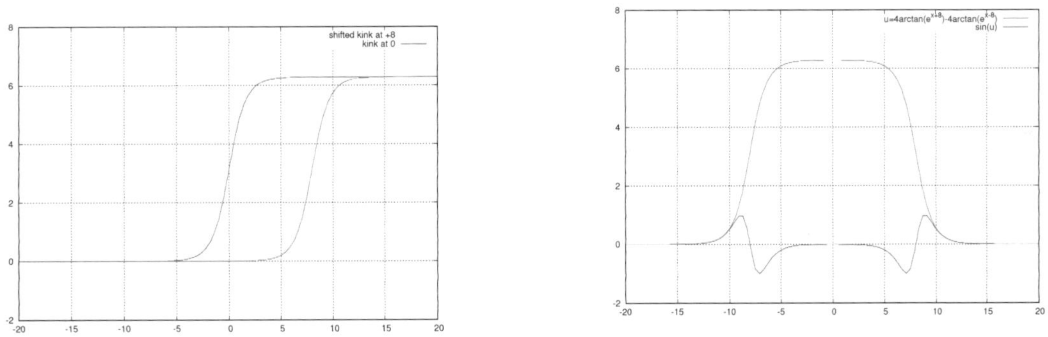

(a) Permanent wave solution as is given in (5).

(b) Combined permanent solutions and its sin term.

Figure 1: The permanent wave solution in (5) in the left panel is not $L^2(-\infty, \infty)$. Combining two solutions though results to the legitimate result, right panel.

In figure 1(a) we show the exact solution and a shifted exact solution, which is a kink type of soliton. That means it can not be developed in a Galerkin type of expansion using the Christov functions as a basis, since are in a different space. We can by pass this obstacle though if we add two solitons, like applying the superposition principle for waves, of this type and the result can be seen in figure 1(b). The two added solutions form an $L^2(-\infty, +\infty)$ function and in the same figure we also show how the nonlinear sine term looks for this particular combined solution. Here we need to comment that in [11] is explained in great detail why for the two soliton case the superposition principle can be applied as in the linear problem which is basically the reason why we can work as we did.

3. Implementation of the spectral expansion

To simulate the KG equation (1) we use a three level Crank–Nicolson type numerical scheme subject to asymptotic boundary conditions $u(x,t) \to 0$ as $|x| \to \infty$. Here we need to remind the reader that we can apply asymptotic boundary conditions since, as explained in the previous section, we consider the superposition of two kink type solutions and if we do that then the solution belongs to $L^2(-\infty,\infty)$. Then,

$$\frac{1}{\tau^2}u^{p+1} - \frac{1}{2}u^{p+1}_{xx} = \frac{2}{\tau^2}u^p - \frac{1}{\tau^2}u^{p-1} + \frac{1}{2}u^{p-1}_{xx} - G^p(u), \tag{28}$$

where $u^p \approx u(x,t^p)$ with $t^p = p\tau$, and τ is the time step.

Before proceeding further, we note that rescaling the spatial variable x does not change the nature of the asymptotic boundary conditions on $(-\infty,+\infty)$. Upon introducing $x = \zeta\hat{x}$ we recast (28) as

$$\frac{1}{\tau^2}u^{p+1} - \zeta^2\frac{1}{2}u^{p+1}_{\hat{x}\hat{x}} = \frac{2}{\tau^2}u^p - \frac{1}{\tau^2}u^{p-1} + \zeta^2\frac{1}{2}u^{p-1}_{\hat{x}\hat{x}} - G^p(u), \tag{29}$$

The scaling parameter ζ can be used to optimize the method in the sense that it allows one to control the typical length scales of the employed system of functions and to also control the width of the sought localized solution. Naturally, such a coordination between the scales will result in a faster convergence of the Fourier–Galerkin series. From this point and on we omit writing $\hat{x}$ and just write x. For further discussion on the use of the scaling parameter ζ, see [4] and [5]. Omitting the ˆ from x we have

$$\frac{1}{\tau^2}u^{p+1} - \zeta^2\frac{1}{2}u^{p+1}_{xx} = \frac{2}{\tau^2}u^p - \frac{1}{\tau^2}u^{p-1} + \zeta^2\frac{1}{2}u^{p-1}_{xx} - G^p(u), \tag{30}$$

In space, we expand the solution in a spectral series using Christov functions:

$$u^n(x) = \sum_{k=0}^{\infty}\left[a_k^n C_k(x) + b_k^n S_k(x)\right]. \tag{31}$$

Now, for the other terms we have:

$$u^p_{xx} = \sum_{k=0}^{\infty}\sum_{m=0}^{\infty}\left[a_k^p\chi_{km}C_m(x) + b_k^p\chi_{km}S_m(x)\right], \quad \sin u^p = \sum_{k=0}^{\infty}\Big[c^p_{1k}C_k(x) + s^p_{1k}S_k(x)\Big], \tag{32a}$$

$$\sin(2u^p) = \sum_{k=0}^{\infty}\Big[c^p_{2k}C_k(x) + s^p_{2k}S_k(x)\Big], \quad \sin(3u^p) = \sum_{k=0}^{\infty}\Big[c^p_{3k}C_k(x) + s^p_{3k}S_k(x)\Big], \tag{32b}$$

$$\begin{aligned}(u^3)^p = \sum_{l=0}^{\infty}\sum_{n=0}^{\infty}\sum_{k=0}^{\infty}\sum_{m=0}^{\infty}\big[&a_l^p a_n^p a_k^p \beta_{lnk,m}C_m(x) + 3b_l^p b_n^p a_k^p \bar{\gamma}_{lnk,m}C_m(x)\\ &+ 3b_l^p a_n^p a_k^p \bar{\bar{\gamma}}_{lnk,m}S_m(x) + b_l^p b_n^p b_k^p \alpha_{lnk,m}S_m(x)\big].\end{aligned} \tag{32c}$$

Here we should note that the term $\beta_{lnk,m}$ is given in (13) but all the coefficients together namely $\beta_{lnk,m}, \alpha_{lnk,m}, \bar{\gamma}_{lnk,m}, \bar{\bar{\gamma}}_{lnk,m}$ are given in [4]. As was aforementioned though, we will not use these expansions since are computationally expensive but we will use instead the result obtained after applying (18).

The next step is to introduce the spectral expansion terms in-to the time-discrete scaled KG equation (30). We multiply the equation with the Christov functions (C_n, S_n) and after integration we arrive to the nonlinear system of equations for the unknown coefficients $\{a_n, b_n\}$:

$$\frac{1}{\tau^2}a_n^{p+1} - \frac{\zeta^2}{2}\sum_{m=0}^{\infty} a_n^{p+1}\chi_{nm} = \frac{2}{\tau^2}a_n^p - \frac{1}{\tau^2}a_n^{p-1} + \frac{\zeta^2}{2}\sum_{m=0}^{\infty} a_n^{p-1}\chi_{nm} - G^p_{\text{even coef.}}(u), \tag{33}$$

$$\frac{1}{\tau^2}b_n^{p+1} - \frac{\zeta^2}{2}\sum_{m=0}^{\infty} b_n^{p+1}\chi_{nm} = \frac{2}{\tau^2}b_n^p - \frac{1}{\tau^2}b_n^{p-1} + \frac{\zeta^2}{2}\sum_{m=0}^{\infty} b_n^{p-1}\chi_{nm} - G^p_{\text{odd coef.}}(u), \tag{34}$$

The initial conditions (for $p = 0$) are obtained by numerical quadrature on the analytical solution. The summations in (33),(34) are all truncated at finite value to solve the system. Here we should explain how the scheme works when the nonlinear term is G_1, (2). Let us assume we have the term $\sin u$. When $p = 0$ we have the analytical solution for u and from that we can compute the $\sin u$ and in turn, after applying numerical quadrature, we get the initial coefficients $\{c^0_{1k}, s^0_{1k}\}$. After the first time step we have the new coefficients $\{a^1_k, b^1_k\}$. Using them we can have u^1. Then applying numerical quadrature again we get the new coefficients $\{c^1_{1k}, s^1_{1k}\}$.

4. Numerical examples and results

In figure 2 we can see the collision of two sech type solutions of the KG equation using G_3 nonlinearity, (4). The initial condition is placed in the bottom of the graph and we plot the evolution of this interaction every 100 time steps. The waves cannot survive the collision and we can see the developments of a complex pattern. Here it should be noted that for all examined cases we used for the time step $\tau = 0.01$, the scaling parameter $\zeta = 0.07$ and 350 Christov functions. The two sech solutions are placed at -15 and 15 respectively and we let them propagate so as to approach each other. Our graph is identical to the graph obtained by Christov and Nicolis in [2].

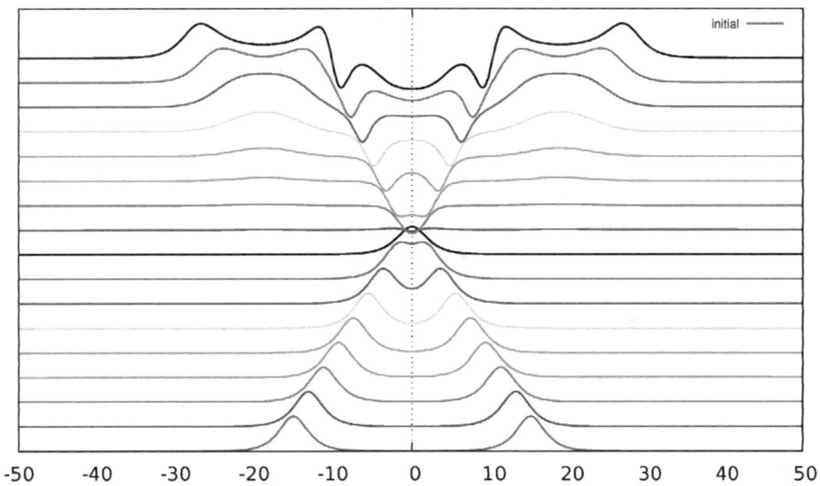

Figure 2: Evolution of a system of two sech-type solutions as given in (8) for G_3, (4). Here $c_l = -c_r = 1.9$ and $\lambda = -1$.

In figure 3 we can see an interaction corresponding to the sine-Gordon equation using G_1 nonlinearity for $\mu_1 = \mu_2 = 0$, (2). The initial condition is placed in the bottom of the graph we let it run and plot the results every 500 time steps. We use the same time step and scaling parameter as in the previous case. Same results were obtained in [1].

Now, as was explained in [2], in the case were $\lambda < 0$ the solution is supersonic where for $\lambda > 0$ is subsonic. A subsonic case is treated in figure 4. Here $\lambda = 1$ and $c_l = c_r = -0.9$. The two sech solutions do not survive the collision.

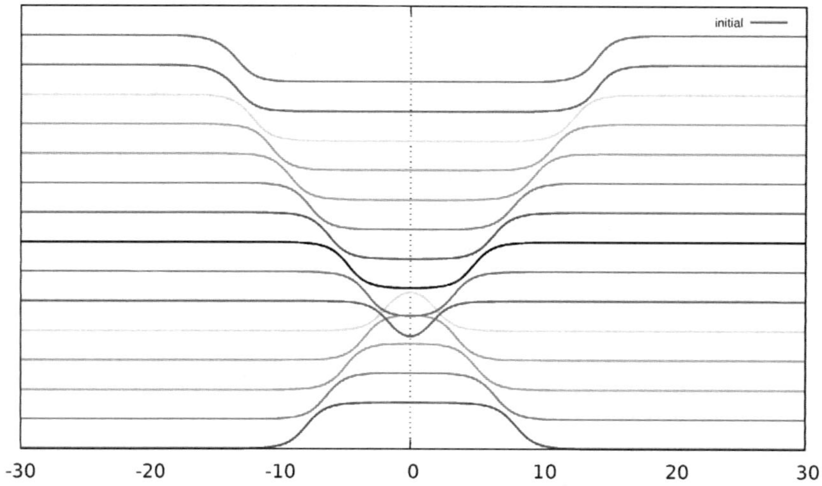

Figure 3: Evolution of a system of two solutions as given in (8) for G_1 when $\mu_1 = \mu_2 = 0$, (2). Here $c_l = -c_r = 0.3$.

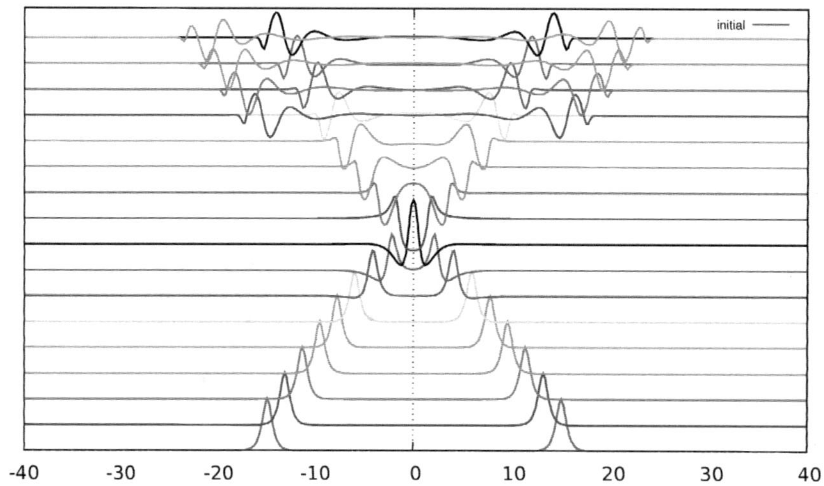

Figure 4: Evolution of a system of two sech-type solutions as given in (8) for G_3, (4). Here $c_l = -c_r = 0.9$ and $\lambda = 1$.

5. Conclusion

In the current work we applied the Christov spectral method in the Klein-Gordon and sine-Gordon equation. These equations posses kink and sech type of solutions. The difficulty here was to develop the kinks in terms of the Christov functions which are functions decaying at infinity, they belong to the $L^2(-\infty, \infty)$ space. To by pass this difficulty we added two solutions and were able to apply the method. We manage to recover results obtained in [1] and [2]. Furthermore, we showed how we can optimize even better the nonlinear terms when we have products such as a cube. That, gave as an extra computational advancement.

In a previous work, [5] we have examined, using the Christov method, the behavior of the method in a nonlinear nonlocal evolution equation, the Benjamin-Ono equation. The results we obtained in the current work and for the Benjamin-Ono equation encourage us to extend it to a more general nonlocal dispersion equation such as the Whitham equation [12] and a more general 2d Klein-Gordon equation.

References

[1] Ablowitz M J, Kruskal M D and Ladik J F 1979 *SIAM* **36(3)** 428–437
[2] Christov C and Nikolis G 1996 *Physica A* **228** 326–343
[3] Liu S, Fu Z and Liu S 2006 *Physics Let. A* **351** 59–63
[4] Christou M A and Christov C I 2002 *J. Comput. Anal. Appl.* **4** 63–77
[5] Christou M A and Christov I C 2023 *J. Phys.:Conf. Ser.* **2675**
[6] Christov C I and Bekyarov K L 1990 *SIAM J. Sci. Stat. Comput.* **11** 631–647
[7] Christov C I 1982 *SIAM J. Appl. Math.* **42** 1337–1344
[8] Wiener N 1964 *Time Series* (Cambridge, MA: MIT Press)
[9] Boyd J P 1990 *J. Approx. Theory* **61** 98–105
[10] Christov C I 1987 *Ann. Univ. Sofia, Fac. Math. Mech.* **76(2)** 87–113
[11] Christov I and Christov C I 2008 *Physics Letters A* **372** 841–848
[12] Whitham G B 1967 *Proc. Royal S. Lond.* **299** 6–25

Analysing the Impact of Second Gas Component on Pulse Pirani Gauge

D Y Dankov*, M K Mironova, P N Gospodinov

Institute of Mechanics, Bulgarian Academy of Sciences, Acad. G. Bonchev str. Bl. 4, 1113 Sofia, Bulgaria
* Corresponding Author E-mail: dankov@imbm.bas.bg

Abstract. This numerical study investigates the influence of a second gas component concentration on Pirani gauge readings. The analysis aims to provide insights into how the presence of additional gas components affects the accuracy and reliability of Pirani gauge measurements. By simulating different gas mixtures and examining the resulting readings, this study sheds light on the complex interactions between gases in Pirani gauge applications.

1. Introduction

The Pirani gauge is a widely used thermal conductivity gauge for measuring vacuum pressures. Its operation relies on the phenomenon that the thermal conductivity of a gas changes with pressure. A heated wire within the gauge cools at a rate dependent on the surrounding gas pressure, allowing for pressure determination through temperature and resistance measurements. Pirani vacuum sensors are suitable for measuring pressures in the range of approximately 10^{-6} bars to 5 bars. They are particularly useful for low to medium vacuum levels, where other types of sensors may not be as effective. The fundamental working principle of a Pirani vacuum gauge is based on the thermal conductivity of gases [5]. The gauge typically consists of a heated filament (often made of tungsten or nickel) suspended in the vacuum. When the filament is heated by an electric current, it loses heat to the surrounding gas molecules. The rate of heat loss depends on the number of gas molecules present, which in turn is related to the pressure of the gas. The gauge uses a heating-cooling cycle of filament reveals significant advantages [9,12]. Numerical calculations for the Pirani gauge for rarefied gas are modelled by the linearized non-stationary Boltzmann equation [15]. Pirani vacuum gauges are essential tools for monitoring and controlling vacuum conditions in various industrial and scientific applications. Their ability to measure a wide range of pressures with good accuracy makes them indispensable in environments where maintaining specific vacuum levels is crucial. Understanding the working principles, components, and limitations of these sensors is key to their effective utilization.

Traditionally, Pirani gauges are calibrated and used with single-component gases. However, many practical applications involve gas mixtures, where the presence of a second gas component can significantly affect gauge readings and accuracy.

In our papers [13, 14] we studied a rarefied monatomic gas between two coaxial, stationary, unconfined cylinders (one dimensional, axis-symmetrical problem). The temperature of the outer cylinder is constant and equal to the gas equilibrium temperature, while the temperature of the inner cylinder varies in time. The inner cylinder (filament) is subjected on periodically heating-cooling cycle.

The primary aim of this numerical study is to analyse the impact of a second gas component on the performance and accuracy of Pirani gauges. Understanding this impact is crucial for improving the

Content from this work may be used under the terms of the Creative Commons Attribution 4.0 licence. Any further distribution of this work must maintain attribution to the author(s) and the title of the work, journal citation and DOI.
Published under licence by IOP Publishing Ltd

reliability of vacuum measurements in mixed gas environments and for developing more accurate calibration methods for Pirani gauges used in industrial and research applications, which can be considered in a single-phase gas environment "slightly" contaminated by the presence of another undesirable component.

To model the thermal and fluid dynamics within the Pirani gauge we use the Taylor-Couette problem, which studies the flow between two coaxial cylinders [7,8,10,11,16,17]. The results are obtained using a continuum model based on the Navier-Stokes-Fourier (NSF) method for one component gas (basis for comparison) and Direct Simulation Monte Carlo (DSMC) method for one component gas and gas mixture.

This study aims to bridge the gap in understanding the behaviour of Pirani gauges in mixed gas environments. The findings will contribute to more precise vacuum measurements and improved calibration techniques, benefiting both industrial applications and scientific research. By leveraging numerical methods, we can predict and mitigate the effects of a second gas component, enhancing the reliability of Pirani gauge readings in complex gas mixtures.

2. Formulation of the problem and methods of solution

The Pirani gauge is modelled by the one dimensional cylindrical Couette flow. The following notations for the physical quantities describing the rarefied gas flow: ρ is the density, T is the temperature, P is the thermodynamic pressure and u is the velocity component in r – direction. The inner cylinder has radius R_1 and temperature T_1 and the outer – R_2 and T_2. Both cylinders are stationary.

2.1 Continuous Model (NSF) for monatomic gas.

The continuous model for monatomic gas is based on the Navier-Stokes-Fourier (NSF) equations for compressible fluid, completed with the equations of continuity and energy transport. The Navier-Stokes equations and the Fourier equation together provide a comprehensive description of fluid flow and heat transfer. The Navier-Stokes equations model the dynamics of fluid motion, while the Fourier equation models the distribution and flow of heat. When coupled, they describe phenomena such as convective heat transfer, where the movement of the fluid significantly influences temperature distribution. The nonzero stress tensor components [6] can be expressed as:

$$\tau_{rr} = -\mu\left(2\frac{\partial u}{\partial r} - \frac{2}{3}\frac{1}{r}\frac{\partial}{\partial r}(ru)\right) \tag{1}$$

$$\tau_{\varphi\varphi} = \mu\left[\frac{2}{3}\left(\frac{1}{r}\frac{\partial}{\partial r}(ru) - 2\left(\frac{u}{r}\right)\right)\right] \tag{2}$$

The dissipation function Φ in the equation of energy transport is

$$\Phi = \frac{4}{3}\mu\left[\left(\frac{\partial u}{\partial r}\right)^2 - \frac{\partial u}{\partial r}\frac{u}{r} + \left(\frac{u}{r}\right)^2\right] \tag{3}$$

For a perfect monatomic gas, the viscosity and the heat transfer coefficients read as [1, 2, 3, 6]:

$$\mu(T) = C_\mu \rho_0 l_0 V_0 \sqrt{T},\ C_\mu = \frac{5}{16}\sqrt{\pi} \tag{4}$$

$$\lambda(T) = C_\lambda c_P \rho_0 l_0 V_0 \sqrt{T},\ C_\lambda = \frac{15}{32}\sqrt{\pi}, \tag{5}$$

The NSF model is completed with the equation of state for an ideal gas:

$$P = \rho R T \tag{6}$$

The governing equations are normalized by using the following scales: for density $\rho_0 = mn_0$ (m is the molecular mass, n_0 – the average number density), for velocity $V_0 = \sqrt{2RT_0}$ – R is the gas constant, for length – the distance between the cylinders $L=R_2-R_1$, for time $t_0=L/V_0$, for temperature T_0 – the wall temperature of outer cylinders. The Knudsen number is $Kn=l_0/L$ where the mean free path is l_0 and $\gamma=c_P/c_V=5/3$ (c_P and c_V are the heat capacities at constant pressure and constant volume respectively). In this way, the characteristic number Kn and the constants C_μ and C_λ take part in the dimensionless model. For brevity, the same symbols r, t, P, T, u and R_i, $i = 1, 2$ are used for the corresponding dimensionless variables.

We use first-order slip boundary conditions at both walls [2,14]. They be written in dimensionless form as follows:

$$u = 0 \quad (7)$$

for the velocity

$$T \pm \zeta_T Kn_{local} \frac{\partial T}{\partial r} = \overline{T_\iota} \quad (8)$$

for the temperature.
at $r=R_i$,: i=1, 2 where for diffuse scattering we have used the temperature jump coefficients $\zeta_T = 2.1904$ calculated, respectively in [12], from the kinetic BGK equation (a form of the collision integral proposed by Bhatnagar, Gross and Krook). The dimensionless wall pulsating temperature on the filament $\overline{T_\iota}$ depends on the dimensionless time harmonically
for $2\pi k \leq \omega t \leq 2\pi(k + 0.5)$ $k = 1,2,3, \ldots.$ $\overline{T_\iota} = T_{i,0}(1 + A|sin\omega t|)$ (9)
where $T_{i,0}$ is the inner cylinder (filament) wall temperature value, and A is the oscillation amplitude. The condition (8)-(9) describes the filament heating stage from the heating-cooling cycle.
For $2\pi(k + 0.5) \leq \omega t \leq 2\pi(k + 1)$ $k = 1,2,3, \ldots.$ instead (8) the adiabatic condition (10) corresponding to the cooling stages of the heating-cooling cycle is applied $\frac{\partial T}{\partial r} = 0.$ $r = R_1$ and T_2 is the outer cylinder wall constant temperature.
The boundary conditions are modeled by using the local Knudsen number Kn_{local}.

$$Kn_{local} = \frac{l}{L} = \left(L\sqrt{2}\pi\sigma^2 \frac{\rho_0}{\rho} n_0\right)^{-1} = \frac{Kn}{\overline{\rho}} \quad (11)$$

In (11) with l is denoted the local mean free path, σ - molecular diameter and ρ – is the dimensional density [10].

Following [9,12,15] we introduce a Stokes number for temperature oscillation $\beta = \sqrt{\frac{\omega L^2}{\nu}}$, where ν is kinematic viscosity and ω temperature circular frequency. The dimensionless pressure P_0 corresponding to initial equilibrium state can be calculated as: $P_0 = 0.5\rho_0 T_0$, ρ_0 and T_0 are the equilibrium dimensionless density and temperature.

2.2 The Direct Simulation Monte Carlo (DSMC) method for gas mixture.

The Direct Simulation Monte Carlo (DSMC) method [1,3,4] is a particle-based simulation technique that models the motion and collisions of individual gas molecules. It uses statistical methods to simulate the behaviour of a large number of particles, providing an approximate solution to the Boltzmann equation, which governs the statistical behaviour of a gas.
Steps in the DSMC Method:
1. Initialize the positions, velocities, and other properties of the simulation particles according to the initial conditions of the problem.
2. Move each particle according to its velocity over the time step Δt.
3. Sort particles into cells to facilitate the collision process. For each cell, determine which pairs of particles will collide based on collision probability. Update the velocities of colliding particles according to the laws of conservation of momentum and energy. We use Bird's scheme "no time counter".
4. Handle interactions with boundaries, such as reflections, absorption, or emission.
5. Collect statistical data for macroscopic properties (e.g., density, velocity, temperature) by averaging over a number of particles and time steps.
6. Repeat the steps for the required number of time steps to simulate the evolution of the system over time.

The Knudsen number calculation: For one component gas Kn $= \frac{l}{L}$, $l = \frac{1}{4\sqrt{2}\pi r^2 n_0}$. For two component gas $l = \sum \frac{n_k}{n} l_k = \frac{1}{(\sum_k n_k)^2} \sum_{k,m} \frac{n_k}{\pi\sqrt{2}(r_k + r_m)^2}$, where l_k, r_k and n_k is respectively the mean free path, the molecular radius and number density of the k-th component. All quantities used are non-dimensional, so that the mean free path at equilibrium is equal to 1.

To understand the pressure in a gas mixture and its impact on a Pirani gauge, we need to consider the individual pressures of each gas component and how they contribute to the total pressure. The macroscopic variables of each species s, namely, the number density n_s, mass density ρ_s, flow velocity u_s, temperatures T_s and pressure tensor P_s are obtained by taking the moments of the respective velocity distribution function. The particle velocity is denoted with $\xi\left(\xi_z, \xi_r, \xi_\varphi\right)$. The flow velocity is calculated by $u_s = \bar{\xi}$, the temperature – $\frac{3}{2}n_s k_B T_s = \frac{1}{2}m_s(\xi - u_s)^2$ and the pressure – $P_s = \frac{1}{3}\rho_s(\xi - u_s)^2$ for component s. The corresponding macrocharacteristics are calculated by formulas: the velocity $\rho u = \sum \rho_s u_s$, the temperature $\frac{3}{2}nk_B T = \sum \frac{3}{2}nk_B T_s + \frac{1}{2}\sum \rho_s(u_s - u)^2$, the pressure $P = \sum P_s + \sum \rho_s(u_s - u)^2$.

Boundary condition: The particle velocity before collision with wall is denoted with $\xi\left(\xi_z, \xi_r, \xi_\varphi\right)$ and after with $\xi^*\left(\xi_z^*, \xi_r^*, \xi_\varphi^*\right)$, respectively the velocity magnitude with $c = \sqrt{\xi_z^2 + \xi_r^2 + \xi_\varphi^2}$ and $c^* = \sqrt{\xi_z^{*2} + \xi_r^{*2} + \xi_\varphi^{*2}}$. This gas-surface interaction model can be describing by keeping the particle velocity magnitude invariant and the velocity direction of the reflected particle set based on isotropic scattering boundary conditions in the half unite sphere. The boundary condition are written in the spherical coordinate system $(c, \alpha_1, \alpha_2), \alpha_1 \in \left[0, \frac{\pi}{2}\right], \alpha_2 \in [0, 2\pi]$ [7,21]. The three velocity components on the inner cylinder for the adiabatic case reflected by:

$$\xi_r^* = c \times cos\alpha_1$$
$$\xi_z^* = c \times sin\alpha_1 \times cos\alpha_2$$
$$\xi_\varphi^* = c \times sin\alpha_1 \times sin\alpha_2$$

and for diffusion case reflected by:

$$\xi_r^* = \sqrt{\frac{2k_B\overline{T_1}}{m}} \times cos\alpha_1$$

$$\xi_z^* = \sqrt{\frac{2k_B\overline{T_1}}{m}} \times sin\alpha_1 \times cos\alpha_2$$

$$\xi_\varphi^* = \sqrt{\frac{2k_B\overline{T_1}}{m}} \times sin\alpha_1 \times sin\alpha_2$$

where k_B is the Boltzmann constant.

The difference between diffusion and adiabatic reflection is only the coefficient. At the diffusion reflection the coefficient $\sqrt{\frac{2k_B\overline{T_1}}{m}}$ depends on the wall temperature $\overline{T_1}$ and the particle mass m, while at the adiabatic reflection the coefficient c is the particle velocity magnitude before collision on the particle with the wall.

3. Numerical results

We study monatomic argon (Ar) gas as a base component and diatomic molecular iodine (I_2) as a additional component with pressure $p_0 = 6.35 \times 10^{-4}(bar)$, temperature $T_0 = 290(K)$, the most portable speed $V_P = 347(m/s)$, the free molecular path $l = 1.23 \times 10^{-5}(m)$, the filament radius $R_1 = 3.07 \times 10^{-5}(m)$, and the outer cylinder radius $R_2 = 64.47 \times 10^{-5}(m)$. The nondimensional flow parameters associate with this case are $Kn = 0.5,\ l = 1, p_0 = 0.5,\ R_1 = 0.05, R_2 = 1.05, A = 0.025$, $\beta = 2$, the particle mass ratio $m_{Ar}: m_{I_2} = 1: 6.35$, and the particle diameter ratio $d_{Ar}: d_{I_2} = 1: 2.53$.

We study three cases for the concentration of additional component (I_2):

(1) $n_{I_2} = 0$ (2) $n_{I_2} = 2.5\% n_{Ar}$ (3) $n_{I_2} = 5\% n_{Ar}$

The figures 1 and 2 show the temperature and the pressure variation in case 1 [13,14]. The differences in numerical results between the two models appear visually large, but as values are below 0.5%. This means that there were no significant differences in numerical results obtained by two methods – NSF and DSMC. The studies have been continued to consider two cases of additional gas iodine (I_2 – a contamination) on the basis of the statistical results for a case 1. Figure 3 presents the variation of filament temperature with time. It is clearly visible the heating-cooling cycle, no differences at heating period and visible differences at cooling period. Small differences and long-time fluctuations in temperature and density are observed when a second type of particle is encountered – Figures 4 and 5.

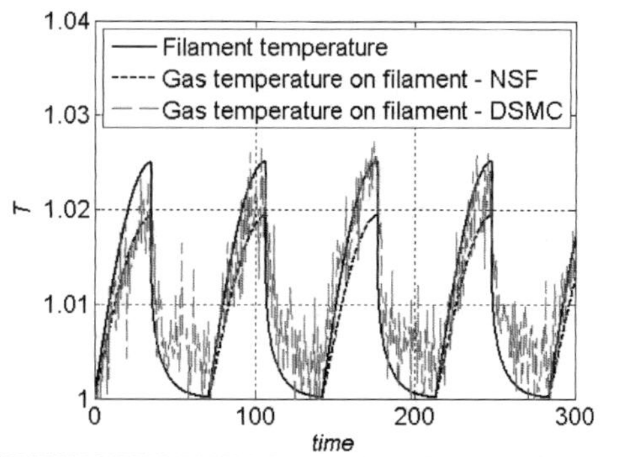

Figure 1. The fibre temperature variation with time in case (1).

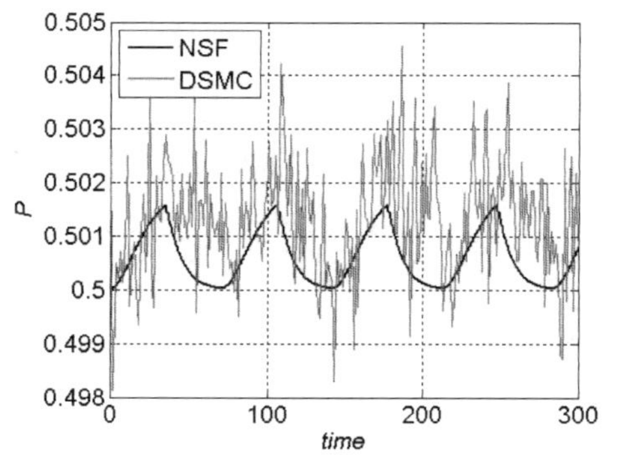

Figure 2. The fibre pressure variation with time in case (1).

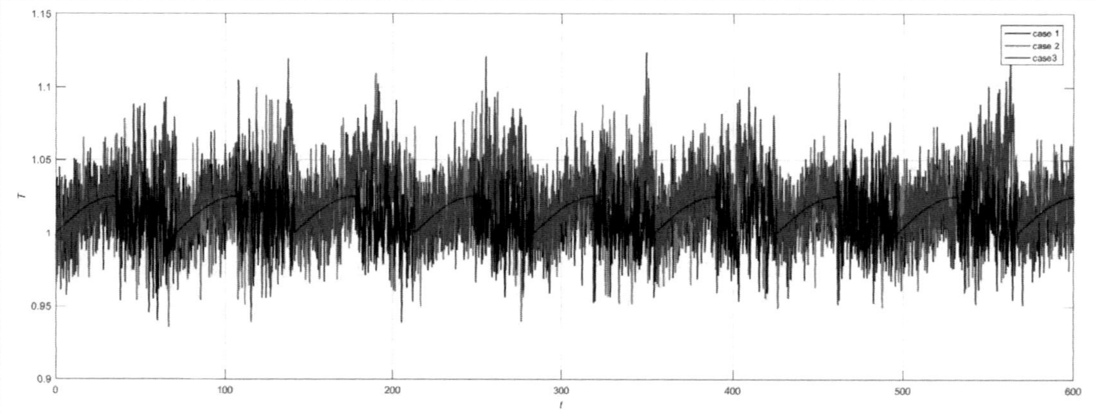

Figure 3. The filament temperature variation with time.

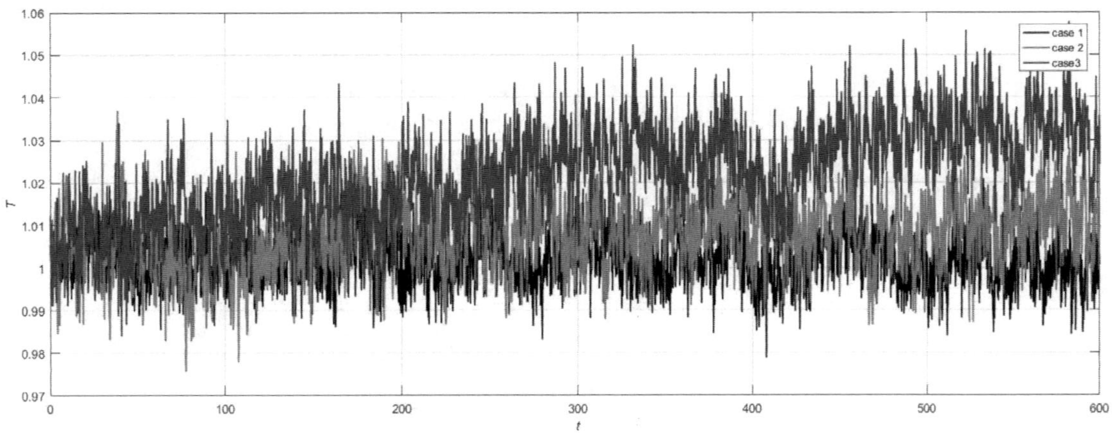

Figure 4. The temperature variation with time at r=0.55.

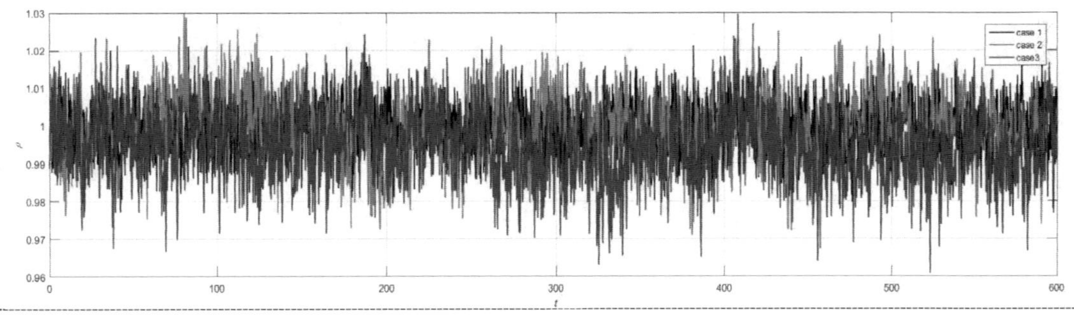

Figure 5. The density variation with time at r=0.55.

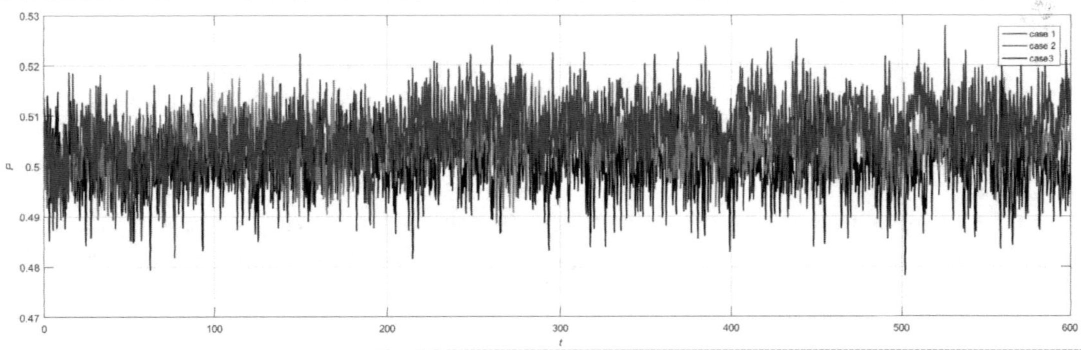

Figure 6. The pressure variation with time at r=0.05.

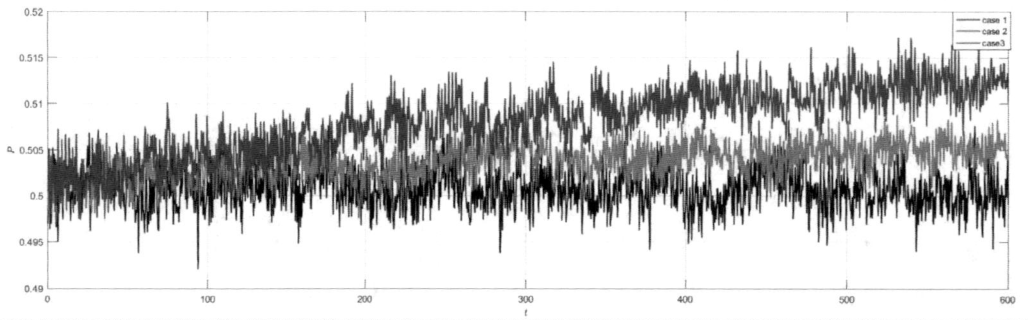

Figure 7. The pressure variation with time at r=0.55.

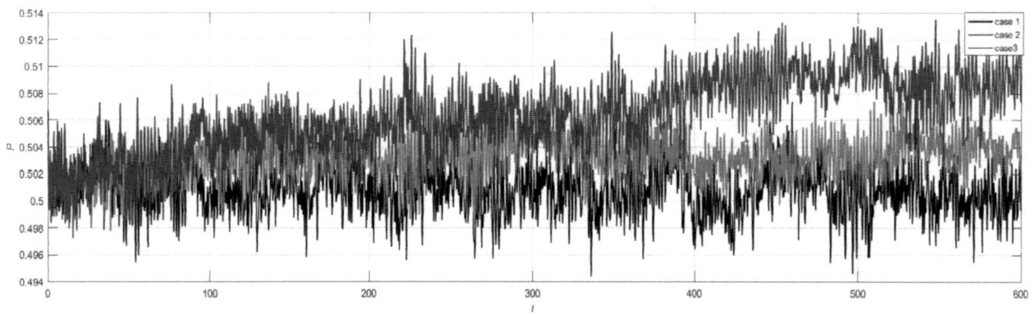

Figure 8. The pressure variation with time at r=1.05.

The performance of a Pirani gauge, which measures vacuum pressure based on thermal conductivity, is significantly influenced by the composition of the gas mixture being measured. The thermal conductivity of a gas mixture is different from that of a single-component gas. Each gas in the mixture contributes differently to the overall thermal conductivity, which the Pirani gauge relies on to infer pressure. The presence of a second gas with different thermal properties can alter the gauge readings. The introduction of a second gas component, especially one with higher molecular weight or higher thermal conductivity, generally leads to increased thermal conductivity of the mixture. This results in the Pirani gauge indicating a higher pressure than the actual value – Figures 6, 7 and 8. Adding a second gas component affects the thermal conductivity of the gas mixture. The presence of a second gas component can lead to longer stabilization times for the Pirani gauge. This is due to the complex interplay between the different molecular masses and thermal conductivities, which affects the stability state of the system. When analysing the impact of a second gas component on the performance of a Pirani gauge, it's important to distinguish between changes due to the actual physical presence of the second gas component and those introduced by numerical effects in the simulation, which is the aim of further research. Pirani gauges need to be calibrated for specific gas mixtures to ensure accurate pressure measurements. Calibration curves for different mixtures can help correct for the changes in thermal conductivity.

4. Conclusion

Understanding the influence of gas mixtures (the second gas can be considered as contamination of the gas medium in a vacuum installation) is vital in applications such as vacuum systems, semiconductor manufacturing, and any process where accurate pressure measurements are critical. In summary, the presence of a pollution has a noticeable impact on the performance of a Pirani gauge. It generally leads to increased pressure readings and longer stabilization times due to changes in thermal conductivity. Accurate modelling and calibration are essential to mitigate these effects and ensure reliable pressure measurements.

5. References

[1] Chapman S and Cowling T. 1952 *The Mathematical Theory of Non-Uniform Gases* (Cambridge University Press Cambridge)

[2] Albertoni S, Cercignani C and Gotusso L 1963 Numerical evaluation of the slip coefficient *Phys. Fluids* 6 pp 993–996

[3] Cercignani C and Sernagiotto F 1967 Cylindrical Couette flow of a rarefied gas, *Physics of Fluids* 10 pp 1200–1204

[4] Bird G 1994 *Molecular Gas Dynamics and the Direct Simulation of Gas Flows* (Oxford University Press Oxford)

[5] Cercignani C 2000 *Rarefied Gas Dynamics, From Basic Concepts to Actual Calculations* (Cambridge University Press Cambridge)

[6] Bird R, Stewart W and Lightfoot E 2002 *Transport Phenomena* (John Wiley New York)

[7] Park J, Bahukudumbi P and Beskok A 2004 Rarefaction effects on shear driven oscillatory gas flows: A DSMC study in the entire Knudsen regime, *Phys. Fluids* 16 p 317

[8] Emerson D, Gu Xiao-Jun, Stefanov S, Yuhong S and Barber W 2007 Nonplanar oscillatory shear flow: From the continuum to the free-molecular regime *Phys. Fluids* 19 107105

[9] Sharipov F 2011 Data on the velocity slip and temperature jump on a gas-solid interface *J. Phys. Chem. Ref. Data* 40 023101

[10] Gospodinov P, Roussinov V and Stefanov S 2012 Nonisothermal oscillatory cylindrical Couette gas flow in the slip regime: A computational study *European Journal of Mechanics B/Fluids* 33, pp 68–77

[11] Kalempa D and Sharipov F 2014 Numerical modeling of thermoacoustic waves in a rarefied gas confined between coaxial cylinders, *Vacuum* 109 pp 326–332

[12] Gospodinov P, Rousinov V and Mironova M 2014 Cylindrical nonisothermal oscillatory Couette gas flow in the slip regime: Wall shear stress and energy transfer, numerical investigation IntellectualArchive.com online publications *Intellectual Archive Bulletin* Feb 19 pp 2–19

[13] Gospodinov P, Dankov D, Roussinov V and Mironova M 2017 Numerical simulation of the pulsed Pirani gauges *AIP Conference Proceedings* 1895, 080003; doi: 10.1063/1.5007397

[14] Gospodinov P, Dankov D, Roussinov V and Mironova M 2018 Thermoacoustic waves in pulsed Pirani gauge. Frequency analysis *AIP Conference Proceedings* 2025, 070002; doi: 10.1063/1.5064914

[15] Oualli H, Boubdallah A, Hanchi S and Gad-Ei-Hak M 2010 Taylor-Couette flow control using a radially oscillating inner cylinder *Geophysical Research Abstracts EGU General Assembly* 2010 12 EGU2010-1252

[16] Oualli H, Mekadem M, Lebbi M and Bouabdallah A 2016 Taylor-Couette flow control by amplitude variation of the inner cylinder cross-section oscillation *Eur. Phys. J. Appl. Phys.* 71 11102

[17] Lalaoua A and Bouabdallah A 2016 On the Onset of Taylor Vortices in Finite-Length Cavity Subject to a Radial Oscillation Motion *Journal of Applied Fluid Mechanics* vol 9 No. 4 pp 1887-1896

doi:10.1088/1742-6596/2910/1/012028

Preconditioned IDR solution method in scientific and industrial applications

Alex Fedoseyev

Ultra Quantum Inc., Huntsville, Alabama, USA

E-mail: af@ultraquantum.com

Abstract. A review of preconditioned solver for large-scale applications in science and industry is presented. The analysis, parallelization, and optimization approach for large unstructured sparse matrices using IDR methods are considered for modern multicore microprocessors.

Performance criteria and methods have been revisited, along with consideration of parallelization involving the solver and preconditioner using the OpenMP environment. Results of the successful implementation for efficient parallelization are presented.

1 Introduction

CNSPACK (Compact Non-symmetric Solver PACKage) is an advanced solver successfully used for the coupled solution of stiff problems arising in multiphysics applications, such as Computational Fluid Dynamics (CFD) for high Reynolds number turbulent flows, turbulent boundary layers, hypersonic and rarefied flows, carrier transport in semiconductors, and kinetic and quantum mechanics problems. CNSPACK employs an iterative IDR (Induced Dimension Reduction) methods with ILU preconditioning (where the user chooses the ILU order).

Originally, CNSPACK was implemented and optimized for early sequential processors, considering their arithmetic and memory-size limitations. In the early 1990s, the first optimization exercise was performed to accelerate the algorithm by 6 times for emerging superscalar microprocessors, such as the Intel i860.

However, there has been a significant shift in processor architectures and computer system organization since that time. Nowadays, desktop computers and cluster nodes utilize high-performance pipelined superscalar multicore processors with out-of-order execution of instructions, deep cache hierarchies, and high-throughput memory capabilities.

As a result, performance criteria and methods have been revisited. The parallelization involving the solver and preconditioner using the OpenMP environment have been implemented.

In this paper we present the analysis, parallelization and optimization approach for the large sparse matrix solver CNSPACK for modern multicore microprocessors.

1.1 CNSPACK application area

Numerical solution of 3D incompressible viscous flow problems by the finite-element method leads to large linear systems of equations with sparse non-symmetric matrices. For the Navier-Stokes equations in primitive variables with the continuity equation, treated directly, the system of related linear equations combines many unwanted properties of algebraic systems. It is non-symmetric, not positive definite and has zero diagonals in rows related to the discretized continuity equation. The advantage of the coupled solution is in the improved stability of the numerical solution for stationary flow at high Reynolds numbers.

Content from this work may be used under the terms of the Creative Commons Attribution 4.0 licence. Any further distribution of this work must maintain attribution to the author(s) and the title of the work, journal citation and DOI.
Published under licence by IOP Publishing Ltd

Also, much larger discrete time steps for transient flow analysis can be used. The drawback of the coupled solution is that it is expensive in terms of computer resources, such as memory and computation time.

Similar situation arises in semiconductor device simulation, where the radiation effects result in multi-scale time dependent problem.

Turbulent flows is another example where the large sparse matrices arise. Hypersonic flow is another challenging problem, that requires solution of large sparse linear system of equations on every time step.

2 Algorithms Used in CNSPACK

Advances in iterative solution techniques resulted in more efficient solvers, based on CGS, BiCGS and IDR [4], [29], [30] and GMRES methods [27]. An excellent review by Benzi [5] provides more details on the current state of art of iterative solution techniques.

A preconditioning technique is used to accelerate the convergence of iterative solution, thus resulting in dramatically fewer iterations, typically 10 to 100 instead of the order of 10^3 iterations without preconditioning for large linear systems. Recurrent formulas have been published to construct a preconditioner for structured (3- or 5-diagonal) matrices [29]. Simple first order preconditioners are typically used for unstructured matrices that are related to unstructured meshes for complicated domain geometry. The efficiency of such preconditioning is low and, in a case of the incompressible Navier-Stokes equations, solved with quadratic (for velocity) and linear (for pressure) finite elements, such a preconditioning may result in a divergence of iterations.

The large sparse matrix linear solver CNSPACK is an iterative solver based on iterative IDR (Induced Dimension Reduction) algorithm with ILU preconditioning. It implements the advanced incomplete LU decomposition preconditioning algorithm of arbitrary order (user chosen). In case of the first order ILU (ID=1), the approximate inverse matrix has non-zero elements only where the original matrix has the non-zero elements. In case of the second order ILU (ID=2), the approximate inverse matrix has additionally one non-zero element in each row from the left and right side of the original matrix non-zero element, etc.

The CNSPACK solver has been successfully implemented and used for coupled solution of stiff problems arising in different multiphysics applications (Computational Fluid Dynamics (CFD), semiconductor transport, kinetic and quantum problems) [9], [10], [11], [12], [13], [14], [15], [17], [18].

Originally, CNSPACK was implemented and optimized for early sequential processors, keeping in mind their arithmetic and memory-size limitations. In the beginning of 1990s, the first optimization exercise was performed in order to accelerate the algorithm for appearing superscalar microprocessors (Intel i860) [11] (see Figure 1 for CNSPACK performance with almost linear CPU time and memory versus N).

Today, we are witnessing significant changes in processor architectures and computer system organization. The desktop computers and cluster nodes use high performance pipelined superscalar multicore processors with out-of-order execution of instructions, deep cache hierarchies and high-throughput memory capabilities. Due to this, performance criteria and methods have been revisited, together with consideration of parallelization. The latter becomes important because now parallelization is no more an option (which requires an access to expensive parallel computer systems). Now it is a must because there is no other way to fully utilize the computational potential of a processor. On the other hand, a multicore processor is a shared memory system rather than a distributed memory cluster. This property simplifies parallelization because the convenient and natural OpenMP environment can be used. This model, of course, can be extended with the distributed memory environment (MPI) for the further parallelization in the context of a cluster built upon several shared memory nodes.

2.1 Numerical experiments with CNSPACK

In [11] we proposed the high order preconditioners for unstructured matrices, which improve the convergence of the iterative solution. In that paper, we analyzed the efficiency of this approach on 2D and 3D flow problems. Numerical experiments showed that the computational complexity of the proposed method appears to be $O(N^{5/4})$. The linear algebraic systems were solved using the IDR iterative method [30] with preconditioning by the incomplete decomposition of the original matrix. Comparing the IDR and GMRES methods [27] in our examples, we found that both methods convergence well, if a good preconditioner is used. The IDR method needs significantly less memory to store (only eight) work vectors. The experiments done in 2001, and illustrated in the Figure 1. The IDR algorithm with preconditioning and iteration termination criteria, the residual $r < 10^{-12}$, demonstrated the following:

- The iterative solution without a preconditioning (ID = 0) is time-expensive, the number of iterations is of the order of the number of unknowns (if converged). The memory requirements are smallest.

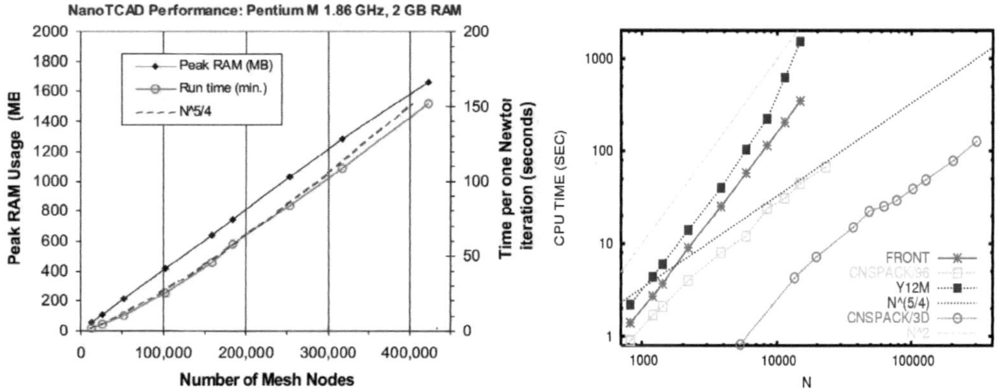

(a) (b)

Figure 1: CNSPACK linear solver performance for electronic device simulation of 3D devices. The dependence on the number of mesh nodes is almost linear for both memory and CPU time, and is close to the optimal theoretical estimate for CPU time, $T = O(N^{5/4})$. (b) Comparison of the CNSPACK solver to other linear solvers for CFD with unstructured sparse matrices shows one to two order faster performance of CNSPACK on an Intel P6 200MHz CPU (done in 2001 [11]): FRONT - direct solution by frontal method; CNSPACK-2D - solution with the preconditioned IDR method [11](P2-P1 finite elements, matrices stored in double precision,incomplete decomposition of the second order); Y12M - solution with Y12M sparse matrix solver package [31]; CNSPACK-3D (lower curve) - solution of 3D problem, matrices are stored in single precision, a first order incomplete decomposition is used. Theoretical asymptotic for direct solution is $T = O(N^2)$, and for the proposed method is $T = O(N^{5/4})$ (these lines are plotted for comparison).

- A first order decomposition for preconditioning (ID = 1), can result in the divergence of the iterations.
- A second order decomposition for preconditioning (ID = 2), dramatically reduced the number of iterations (for $N = 100000$ the number of iteration did not exceed 100). The memory requirements were about twice as high as for no preconditioning (ID = 0).
- Solving the Generalized Hydrodynamic Equations, Eq.(2, 3) [2, 3] instead of Navier-Stokes equations for 3D problems [17, 18] resulted in a much faster convergence: for $N = 300,000$ the number of iterations to converge was 30 to 50.
- We compared the solution time for the same linear system by different methods: direct solution by frontal method, direct solution with Y12M sparse matrix package [31], and by the CNSPACK software (Figure 1). We have been unable to use the SPARSKIT [26] and SPARSE [28] solvers, as those failed for zero matrix diagonal elements.
- The proposed method IDR method with ILU preconditioning [11] can solve the linear system with a high accuracy, much higher than one obtained by a direct solution.
- The node/equation numbering in finite-element mesh is of great importance for the rate of convergence [7, 24]. We found that the optimal node or equation renumbering is quite necessary for unstructured meshes, and the second order incomplete decomposition (ID = 2) should be used for faster convergence and smaller CPU time.
- It appears (Figure 1), that the computational complexity of the proposed method is $T = O(N^{5/4})$.

2.2 Details of implementation

To reduce the memory requirements in CNSPACK, a compact storage scheme for matrices was used. We stored only the nonzero entries. The incomplete decomposition (ID) was a product of triangular and diagonal matrices, $P = LDU$. To avoid a diagonal pivot degeneration we used the Kershaw diagonal modifications [25]: if the value of diagonal element was small, i.e.

$$|a_{ii}| < \alpha = \sqrt{2^{-t}\sigma\mu}, \tag{1}$$

the diagonal was replaced by α . Here σ and μ are the maximum magnitudes of row and column elements, and 2^{-t} is a machine precision (t bits in mantissa, see for the choice of minimum α and other details in [25]).

Aspects of the efficiency of the method on superscalar pipelined microprocessors and parallel architectures were considered. We found that the solution time can be reduced by a factor of three by using special optimization techniques for a superscalar pipelined architecture.

2.3 Convergence of iterations for different problems

Figure 2 shows residuals versus the iteration number for several problems: (a) 3D turbulent flow of incompressible fluid, the number of unknowns $N = 2.5 \cdot 10^6$, (b) 3D axisymmetric hypersonic flow around missile at the altitude of 130km, the number of unknowns $N = 250,000$, and (3) (b) 3D axisymmetric hypersonic flow around missile at the altitude of 110km, the number of unknowns $N = 250,000$.

The solution time on the Dell Precision T5500 computer with Intel Xeon CPU X5550 processors at 2.67GHz was $T = 40.5s$ for the case (a), and $T = 3.05s$ for the case (b) with scalar version of CNSPACK software. The parallel version of CNSPACK resulted in $T = 14.0s$ for the case (a), and $T = 1.05s$ for the case (b).

2.4 Generalized Hydrodynamic Equations

In cases of turbulent flows, hypersonic flow, the Generalized Hydrodynamic Equations by Alexeev [2, 3] were solved, which were presented first in [18] and provided here for the incompressible fluid:

$$\frac{\partial \mathbf{V}}{\partial \mathrm{t}} + (\mathbf{V}\nabla)\mathbf{V} - Re^{-1}\nabla^2\mathbf{V} + \nabla \mathrm{p} - \mathbf{F} = \tau\left\{\frac{\partial^2 \mathbf{V}}{\partial \mathrm{t}^2} + 2\frac{\partial}{\partial \mathrm{t}}(\nabla \mathrm{p}) + \nabla^2(\mathrm{p}\mathbf{V}) + \nabla(\nabla \cdot (\mathrm{p}\mathbf{V}))\right\} \tag{2}$$

while continuity equation is

$$\nabla \cdot \mathbf{V} = \tau\left\{\mathbf{2}\frac{\partial}{\partial \mathbf{t}}(\nabla \cdot \mathbf{V}) + \nabla \cdot (\mathbf{V}\nabla)\mathbf{V} + \nabla^{\mathbf{2}}\mathrm{p} - \nabla \cdot \mathbf{F}\right\} \tag{3}$$

where $\mathbf{V}$ and p are nondimensional velocity and pressure, $Re = V_0 L_0/\nu$ - the Reynolds number, V_0 - velocity scale, L_0 - hydrodynamic length scale, ν - kinematic viscosity, $\mathbf{F}$ is a body force and a nondimensional τ is

$$\tau = \tau^* L_0^{-1} V_0 = l^2/L_0^2 Re = \delta^2 Re, \tag{4}$$

where τ^* is a material property.

Note that the right-hand side of Eq.(3) is the divergence of the fluctuation part of flow velocity, expressed explicitly through original primitive variables according to [2]. The parameter τ is assumed to be constant. No assumption of slow flow variation is made, so the second derivatives in time are kept.

Additional boundary conditions on walls are for fluctuations to be zero. The following boundary condition for pressure on walls is

$$(\nabla \mathrm{p} - \mathbf{F}) \cdot \mathbf{n} = 0, \tag{5}$$

where $\mathbf{n}$ is a wall normal. The formulation is completed by providing the boundary conditions:

$$\mathbf{V} = 0 \text{ on walls, and } \mathbf{V} = \mathbf{V_{in}}(\mathbf{r}), \text{ and } \frac{\partial p}{\partial \mathbf{n}} = 0 \text{ on the inlet}, \tag{6}$$

$$\frac{\partial \mathbf{V}}{\partial \mathbf{n}} = 0, \text{ and } p = 0 \text{ on the outlet}, \tag{7}$$

and the initial conditions:

$$\mathbf{V}(t = 0) = \mathbf{V_0}(\mathbf{r}), \text{ and } p(t = 0) = p(\mathbf{r}), \tag{8}$$

where $\mathbf{r}$ is a vector of a point (x, y, z).

The governing equations for the hypersonic flows are provided in [19, 20, 21], where a number of application examples are presented.

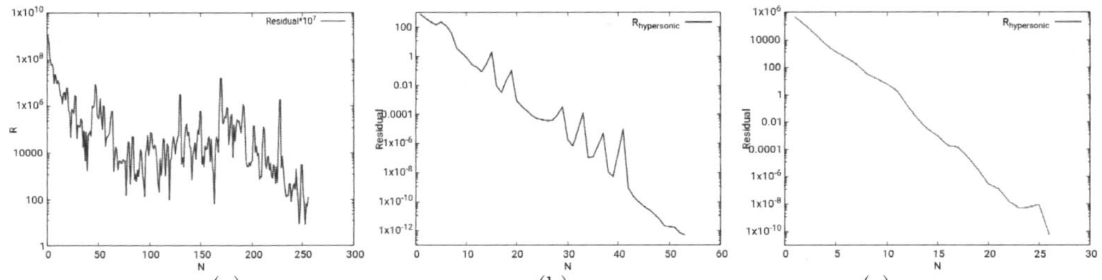

(a) (b) (c)

Figure 2: CNSPACK solution convergence, the residual versus the iteration number for (a) 3D turbulent flow solution of incompressible fluid (the number of unknowns N=2,500,000); (c) hypersonic flow around missile at the altitude of 130km (the number of unknowns N=200,000); (c) hypersonic flow around missile at the altitude of 110km (the number of unknowns N=200,000).

2.5 Parallelization of CNSPACK

It was initially known that the CNSPACK algorithm requires both the speed of arithmetic calculations and the throughput of memory accesses. Because of the computation-memory disbalance of modern processors, the latter property becomes even more important (i.e. the algorithm is memory-bound rather than computational-bound). The above determines the choice of a computer system used for the computations. On the other hand, there are no more memory-size limitations, and algorithms can be now re-arranged correspondingly in favor of the computational speed.

In order to implement a parallel variant of the iterative procedure, it was necessary to check some storage schemes suitable for parallelization. First, the reverse-numbered storage scheme was implemented and tested for convergence. Test results have demonstrated good convergence properties and accuracy of the solution vector. This scheme was then incorporated into the more complicated twisted scheme for parallelization (considered in the next section).

Next, the special storage scheme was implemented which stores elements by rows both in lower and upper parts of the original matrix A. This scheme is very convenient for parallelization of the multiplication routine (CNSMUL) because it makes splitting of the matrix and vectors in the matrix-by-vector expression Y = A X very natural: lines of both matrix parts, L and U, are used to update a single elements in vector Y (while in the standard storage scheme, a sparse group of elements in Y is updated through the processing of a column in U). As a result, parallelization of CNSMUL for any number of threads becomes possible without any additional care.

New advanced block-pipelined parallelization method of ILU-preconditioning was finally developed and implemented based on our preliminary experience described above. The idea of this method is to split a matrix half-band into pairs of adjacent trapezoidal blocks that have no mutual data dependence and can therefore be processed in parallel. As a result, parallelization of the preconditioning routine for 4 threads was implemented. For this method, the new blocked storage scheme was designed. Pipelined parallelization methods are used in situations when the natural parallelization potential of a recursive method is exhausted [1].

These methods are based on the ability to split a sequence of dependent steps into substeps in such a way, that substeps of each step are executed in different iterations of the algorithm (graphically it can be represented as stairs, therefore the method itself is sometimes called as "staircase method"). In the current algorithm, each step is split into two substeps. In the first iteration, only the first substep is executed. Then, when the first substep processes iteration k, the second substep processes iteration k-1. At the very end, only the second substep is executed. Similarly, a multi-substep method can be described [6].

As was analyzed above, parallelization potential of the ILU-decomposition (in particular, in solving a preconditioning system) is limited by two threads: it is possible to eliminate elements by moving from both ends of a banded matrix to the center, but it is not generally possible to eliminate elements in a particular row before finishing elimination in the previous row. However, we can try to split a matrix half-band into blocks in such a way that dependencies between some neighbor blocks are removed. This splitting is shown on Figure 3 (at the center) [6].

The experiments with CFD matrix for incompressible flow, of the order N=78,000, using the target computer system Intel Core i7-920 (2.66 GHz, 3 memory channels DDR3-1333) [6], are shown in Table 1.

	Serial 1 thread	Parallel 2 threads	Parallel 4 threads	Pipelined 4 threads
Time 1 iter	89.1 ms	51.3 ms	41.6 ms	32.05 ms
Acceleration	1.00	1.74	2.14	2.78
real*4 /real*8	1.00	1.08	1.16	1.28

Table 1: Parallelization performance results for the CFD matrix of the order N=78,000 (real*4).

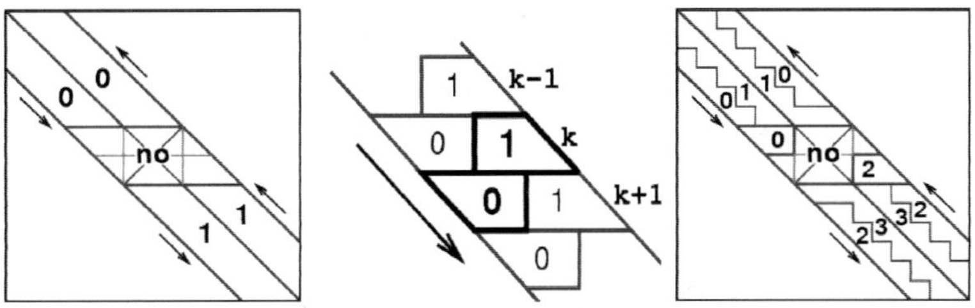

Figure 3: Example of the splitting of the matrix L (center) and illustration of the implementation of the pipelined parallelization during the elimination process and the backsubstitution (right). The original twisted factorization algorithm is shown for the comparison (left).

2.6 Summary of CNSPACK performance
Here we provide a brief summary of CNSPACK convergence, properties and performance:

- The CNSPACK solver has been tested on a variety of unstructured large sparse matrices arising from diverse application areas.
- The computational complexity of CNSPACK is $T = O(N^{5/4})$.
- The CPU time and memory requirement of CNSPACK were 10 to 100 times less compared to best industrial linear solvers.
- CNSPACK solver demonstrated convergence for any problem we have solved. Sometimes the order of preconditioner need to be increased (up to ID=4) to get the convergence. This was in the case of MOSFET2 matrix from semiconductor transport equations (high energy ion strike on MOSFET). Any other solver we tried (including the Hypre software library [8]) did not converge for this problem.
- The pipelined parallel version of CNSPACK is about three times faster than a superscalar version.
- CNSPACK solver has several interfaces to easy be included in various application packages: (i) finite element method interface; (ii) sparse matrix input; (iii) element by element matrix input; (iv) any other formats can be easily adjusted to CNSPACK matrix format.
- The order of preconditioner is a user choice.

Availability: The CNSPACK package is commercially available from Ultra Quantum Inc. Requests can be sent directly to the author.

3 Applications
In this section, we present examples of CNSPACK applied to the solution of large sparse non-symmetric linear systems across various application areas.

3.1 Scientific applications
Figure 4 shows 3D simulation result of internal combustion engine performed using FEMINA-3D multiphysics solver with CNSPACK: the pressure field (a) and the magnitude of velocity in a cross-section (b).

Figure 5 shows the 1D, 2D, and 3D simulations of a car radiator performed using the FEMINA-3D multiphysics solver with CNSPACK.

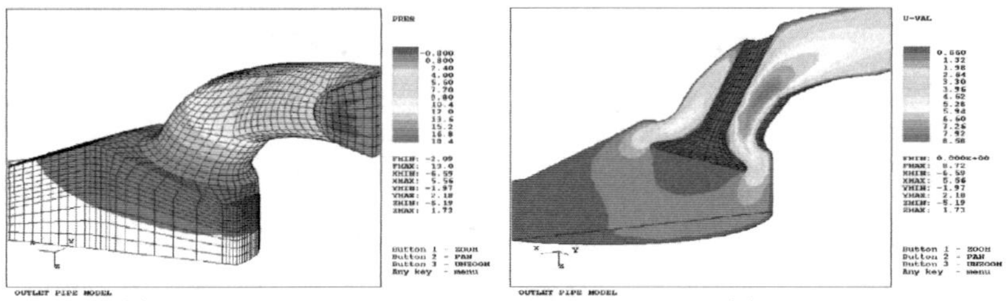

(a) (b)

Figure 4: 3D simulations of of an internal combustion engine performed using the FEMINA-3D multiphysics solver with CNSPACK: the pressure field (a) and the velocity magnitude in a cross-section (b).

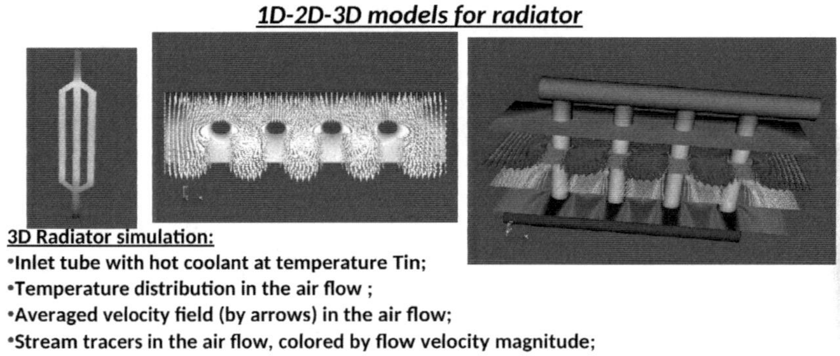

Figure 5: 1D, 2D and 3D simulations of car radiator performed by FEMINA-3D multiphysics solver with CNSPACK.

Figures 6, 7 show the 3D simulations of MOSFET subject to high-energy ion strike (semiconductor application solver (TCAD)) with CNSPACK [15].

3.2 Industrial applications

The industrial applications include large-scale simulations of hypersonic flows at high altitudes and turbulent flows.

Figures 8(a, b) show 3D axisymmetric simulations of hypersonic flows for a 22/55° bi-conic geometry at Ma = 15.6 and a 70° blunt cone flow at Ma = 20.2, performed using the FEMINA-3D/CNSPACK software. The number of unknowns is approximately N = 250,000 [20, 21].

Figures 9(a,b) show the 3D axisymmetric simulations of hypersonic flows around missile at the altitude of 160km (comparison to DSMC simulation, the number of unknowns is approximately N=200000), and a hypersonic flow around 2D cylinder conducted with FEMINA-3D/CNSPACK software [20, 21].

Figures 10(a,b) show the 3D simulations of turbulent flow in a lid-driven cavity at Re=10,000, compared with the experiment by Koseff and Street (1984), and a 2D simulation of backward-facing step flow at Re=132,000, compared with the experiment by Kim (1984) with FEMINA-3D/CNSPACK software [17, 18, 23].

Figures 11(a,b) show the 2D simulation of turbulent flow in channel at Re=14914 (number of unknowns is approximately N=60000), and 3D simulation of backward-facing step turbulent flow at Re=55,000 (the number of unknowns is approximately N=2,500,000) performed with FEMINA-3D/CNSPACK software.

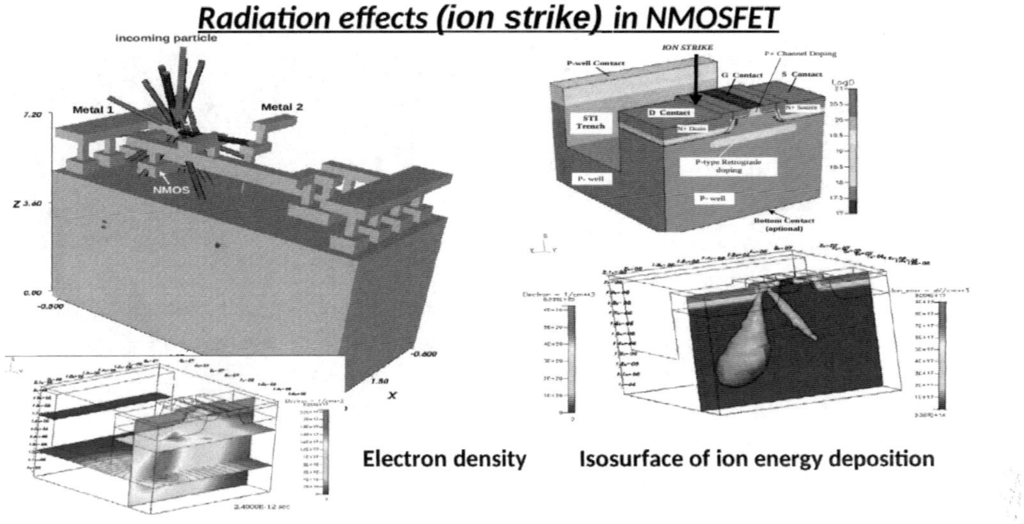

Figure 6: 3D simulations of MOSFET subject to high-energy ion strike (semiconductor application TCAD with CNSPACK).

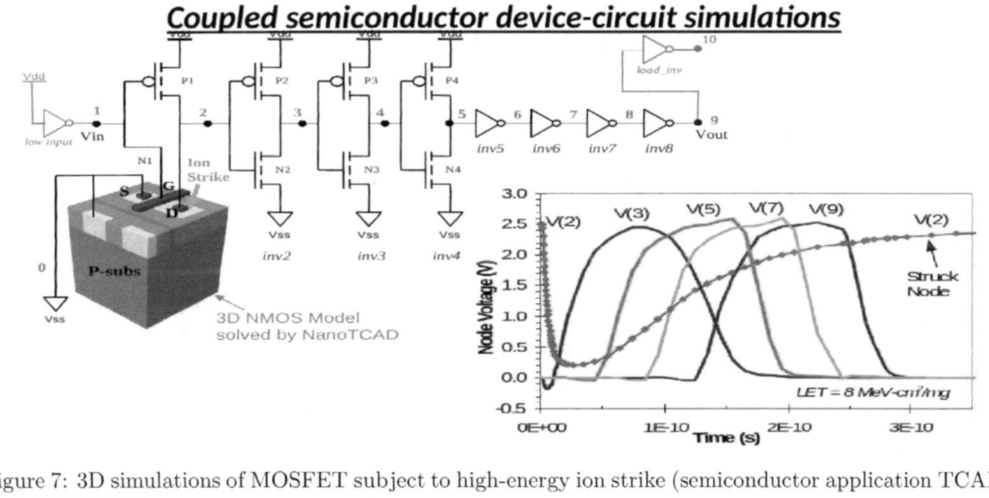

Figure 7: 3D simulations of MOSFET subject to high-energy ion strike (semiconductor application TCAD with CNSPACK).

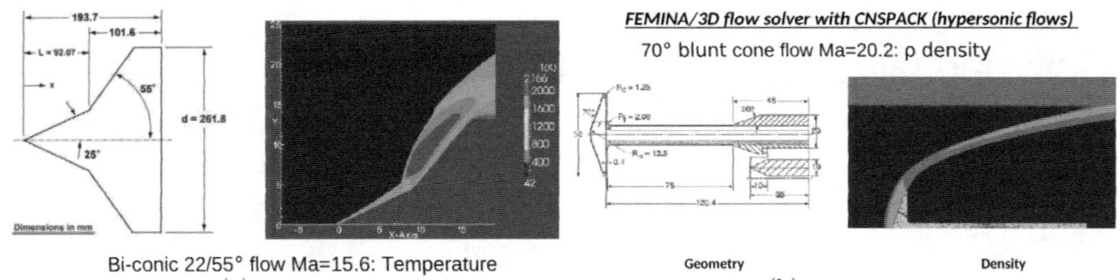

(a) (b)

Figure 8: (a) 3D axisymmetric simulations of hypersonic flow around bi-conic $22/55^o$ geometry at Ma=15.6, and (b) 70° blunt cone flow at Ma=20.2 with FEMINA-3D/CNSPACK, the number of unknowns is approximately N=250,000.

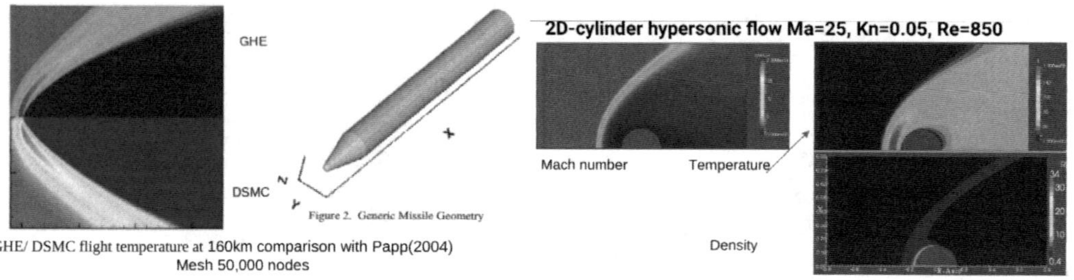

(a) (b)

Figure 9: (a) 3D axisymmetric simulations of hypersonic flow around missile at the altitude of 160km (comparison to DSMC simulation, the number of unknowns is approximately N=200,000), and (b) hypersonic flow around 2D cylinder performed with FEMINA-3D/CNSPACK).

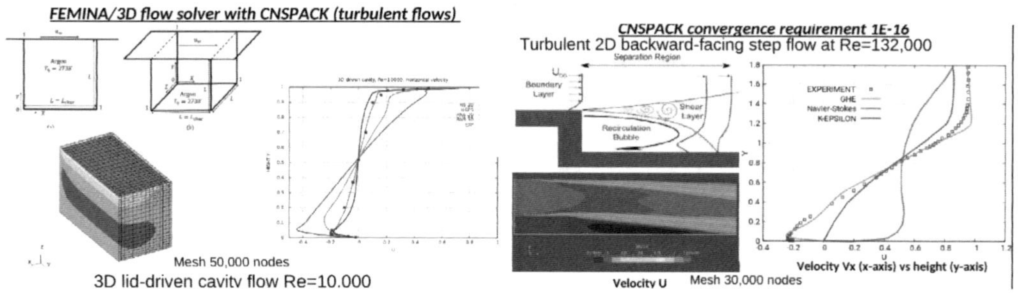

(a) (b)

Figure 10: (a) 3D simulations of turbulent flow in a lid-driven cavity at Re=10,000, compared with the experiment by Koseff and Street (1984), and (b) 2D simulation of backward-facing step flow at Re=132,000, compared with the experiment by Kim (1984) with FEMINA-3D/CNSPACK software.

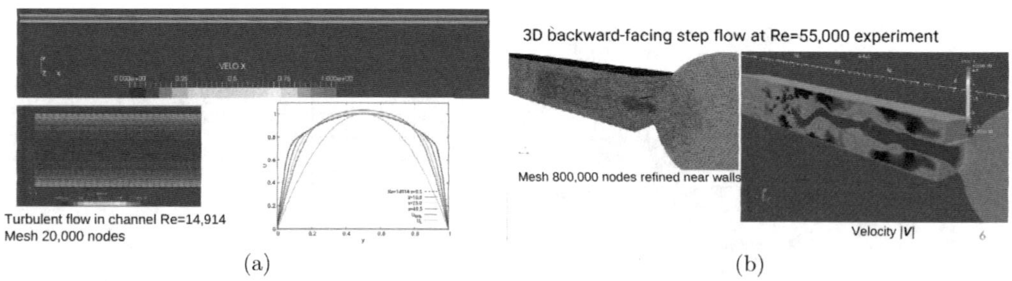

(a) (b)

Figure 11: (a) 2D simulations of turbulent flow in channel at Re=14,914, and (b) 3D simulation of backward-facing step flow at Re=55,000 performed with FEMINA-3D/CNSPACK software, the number of unknowns is approximately N=2,500,000.

Conclusions

Recent advancements in the universal iterative linear solver CNSPACK, developed for large-scale scientific and industrial applications, have been presented. The performance of the CNSPACK solver is significantly improved through preconditioning by incomplete decomposition of arbitrary (user-chosen) order and approaches the theoretical time complexity of $T = O(N^{5/4})$. It achieves minimal storage and CPU time usage across a wide variety of problems, including incompressible fluid flows, turbulent flows, semiconductor device carrier transport, optoelectronics and nano-device carrier transport, and hypersonic flows..

CNSPACK is the only parallel solver currently used for semiconductor applications (TCAD).

The performance of the parallel version of CNSPACK is three times faster than that of the superscalar version.

References

[1] G. Accary, O. Bessonov, D. Fougere, K. Gavrilov, S. Meradji, and D. Morvan. 2009 Efficient parallelization of the preconditioned Conjugate gradient method. Lecture Notes in Computer Science 5698, 60-72 (2009).

[2] Alexeev B.V., 1994 The generalized Boltzmann equation, generalized hydrodynamic equations and their applications. Phil. Trans. Roy. Soc. London, A. 349, 417-443.

[3] Alexeev B.V., 2004 Generalized Boltzmann Physical Kinetics, Elsevier.

[4] R. Barrett, M. Berry, T. Chan, J. Demmel, J. Donato, J. Dongarra,V. Eijkhout, R. Pozo, C. Romine, and H. van der Vorst. 1994 Templates for the Solution of Linear Systems: Building Blocks for Iterative Methods, Philadelphia: SIAM, 1994.

[5] M. Benzi, 2002 Preconditioning Techniques for Large Linear Systems: A Survey, J. Comp. Phys 182, 418-477.

[6] O. A. Bessonov and A. I. Fedoseyev, 2012 Parallelization of the preconditioned IDR solver for modern multicore computer systems, AIP Conf. Proc. 1487, 314 (2012); doi: 10.1063/1.4758973

[7] L. C. Dutto. 1993 The Effect of Ordering on Preconditioned GMRES Algorithm, for Solving the Compressible Navier-Stokes Equations. Int. J. Numer. Meth. Eng., 36 (1993) 457-497.

[8] Falgout R.D., Jones J.E., Yang U.M., 2006 The Design and Implementation of hypre, a Library of Parallel High Performance Preconditioners, Numerical Solution of Partial Differential Equations on Parallel Computers, 267-294, Springer 2006.

[9] Fedoseyev A.I., Alexeev, B.V., 1998 Mathematical model for viscous incompressible fluid flow using Alexeev equations and comparison with experimental data. In: Dey, S.K., Ziebarth, G., Ferrandiz (Eds.), Proceedings of Advances in Scientific Computing Modelling (Special Proceedings of IMACS'98). Alicante, Spain, 1998, 158-163.

[10] Fedoseyev A.I., Kansa E.J., C. Marin, M. Volz, and A. G. Ostrogorsky, 2000 AIAA Paper 2000-0698.

[11] Fedoseyev A., Bessonov O., 2001 Iterative solution of large linear systems for unstructured meshes with preconditioning by high order incomplete decomposition, Japanese Computational Fluid Dynamics Journal 10, 299-303.

[12] Fedoseyev A., 2001a A regularization approach to solving the Navier-Stokes Equations for Problems with Boundary Layer, Comput. Fluid Dynamics J., 9, 317-324.

[13] Fedoseyev A.I., Kansa E.J., C. Marin, and A.G. Ostrogorsky, 2001b Magnetic field suppression of semiconductor melt flow in crystal growth: comparison of three methods for numerical modeling, Japanese Computational Fluid Dynamics Journal, 10(3), 325-333.

[14] Fedoseyev A., M. Turowski, M. Wartak, 2007 Kinetic and Quantum Models for Nanoelectronic and Optoelectronic Device Simulation, Journal of Nanoelectronics and Optoelectronics, 2(3):234-256

[15] Fedoseyev A., M. Turowski, L. Alles, and R. A. Weller, 2008 Math. and Computers in Simulation 79, 1086-1096.

[16] Fedoseyev A.I, M. Turowski, M. Alles, R. A. Weller, 2006 Accurate Numerical Models for Simulation of Radiation Events in Nano-Scale Semiconductor Devices, Mathematics and Computters in Simulation, 79, Issue 4, 1086-1095.

[17] Fedoseyev A., Alexeev, B.V., 2010 Simulation of viscous flows with boundary layers within multiscale model using generalized hydrodynamics equations, Procedia Computer Science, 1, 665-672.

[18] Fedoseyev A., Alexeev B.V.,2012 Generalized hydrodynamic equationsfor viscous flows-simulation versus experimental data,in AMiTaNS-12, American Institute of Physics AIP CP 1487, pp.241-247.

[19] Fedoseyev A., 2020 Finite element method stabilization for supersonic flows with flux correction transport method, AIP CP 2302, 120003.

[20] Fedoseyev A., Griaznov V., Simulation of Rarefied Hypersonic Gas Flow and Comparison with Experimental Data, in AMiTaNS-2021, Conf. Proc., AIP CP 2522, 100003, 2021, Ed. M.Todorov.

[21] Fedoseyev A., Griaznov V., Ouazzani J., Simulation of rarefied hypersonic gas flow and comparison with experimental data II, Proc. AMITANS-2022 Conf., AIP CP 2953, 2023, Ed. M.Todorov.

[22] Fedoseyev A. 2023 Approximate Analytical Solution for Turbulent Flow in Channel, J. Phys.: Conf. Ser. 2675 012011.

[23] Fedoseyev A. 2024 Modeling of Turbulent Flow over 2D Backward-Facing Step Using Generalized Hydrodynamic Equations, in: L. Franco et al. (Eds.): ICCS 2024, LNCS 14838, pp. 63-69, Springer, 2024.

[24] N.E. Gibbs, W.G. Pool, Jr. and P.K. Stockmeyer. 1976 An algorithm for reducing the bandwidth and profile of a sparse matrix, SIAM J. Numer. Anal., 13 (1976) 236-250.

[25] D. S. Kershaw. 1980 On the Problem of Unstable Pivots in the Incomplete LU-Conjugate Gradient Method. J. Comput. Phys., 38 (1980) 114-123.

[26] K. S. Kundert, A. S. Sangiovanni-Vincentelli. 1988 A Sparse linear equation solver, Manual version 1.3a. University of California at Berkeley, 1988,68p.

[27] Y. Saad. 1986 GMRES: A generalized minimal residual algorithm for solving nonsymmetric linear systems, SIAM J. Sci. Statist. Comput., 7 (1986) 856-869.

[28] T. Skalicky. 1995 LASPACK user manual, version 1.12.2. Technical University of Dresden, 1995. http: //www.tu-dresden.de/mwism/skalicky/laspack/laspack.html

[29] H. Van Der Vorst, 1981 Iterative solution methods for certain sparse linear systems with a non-symmetric matrix arising from PDE-problems. J. Comput. Phys., 44 (1981) 1-19.

[30] P. Wesseling, P. Sonneveld. 1980 Numerical experiments with a multiple grid and a preconditionned Lanczos type methods. Lecture Notes in Mathematics, 771, Berlin: Springer, 1980, 543-562.

[31] Z. Zlatev et al. 1981 Y12M solution of large and sparse systems of linear algebraic equations. Lecture Notes in Computer Science, 121, Springer, Berlin-Heidelberg-New York, 1981.

Mathematical modeling of the vortices couple dynamics in a stratified fluid

V A Gushchin[1] and I A Smirnova[2]

[1] Professor, Institute for Computer Aided Design RAS, 84/32-14, Profsoyuznaya str., Moscow, Russia

[2] Junior researcher, Institute for Computer Aided Design RAS, 84/32-14, Profsoyuznaya str., Moscow, Russia

E-mail: gushchin47@mail.ru

Abstract. The problem of the dynamics of the vortex couple in stratified fluid is considered. The couple is two vortexes, arranged symmetrically relative to the vertical axis, but with the opposite direction of rotation, so that it begins to rise upwards. The mathematical model of a couple of vortices in a viscous stratified medium was studied. Salinity is chosen as a stratifying component, just as it can be implemented in the laboratory. The task is described by the Navier-Stokes equations in the Boussinesq approximation. To solve the problem, the SMIF method is used, of course, the finite different scheme of which has such properties as the second order of approximation on spatial variables, minimal scheme dissipation and dispersion, performance in wide range of Reynolds and Froude numbers and, most importantly, when solving wave processes are a property of monotony. The dynamics of such a couple are studied depending on its geometric sizes, the intensity of the vortexes and the depth of their initial immersion.

1. Introduction

In this work, the mathematical model of the dynamics of a vortices couple in a viscous stratified medium was studied. Vortices in rotating, stratified fluids are subject to subtle physical processes which are as yet not fully understood. The stability conditions (baroclinic instability) of such vortices depend on a number of parameters, which makes it difficult to derive a simple criterion. The interaction properties raise even more questions [1]. The aviation security tasks is to explore the movement of vortexes that break from the aircraft surface. It also explores the aero acoustic noises generated by such vortexes [2]. In studying the pollution, that arises on the free surface of large reservoirs, the tasks of interaction of vortex structures with such boundaries are also considered [3]. The tasks of modeling the non acoustic signs of dynamics of underwater objects also based on the movement of vortices and there interaction with a free surface or a pycnocline [4]. Similar challenges arise in meteorology, when the equations take into account the forces of Coriolis.

The interaction between two vortices depends on whether they are homostrophic or heterostrophic (of the same sign or of the opposite sign). Let's call two vortices of opposite sign a couple, and two vortices of the same sign a pair. A mathematical model of the dynamics of a vortices couple in a viscous stratified fluid is proposed. The dynamics of a vortex couple is investigated. This task is of undoubted significance from a practical point of view.

2. Statement of the problem

Consider two vortexes of the radius r_0 and the same intensity of Γ, but with the opposite directions of rotation, are arranged symmetrically relative to the vertical axis (Figure 1). Distance between the centers of vortexes is b. Due to the symmetry of the task it is possible to consider the flow in

Content from this work may be used under the terms of the Creative Commons Attribution 4.0 licence. Any further distribution of this work must maintain attribution to the author(s) and the title of the work, journal citation and DOI.
Published under licence by IOP Publishing Ltd

the right half-plain only. For certainty, let the right vortex rotates clockwise. We consider here 2D problem.

It is known [5] that such a vortex couple will rise up at speed $U = \Gamma/2\pi b$, where Γ is intensity of the vortex $\Gamma(\mathbf{v}) = \oint \mathbf{v}dl = \oint \omega r_0^2 d\varphi = 2\pi\omega r_0^2$, ω – the magnitude of the vorticity.

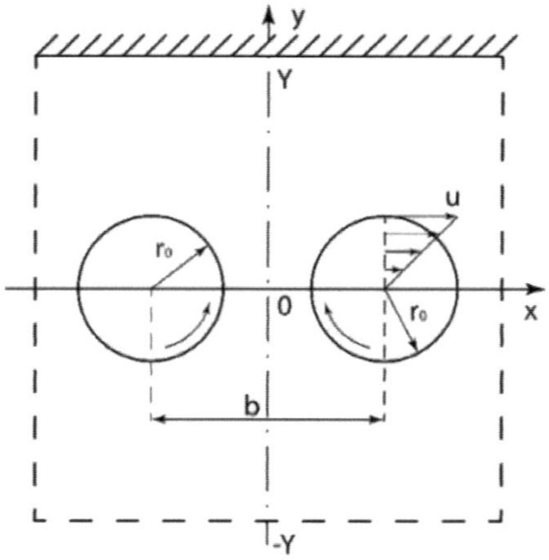

Figure 1. Statement of the vortex couple dynamics problem.

The task is described by the equations (1),

$$\begin{aligned} &\frac{\partial \mathbf{v}}{\partial t} + (\mathbf{v}\cdot\nabla)\mathbf{v} = -\nabla p + \frac{1}{Re}\Delta\mathbf{v} + \frac{1}{Fr}\frac{\mathbf{g}}{g}, \\ &\frac{\partial s}{\partial t} + (\mathbf{v}\cdot\nabla)s = \frac{1}{Sc\cdot Re}\Delta s + v\cdot Fr, \\ &\nabla\cdot\mathbf{v} = 0 \end{aligned} \tag{1}$$

where **v** is velocity vector with constituents u, v, accordingly, along the axes x and y at the Cartesian coordinate system, chosen as indicated in Figure 1, p is pressure, $Re = r_0 U_0/\nu$ is the Reynolds number, $Fr = U_0^2/(g\ r_0)$ is the Froude number, r_0 is radius of vortex at t=0, U_0 is velocity of vortex rotation at t=0, ν is the kinematic viscosity coefficient, and $\mathbf{g} = (0, -g)$, where g is the acceleration due to gravity. We intentionally use equations in primitive variables ($\mathbf{v}$, p), to summarize the stratified fluid approach in the future.

The initial conditions. Assume that at the initial moment of time $t = 0$ inside the area of the initial vortex A particles of fluid move at speed $\mathbf{v} = (u, v)$, described by equations (2), and in the rest of the area, the fluid is at rest $\mathbf{v} = 0$ (no motion).

$$\begin{aligned} &\sqrt{(x-b/2)^2+y^2} \le 1, \ (x,y)\in A, \\ &u(x,y) = \omega\sqrt{(x-b/2)^2+y^2}\cdot sign(y), \\ &v(x,y) = -\omega\sqrt{(x-b/2)^2+y^2}\cdot sign(x-b/2). \end{aligned} \tag{2}$$

The boundary conditions. On the left boundary ($x = 0$) the conditions of symmetry are used. At the upper boundary is a condition taken in oceanology as an approximation of a solid cover (a hard impenetrable wall with slippage). On the lower and right borders, considering them quite remote from the vortex, conditions of rest are put.

3. Solution method

For mathematical modeling of the dynamics of a vortices couple, one of the latest versions of the Splitting Method for Incompressible Fluid (SMIF) [6] developed by the authors is used to solve the problem. The finite difference scheme of the method has the following properties: the second order of approximation on spatial variables, minimal scheme viscosity and dispersion, monotonicity.

Splitting scheme. Suppose that at a time $t_n = n\tau$, where τ is the time step and n is the number of steps, the fields of velocity $\mathbf{v}$, pressure p are known. Then, the scheme for finding the unknown functions at the time $t_{n+1} = (n + 1)\cdot\tau$ can be written as

$$\frac{\tilde{\mathbf{v}} - \mathbf{v}^n}{\tau} = -(\mathbf{v}^n \cdot \nabla)\mathbf{v}^n + \frac{1}{Re}\Delta\mathbf{v}^n + \frac{1}{Fr}\frac{\mathbf{g}}{g}, \tag{3.1}$$

$$\tau\Delta p = \nabla \cdot \tilde{\mathbf{v}}, \tag{3.2}$$

$$\frac{\mathbf{v}^{n+1} - \tilde{\mathbf{v}}}{\tau} = -\nabla p, \tag{3.3}$$

$$\frac{s^{n+1} - s^n}{\tau} = -(\mathbf{v}^{n+1} \cdot \nabla)s^n + \frac{1}{Sc \cdot Re}\Delta s^n + \frac{v^{n+1}}{C}. \tag{3.4}$$

At Stage I (3.1), it is assumed that the transfer of momentum occurs only due to convection, diffusion, and buoyancy forces. At Stage II (3.2), given the intermediate velocity field $\tilde{\mathbf{v}}$ found earlier and taking into account the solenoidity of the velocity vector $\mathbf{v}^{n+1}$, the pressure field is found by solving the Poisson equation (over-relaxation method is used here). At Stage III (3.3), it is assumed that the transfer of momentum occurs only due to the pressure gradient. At Stage IV (3.4), on found velocity field, the salinity perturbation is calculated.

Finite difference scheme. The flow region under examination is covered by grid of cells (the grid is uniform with respect to x and y)

$$\Omega = \begin{cases} x_{i+1/2} = ih_x, & h_x > 0, \quad i = 0,1,\dots,L; \quad Lh_x = X, \\ y_{j+1/2} = jh_y, & h_y > 0, \quad j = 0,1,\dots,M; \quad Mh_y = 2Y, \end{cases}$$

where h_x and h_y are the grid sizes, and L and M are the numbers of grid cells in the directions x and y, respectively (the point with the coordinates (i, j) coincides with the cell center). Here, as in the original splitting method, the staggered grid is used, *i.e.*, the coordinates of the grid functions are spaced apart. This allows us to visually interpret each cell as an element of the medium volume; such an element is characterized by the pressure $p_{i,j}$, (and possibly by the temperature, energy, *etc.*), and the divergence $D_{i,j}$ (D determines the presence of a source or sink in the volume, depending on its sign) computed at the cell center. The knowledge of the normal velocity component on a cell side makes it possible to directly calculate the momentum flux through this side. The finite difference scheme for the problem of spot collapse in stratified fluid for 2D case was published in details in [7].

4. Calculation results

The results of numerical study of the vortex couple dynamics in a viscous stratified medium are presented. A study was carried out of Reynolds number influence on the movement of a vortex couple. One of the latest variants of the Splitting Method for Incompressible Fluid (SMIF) [6] developed by the authors is used to solve the problem. In this case, a series of calculations was carried out, presented in Table 1.

Table 1. The series of calculations

	1	2	3	4	5	6	7
Re	10	$10\sqrt{10}$	100	500	2 000	5 000	10 000
Fr	100	100	100	100	100	100	100

Computation area is $0 < x < 4$, $-4 < y < 4$. The rectangular computational grid has a computational step $h = 0.02$. The size of the computational grid is 200×400. The coefficients and parameters are: $\mu/\rho_0 = 0.01$ cm²/s, $k_s = 1.41\cdot10^{-5}$ cm²/s, $T_b = 2\pi$ s, $\Lambda = 10$ cm, Schmidt number $Sc = 709.2$. The computational time step is $\tau = 0.001$. At the initial moment of time, the vortices is round and has a radius $r_0 = 1$, angular velocity $\omega = 6$. The distance between the centers of the vortices is b = 2 (Figure 2). On the left boundary $x = 0$ the symmetry conditions are set. At the upper boundary $y = 4$, the conditions of a solid cover are set. At the lower boundary $y = -4$ and on the right boundary $x = 4$, the fluid is at rest.

The characteristic calculated variant of the vortex couple dynamics where $Re = 2000$, $Fr = 100$ is investigated (variant 5 in the Table 1).

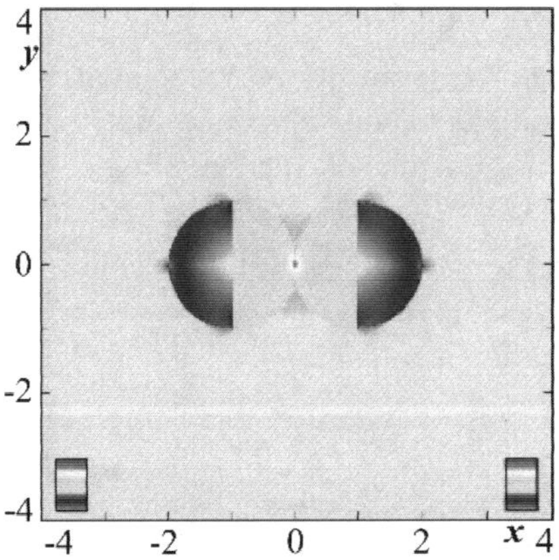

Figure 2. The salinity perturbation isoclines, $t = 0$

At the initial moment of time, a couple of vortices forms a pair of salinity perturbations in the salinity perturbation field, which at the initial moment of time have a circular shape. Figure 3.a shows one of the salinity perturbations pair. Due to the interaction of the couple and the influence of gravity, each area of salinity perturbations splits into two areas. The salinity perturbation area is deformed as the rotation slows down (Figures 3.b-c).

By $t = 0.5$, the rotation stops, two perturbation areas are formed: with minimum and maximum values (Figure 3.d). By time $t = 1$, there is a tendency for the areas to separate; the area with maximum values of salinity perturbations tends upward, and the area with minimum values of salinity perturbations tends downward (Figure 3.e). By this time, the left and symmetrical right parts that make up the pair merge. One area of salinity perturbations moves up, the second area moves down.

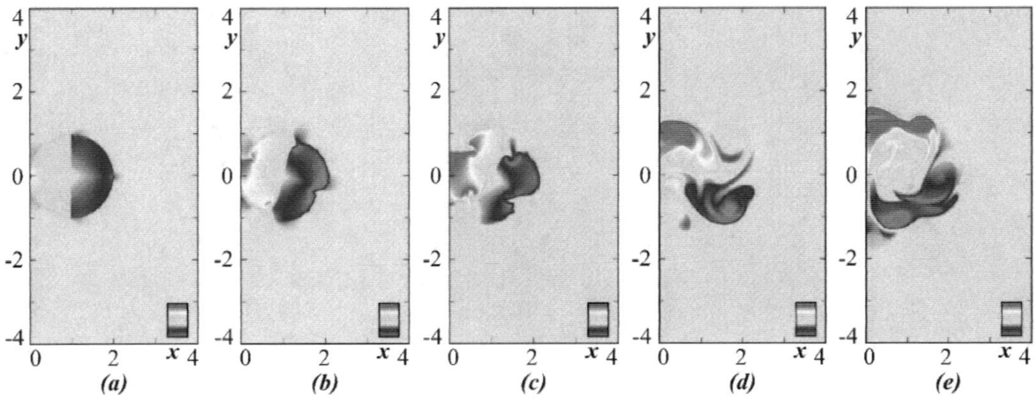

Figure 3. The salinity perturbation isoclines for: Re = 2000, Fr = 100, Sc = 709.2, μ/ρ_0 = 0.01 cm^2/s, k_s = 1.41 10^{-5} cm^2/s, computation area is 4×8 (**a** - t=0, **b** - t=0.05, **c** - t=0.1, **d** - t=0.5, **e** - t=1)

At t = 1.5, the areas of salinity perturbations acquire a flattened shape (Figure 4.a). At t = 2, the shape of the salinity perturbations becomes more circular (Figure 4.b). The two areas of salinity perturbations are moving away from each other. Minor salinity perturbations remain between them. At t = 3, one area of salinity perturbations reaches the upper boundary, the second area approaches the lower boundary (Figure 4.c). Then the upper area of salinity perturbations splits into two areas. At t > 3, the area of salinity perturbations moves along the upper boundary (Figures 4.d-e). At this time, the lower area of salinity perturbations decreases with time. A small perturbation remains in the center of the computational area.

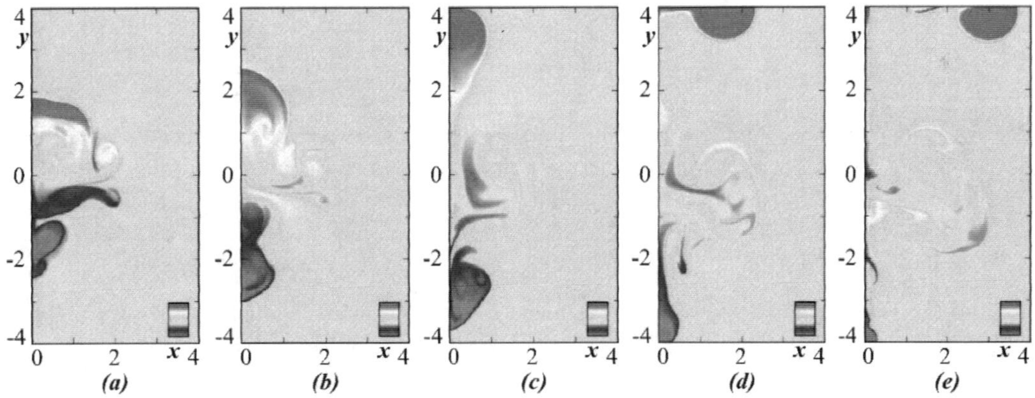

Figure 4. The salinity perturbation isoclines for: Re = 2000, Fr = 100, Sc = 709.2, μ/ρ_0 = 0.01 cm^2/s, k_s = 1.41 10^{-5} cm^2/s, computation area is 4×8 (**a** - t=1.5, **b** - t=2, **c** - t=3, **d** - t=4, **e** - t=5)

Figure 5 shows the complete field of salinity perturbations for vortices couple at t = 0, t = 2.5, t = 4. Here it can be seen that a couple of salinity perturbations is first transformed into two new areas of salinity perturbations, which begin to move away from each other in the vertical direction, then the upper area splits into two areas that move along the horizontal boundary. Gradually the size of the areas decreases.

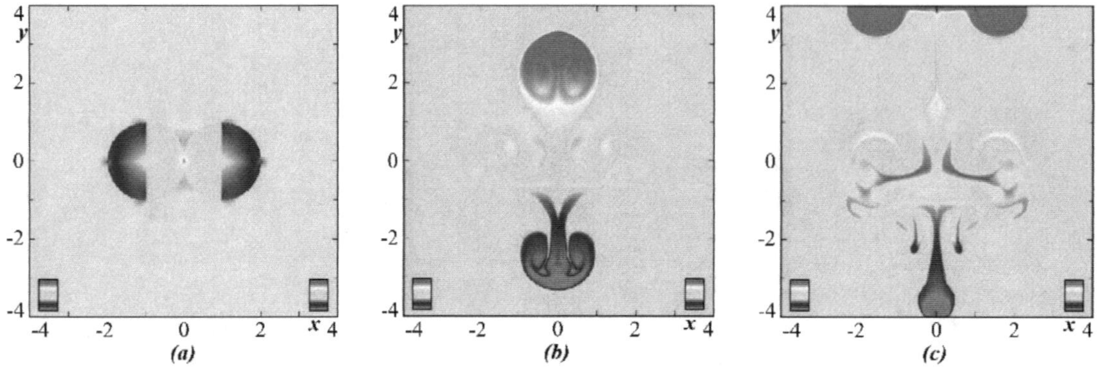

Figure 5. The salinity perturbation isoclines for: *Re* = 2000, *Fr* = 100, *Sc* = 709.2, μ/ρ_0 = 0.01 cm²/s, k_s = 1.41 10^{-5} cm²/s, computation area is 8×8 (**a** - *t*=0, **b** - *t*=2.5, **c** - *t*=4)

At the initial moment of time, the couple of vortices rotates in the opposite direction. One of the vortices of the couple is located in the computational area. At $t < 1$, the vortex continues to rotate, slightly increasing in size (Figures 6.a-c). At $t > 1$, the vortex transforms into two vortices – a new pair of vortices is formed (Figure 6.d). Salinity perturbations behave synchronously.

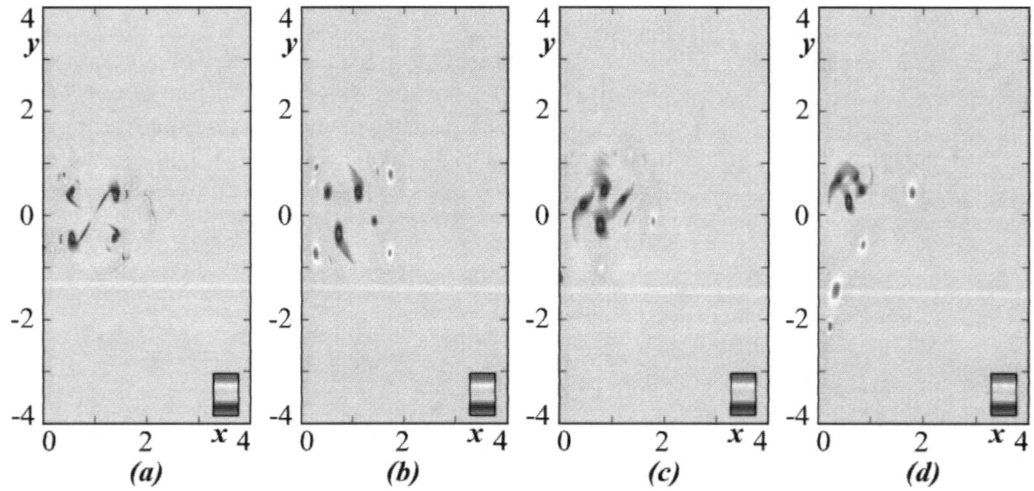

Figure 6. The isoclines of vorticity for: *Re* = 2000, *Fr* = 100, *Sc* = 709.2, μ/ρ_0 = 0.01 cm²/s, k_s = 1.41 10^{-5} cm²/s, computation area is 4×8 (**a** - *t*=0.1, **b** - *t*=0.5, **c** - *t*=1, **d** - *t*=1.5).

At $t = 2$, the vortices increase in size, one vortex rises, the second descends (Figure 7.a). A small vortex is formed in the center of the computational area. At $t = 3$, the pair of vortices reaches horizontal boundaries (Figure 7.b). One vortex rises to the upper boundary. The second vortex approaches the lower boundary. The intensity of the vortex pair increases. At $t = 3$, the salinity perturbation areas also approach the upper and lower boundaries.

At $t > 3$, the upper vortex moves along the horizontal upper boundary, and the lower vortex reaches the lower boundary (Figures 7.c-d). Over time, the intensity of the vortex pair decreases, as do the areas of salinity perturbation.

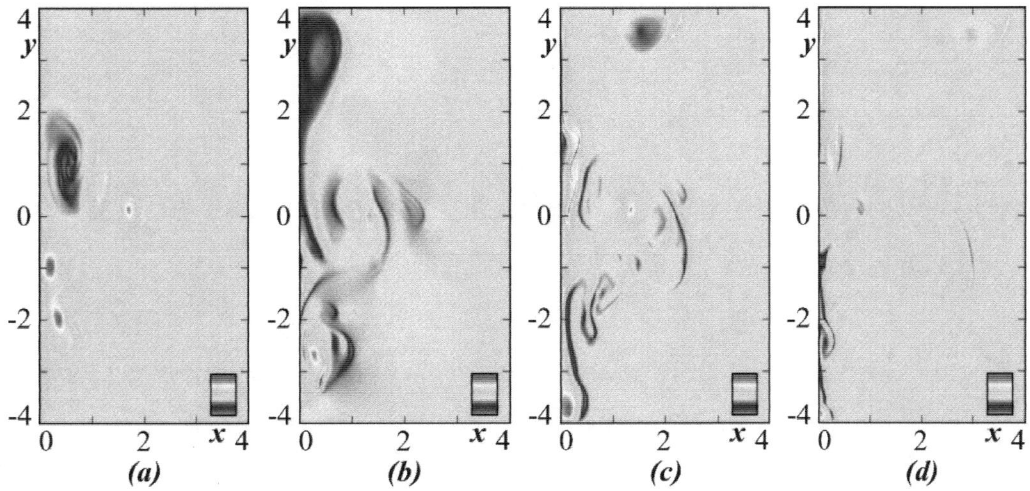

Figure 7. The isoclines of vorticity for: Re = 2000, Fr = 100, Sc = 709.2, μ/ρ_0 = 0.01 cm²/s, k_s = 1.41 10-5 cm²/s, computation area is 4×8 (**a** - t=2, **b** - t=3, **c** - t=4, **d** - t=5).

5. Conclusion

Due to the interaction of the couple and the influence of gravity, each initial vortex splits into two vortices, and synchronously each initial area of salinity perturbations splits into two areas. One vortex rises and the second vortex descends. One area of salinity perturbations rises and the second area falls. The upper vortex moves along the boundary of the computational area. The intensity of the vortex increases as the vortex moves along the vertical boundary. The intensity decreases as the vortex moves along the upper boundary.

References

[1] Hopfinger E J and van Heijst G J F 1993 Vortices in rotating fluids *Annu. Rev. Fluid Mech.* **25** pp. 241-289

[2] Nakashima Y 2008 Sound generation by head-on and oblique collisions of two vortex rings *Physics of Fluids* **20**(5) pp. 056-102

[3] Wu-Ting T and Dick K P Y 1995 Effects of soluble and insoluble surfactant on laminar interactions of vortical flows with a free surface *J. Fluid Mech.* **289** pp. 315-349

[4] Sarpkaya T and Suthon P 1991 Interaction of a vortex couple with a free surface *Experiments in Fluids* **11** pp. 205-217

[5] Batchelor G K 1967 *An Introduction to Fluid Dynamics* (Camb. Univ. Press)

[6] Gushchin V A 2016 On a one family of quasi monotone finite-difference schemes of the second order of approximation *Math. Models Comput. Simulation* **8** pp. 487–496

[7] Gushchin V A and Smirnova I A 2020 On a Vortex Couple Dynamics in Fluid *Apllication of Math. in Technical and Natural Siences* (AIP Conf. Proc.)

Instability of three-body periodic collisionless equal-mass free-fall orbits

Ivan Hristov[1], **Radoslava Hristova**[1], **Veljko Dmitrašinović**[2], **Kiyotaka Tanikawa**[3]

[1] Faculty of Mathematics and Informatics, Sofia University, 5 James Bourchier blvd., Sofia 1136, Bulgaria
[2] Institute of Physics Belgrade, University of Belgrade, Pregrevica 118, Zemun, P.O.Box 57, 11080 Beograd, Serbia
[3] National Astronomical Observatory of Japan, Mitaka, Tokyo 181-8588, Japan

E-mail: ivanh@fmi.uni-sofia.bg, radoslava@fmi.uni-sofia.bg, ntkaos@gmail.com

Abstract. This work is a continuation of our recent work *Celest Mech Dyn Astron 136, 7, (2024)*. Here we give more details about the high-precision numerical methods used and present the results for the linear stability investigation of the found periodic orbits. The eigenvalues of the monodromy matrices for all periodic orbits are computed with high-precision and given with 30 correct digits. All found periodic orbits are unstable, more precisely they are of a hyperbolic-hyperbolic or a hyperbolic-elliptic type. This result gives rise to a hypothesis that the three-body periodic collisionless equal-mass free-fall orbits are unstable ones. The additionally made high-precision computations for the escape-times in the two dimensional initial conditions' domain also support this hypothesis. A discussion in relation to chaos theory is held.

1 Introduction

Using a new search strategy and high-precision algorithms we have increased in [1] more than 400-fold the number of three-body periodic collisionless equal-mass free-fall orbits known from [2]. A natural continuation of our recent work [1], which is intended to give more completeness of it, is to present more details about the numerical methods used and to investigate as well the discovered solutions for linear stability. The present work sets precisely these two goals. The previous computational difficulties in the numerical search for free-fall periodic orbits and the small known number of them can be partially explained by the main result concluded in this paper, namely that all found orbits in [1] are unstable.

The paper falls into 6 sections. After a short introduction, in Sect. 2 we present the differential equations describing the motion of the three bodies and the two dimensional initial conditions' domain for the free-fall problem. Sect. 3 defines the proximity functions. In Sect. 4 we explain the main idea of the new search strategy, the numerical methods used and the stages of the search. The definition of the monodromy matrix for a given periodic orbit and the interpretation of its eigenvalues with respect to stability is given in Sect. 5. The linear stability results and the computed escape-times in the initial conditions' domain are also analyzed in Sect. 5. In Sect. 6 we draw conclusions.

Content from this work may be used under the terms of the Creative Commons Attribution 4.0 licence. Any further distribution of this work must maintain attribution to the author(s) and the title of the work, journal citation and DOI.
Published under licence by IOP Publishing Ltd

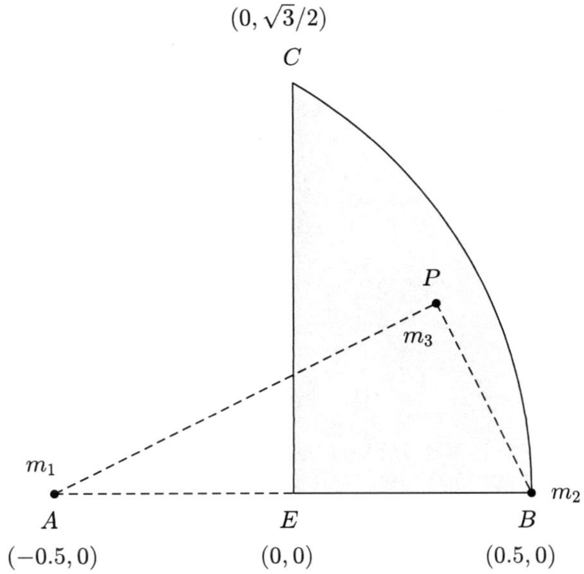

Figure 1: Agekyan-Anosova's domain (in yellow)

2 Differential equations and initial conditions

The movement of three bodies in space under their mutual gravitational attraction is described by the following second order system of ODEs:

$$m_i \ddot{r}_i = \sum_{j=1, j\neq i}^{3} G m_i m_j \frac{(r_j - r_i)}{\|r_j - r_i\|^3}, i = 1, 2, 3. \tag{1}$$

In system (1) r_i is the vector of positions of the i-th body, m_i is the mass of the i-th body, G is the gravitational constant. The derivatives are with respect to time t. At the left side of equations (1) is the force imposed on the i-th body given by the second Newton's law. At the right side is the sum of mutual forces imposed on the i-th body from each other body given by Newton's law of gravitation.

We consider free-fall initial conditions, meaning that the bodies are positioned at the vertices of a given triangle with zero velocities at $t = 0$. It follows from these special initial conditions that the movement of the bodies stays in the plane determined by its initial positions, i.e. the motion is planar. Thus, the vectors r_i, $\dot{r}_i$ have two components - $r_i = (x_i, y_i)$, $\dot{r}_i = (\dot{x}_i, \dot{y}_i)$. Additionally, in this paper equal masses $m_1 = m_2 = m_3$ and normalization with $G = 1$ and $m_i = 1$ are considered. If the dependent variables vx_i and vy_i are introduced, so that $vx_i = \dot{x}_i, vy_i = \dot{y}_i$, then the second order system (1) can be written as a first order one:

$$\dot{x}_i = vx_i,\ \dot{y}_i = vy_i,\ \dot{vx}_i = \sum_{j=1,j\neq i}^{3} \frac{(x_j - x_i)}{\|r_j - r_i\|^3},\ \dot{vy}_i = \sum_{j=1,j\neq i}^{3} \frac{(y_j - y_i)}{\|r_j - r_i\|^3},\ i = 1, 2, 3 \tag{2}$$

This first order form of the second order system (1) corresponds to the 12-component unknown vector

$$X(t) = (x_1(t), y_1(t), x_2(t), y_2(t), x_3(t), y_3(t), vx_1(t), vy_1(t), vx_2(t), vy_2(t), vx_3(t), vy_3(t))^\top$$

and is more convenient for study of the 3-body system and particularly for its numerical solving.

For the equal-mass case all different shapes of the initial triangle at $t = 0$ can be generated in the following way. We put body 1 and body 2 as fixed at points $A = (-0.5, 0)$ and $B = (0.5, 0)$ respectively and then body 3 is placed at all possible positions of point $P(p_x, p_y)$ in the Agekyan-Anosova's domain (the yellow domain in Fig. 1). More precisely, the Agekyan-Anosova's domain is bounded by the x and y axes and the arc of a circle with a center A and a radius of 1 (see [3] for details). So, the free-fall initial conditions are:

$$\begin{aligned}(x_1(0), y_1(0)) = (-0.5, 0),\ \ (x_2(0), y_2(0)) = (0.5, 0),\ \ (x_3(0), y_3(0)) = (p_x, p_y) \\ (vx_1(0), vy_1(0)) = (vx_2(0), vy_2(0)) = (vx_3(0), vy_3(0)) = (0, 0)\end{aligned} \tag{3}$$

3 Return proximity function. Stopping proximity function.

Let us denote the solution of (2) with initial conditions (3) with $X(p_x, p_y, t)$. Finding a periodic solution with period $T > 0$ means to find a triplet (p_x, p_y, T) for which the periodicity condition (equation):

$$X(p_x, p_y, T) = X(p_x, p_y, 0) \tag{4}$$

is satisfied. The function:

$$R(t) = \|X(p_x, p_y, t) - X(p_x, p_y, 0)\|_2,\ \ t > 0$$

is called the return proximity function. The triplet (p_x, p_y, T), $T > 0$ corresponds to a periodic solution with period T only if $R(T) = 0$. For an approximate solution with approximations (p_x, p_y, T), $R(T)$ measures how close we are to a periodic solution. $R(T)$ is the residual that we use, when we solve the nonlinear equation (4) with Newton's method and the convergence in this case is in terms of this function.

We know from theory that periodic free-fall orbits have the property to shuttle back and forth between two triangles, at which triangles the bodies stop (the bodies have zero velocities) (see for example [4]). The first triangle is at $t = 0$ and the second is at $t = T/2$. The new search strategy for free-fall periodic orbits used in [1] is the following. Instead of searching for orbits that return to the first (initial) triangle, we search for orbits for which the bodies stop at a later time. This time is an approximation of $T/2$. If we introduce the vector of velocities $V(t) = (vx_1(t), vy_1(t), vx_2(t), vy_2(t), vx_3(t), vy_3(t))^\top$ (this is the second half of $X(t)$), the stopping condition (equation) becomes:

$$V(p_x, p_y, T/2) = V(0) \tag{5}$$

The stopping proximity function (the proximity function for stopping condition (5)) is:

$$R_{vel}(t) = \|V(p_x, p_y, t) - V(p_x, p_y, 0)\|_2 = \|V(p_x, p_y, t)\|_2,\ \ t > 0$$

For a free-fall periodic orbit with period T in addition to equality $R(T) = 0$, the equality $R_{vel}(T/2) = 0$ is also satisfied. Let $\overline{T}$ be an approximation of $T/2$. For an approximate solution with approximations $(p_x, p_y, \overline{T})$, $R_{vel}(\overline{T})$ measures how well we satisfy the stopping condition (5). The value of stopping proximity function $R_{vel}(\overline{T})$ is the residual that we use when solving the nonlinear equation (5) with Newton's method and the convergence in this case is in terms of this function.

4 Search strategy and numerical methods

4.1 The essence of the search strategy

The two key works [5, 6] presented the main concepts in the numerical search for three-body periodic orbits and what is also important to remark, they found some new stable periodic orbits. The use of the standard double precision arithmetic in [5, 6], however, restricted the number of the found periodic orbits, because of the sensitive dependence of the solution on the initial conditions. This sensitivity is an inherent property of the three-body problem and is related to its chaotic nature (see the discussion about chaos in subsection 5.3).

Relying on works [5, 6], a big progress was made in [2, 7, 8], where the grid search method combined with Newton's method, were successfully implemented for the discovery of many new periodic orbits. The success of [2, 7, 8] lies in the replacement of the standard double precision arithmetic with a higher precision. The elements of the linear systems for each Newton's iteration were computed with the high order and high precision Taylor series method [9–12]. The latter allows the trajectories of the bodies to be followed accurately for a sufficiently long time regardless of their sensitivity on the initial conditions.

The choice of a suitable ODE-solver that works with high precision is an important and necessary step, but it is not all that we need in order to make the grid search method in combination with Newton's method to work efficiently. For a given periodic orbit, the properties of Newton's iterations and particularly the region of convergence of Newton's method depends on the nonlinear equation we solve. As numerous numerical experiments show, by solving the stopping equation (5) instead of the standard periodicity equation (4), we capture much more periodic orbits for a given search grid in Agekyan-Anosova's domain. The reason is that the regions of convergence of Newton's method around the periodic orbits are enlarged, which is presumably connected to the smaller divergence of nearby trajectories when the integration time is shortened by half.

So, let a search grid be chosen. Our new search strategy consists of applying Newton's method not to the nonlinear equation (4), but to the nonlinear equation (5) instead. Of course, each captured solution is tested after that for convergence with Newton's method for the standard periodicity equation (4).

4.2 Newton's method for the periodicity equation. Newton's method for the stopping equation.
Now we will give the formulas for the linear (algebraic) equations at each Newton's iteration for solving (4) and (5). Let us agree to use the same notation for (p_x, p_y, T) and for their approximations and to not index the number of the iteration. For nonlinear equation (4) the approximations (p_x, p_y, T) are improved with corrections $\Delta p_x, \Delta p_y, \Delta T$ by expanding the periodicity condition:

$$X(p_x + \Delta p_x, p_y + \Delta p_y, T + \Delta T) = X(p_x + \Delta p_x, p_y + \Delta p_y, 0)$$

in a multivariable linear approximation:

$$\begin{pmatrix} x_1(T) \\ y_1(T) \\ x_2(T) \\ y_2(T) \\ x_3(T) \\ y_3(T) \\ vx_1(T) \\ vy_1(T) \\ vx_2(T) \\ vy_2(T) \\ vx_3(T) \\ vy_3(T) \end{pmatrix} + \begin{pmatrix} \frac{\partial x_1}{\partial p_x}(T) & \frac{\partial x_1}{\partial p_y}(T) & \dot{x}_1(T) \\ \frac{\partial y_1}{\partial p_x}(T) & \frac{\partial y_1}{\partial p_y}(T) & \dot{y}_1(T) \\ \frac{\partial x_2}{\partial p_x}(T) & \frac{\partial x_2}{\partial p_y}(T) & \dot{x}_2(T) \\ \frac{\partial y_2}{\partial p_x}(T) & \frac{\partial y_2}{\partial p_y}(T) & \dot{y}_2(T) \\ \frac{\partial x_3}{\partial p_x}(T) & \frac{\partial x_3}{\partial p_y}(T) & \dot{x}_3(T) \\ \frac{\partial y_3}{\partial p_x}(T) & \frac{\partial y_3}{\partial p_y}(T) & \dot{y}_3(T) \\ \frac{\partial vx_1}{\partial p_x}(T) & \frac{\partial vx_1}{\partial p_y}(T) & \dot{vx}_1(T) \\ \frac{\partial vy_1}{\partial p_x}(T) & \frac{\partial vy_1}{\partial p_y}(T) & \dot{vy}_1(T) \\ \frac{\partial vx_2}{\partial p_x}(T) & \frac{\partial vx_2}{\partial p_y}(T) & \dot{vx}_2(T) \\ \frac{\partial vy_2}{\partial p_x}(T) & \frac{\partial vy_2}{\partial p_y}(T) & \dot{vy}_2(T) \\ \frac{\partial vx_3}{\partial p_x}(T) & \frac{\partial vx_3}{\partial p_y}(T) & \dot{vx}_3(T) \\ \frac{\partial vy_3}{\partial p_x}(T) & \frac{\partial vy_3}{\partial p_y}(T) & \dot{vy}_3(T) \end{pmatrix} \begin{pmatrix} \Delta p_x \\ \Delta p_y \\ \Delta T \end{pmatrix} = \begin{pmatrix} x_1(0) \\ y_1(0) \\ x_2(0) \\ y_2(0) \\ x_3(0) + \Delta p_x \\ y_3(0) + \Delta p_y \\ vx_1(0) \\ vy_1(0) \\ vx_2(0) \\ vy_2(0) \\ vx_3(0) \\ vy_3(0) \end{pmatrix}$$

After reordering this equation we finally obtain the following linear system with a 12×3 matrix with respect to $(\Delta p_x, \Delta p_y, \Delta T)^\top$.

$$\begin{pmatrix} \frac{\partial x_1}{\partial p_x}(T) & \frac{\partial x_1}{\partial p_y}(T) & \dot{x}_1(T) \\ \frac{\partial y_1}{\partial p_x}(T) & \frac{\partial y_1}{\partial p_y}(T) & \dot{y}_1(T) \\ \frac{\partial x_2}{\partial p_x}(T) & \frac{\partial x_2}{\partial p_y}(T) & \dot{x}_2(T) \\ \frac{\partial y_2}{\partial p_x}(T) & \frac{\partial y_2}{\partial p_y}(T) & \dot{y}_2(T) \\ \frac{\partial x_3}{\partial p_x}(T) - 1 & \frac{\partial x_3}{\partial p_y}(T) & \dot{x}_3(T) \\ \frac{\partial y_3}{\partial p_x}(T) & \frac{\partial y_3}{\partial p_y}(T) - 1 & \dot{y}_3(T) \\ \frac{\partial vx_1}{\partial p_x}(T) & \frac{\partial vx_1}{\partial p_y}(T) & \dot{vx}_1(T) \\ \frac{\partial vy_1}{\partial p_x}(T) & \frac{\partial vy_1}{\partial p_y}(T) & \dot{vy}_1(T) \\ \frac{\partial vx_2}{\partial p_x}(T) & \frac{\partial vx_2}{\partial p_y}(T) & \dot{vx}_2(T) \\ \frac{\partial vy_2}{\partial p_x}(T) & \frac{\partial vy_2}{\partial p_y}(T) & \dot{vy}_2(T) \\ \frac{\partial vx_3}{\partial p_x}(T) & \frac{\partial vx_3}{\partial p_y}(T) & \dot{vx}_3(T) \\ \frac{\partial vy_3}{\partial p_x}(T) & \frac{\partial vy_3}{\partial p_y}(T) & \dot{vy}_3(T) \end{pmatrix} \begin{pmatrix} \Delta p_x \\ \Delta p_y \\ \Delta T \end{pmatrix} = \begin{pmatrix} x_1(0) - x_1(T) \\ y_1(0) - y_1(T) \\ x_2(0) - x_2(T) \\ y_2(0) - y_2(T) \\ x_3(0) - x_3(T) \\ y_3(0) - y_3(T) \\ vx_1(0) - vx_1(T) \\ vy_1(0) - vy_1(T) \\ vx_2(0) - vx_2(T) \\ vy_2(0) - vy_2(T) \\ vx_3(0) - vx_3(T) \\ vy_3(0) - vy_3(T) \end{pmatrix} \quad (6)$$

Let us recall that with $\overline{T}$ we denote the approximation of $T/2$. For the nonlinear equation (5) the approximations $(p_x, p_y, \overline{T})$ are improved with corrections $\Delta p_x, \Delta p_y, \Delta\overline{T}$ by expanding the stopping condition:

$$V(p_x + \Delta p_x, p_y + \Delta p_y, \overline{T} + \Delta\overline{T}) = V(0)$$

in a multivariable linear approximation:

$$\begin{pmatrix} vx_1(\overline{T}) \\ vy_1(\overline{T}) \\ vx_2(\overline{T}) \\ vy_2(\overline{T}) \\ vx_3(\overline{T}) \\ vy_3(\overline{T}) \end{pmatrix} + \begin{pmatrix} \frac{\partial vx_1}{\partial p_x}(\overline{T}) & \frac{\partial vx_1}{\partial p_y}(\overline{T}) & \dot{vx}_1(\overline{T}) \\ \frac{\partial vy_1}{\partial p_x}(\overline{T}) & \frac{\partial vy_1}{\partial p_y}(\overline{T}) & \dot{vy}_1(\overline{T}) \\ \frac{\partial vx_2}{\partial p_x}(\overline{T}) & \frac{\partial vx_2}{\partial p_y}(\overline{T}) & \dot{vx}_2(\overline{T}) \\ \frac{\partial vy_2}{\partial p_x}(\overline{T}) & \frac{\partial vy_2}{\partial p_y}(\overline{T}) & \dot{vy}_2(\overline{T}) \\ \frac{\partial vx_3}{\partial p_x}(\overline{T}) & \frac{\partial vx_3}{\partial p_y}(\overline{T}) & \dot{vx}_3(\overline{T}) \\ \frac{\partial vy_3}{\partial p_x}(\overline{T}) & \frac{\partial vy_3}{\partial p_y}(\overline{T}) & \dot{vy}_3(\overline{T}) \end{pmatrix} \begin{pmatrix} \Delta p_x \\ \Delta p_y \\ \Delta \overline{T} \end{pmatrix} = \begin{pmatrix} vx_1(0) \\ vy_1(0) \\ vx_2(0) \\ vy_2(0) \\ vx_3(0) \\ vy_3(0) \end{pmatrix}$$

Now we have to solve at each Newton's iteration the following system with 6×3 matrix with respect to $(\Delta p_x, \Delta p_y, \Delta\overline{T})^\top$.

$$\begin{pmatrix} \frac{\partial vx_1}{\partial p_x}(\overline{T}) & \frac{\partial vx_1}{\partial p_y}(\overline{T}) & \dot{vx}_1(\overline{T}) \\ \frac{\partial vy_1}{\partial p_x}(\overline{T}) & \frac{\partial vy_1}{\partial p_y}(\overline{T}) & \dot{vy}_1(\overline{T}) \\ \frac{\partial vx_2}{\partial p_x}(\overline{T}) & \frac{\partial vx_2}{\partial p_y}(\overline{T}) & \dot{vx}_2(\overline{T}) \\ \frac{\partial vy_2}{\partial p_x}(\overline{T}) & \frac{\partial vy_2}{\partial p_y}(\overline{T}) & \dot{vy}_2(\overline{T}) \\ \frac{\partial vx_3}{\partial p_x}(\overline{T}) & \frac{\partial vx_3}{\partial p_y}(\overline{T}) & \dot{vx}_3(\overline{T}) \\ \frac{\partial vy_3}{\partial p_x}(\overline{T}) & \frac{\partial vy_3}{\partial p_y}(\overline{T}) & \dot{vy}_3(\overline{T}) \end{pmatrix} \begin{pmatrix} \Delta p_x \\ \Delta p_y \\ \Delta \overline{T} \end{pmatrix} = \begin{pmatrix} vx_1(0) - vx_1(\overline{T}) \\ vy_1(0) - vy_1(\overline{T}) \\ vx_2(0) - vx_2(\overline{T}) \\ vy_2(0) - vy_2(\overline{T}) \\ vx_3(0) - vx_3(\overline{T}) \\ vy_3(0) - vy_3(\overline{T}) \end{pmatrix} \tag{7}$$

The linear algebraic systems (6) and (7) are with rectangular matrices and are solved in least square sense by QR-decomposition [13].

4.3 Modified Newton's method for the stopping equation

For enlarging the region of convergence of Newton's method when applying it in the solving of a given nonlinear equation one can use a modification based on the continuous analogue of Newton's method [14]. Of course, this comes at the cost of a certain slowing of the method (more iterations are needed for a given accuracy). With the (classic) Newton's method for equation (5) the corrections at each iteration are:

$$p_x := p_x + \Delta p_x, \quad p_y := p_y + \Delta p_y, \quad \overline{T} := \overline{T} + \Delta\overline{T}$$

To modify the method, we introduce a parameter $0 < \tau_k <= 1$, where k is the number of the iteration. We correct this way:

$$p_x := p_x + \tau_k\Delta p_x, \quad p_y := p_y + \tau_k\Delta p_y, \quad \overline{T} := \overline{T} + \tau_k\Delta\overline{T}$$

Let us denote with R_k the stopping proximity $R_{vel}(\overline{T})$ at the k-th iteration. For a given τ_0 the next τ_k, $k = 1, 2, ...$ is computed with the following algorithm [14]:

$$\tau_k = \begin{cases} \min(1,\ \tau_{k-1}R_{k-1}/R_k), & R_k \le R_{k-1}, \\ \max(\tau_0,\ \tau_{k-1}R_{k-1}/R_k), & R_k > R_{k-1}. \end{cases} \tag{8}$$

We usually take $\tau_0 = 0.2$. The numerical experience shows that it is more efficient to use the modified method for a coarser search grid than to obtain a similar result with the classic method for a finer grid. However, the crucial decision in [1] is not to use the modified Newton's method when capturing new periodic orbits, but to solve equation (5) instead of (4). Let us also note that we have no additional computational difficulties when we use the modified method, as we basically need to compute the same linear systems.

4.4 The choice of ODE solver - Taylor series method

The choice of a suitable high precision ODE-solver for solving the differential problem (2), (3) and also for computing the partial derivatives for (6) and (7) is very important. We chose (as in [2, 7, 8]) the high order and high precision Taylor series method [9–12]. An adaptive time stepsize strategy is used for accurate handling of the close approaches of the bodies. We determine the time stepsize according to [9] by setting the absolute and the relative tolerances from this paper to be equal. We use a fixed order Taylor series method, meaning that the order is determined by the used precision. As the numerical experiments and computational experience show, 22 order for each 64 bit of precision is an optimal choice for the used time stepsize strategy.

An important property of Taylor series method is that the automatic differentiation formulas for computing the Taylor coefficients (the so called normalized derivatives) of the original 12 unknowns are easily generalized for their partial derivatives with respect to p_x and p_y (see [9–12]). Thus, Taylor series method and Newton's method complement each other very well. The details about the Taylor series method, including the formula for the time stepsize and the formulas for computing the Taylor coefficients with the rules of automatic differentiation can be seen in [15]. Overall, Taylor series method handles both the sensitivity of the solution on the initial conditions and the possible close approaches of the bodies very well.

4.5 Stages of the numerical search

The numerical search consists of three stages: stage (I) - scanning stage, stage (II) - capturing stage, stage (III) - verification stage with computing the solution with many correct digits. In general, the verification stage may consist of several substages, depending on the accuracy we want.

At stage (I) we introduce in Agekyan-Anosova's domain a grid with size $2^{-14}(\approx 6.1 * 10^{-5})$ for both parameters p_x and p_y. At each grid point (p_x, p_y) periods less than $T_0(p_x, p_y) = \min(7, 80/|E(p_x, p_y)|^{3/2})$ are considered. Here $E(p_x, p_y)$ is the energy, which depends on the point. We simulate the problem (2), (3) up to $T_0/2$ with precision of 128 bit ($\approx$ 38.5 decimal digits) and 44 Taylor's order. Then we compute the time $\overline{T}(p_x, p_y)$ at which the minimum $d(p_x, p_y) = \min_{t \le T_0/2} R_{vel}(t)$ is obtained. The triplet $(p_x, p_y, \overline{T})$ becomes a candidate for a periodic orbit if $d(p_x, p_y)$ has a local minimum on the grid at the point (p_x, p_y) and $d(p_x, p_y) < 0.3$.

At Stage (II) the stopping equation (5) is solved. The modified Newton's method starting with the initial approximations from stage (I) is applied. The convergence is in terms of the stopping proximity function $R_{vel}(\overline{T})$. 192-bit precision ($\approx$ 57.8 decimal digits) and 66 Taylor's order is used at this stage. A periodic solution is captured if $R_{vel}(\overline{T}) < 10^{-20}$.

At Stage (III) we solve the standard periodicity equation (4). We use only the classic Newton's method, because the approximations $(p_x, p_y, T) = (p_x, p_y, 2\overline{T})$ are already good enough. We gradually increase the precision and Taylor's order at each subsequent substage starting with 384-bit precision and Taylor's order 132. As initial approximations at each substage we take the results from the previous substage. The convergence is in terms of the standard return proximity $R(T)$. We iterate until convergence (until $R(T)$ stops to decrease). We verify the results by comparing the first digits for (p_x, p_y, T) for different Taylor's order and precision and also by comparing the results after dividing the prescribed time stepsize by 2 and 4. It is worth noting that for collisionless orbits the convergence of Newton's method has the property that the number of time steps at each Newton's iteration becomes fixed when we are sufficiently close to a periodic solution. This property is checked for each substage of stage (III). The observed convergence of Newton's method in terms of $R(T)$ is also checked. It agrees very well with the theoretical quadratic convergence.

5 Linear stability and linear stability results

5.1 The monodromy matrix and its eigenvalues

Let $X(X^0, t), 0 \le t \le T$, be a periodic orbit with period T and initial condition $X(0) = X^0$. The stability of $X(X^0, t)$ is manifested by the behaviour of the neighboring trajectories, i.e. whether the distance between them grows or not. If it is not growing, the orbit is stable, otherwise the orbit is unstable. Let the initial vector X^0 be perturbed on each of its 12 components. A trajectory that starts from the perturbed initial vector $X^0 + \delta X^0$ will be displaced after one period T by:

$$\delta X(X^0, T) = X(X^0 + \delta X^0, T) - X(X^0, T)$$

Expanding this equation in a multivariable linear approximation gives:

$$\delta X(X^0, T) \approx M(X^0, T)\delta X^0$$

Here the 12×12 matrix $M(X^0, T)$ with elements $M_{ij}(X^0, T) = \frac{\partial X_i(X^0, T)}{\partial {X_j}^0}$ is the so called monodromy matrix [16]. Note that the monodromy matrix $M(X^0, T)$ is a matrix of the same type (Jacobian type) as the matrices in linear systems for Newton's method. Hence its elements M_{ij} are computed in the same way – with the high order high precision Taylor series method using the rules of automatic differentiation [15].

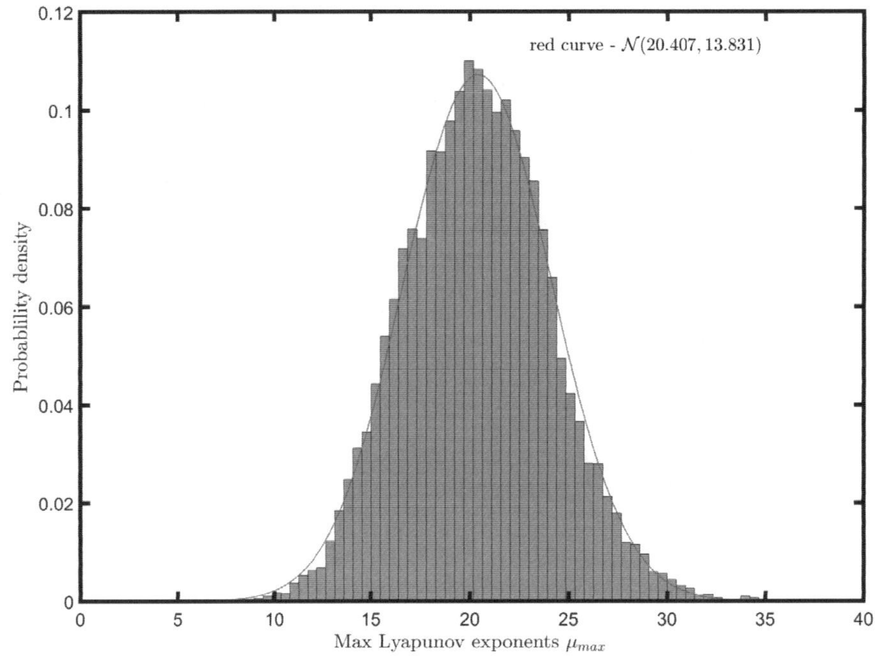

Figure 2: Distribution of the maximal Lyapunov exponents μ_{max}

When we compute $M(X^0, T)$ the initial conditions that have to be set are $M(X^0, 0) = I$, where I is the identity matrix. The linear stability of a given periodic solution $X(X^0, t)$ with period T is determined by the eigenvalues γ of $M(X^0, T)$ [16], since their values determine whether the perturbation applied in the direction of the corresponding eigenvectors grows or not. The eigenvalues γ of $M(X^0, T)$ (generally complex numbers) are in pairs or quadruplets [16]: $(\gamma, \gamma^{-1}, \gamma^*, \gamma^{*-1})$ (with a star we denote the complex conjugation). The eigenvalues are of four types:

(1) Elliptically stable - $\gamma = e^{\pm 2\pi i\nu}$, where $\nu > 0$ (real) is the so called stability angle. The eigenvalues are on the unit circle. The angle ν describes the stable revolution of adjacent trajectories around a periodic orbit.

(2) Marginally stable - $\gamma = \pm 1$.

(3) Hyperbolic (unstable) - $\gamma = \pm e^{\pm\mu}$, where $\mu > 0$ (real) is the Lyapunov exponent.

(4) Loxodromic (unstable) - $\gamma = e^{\pm\mu \pm i\nu}$, $\mu, \nu > 0$ (real).

It is known that eight of the eigenvalues of M are equal to 1 [16]. The linear stability is determined by the other four. The monodromy matrices are computed by taking the most accurate values of the initial conditions and periods obtained after the last substage of stage (III) of the search. Then their eigenvalues are computed with a Multiprecision Computing Toolbox [17] for MATLAB® [18]. For verification of the computed eigenvalues we do the computations with the toolbox using different precisions and comparing the results. In this paper we present the results with 30 correct digits.

5.2 The computed eigenvalues. Instability of free-fall periodic orbits.

All found orbits in [1] are unstable (23969 hyperbolic-hyperbolic and 613 hyperbolic-elliptic orbits' initial conditions). A hyperbolic-hyperbolic orbit means that there are two hyperbolic pairs of eigenvalues, while a hyperbolic-elliptic orbit means that there is one hyperbolic and one elliptic pair. No other types of orbits are found. Let us denote with μ_{max} the maximal Lyapunov exponents. They are scale invariant quantities. Our very accurate computations allow to compare the values $e^{\mu_{max}}$ with the simulated by Taylor's method multiplication factors of the separation $\delta(t)$ between trajectories for one period. These values match very well, meaning that closely to the periodic orbits, the approximate low of separation

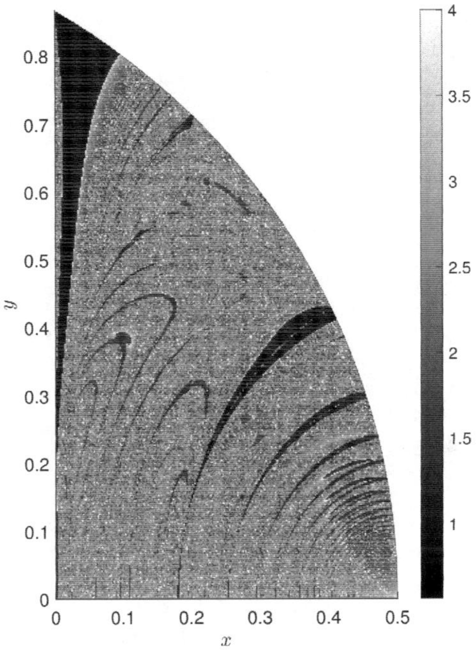

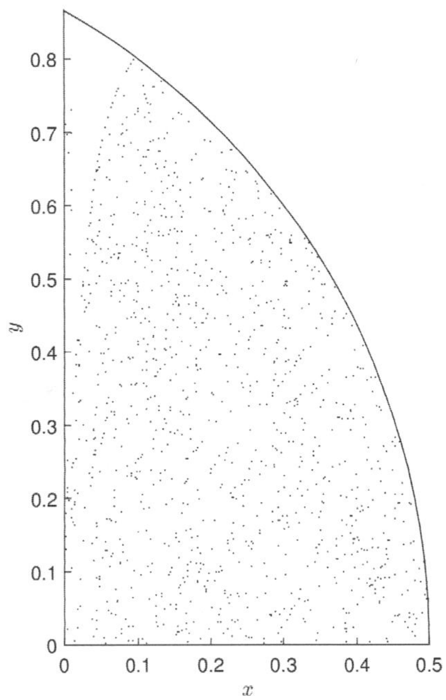

Figure 3: Left – Color map of $\log(t^*_{esc})$ Right – the points for $t^*_{esc} > 10,000$

$\|\delta(t)\| \approx \|\delta(0)\|e^{\lambda t}$ is satisfied with $\lambda = \mu_{max}/T$. The four eigenvalues which determine the stability and their absolute values with 30 correct digits for all 24,582 initial conditions can be downloaded from [19] (the same site where the data from [1] is placed). The maximal Lyapunov exponents μ_{max} with 30 correct digits for all solutions are also there. The distribution of the maximal Lyapunov exponents μ_{max} can be seen in Figure 2. It very well fits with a normal distribution with mean ≈ 20.407 and standard deviation ≈ 13.831. The minimal value of μ_{max} is ≈ 5.376, the maximal is ≈ 40.25. It should be noted however, that these results correspond to only a restricted set of periods - with scale invariant periods $T^* = T|E|^{3/2} < 80$.

It is unlikely that we have missed a stable solution for the periods that were taken under consideration, because the stable orbits are much easier to be captured numerically. The hypothesis that the three-body periodic collisionless equal-mass free-fall orbits are unstable arises naturally. Since it is computationally very laborious (impossible) to search for periodic orbits for very long (arbitrary long) periods, it would be a good idea to confirm this hypothesis by analyzing the escape-times (see [20]) in the Agekyan-Anosova's domain in the context of stability. Such escape-time computations are already done in [21]. According to KAM-theory stable periodic orbits have to be surrounded by sets with nonzero measure consisting of orbits with bounded motions [22,23]. However, no continuous regions of points on the computational grid with escape times larger than some very large (but bounded) time are observed in [21]. These points are "roughly uniformly and randomly distributed over the Agekyan-Anosova's domain". Thus, this result in [21] supports the hypothesis of instability of the free-fall periodic orbits. In contrast to the free-fall initial conditions' domain, for another two dimensional initial conditions' domain (those corresponding to Eulerian initial conditions [5–7]) "stability regions" are observed according to [23]. Of course, the latter result in [23] is in agreement with the fact that many linearly stable orbits for Eulerian initial conditions are known [24–26].

To make our study more complete, we additionally computed the escape-times with the relatively large grid size 2^{-9} ($\approx 80,000$ points inside the Agekyan-Anosova's domain) and with a scale invariant escape-times $t^*_{esc} \leq 10,000$. We use the escape condition from [20]. Our results for the escape-times are shown in Figure 3 (left) and are qualitatively the same as those in [21, 27, 28]. The points with $t^*_{esc} > 10,000$ (the point for which the simulation time is not enough for an escape) are shown in Figure 3 (right). Note that these points are in the "maximal" white color in the color map at left. The picture is qualitatively the same as that in [21], meaning that no "stability regions" are observed in Figure 3 (right). Using the terminology from [21], the points are "roughly uniformly and randomly distributed over the domain". We shall add that this excludes the black regions in Figure 3 (left) that correspond to very short escape-times. It is worth noting that the data for Figure 3 is verified with our high precision ODE-solver by comparing the results for gradually increased precision and also with a reversibility test. For the simulations on the considered grid with $t^* \leq 10,000$ there is a small number of points for which the (quite high) 768-bit precision and 264 Taylor's order are needed for reliable numerical results.

5.3 Chaos in the three-body equal-mass free-fall problem

The three-body system is habitually described as chaotic [29]. However, there are plenty of stable periodic equal-mass three-body orbits, which contradicts the requirement that a chaotic system has only unstable periodic orbits [30]. Manifestly, only a subset of the full equal-mass three-body problem can be chaotic. Also note that chaotic dynamical systems are expected to have an infinite number of (dense) periodic orbits, whereas the heretofore known periodic three-body equal-mass free-fall orbits were very few (30 in number).

Let us use a widely accepted definition of chaos given by Robert L. Devaney [30, 31] and test it with the three-body equal-mass free-fall problem. According to this definition a dynamical system can be classified as chaotic if the following three properties are satisfied: (1) It must be sensitive to initial conditions, (2) It must have dense periodic orbits, (3) It must have topological mixing. Property (1) is shown in the previous subsection 5.2 by presenting the instability of the periodic orbits. Property (2) may yet be satisfied, as there are many more periodic orbits that can be discovered with our highly efficient approach, when finer search grid and longer periods are considered. When doing this, we expect that the Agekyan-Anosova's domain (with the exception of the black regions with very short escape-times in Figure 3 (left)) will be increasingly filled with periodic orbits' points. Property (3), which is often omitted from popular accounts of chaos, has to be studied separately in the future, but even without it, we have enough confidence from all the results obtained in this work to draw the conclusion that the three-body equal-mass free-fall problem is chaotic.

Comparing the two different two-dimensional sectors of the three-body equal-mass problem studied so far - the one with Eulerian initial conditions and the one with free-fall initial conditions, we see that the first problem can be regular (though not everywhere), with many stable orbits and the second is chaotic without a single stable orbit. Therefore one should recognize the free-fall three-body problem as the "chaotic" face of the three-body problem. These results give credence to the oft-repeated, but not fully justified mantra that the three-body problem is chaotic: Well, it is if you look at it from one end, but not otherwise.

6 Conclusions

The search strategy and numerical methods used in [1] are presented in more detail. The found orbits are investigated for linear stability by computing the eigenvalues of the monodromy matrices. All found free-fall collisionless periodic orbits are unstable. The numerical results are presented as a large 30 correct digits data base of the eigenvalues and the maximal Lyapunov exponents for all 24,582 initial conditions. The hypothesis that three-body periodic collisionless equal-mass free-fall orbits are unstable is additionally supported by escape-time computations in the Agekyan-Anosova's domain. Taking everything into consideration, the results of this work reveal that the found periodic free-fall orbits cannot be observed in reality due to their instability, yet their discovery is of significance to chaos theory.

Acknowledgement

The authors acknowledge the access to the Nestum cluster at HPC Laboratory, Research and Development and Innovation Consortium, Sofia Tech Park. The work of R.H. is partially supported by the European Union-NextGenerationEU, through the National Recovery and Resilience Plan of the Republic of Bulgaria, project number BG-RRP-2.004-0008-C01.

References

[1] I. Hristov, R. Hristova, V. Dmitrašinović, K. Tanikawa. "Three-body periodic collisionless equal-mass free-fall orbits revisited." Celestial Mechanics and Dynamical Astronomy, 136, 7, (2024).

[2] X. Li, S. Liao. "Collisionless periodic orbits in the free-fall three-body problem." New Astronomy 70, 22-26, (2019).

[3] T. Agekyan, Z. Anosova. "A study of the dynamics of triple systems by means of statistical sampling." Astronomicheskii Zhurnal, 44, 1261 (1967) (in Russian); Soviet Physics - Astronomy, 11, 1006 (1968).

[4] R. Montgomery. “Dropping bodies.” The Mathematical Intelligencer 45.2, 168-174, (2023).

[5] M. Šuvakov, V. Dmitrašinović. “Three classes of Newtonian three-body planar periodic orbits.” Phys. Rev. Lett. 110.11, 114301 (2013).

[6] M. Šuvakov, V. Dmitrašinović. “A guide to hunting periodic three-body orbits.” American Journal of Physics 82.6, 609-619, (2014).

[7] X. Li, S. Liao. “More than six hundred new families of Newtonian periodic planar collisionless three-body orbits.” Science China Physics, Mechanics & Astronomy 60, 1-7, (2017).

[8] X. Li, Y. Jing, S. Liao. “Over a thousand new periodic orbits of a planar three-body system with unequal masses.” Publications of the Astronomical Society of Japan 70.4, 64, (2018).

[9] A. Jorba, M. Zou. “A software package for the numerical integration of ODEs by means of high-order Taylor methods.” Experimental Mathematics 14.1, 99-117, (2005).

[10] R. Barrio, “Sensitivity analysis of ODEs/DAEs using the Taylor series method.” SIAM Journal on Scientific Computing 27.6, 1929-1947, (2006).

[11] R. Barrio, M. Rodriguez, A. Abad, and F. Blesa. “Breaking the limits: the Taylor series method”. Applied mathematics and computation 217.20, 7940-7954, (2011).

[12] F. Biscani, D. Izzo. “Revisiting high-order Taylor methods for astrodynamics and celestial mechanics.” Monthly Notices of the Royal Astronomical Society 504.2, 2614-2628, (2021).

[13] J. Demmel, “Applied numerical linear algebra”. Society for Industrial and Applied Mathematics, (1997).

[14] A. Abrashkevich, I. Puzynin, “CANM, a program for numerical solution of a system of nonlinear equations using the continuous analog of Newton’s method”. Computer physics communications, 156(2), 154-170, (2004).

[15] I. Hristov, R. Hristova, I. Puzynin, T. Puzynina, Z. Sharipov, Z. Tukhliev, “Newton’s method for computing periodic orbits of the planar three-body problem.” arXiv preprint arXiv:2111.10839, (2021).

[16] G. Roberts, “Linear stability analysis of the figure-eight orbit in the three-body problem.” Ergodic Theory and Dynamical Systems, 27.6, 1947-1963, (2007).

[17] Advanpix LLC., Multiprecision Computing Toolbox for MATLAB, `http://www.advanpix.com/`, Version 4.9.0 Build 14753, 2022-08-09.

[18] MATLAB version 9.12.0.1884302 (R2022a), The Mathworks, Inc., Natick, Massachusetts, (2021).

[19] `http://db2.fmi.uni-sofia.bg/3bodyfree/`

[20] E. Standish, “Sufficient conditions for escape in the three-body problem.” Celestial mechanics 4, 44-48, (1971).

[21] A. Martynova, V. Orlov. “The decay of triple systems.” Astronomy Reports 58, 756-766, (2014).

[22] V. Arnold, V. Kozlov, A. Neishtadt. “Mathematical aspects of classical and celestial mechanics.” Vol. 3. Berlin: Springer, (2006).

[23] A. Martynova, V. Orlov, A. Rubinov. “The structure of non-hierarchical triple system stability regions.” Astronomy reports 53, 710-721, (2009).

[24] V. Dmitrašinović, A. Hudomal, M. Shibayama, A. Sugita, “Linear stability of periodic three-body orbits with zero angular momentum and topological dependence of Kepler’s third law: a numerical test.” Journal of Physics A: Mathematical and Theoretical, 51(31), 315101, (2018).

[25] I. Hristov, R. Hristova, I. Puzynin, T. Puzynina, Z. Sharipov, Z. Tukhliev, “Hundreds of new satellites of figure-eight orbit computed with high precision.” In AIP Conference Proceedings (Vol. 2953, No. 1). AIP Publishing. (2023).

Journal of Physics: Conference Series **2910** (2024) 012030 doi:10.1088/1742-6596/2910/1/012030

[26] I. Hristov, R. Hristova, I. Puzynin, T. Puzynina, Z. Sharipov, Z. Tukhliev, "A database of high precision trivial choreographies for the planar three-body problem." In International Conference on Numerical Methods and Applications (pp. 171-180). Cham: Springer Nature Switzerland, (2022).

[27] H. Lehto, S. Kotiranta, M. Valtonen, P. Heinämäki, S. Mikkola, A. Chernin, "Mapping the three-body system–decay time and reversibility." Monthly Notices of the Royal Astronomical Society, 388(3), 965-970, (2008).

[28] T. Boekholt, S. Portegies Zwart, M. Valtonen, "Gargantuan chaotic gravitational three-body systems and their irreversibility to the Planck length." Monthly Notices of the Royal Astronomical Society, 493(3), 3932-3937, (2020).

[29] Encyclopedia Brittanica, `https://www.britannica.com/science/three-body-problem`

[30] Wikipedia, `https://en.wikipedia.org/wiki/Chaos_theory`

[31] B. Hasselblatt, A. Katok. "A first course in dynamics: with a panorama of recent developments". Cambridge University Press, (2003).

Noise as a Factor of Useful Signal Amplification Based on the Effect of Stochastic Resonance

Oksana Kharchenko [1,3], Zlatinka Kovacheva [1,2*] and Volodymyr Chumakov [3]

[1] Institute of Mathematics and Informatics, Bulgarian Academy of Sciences: Sofia, Bulgaria
[2] University of Mining and Geology, Sofia, Bulgaria
[3] Kharkiv National University of Radio Electronics, Kharkiv, Ukraine

*E-mail: zkovacheva@math.bas.bg

Abstract. The effect of stochastic resonance is investigated. The principles of the effect are considered on the basis of a Brownian particle moving in a system with a symmetric bistable potential. Different from the widely existing noise elimination methods, stochastic resonance enables a new method to utilize the energy of noises, nonlinear system, and signal frequency to reach certain weak signal enhancement effects that noise elimination cannot perform well. The presence of noise at the input of nonlinear systems possessing the effect of the stochastic resonance allows to stand out a weak signal from an additive mixture with white Gaussian noise. The equation and structural scheme of stochastic resonance are considered The conditions for the occurrence of stochastic resonance are formulated and the determining role of noise variance in the realization of this effect is shown. For the first time, a numerical calculation of standing out of harmonic oscillation from an additive mixture with white Gaussian noise based on the stochastic resonance effect is carried out. The output signal-to-noise ratio dependence of the stochastic resonator on the intensity of input noise and the input signal frequency is investigated. The frequency dependencies of the output signal are investigated and the components of this signal are analysed. It is shown that the stochastic resonator acts as a low-pass filter and reduces the output noise level. Numerical modelling of the stochastic resonance effect also shows that the output signal of the stochastic resonator contains odd harmonics of the input oscillation, which is characteristic of nonlinear systems. It is shown that the stochastic resonance effect can be widely used in radio engineering and telecommunications to amplify the useful signal.

1. Introduction

The main task in signal receiving is reduced to the best recovery of the useful information from the signal distorted during propagation and received together with interference. Standing out weak signals from noise is a pressing problem in many practical situations. The presence of noise reduces the likelihood of correct signal receiving. A quantitative measure of signal receiving is the signal-to-noise ratio (SNR), which have to exceed a certain value. Thus, noise reducing is the main task of radio engineering and telecommunication [1-4].

Accordingly, signal processing developing of new methods and mathematical models for useful signal standing out from mixture with intensive noise is relevant.

At the same time, the investigations carried out at the end of XX century [5-11] led to paradoxical conclusions. The presence of noise at the input of nonlinear systems possessing the

Content from this work may be used under the terms of the Creative Commons Attribution 4.0 licence. Any further distribution of this work must maintain attribution to the author(s) and the title of the work, journal citation and DOI.
Published under licence by IOP Publishing Ltd

effect of the so-called stochastic resonance (SR) allows to stand out a weak signal from an additive mixture with white Gaussian noise.

The SR effect characterizes the output of a nonlinear system to a weak input signal. In this case, the output parameters of the nonlinear system, such as gain and SNR, have a maximum under certain conditions [12].

As it is known, the concept of SR was introduced by physicists investigating the occurrence of ice ages [13-15].

Nowadays, this phenomenon is fundamental and takes place in various fields of science (biology, physics, chemistry, medicine, sociology) [12, 16-21], but in radio engineering and telecommunication SR effect is just beginning to develop [22].

2. Generalized model of stochastic resonance

The physical sense of the SR effect is described in [5]. As an example, we consider the behavior of a Brownian particle moving in a system with a symmetric bistable potential:

$$U(x) = -ax^2/2 + bx^4/4 , \tag{1}$$

where x is a particle coordinate.

The particle is moving under the weak periodic signal influence $A\cos(\omega_0 t)$ [9,12]. In the absence of noise, the signal energy is not enough to overcome the potential barrier and move from one potential pit to another. A completely different picture is observed when, at the same time as a periodic signal, a particle is exposed to white Gaussian noise. Thus, at a certain value of noise variance, the particle begins to make regular jumps from one potential pit to another, and the frequency of these jumps coincides with the frequency of a weak periodic signal. Note that at a certain value of noise variance, the amplitude of the particle's periodic motion reaches its maximum, which is the SR condition.

The potential function is given in Figure 1, where minima are located at points $x = \pm\sqrt{a/b}$, and the barrier height is equal to $\Delta U = a^2/4b$.

The equation describing the Brownian particle behaviour looks like [9]:

$$x'(t) = - U'(t) + A\cos(\omega_0 t + \varphi_0) + n(t) \tag{2}$$

After substituting the function U(t) equation (2) can be rewritten as

$$x'(t) = ax - bx^3 + A\cos(\omega_0 t + \varphi_0) + n(t) \tag{3}$$

where a and b are system parameters;

A – signal amplitude;

$n(t)$ – white Gaussian noise with zero mean value and variance D.

Thus, the SR effect is possible if three conditions are met simultaneously:

1) There is a weak harmonic signal.
2) There is noise of a certain intensity.
3) There is a bisquare nonlinear potential, which describes the system nonlinearity.

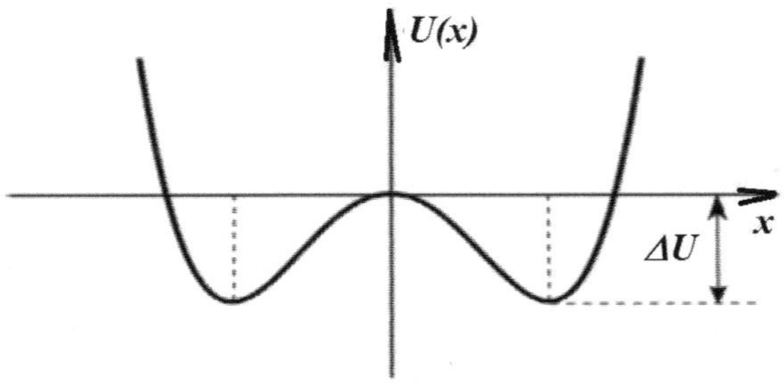

Figure 1. Classical bistable potential function $U(x)$.

In radio-engineering and telecommunication, physical devices in which the described effect is realized, are called stochastic resonators [23]. The structural scheme of the stochastic resonator is shown in Figure 2.

It should be emphasized once again that in all cases the amplitude of the input periodic effect is assumed to be so small that it excludes the presence of an output oscillation of this frequency at the stochastic resonator output. [24].

3. Determination of the output signal of a stochastic resonator in the case of an additive mixture of a harmonic signal and white Gaussian noise at the input

We consider the case when the input signal of the stochastic resonator is an additive sum of a sinusoidal signal and white Gaussian noise (3).

This equation is an Abel equation of first kind and has no exact analytical solution [25]. Even in the absence of an external harmonic signal, it is impossible to find an exact solution to the Fokker-Planck equation for two-dimensional probability densities of the output signal [3,1]. Consequently, the autocorrelation function and spectral density of the output signal cannot be determined accurately. Naturally, with the inclusion of a harmonic signal, additional difficulties arise [12,26].

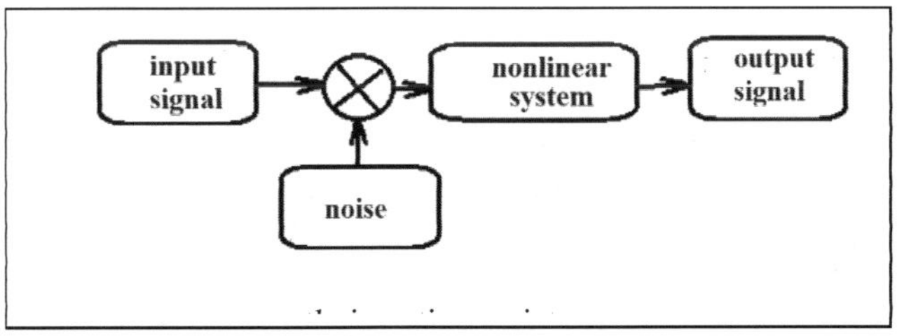

Figure 2. Structural scheme of stochastic resonator

The results of numerical modeling of the response of the stochastic resonator to an additive mixture of a harmonic signal and white Gaussian noise are shown in Figure 3. The characteristics of the input signal are as follows: frequency of the harmonic signal $f=0.125$ Hz, input signal-to-noise ratio $SNR=3$ dB. A significant increasing of the output SNR is shown in this figure. The amplitude spectra of the input and output signals are shown in Figure 4. The input harmonic signal has a finite spectrum width because its length is limited to 55 s.

The enrichment of the output spectrum with harmonics is a natural result of nonlinear transformation. In this figure, the fundamental feature of the SR effect has been reflected, which consist of the harmonic signal power increasing.

Thus, effective noise suppression is provided. Figure 4 also shows that the stochastic resonator works as a low-pass filter. Figure 5 shows the dependence of the SNR at the output on the variance of the input noise D. This dependence exhibits a local maximum at a point D=0.4.

4. Signal-to-noise ratio at the output of a stochastic resonator in the case of an additive mixture of a harmonic signal and white Gaussian noise at the input

Let us determine the output SNR of the stochastic resonator depending on the frequency of the input oscillation *f* and the input SNR. The frequency range here is shown to equal from 0.05 to 10 Hz. The input SNR values are chosen 0.005; 0.02; 0.5 (in power), respectively. The calculation results are shown in Figure. 6.

As Figure 6 shows, the SR effect is more pronounced at low frequencies, i.e., the stochastic resonator in this case is a low-pass filter. Obviously, the output SNR of the stochastic resonator is controlled by the intensity of the external noise and the input signal power. The maximum of the stochastic resonator output SNR is a function of frequency and input SNR.

Thus, we can see that at the point of *f*=0.4 Hz there is a resonance for two curves (SNR_{INPUT}=0.02; SNR_{INPUT}=0.5), and for the third curve (SNR_{INPUT}=0.05) resonance appears at the

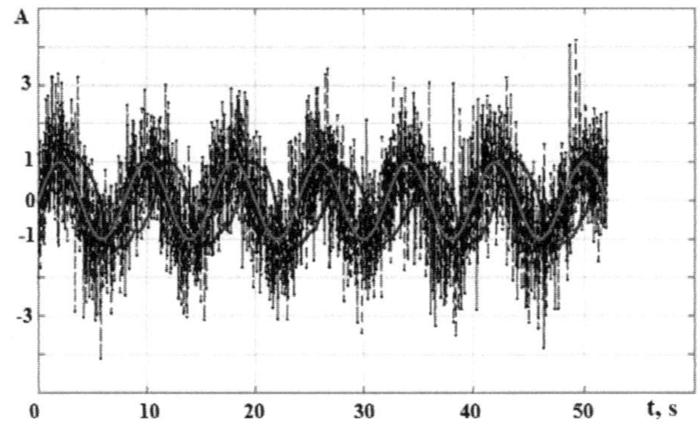

Figure 3. Signal extraction using SR effect (red line – harmonic input signal with frequency of *f=0.125* Hz, black dotted line – additive mixture of harmonic signal and white at the input with variance of *D=1*, blue line – signal at the output of nonlinear device)

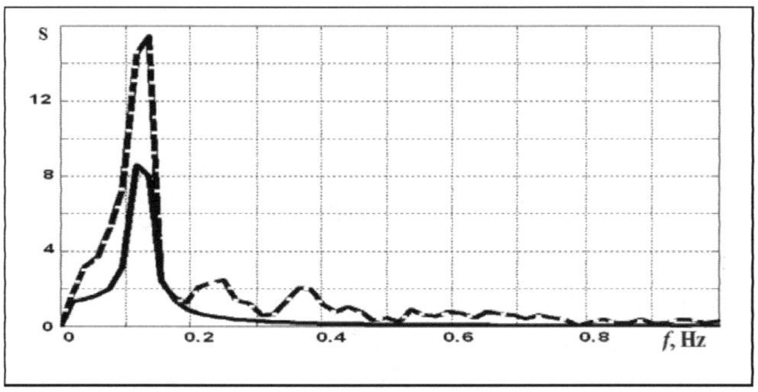

Figure 4. Amplitude spectrum of the input harmonic signal (solid line) and the output signal (dotted line)

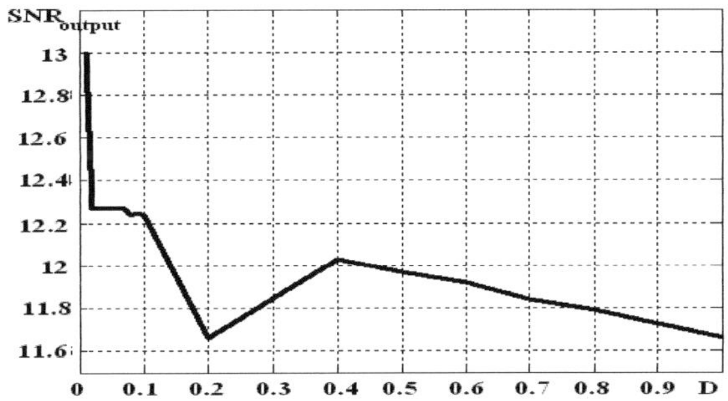

Figure 5. The output signal-to-noise ratio (SNR out) dependence on the input noise variance D (frequency of the input harmonic signal f=0.125 Hz)

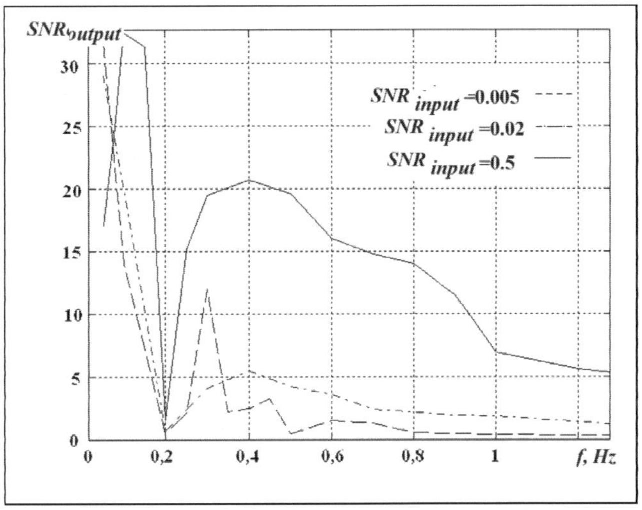

Figure 6. The output signal-to-noise ratio (SNRoutput) dependences on the signal-to-noise ratio (SNRinput) and on the input oscillation frequency

point of f=0.3 Hz. Notice that an antiresonance appears at the point of f=0.2 Hz, and this effect appears at the all input SNR values.

At the same time, there is a low-power third harmonic in the signal spectrum (Figure 7). But as the input harmonic signal frequency increases, powerful low-frequency components appear in the output spectrum (Figure 8).

Figure 7 shows that the low-frequency components are much more powerful than the component of the input signal, hence there is no signal standing out effect in this case.

It should be noted that all the above calculation results were performed for signal frequencies of the order of units and fractions of Hz. However, this does not limit the use of the SR effect in the analysis of high-frequency radio signals. Such calculations can be performed by the scaling method, which will provide results for the problem of real signal standing out in radio engineering and telecommunication systems.

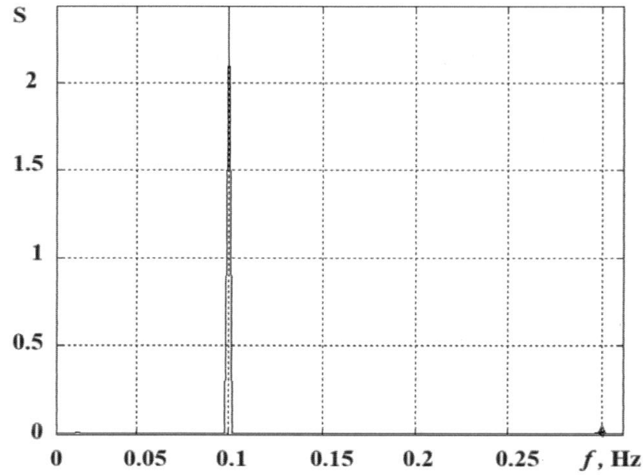

Figure 7. Amplitude spectrum of output signal (SNRinput =0.02; f=0.1 Hz; third harmonic at frequency f=0.3 Hz)

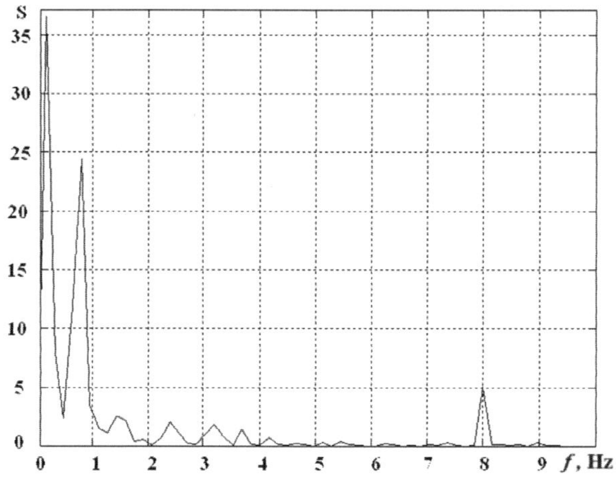

Figure 8. Amplitude spectrum of the output signal (SNRinput =0.02; f=8 Hz)

5. Conclusion

The SR effect is considered, which allows to stand out a weak (compared to noise) signal from an additive mixture of a harmonic signal with white Gaussian noise.

It is shown that the SR effect opens an opportunity for its usage in radio engineering and telecommunications to amplify the useful signal. This effect manifests itself in a non-monotone dependence of the output signal on input signal amplitude and noise intensity.

Numerical modelling of the SR effect shows that the stochastic resonator acts as a low-pass filter and it provides a significant reduction of the output noise level. It also shows that the output signal of the stochastic resonator contains odd harmonics of the input oscillation, which is characteristic of nonlinear systems.

The output SNR of the stochastic resonator is controlled by the input noise intensity and the input signal power.

Acknowledgment

This work is supported by a grant from The Simons Foundation International (grant no. 992227, IMI-BAS).

References

[1] Ya. D. Shirman et al, Theoretical bases of radar (selected pages) Transl. into ENGLISH from Teoreticheskiye Osnovy Radiolokatsii, Moscow, USSR, 1970.

[2] Ya. D. Shirman et al, Computer Simulation of Aerial Target Radar Scattering, Recognition, Detection, and Tracking, CD-ROM plus User's Manual, IEEE Aerospace and Electronic Systems Magazine, vol. 18, Issue: 5, 2003, pp. 40-43.

[3] David Middleton, An Introduction to Statistical Communication Theory: An IEEE Press Classic Reissue, **1996.**

[4] Bernard Sklar, Digital Communications. Fundamentals and Applications Second Edition. Communications Engineering Services, Tarzana, CaliforniaandUniversity of California, Los Angeles, 2001.

[5] Jerome Moses M., Ramakalyan Ayyagari, "A Brief Survey, of Stochastic Resonance and Its Application to Control", Third International Conference on Advances in Control and Optimization of Dynamical Systems March 13-15, 2014.

[6] Luca Gammaitoni, Peter Hanggi, Peter Jung "Stochastic resonance", Reviews of Modern Physics, vol. 70, No. 1, 1998, pp. 224-283.

[7] Zhi-hui Lai, Yong-gang Leng, "Weak-signal detection based on the stochastic resonance of bistable Duffing oscillator and itsapplication in incipient fault diagnosis", Mechanical Systems and Signal Processing, vol. 81, 2016, pp. 60–74.

[8] Siliang Lu, Qingbo He, Fanrang Kong, "Stochastic resonance with Woods–Saxon potential lforrolling element bearing fault diagnosis", Mechanical Systems and Signal Processing, vol. 45, 2014, pp. 488–503.

[9] Xiaofei Zhang, Niaoqing Hu, Lei Hu, Zhe Cheng "Stochastic resonance in multi-scale bistable array", Physics Letters vol. A 377, 2013, pp. 981–984.

[10] Xi-hui Chen, Gang Cheng, Xian-lei Shan, Xiao Hu, Qiang Guo, Hou-guang Liu "Research of weak fault feature information extraction of planetary gear based on ensemble empirical mode decomposition and adaptive stochastic resonance", Measurement vol. 73, 2015, pp. 55–67.

[11] Jimeng Li,Xuefeng Chen, Zhengjia He, "Multi-stable stochastic resonance and its application research on mechanicalfaultdiagnosis", Journal of Sound and Vibration, vol. 332, 2013, pp. 5999–6015.

[12] V S Anishchenko, A B Neiman, F Moss, L Schimansky-Geier, "Stochastic resonance: noise-enhanced order" Uspekhi Fizicheskikh Nauk, Russian Academy of Sciences, vol. 42(1), pp.7-34.

[13] Benzi R., Sutera A., Vulpiani A., "The mechanism of stochastic resonance", Journal of Physics A: mathematical and general, vol.14 (11), 1981, p. 1453-1462.

[14] Benzi R., Parisi G., Sutera A., Vulpiani A., "Stochastic resonance in climatic change", Tellus, vol.34(1), 1982, pp. 10-16.

[15] C. Nicolus, "Stochastic aspects of climatic transition – response to a periodic forcing", Tellus, vol. 34(1), 1982, pp. 1-9.

[16] A. Bruno, S. Graziani, "Stochastic resonance. Theory and application", Kluwer Academic Publishers, 2000.

[17], Peter Hänggi, "Stochastic resonance in biology. How noise can enhance detection of weak signals and help improve biological information processing", Chemphyschem ;vol. 3(3), 2002, pp. 90-101.

[18] Honggang Hu et al., "An experimental study of stochastic resonance ina bistable mechanical system", Journal of Physics: Conference Series382(2012) 012024, 2012, pp. 1-6.

[19] A. Guderian et al, "Stochastic Resonance in Chemistry. 1. The Belousov–Zhabotinsky Reaction", J. Phys. Chem. 100, vol. 11, 1996, pp. 4437–4441.

[20] *O. Kaut, "Stochastic resonance* therapy induces increased movement related caudate nucleus activity", Journal of rehabilitation medicine, vol. 48(9), 2014, pp. 815-818.

[21] P. Babinec, "Stochastic resonance in the Weidlich model of public opinion formation", Physics Letters A, vol. 225, Issues 1–3, 1997, pp. 179-181.

[22] Oksana Kharchenko, Zlatinka Kovacheva, Volodymyr Chumakov, "The Calculation of the Stochastic Resonator Output Driven by Signal plus Noise "// TEM Journal, 12(2), 2023, pp.585-589.

[23] Hao Chen, Pramod K. Varshney, Steven M. Kay and James H. Michels, "Theory of the Stochastic Resonance Effect in Signal Detection: Part I—Fixed Detectors", IEEE transactions on signal processing, vol. 55(7), July 2007. pp. 3172-3184.

[24] Xiaofei Zhang, Niaoqing Hu, Zhe Cheng, «Stochastic resonance in multi-scale bistable array», Physics Letter A, , vol. 377, 2013, pp. 981-984.

[25] Erich Kamke, Differentialgleichungen. B.1: Ge-woehnliche Differentialgleichungen. 1943. (in German).

[26] O.I. Kharchenko, A.M.Gorban, «Non-linear filtering of pulse signals in case of high intensity noise», Problems of Atomic Science and Technology, Series: Nuclear Physics Investigations , №6 (112), 2017, pp.113-116.

Spreading and leveling of finite thin films with elastic interfaces: An eigenfunction expansion approach

N C Papanicolaou[1] and I C Christov[2]

[1] Department of Computer Science, University of Nicosia, 46 Makedonitissas Avenue, CY-2417 Nicosia, Cyprus

[2] School of Mechanical Engineering, Purdue University, West Lafayette, Indiana 47907, USA

E-mail: papanicolaou.n@unic.ac.cy, christov@purdue.edu

Abstract. The flow of a thin viscous liquid layer under an elastic film arises in natural processes, such as magmatic intrusions between rock strata, and industrial applications, such as coating surfaces with cured polymeric films. We study the linear dynamics of small perturbations to the equilibrium state of the film in a closed trough (finite film). Specifically, we are interested in the spreading (early-time) and leveling (late-time) dynamics as the film adjusts to equilibrium, starting from different initial perturbations. We consider both smooth and non-smooth spatially symmetric and localized initial conditions (perturbations). We find the exact series solutions for the film height, using the sixth-order complete orthonormal eigenfunctions associated with the posed initial-boundary-value problem derived in our previous work [Papanicolaou N C and Christov I C 2023 *J. Phys.: Conf. Ser.* **2675** 012016]. We show that the evolution of the perturbations begins with the spreading of the localized perturbations, followed by their mutual interaction as they spread, and finally, interactions with the confining lateral boundaries of the domain as the perturbations level. In particular, we highlight how the leading eigenvalues of the problem determine the scalings of certain figures of merit with time.

1. Introduction

The flow of a thin viscous liquid layer under an elastic film [1, 2] arises in natural processes, such as magmatic intrusions between rock strata [3, 4, 5], and industrial applications, such as coating of surfaces with cured polymeric films [6, 7], growth of epitaxial films for optoelectronics [8] and semiconductors [9], microfluidic actuators [10], impact mitigation [11, 12], and adhesion [13, 14], to name a few. In our previous work [15], we considered the linear dynamics of a *finite* thin film with an elastic interface. Specifically, we derived the set of complete sixth-order orthonormal eigenfunctions associated with the initial boundary value problem. In this work, we continue our studies of the linear dynamics of small perturbations to the equilibrium state of the film in a closed trough (finite film).

Specifically, we are interested in the spreading (early time) and leveling (late time) dynamics as the film adjusts to equilibrium, starting from different initial perturbations. The motivation for the present study is the recent work of Sæter et al. [16], who performed a combined experimental-numerical study of the fate of localized *blisters* [17] on the surface of a laterally unbounded elastic-plated film. They focused on the nonlinear regime of coalescence and its self-similar dynamics, noting an "anomalous exponential growth with time, contrasting the power-law solutions usually describing the self-similar behavior of lubrication flows" (e.g., [18, 19]).

To further understand this behavior, we analyze the *linear* dynamics of blisters on the elastic surface of an *confined* thin film, the governing equations of which are reviewed in section 2. Using the complete

Content from this work may be used under the terms of the Creative Commons Attribution 4.0 licence. Any further distribution of this work must maintain attribution to the author(s) and the title of the work, journal citation and DOI.
Published under licence by IOP Publishing Ltd

sixth-order orthonormal eigenfunctions from our previous work [15], which are reviewed in section 3, we find the exact series solution for the film height by separation of variables in section 4. Next, in section 5, we consider non-smooth (section 5.1) and smooth (section 5.2) forms of the spatially symmetric initial conditions representing the localized perturbations that create blisters on the surface of the film. We show that the evolution of the perturbations begins with the spreading of the localized perturbations, followed by their mutual interaction as they spread, and finally, interactions with the finite lateral boundaries of the domain as the perturbations level. In particular, we highlight the exponential decay in time of various relevant quantities, whose time scales are set by the leading eigenvalues of the problem. Conclusions are stated in section 6.

2. Governing equations

Using lubrication theory (i.e., a long-wave approximation) [20, 21, 22], it can be shown [1, 23, 15] that a slender, elastic-plated film's height $h(x,t)$ obeys the sixth-order *thin-film equation*

$$\frac{\partial h}{\partial t} = \frac{\partial}{\partial x}\left\{\frac{h^3}{12\mu_f}\left[\rho_f g\frac{\partial h}{\partial x} + B\frac{\partial^5 h}{\partial x^5}\right]\right\}, \tag{2.1}$$

where μ_f is the (Newtonian) fluid's dynamic viscosity, ρ_f is its density, and B is the elastic interface's bending rigidity. The equilibrium film height is h_0, the trough has an axial length $2\ell \gg h_0$, and the acceleration due to gravity is denoted by g. A schematic is shown in figure 1.

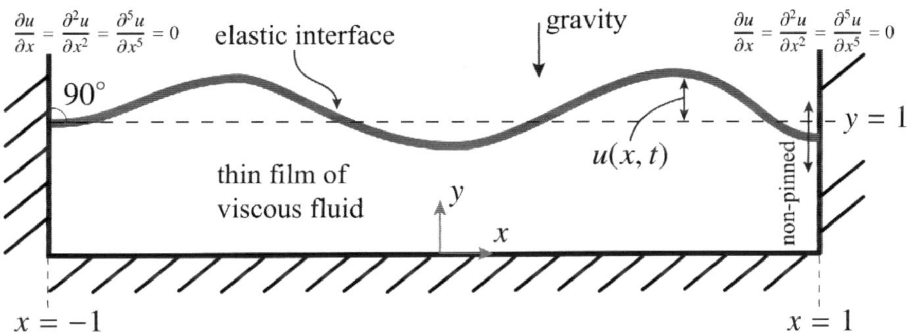

Figure 1. Schematic of a finite thin film with an elastic interface in terms of the dimensionless variables of the problem. The perturbation of the flat state is denoted $u(x,t)$. The domain and boundary conditions on u at $x = \pm 1$ (non-pinned, 90° contact angle, and no flux through the vertical confining walls) are indicated.

Making the problem dimensionless through the transformations $x \mapsto \ell x$, $h \mapsto h_0 h$, $t \mapsto t_c t$ and choosing $t_c = 12\mu_f \ell^6/Bh_0^3$ to be the so-called "viscous-elastic" timescale [24], (2.1) becomes

$$\frac{\partial h}{\partial t} = \frac{\partial}{\partial x}\left[h^3\left(\mathrm{Bo}\frac{\partial h}{\partial x} + \frac{\partial^5 h}{\partial x^5}\right)\right], \tag{2.2}$$

where the *Bond number*

$$\mathrm{Bo} = \frac{\rho_f g \ell^4}{B} \tag{2.3}$$

quantifies the effect of gravity relative to the bending resistance.

As in [15, 19], we are interested in *small perturbations* to the flat state: $h(x,t) = 1 + u(x,t)$, where $\max_{x,t}|u| \ll 1$. Then, to leading-order in u, (2.2) becomes

$$\frac{\partial u}{\partial t} = \mathrm{Bo}\frac{\partial^2 u}{\partial x^2} + \frac{\partial^6 u}{\partial x^6}, \tag{2.4}$$

which is a linear sixth-order partial differential equation of "parabolic" type [19]. As discussed in [15, 19], for a free (non-pinned) film in a closed trough, as shown in figure 1, the boundary conditions for (2.4) are

$$\frac{\partial u}{\partial x} = \frac{\partial^2 u}{\partial x^2} = \frac{\partial^5 u}{\partial x^5} = 0 \quad \text{at} \quad x = \pm 1. \tag{2.5}$$

Here, we consider the simplest form of (2.4), which occurs when Bo = 0 (i.e., gravity is negligible).

3. The complete orthonormal set of sixth-order eigenfunctions

Based on the above discussion, consider the following sixth-order initial boundary value problem (IBVP) for $u = u(x, t)$:

$$\begin{cases} \dfrac{\partial u}{\partial t} - \dfrac{\partial^6 u}{\partial x^6} = 0, & (x,t) \in (-1,1) \times \in (0,\infty), \quad (3.1a) \\ \left.\dfrac{\partial u}{\partial x}\right|_{x=\pm 1} = \left.\dfrac{\partial^2 u}{\partial x^2}\right|_{x=\pm 1} = \left.\dfrac{\partial^5 u}{\partial x^5}\right|_{x=\pm 1} = 0, & t \in (0,\infty), \quad (3.1b) \\ u(x,0) = u^0(x), & x \in (-1,1), \quad (3.1c) \end{cases}$$

for a given $u^0 \in L^2[-1, 1]$.

To solve IBVP (3.1) by separation of variables, we must solve the *associated Sturm–Liouville (SL) eigenvalue problem (EVP)*:

$$\begin{cases} -\dfrac{d^6\psi}{dx^6} = \lambda^6 \psi, & x \in (-1,1), \quad (3.2a) \\ \left.\dfrac{d\psi}{dx}\right|_{x=\pm 1} = \left.\dfrac{d^2\psi}{dx^2}\right|_{x=\pm 1} = \left.\dfrac{d^5\psi}{dx^5}\right|_{x=\pm 1} = 0. & (3.2b) \end{cases}$$

Although SL theory was developed for second-order problems [25], it is customary in the literature to also refer to higher-order self-adjoint EVPs as "SL EVPs" [26, 27].

As shown in [15], SL EVP (3.2) has a countable set of solutions $\{\psi_m(x)\}_{m=1,2,\ldots}^{\infty}$, which can be split into even ("cosine") $\{\psi_m^c(x)\}_{m=1,2,\ldots}^{\infty}$ and odd ("sine") $\{\psi_m^s(x)\}_{m=1,2,\ldots}^{\infty}$ subsets:

$$\begin{aligned} \psi_m^c(x) = c_m^c \Bigg\{ & \frac{4\sin\lambda_m^c}{\cos\lambda_m^c - \cosh(\sqrt{3}\,\lambda_m^c)} \Bigg[-\cos\left(\tfrac{\lambda_m^c}{2}\right)\sinh\left(\tfrac{\sqrt{3}\,\lambda_m^c}{2}\right)\sin\left(\tfrac{\lambda_m^c}{2}x\right)\sinh\left(\tfrac{\sqrt{3}\,\lambda_m^c}{2}x\right) \\ & + \sin\left(\tfrac{\lambda_m^c}{2}\right)\cosh\left(\tfrac{\sqrt{3}\,\lambda_m^c}{2}\right)\cos\left(\tfrac{\lambda_m^c}{2}x\right)\cosh\left(\tfrac{\sqrt{3}\,\lambda_m^c}{2}x\right)\Bigg] + \cos(\lambda_m^c x)\Bigg\}, \end{aligned} \tag{3.3a}$$

$$\begin{aligned} \psi_m^s(x) = c_m^s \Bigg\{ & \frac{4\cos\lambda_m^s}{\cos\lambda_m^s + \cosh(\sqrt{3}\,\lambda_m^s)} \Bigg[-\cos\left(\tfrac{\lambda_m^s}{2}\right)\cosh\left(\tfrac{\sqrt{3}\,\lambda_m^s}{2}\right)\sin\left(\tfrac{\lambda_m^s}{2}x\right)\cosh\left(\tfrac{\sqrt{3}\,\lambda_m^s}{2}x\right) \\ & + \sin\left(\tfrac{\lambda_m^s}{2}\right)\sinh\left(\tfrac{\sqrt{3}\,\lambda_m^s}{2}\right)\cos\left(\tfrac{\lambda_m^s}{2}x\right)\sinh\left(\tfrac{\sqrt{3}\,\lambda_m^s}{2}x\right)\Bigg] + \sin(\lambda_m^s x)\Bigg\}, \end{aligned} \tag{3.3b}$$

where the lengthy expressions for the normalization coefficients c_m^c and c_m^s are given in [15].

The eigenvalues λ_m^c correspond to the ψ_m^c and are the solutions of the transcendental equation (3.4a) whereas the λ_m^s correspond to the ψ_m^s and are the solutions of (3.4b):

$$\text{even}: \quad \cos(2\lambda_m^c) + \sqrt{3}\sin\lambda_m^c \sinh(\sqrt{3}\,\lambda_m^c) - \cos\lambda_m^c \cosh(\sqrt{3}\,\lambda_m^c) = 0\,, \tag{3.4a}$$

$$\text{odd}: \quad \sin(2\lambda_m^s) + \sqrt{3}\cos\lambda_m^s \sinh(\sqrt{3}\,\lambda_m^s) + \sin\lambda_m^s \cosh(\sqrt{3}\,\lambda_m^s) = 0\,. \tag{3.4b}$$

Remark 3.1. $\psi_0^c(x) = 1$ satisfies the EVP (3.2) with $\lambda_0 = 0$.

The two sets of solutions ψ^c_m and ψ^s_m of the SL EVP (3.2) given by (3.3) and supplemented by $\psi^c_0 = 1$ form a *complete orthonormal* (CON) set of functions on $[-1, 1]$. Thus, given any function $u(x) \in L^2[-1, 1]$, we can write its spectral expansion in this CON basis as

$$u(x) = u^c_0\psi^c_0(x) + \sum_{m=1}^{\infty} u^c_m\psi^c_m(x) + u^s_m\psi^s_m(x), \tag{3.5a}$$

$$u^c_0 = \frac{1}{2}\int_{-1}^{1} u(x)\,\mathrm{d}x, \tag{3.5b}$$

$$u^c_m = \int_{-1}^{1} u(x)\psi^c_m(x)\,\mathrm{d}x, \tag{3.5c}$$

$$u^s_m = \int_{-1}^{1} u(x)\psi^s_m(x)\,\mathrm{d}x, \tag{3.5d}$$

where the series (3.5a) converges to $u(x)$ in the mean. For a sketch of the proof, see [15] and the references therein.

4. Solution method (separation of variables)

In this work, we restrict ourselves to initial conditions that are even in space, namely $u^0(x) = u^0(-x)$ on $[-1, 1]$. It is easy to show from (3.1) that it follows that the solution is also even, $u(x, t) = u(-x, t)\ \forall t > 0$.

We solve PDE (3.1a) using an eigenfunction expansion based on sixth-order even eigenfunctions (3.3). Specifically, we write the sought function as $u(x, t) = X(x)T(t)$. Substituting the latter into PDE (3.1a) leads to

$$\frac{T'(t)}{T(t)} = \frac{X^{(6)}(x)}{X(x)} = -\lambda^6, \tag{4.1}$$

where λ must be a constant.

We observe that $X^{(6)}(x) = -\lambda^6 X(x)$ is precisely (3.2a). Subject to the BCs (3.1b), we have shown that it is solved by the even eigenfunctions ψ^c_m corresponding to λ^c_m, i.e., $X(x) = \psi^c_m(x)$ and $\lambda = \lambda^c_m$, $m = 0, 1, 2, \ldots$. Therefore, the general solution is of the form

$$X(x) = \sum_{m=0}^{\infty} u_m\psi^c_m(x), \tag{4.2}$$

where the coefficients $\{u_m\}_{m=0}^{\infty}$ are to be determined. Meanwhile, the IVP $T'(t) = -\lambda^6 T(t)$, $T(0) = T_0$ has solution

$$T(t) = T_0 \mathrm{e}^{-(\lambda^c_m)^6 t}. \tag{4.3}$$

Combining (4.2) and (4.3), using the initial condition $u(x, 0) = u^0(x)$ and recalling remark 3.1, we obtain

$$u(x, t) = u^0_0\psi^c_0(x) + \sum_{m=1}^{\infty} \mathrm{e}^{-(\lambda^c_m)^6 t} u^0_m\psi^c_m(x), \tag{4.4a}$$

where $\{u^0_m\}_{m=0}^{\infty}$ are the spectral expansion coefficients of the (even) initial condition $u^0(x)$. From (3.5), we know these coefficients can be obtained by calculating the integrals

$$u^0_0 = \frac{1}{2}\int_{-1}^{1} u^0(x)\,\mathrm{d}x, \tag{4.4b}$$

$$u^0_m = \int_{-1}^{1} u^0(x)\psi^c_m(x)\,\mathrm{d}x, \qquad m = 1, 2, \ldots, \tag{4.4c}$$

which can be done *exactly* for certain functions $u^0(x)$. When evaluating solutions to specific problems, the infinite sum (4.4a) is truncated at some upper limit $m = M$.

Remark 4.1. The eigenfunctions $\{\psi_m^c(x)\}_{m=0,1,2,...}$ intrinsically satisfy the BCs (3.1b). This guarantees that the solution, as given by (4.4), will also satisfy the BCs (3.1b).

5. Spreading and leveling of finite thin films with elastic interfaces

5.1. Two point loads

Consider IBVP (3.1) with the initial condition

$$u(x,0) = u^0(x) = \frac{1}{2}\left[\delta(x-\tfrac{1}{2}) + \delta(x+\tfrac{1}{2})\right], \qquad x \in (-1,1), \tag{5.1}$$

where $\delta(x)$ is the Dirac delta functional (i.e., the unit impulse) centered at $x = 0$.

The IC (5.1) represents two point loads at an equal distance of 1/2 from the center of the domain. This IC has a total "volume"

$$\int_{-1}^{1} u^0(x)\,\mathrm{d}x = 1. \tag{5.2}$$

Although u represents the deviation from a flat interface (see figure 1), u does not have to decay to 0 as $t \to \infty$, since we have free BCs at $x = \pm 1$ that allow for steady states $u(x, t \to \infty) = const. \neq 0$. For an initial perturbation that maintains $u = 0$ at $x = \pm 1$, but is nonvanishing on $(-1, 1)$, some "volume" is added according to (5.2).

Employing well-known properties of the Dirac delta functional (see, e.g., [28]), from (4.4b), (4.4c), and (5.1) we find:

$$u_0^0 = \frac{1}{2}, \tag{5.3a}$$

$$u_m^0 = \frac{1}{2}\left[\psi_m^c(\tfrac{1}{2}) + \psi_m^c(-\tfrac{1}{2})\right], \qquad m = 1, 2, \ldots. \tag{5.3b}$$

The spectral expansion of IC (5.1) is shown in figure 2. The strong Gibbs effects observed are expected since (5.1) is not only discontinuous at $x = \pm 1/2$, but is not even a "proper function". The question of whether or not the Gibbs effects in the series expansion affect the accuracy of the spectral solution is addressed in section 5.2.

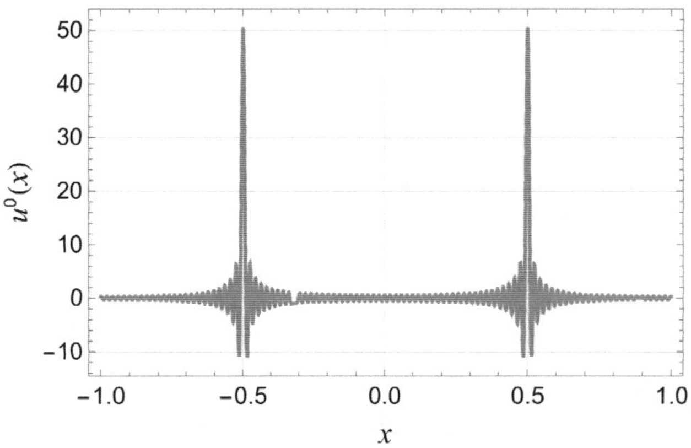

Figure 2. Profile of the spectral representation, which follows from (3.5), of the IC (5.1). The series is truncated at $m = M = 100$ (i.e., 101 terms). Strong Gibbs effects are observed, especially near the discontinuities at $x = \pm 1/2$.

Figure 3 depicts selected snapshots of the solution of this IBVP. Figure 3(a) shows the evolution of the initial disturbance (5.1) from early to intermediate times, whilst figure 3(b) shows how it evolves from

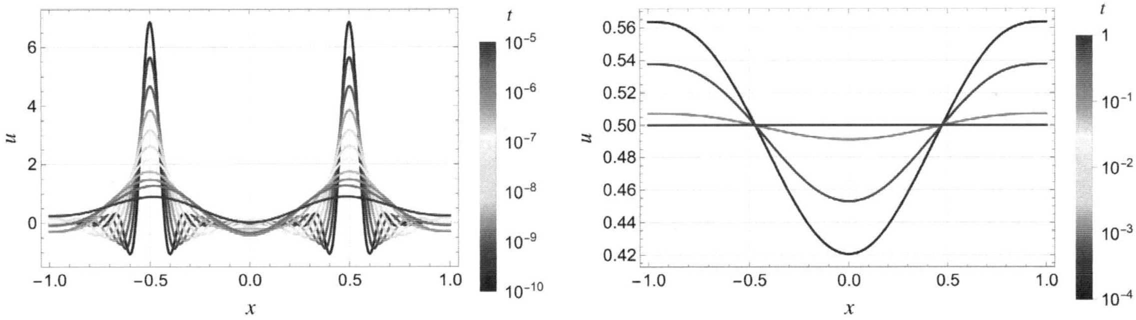

(a) (b)

Figure 3. Solution snapshots for IBVP (3.1) subject to IC (5.1). An eigenfunction expansion (4.4a) with 101 terms is used (i.e., truncated at $m = M = 100$). (a) Spreading of the initial disturbance (early to intermediate times). (b) Leveling of the initial disturbance (intermediate to large times). The color scheme represents time on a logarithmic scale, according to each color bar.

intermediate to large times. Of course, the definition of "early" and "late" is not quantitative. Rather, the classification is based on the dynamics observed, which we now analyze.

From figure 3(a), we observe that the point loads create smooth blisters, which spread horizontally. Their amplitudes decay with time. On the other hand, from figure 3(b), it is evident that for $t \geq 10^{-4}$, the solution has spread to the boundaries, with the two blisters coalescing into a profile that exhibits a single local minimum at $x = 0$. The film is now leveling. For $t \geq 10^{-2}$, the solution profile is essentially flat. Importantly, this flat profile has a "volume" equal to one, which is exactly equal to the "volume" (5.2) of the initial condition. This shows that our solution method preserves the steady state.

An important feature of the solution is the *bridge*, namely the point where the two blisters make "contact", following the terminology of Sæter et al. [16]. Here, this point is $x = 0$. Therefore, the bridge height is represented by $u(0, t)$, which is easily formed by evaluating the obtained spectral solution (4.4a) at $x = 0$. Studying the time evolution of the bridge height provides information on how the two blisters "interact". The bridge height $u(0, t)$ is plotted as a function of time in figure 4. Observing figure 4(a), we see that for very early times, $u(0, t)$ oscillates rapidly. However, these oscillations are between relatively small values, namely $-0.017 \leq u(0, t) \leq 0.007$. Concentrating on early to intermediate times in figure 4(b), the bridge initially rises to a maximum value and then decreases, dropping below $u = 0$ and eventually plateauing at a negative lower bound. The maximum and minimum values of $u(0, t)$ over this time range have absolute values that are substantially larger compared to their counterparts in figure 4(a).

A completely different picture is painted by figures 4(c) and 4(d) for $t > 10^{-5}$. Here, later times are presented using a linear abscissa scale in figure 4(c) and a logarithmic abscissa scale in figure 4(d). The length of the time interval here is orders of magnitude larger compared to figures 4(a) and 4(b). The bridge height increases, reaching the equilibrium value $u = 1/2$ at $t \approx 0.015$. Furthermore, figure 4(d) reveals three different regimes according to the rate of increase of $u(0, t)$ with t. These regimes occur over three different time intervals and correspond to different (positive) slopes of the graph of $u(0, t)$ vs t. First, we observe a region of rapid increase, then a region of intermediate positive slope, and finally, a region of almost zero slope corresponding to the equilibration to the final (flat) state.

Now, we quantitatively investigate the asymptotic behavior of the bridge height (as a function of time) for intermediate to large times. Using the obtained solution for the bridge height, $n = 21$ data points $\big(t_i, (u(0, t_i)\big)$, $i = 1, 2, \ldots, n$ were selected for times $t_i \in [10^{-5}, 10^{-1}]$ and fitted to a general exponential function with the aid of Mathematica's [29] `NonlinearModelFit` routine. The results are presented using a linear abscissa scale in figure 5(a) and on a logarithmic abscissa scale in figure 5(b). The best-fit curve was found to be $u(0, t) \simeq \frac{1}{2} - 0.101127\, e^{-(\lambda_1^c)^6 t} - 1.019245\, e^{-(\lambda_2^c)^6 t}$, where $\lambda_1^c \approx 3.66606496814$ and

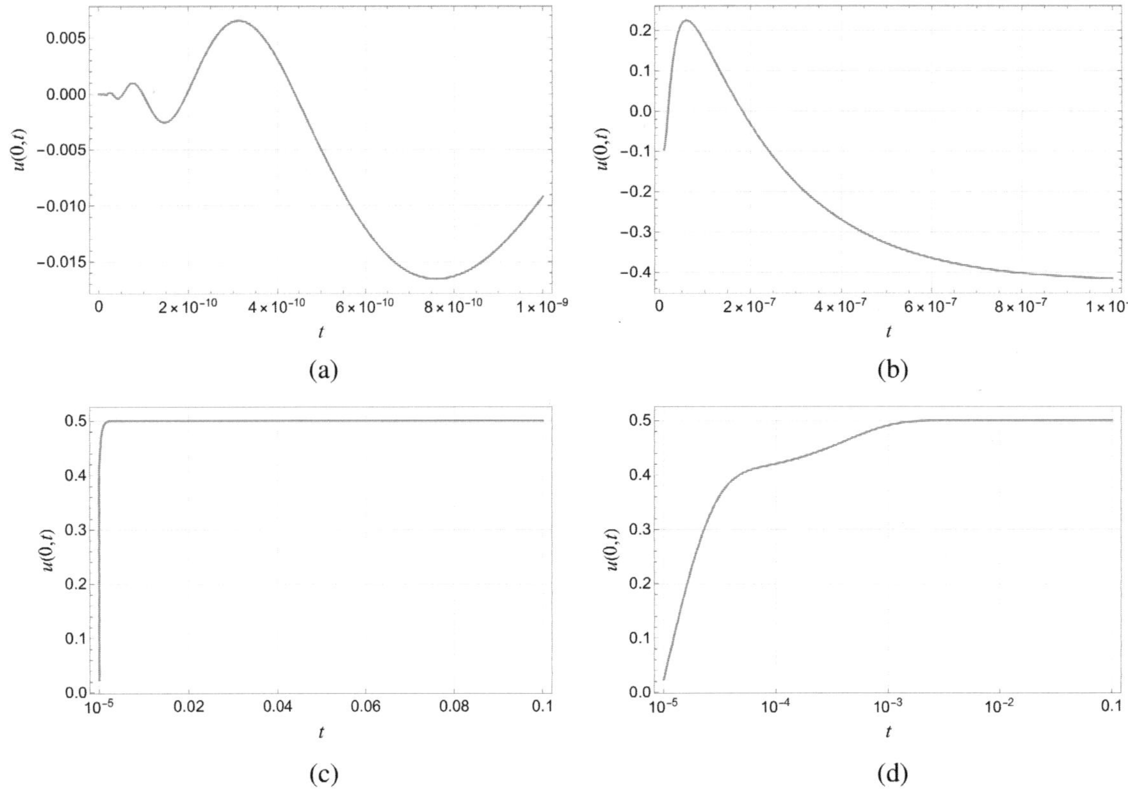

Figure 4. Evolution of the bridge height $u(0,t)$ for IBVP (3.1) subject to IC (5.1) as a function of time: (a) Early times. (b) Intermediate times. (c) Late times. (d) Late times on a log-linear scale.

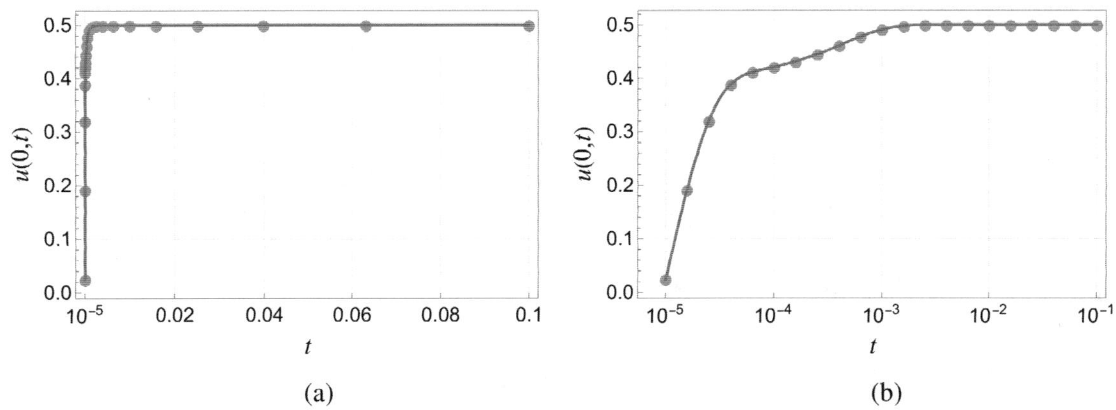

Figure 5. Asymptotic behavior of the bridge height $u(0,t)$ for IBVP (3.1) subject to IC (5.1) for intermediate to large times on (a) linear scale and (b) log-linear scale. Symbols: computed bridge height $u(0,t)$ at the sampling points. Solid: best-fit curve $u(0,t) \simeq \frac{1}{2} - 0.101127\,\mathrm{e}^{-(\lambda_1^c)^6 t} - 1.019245\,\mathrm{e}^{-(\lambda_2^c)^6 t}$.

$\lambda_2^c \approx 6.80678029161$ are the first two nontrivial even eigenvalues of SL EVP (3.2) [15]. Interestingly, this exponential asymptotic behavior of the bridge is similar to what was reported in [16], despite the differences in the two problems. Sæter et al. [16] found the exponential behavior in a *nonlinear* version of the problem on an unconfined domain by seeking a self-similar solution. Here, unlike in [16], the exponential decay is a direct consequence of solving a linear first-order-in-time PDE (3.1a).

Next, we consider the evolution of the peaks of the blisters, namely $u(\pm 1/2, t)$, as functions of time. The points $x = \pm 1/2$ are the coordinates of the centers of the Dirac deltas (point loads), which appear in IC (5.1). Due to the symmetry of the problem (see also figure 3), $u(-1/2, t) = u(1/2, t)$, $\forall t \geq 0$. Therefore, it suffices to consider only $u(1/2, t)$. Now, using the same sampling points $\{t_i\}_{i=1}^{21}$ as those selected for fitting the bridge height, as well as the same computational routine, we find the best-fit curve $u(1/2, t) \simeq \frac{1}{2} + 1.066850\, e^{-(\lambda_2^c)^6 t}$. Therefore, in contrast to our findings for the bridge height, the eigenmode corresponding to λ_1^c does not participate in the dynamics of the blister height decay. The results for $t \geq 10^{-5}$ are presented using a linear abscissa scale in figure 6(a) and using a logarithmic abscissa scale in figure 6(b). Indeed, we observe that the decay of $u(1/2, t)$ has only two distinct regimes: (i) a region of rapid decrease for $10^{-5} \leq t \leq 10^{-4}$ and (ii) a region of almost zero slope corresponding to the equilibration to the final (flat) state ($u = 1/2$) for $t > 10^{-4}$.

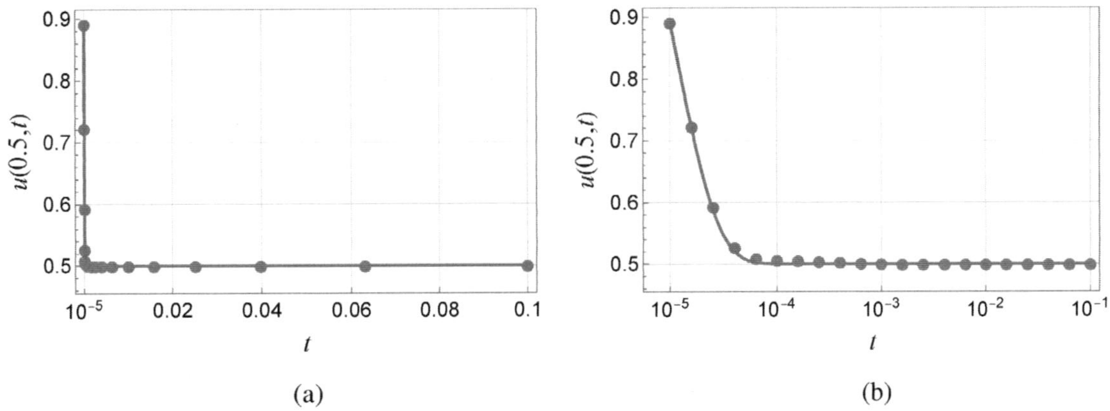

Figure 6. Asymptotic behavior of the blister heights $u(1/2, t)$ for IBVP (3.1) subject to IC (5.1) for intermediate to large times on (a) linear scale and (b) log-linear scale. Symbols: computed bridge height $u(1/2, t)$ at the sampling points. Solid: best-fit curve $u(1/2, t) \simeq \frac{1}{2} + 1.066850\, e^{-(\lambda_2^c)^6 t}$.

5.2. Two smooth bumps

Now, consider IBVP (3.1) subject to the (continuous) piecewise polynomial initial condition

$$u_0(x) = \begin{cases} \frac{15}{8}\left[1 - \frac{(x+\frac{1}{2})^2}{(\frac{1}{4})^2}\right]^2 & \text{if } -\frac{3}{4} \leq x \leq -\frac{1}{4}, \\ \frac{15}{8}\left[1 - \frac{(x-\frac{1}{2})^2}{(\frac{1}{4})^2}\right]^2 & \text{if } \quad \frac{1}{4} \leq x \leq \frac{3}{4}, \\ 0 & \text{otherwise}. \end{cases} \tag{5.4}$$

This IC (5.4) is an adaptation of the function used by Sæter et al. [16] to model the two elastic blisters of their experimental study. In the present context, the shape function has been reduced to a function of a single spatial variable, x, and normalized to have unit "volume" on $x \in [-1, 1]$. It is important to note that (5.4) is a smooth and even function of x, exhibiting no discontinuities and satisfying BCs (3.2b).

The eigenfunction expansion of IC (5.4) is truncated at $M = 100$ and is compared to the function it approximates in figure 7(a). As we have already noted, this IC is a smooth function satisfying the same BCs as the ψ_m^c eigenfunctions. Hence, there is an excellent agreement between (5.4) and its spectral expansion (the profiles in the figure are essentially indistinguishable), which should be expected. In figure 7(b), we further present the spatial variation of the absolute difference between the IC and its

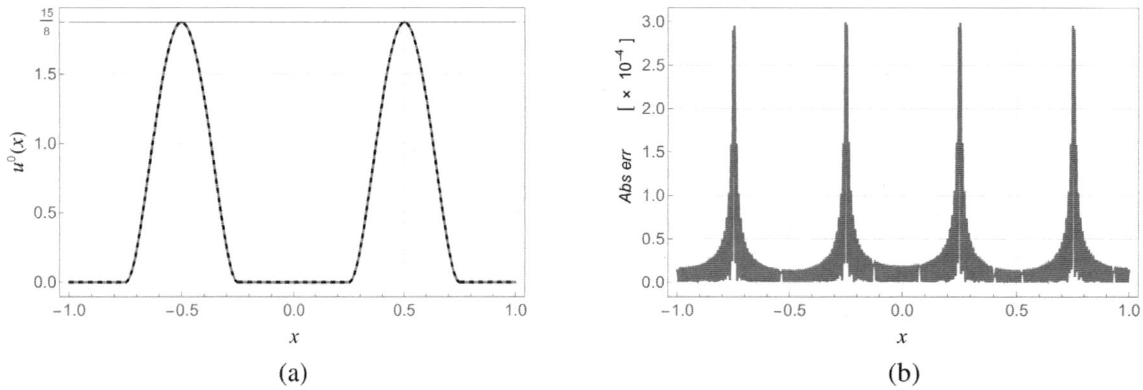

Figure 7. (a) Solid: profile of the IC as given by (5.4). Dashed: its spectral approximation, which follows from (3.5), truncated at $m = M = 100$. (b) The absolute error between $u^0(x)$ and its spectral expansion.

spectral representation. We observe that the maximum absolute error is $< 3 \times 10^{-4}$ and that there is no visible Gibbs effects.

As before, the solution $u(x, t)$ of IBVP (3.1) subject to IC (5.4) can be written as (4.4a), however evaluating the expansion coefficients, via (4.4b) and (4.4c), based on the IC (5.4), is more tedious. Thus, the lengthy expressions for the coefficients $\{u_m^0\}_{m=0,1,2,...}$ are given by (A.1) in the Appendix.

In figure 8, we present snapshots of the eigenfunction solution of IBVP (3.1) subject to IC (5.4). From figure 8(a) (early to intermediate times), we see that the blisters centered at $x = \pm 1/2$ decrease in height, with the solution spreading out. In figure 8(b) (intermediate to later times), the blisters have already coalesced into a single smooth bump centered at $x = 0$ by $t \approx 10^{-4}$. As time increases, the (absolute) height of this bump, namely $|u(0, t)|$, reduces, with the solution essentially reaching the flat equilibrium state ($u = 1/2$) at $t \approx 10^{-2}$. Once again, the "volume" of the equilibrium state is conserved. The behavior exhibited here is *qualitatively* (but not quantitatively) similar to that shown in figure 3, highlighting the universal aspects of spreading and leveling of finite thin films with elastic interfaces.

The evolution of the bridge height $u(0, t)$ with time for IBVP (3.1) subject to IC (5.4) is illustrated in figure 9. The time intervals and scales used in figures 9(a)-(d) are the same as their counterparts in figures 4(a)-(d). Comparing the two figures, we observe that the bridge heights corresponding to blisters emerging from the two different ICs exhibit similar behavior. The same remarks and observations made

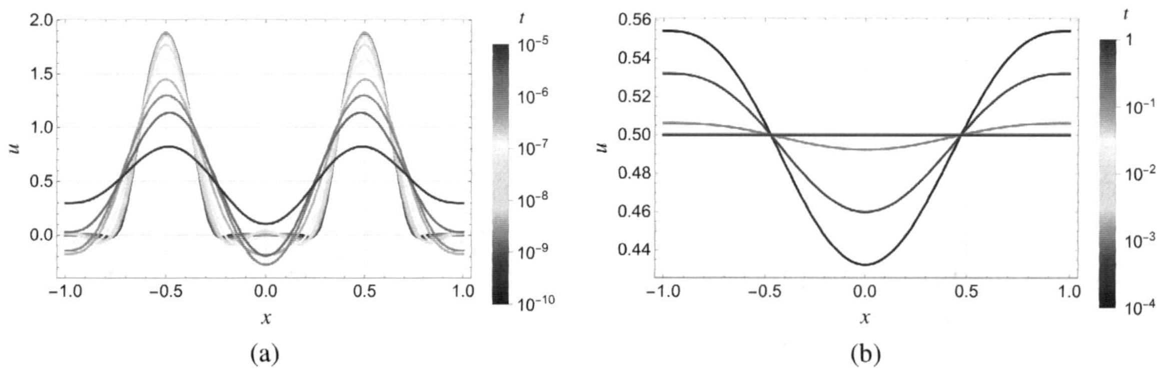

Figure 8. Solution snapshots for IBVP (3.1) subject to IC (5.4). (a) spreading of the initial disturbance (early to intermediate times). (b) leveling of the initial disturbance (intermediate to large times). Notice the similarity of the leveling curves between the two different ICs.

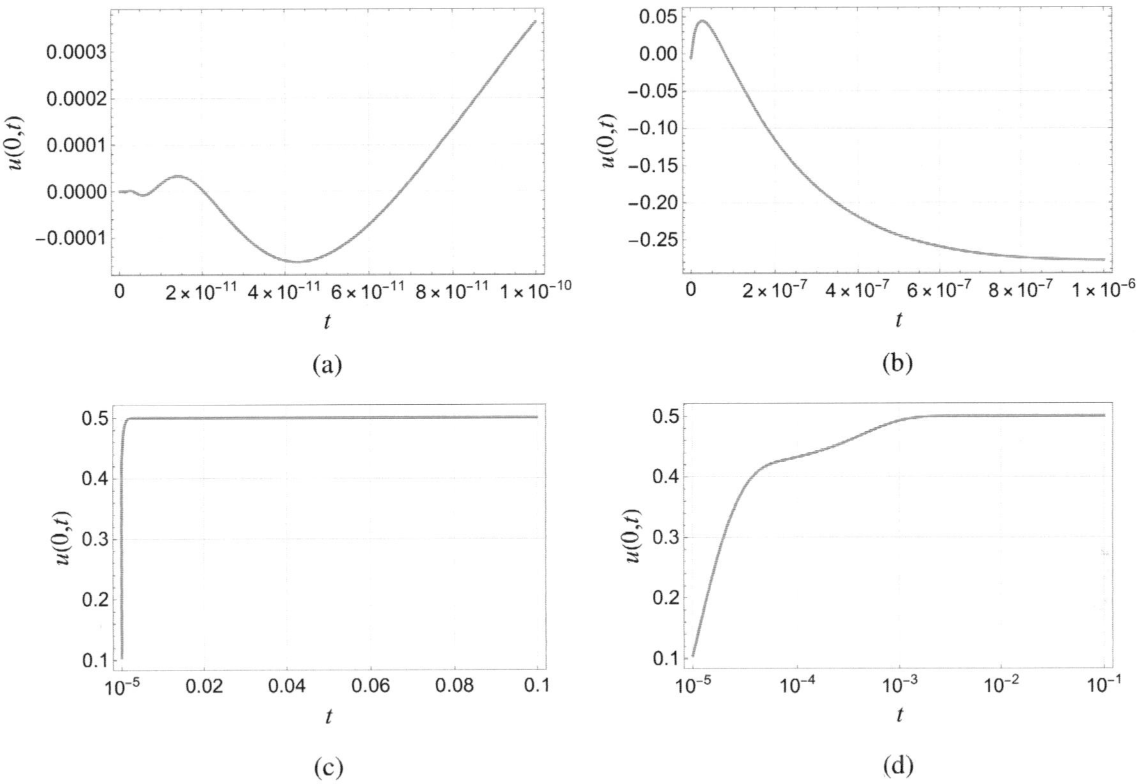

Figure 9. Evolution of the bridge height $u(0,t)$ of the solution of IBVP (3.1) subject to IC (5.4) as a function of time: (a) Early times. (b) Intermediate times. (c) Late times. (d) Late times on a log-linear scale.

when discussing figures 4(a)-(d) in the previous subsection also hold here. Other than the expected quantitative differences, perhaps the only other difference of note is that for very early times, the bridge height in figure 9(a) is slightly less oscillatory compared to that in figure 4(a).

We may, therefore, conclude that the small-amplitude oscillations observed at very early times are not an artifact of the solution method (for instance, a result of the Gibbs effects) but a characteristic of the problem since they are also present when our IC is smooth and has an accurate well-behaved eigenfunction expansion. In addition, *both* figures 9(b),(d) and figures 4(b),(d) show that for the time interval $t \in [2 \times 10^{-7}, 10^{-6}]$ the bridge height is negative. This means that following some very small (in amplitude) oscillations at very early times and after contact between the two blisters is established, the elastic interface in the vicinity of the point of contact drops below the initial equilibrium film height (the flat state $u = 1/2$) before it eventually rises back to it as $t \to \infty$. These negative values were not observed in [16] since that work concentrated on intermediate to later times, with the computational time being reset to zero once contact was made.

In figure 10, we examine the asymptotic behavior of the bridge height evolving from the two bumps IC (5.4) following the same method as that of figure 5 for the two point loads IC. Once again, we find that the best-fit function (obtained from Mathematica) is exponential and that it involves the first two nontrivial even eigenvalues, namely $u(0,t) \simeq \frac{1}{2} - 0.086219\, e^{-(\lambda_1^c)^6 t} - 0.840805\, e^{-(\lambda_2^c)^6 t}$. The results are presented using a linear abscissa scale in figure 10 (a) and using a logarithmic abscissa scale in figure 10(b).

Finally, in figure 11, we study the time-evolution of the blister heights $u(1/2,t)$ in the same way as in

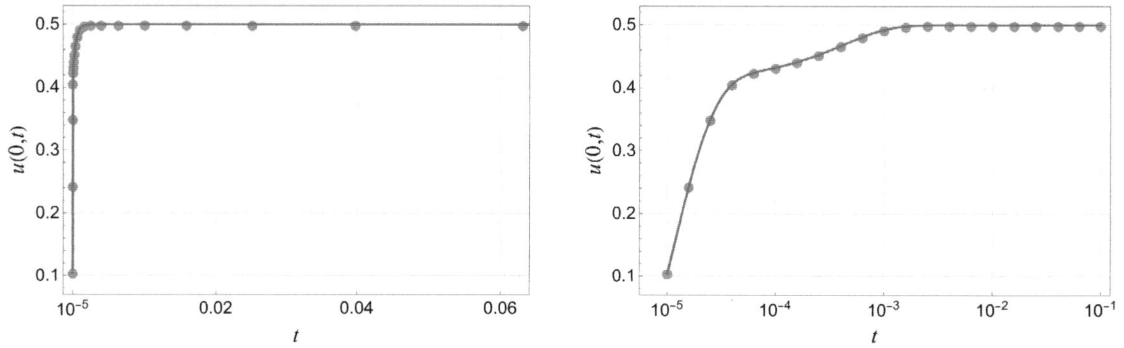

(a) (b)

Figure 10. Asymptotic behavior of the bridge heighfor IBVP (3.1) subject to IC (5.4) for intermediate to large times on (a) linear scale and (b) log-linear scale. Symbols: computed bridge height $u(0,t)$. Solid: best-fit curve $u(0,t) \simeq \frac{1}{2} - 0.086219\,\mathrm{e}^{-(\lambda_1^{\varsigma})^6 t} - 0.840805\,\mathrm{e}^{-(\lambda_2^{\varsigma})^6 t}$.

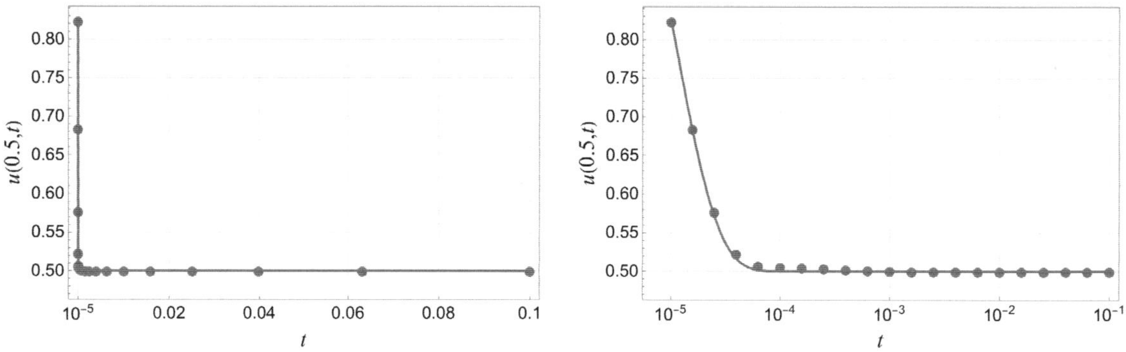

(a) (b)

Figure 11. Asymptotic behavior of the bumps' heights $u(\pm 1/2,t)$ for IBVP (3.1) subject to IC (5.4) for intermediate to large times on a (a) linear scale and (b) log-linear scale. Symbols: computed bridge height $u(1/2,t)$. Solid: best-fit curve $u(1/2,t) \simeq \frac{1}{2} + 0.880922\,\mathrm{e}^{-(\lambda_2^{\varsigma})^6 t}$.

section 5.1. Once again, the best-fit function is exponential, involving only the zeroth and second, but not first, eigenvalues of the problem. The best-fit curve is now $u(1/2,t) \simeq \frac{1}{2} + 0.880922\,\mathrm{e}^{-(\lambda_2^{\varsigma})^6 t}$. In addition, we observe the same two regimes of decay of $u(1/2,t)$, as we previously did in figure 6, indicative of (i) spreading and (ii) leveling of the film.

6. Conclusion

We studied the linearized dynamics of spreading and leveling of thin films with elastic interfaces laterally confined in finite-length troughs in the absence of gravity. The governing sixth-order IBVP was solved using separation of variables and the even functions of the complete orthonormal (CON) basis previously developed in [15].

Two exemplar problems with even ICs were solved to demonstrate how initial perturbations of the film height lead to the development, spreading, and leveling of blisters on the film. In the first case, the non-smooth IC was the semi-sum of two translated delta "functions" equidistant from the origin, and in the second, the smooth IC was the sum of two quartics with finite support and vertices equidistant from the origin. In both cases, the coefficients of the spectral expansions were found exactly (analytically).

We observed that the leveling of the two different initial disturbances considered was qualitatively the same, suggesting that any Gibbs' effects in the expansion of non-smooth ICs do not affect the dynamics. Our observations also highlight the universal aspects of spreading and leveling under a linear sixth-order equation of "parabolic" type. We found that in both cases, the time-dependence of the bridge height for intermediate to large times is *exponential* and determined by the first two non-trivial modes' eigenvalues, λ_1^c and λ_2^c, of the problem. In both cases, the time dependence of the blisters' heights was also found to be exponential for intermediate to large times. However, their evolution is set by *only* λ_2^c.

Acknowledgments
ICC would like to acknowledge the US National Science Foundation, which supports his research on interfacial dynamics under grant CMMI-2029540.

Appendix
The expressions for the coefficients u_m^0 used in the expansion of the initial condition (5.4) into even eigenfunctions ψ_m^c from (3.3a) are given by:

$$u_0^0 = 1, \tag{A.1a}$$

$$u_m^0 = \frac{n_m^c}{\sqrt{d_m^c}}, \qquad \text{where} \tag{A.1b}$$

$$\begin{aligned} n_m^c = 960\Big\{ & 2\cos\tfrac{\lambda_m^c}{2}\Big[12\lambda_m^c \cos\tfrac{\lambda_m^c}{4}\cosh\sqrt{3}\lambda_m^c + (-48 + (\lambda_m^c)^2)(-\cos\lambda_m^c + \cosh\sqrt{3}\lambda_m^c)\sin\tfrac{\lambda_m^c}{4}\Big] \\ & - 6\lambda_m^c \cos\lambda_m^c \csc\tfrac{\lambda_m^c}{4}\sin\lambda_m^c \\ & +2(\cos\tfrac{\lambda_m^c}{2} + \cosh\tfrac{\sqrt{3}}{2}\lambda_m^c)\sin\lambda_m^c \times \Big[-2(24 + (\lambda_m^c)^2)\cos\tfrac{3}{8}\lambda_m^c \cosh\tfrac{\sqrt{3}}{8}\lambda_m^c + 48\sqrt{3}\sin\tfrac{3}{8}\lambda_m^c \sinh\tfrac{\sqrt{3}}{8}\lambda_m^c \\ & + 12\lambda_m^c\Big(\cosh\tfrac{\sqrt{3}}{8}\lambda_m^c \sin\tfrac{3}{8}\lambda_m^c - \sqrt{3}\sinh\tfrac{\sqrt{3}}{8}\lambda_m^c \cos\tfrac{3}{8}\lambda_m^c\Big) \\ & - 12\sqrt{3}(\lambda_m^c\cos\tfrac{\lambda_m^c}{8} + 4\sin\tfrac{\lambda_m^c}{8})\sinh\tfrac{3\sqrt{3}}{8}\lambda_m^c + 2\cosh\tfrac{3\sqrt{3}}{8}\lambda_m^c\Big((24 + (\lambda_m^c)^2)\cos\tfrac{\lambda_m^c}{8}\Big]\Big\}, \end{aligned} \tag{A.1c}$$

$$\begin{aligned} d_m^c = (\lambda_m^c)^9\Big\{ & (-6\lambda_m^c(-2 + \cos 2\lambda_m^c) + 2\lambda_m^c \cosh 2\sqrt{3}\lambda_m^c + 2\cosh^2\sqrt{3}\lambda_m^c \sin 2\lambda_m^c + \sin 4\lambda_m^c \\ & + \cosh\sqrt{3}\lambda_m^c\Big(4\lambda_m^c(-3\cos\lambda_m^c + \cos 3\lambda_m^c) + \sin\lambda_m^c - 3\sin 3\lambda_m^c\Big) \\ & + 4\sqrt{3}(\cos\lambda_m^c - \cosh\sqrt{3}\lambda_m^c)\sin^2\lambda_m^c \sinh\sqrt{3}\lambda_m^c)\Big\}, \qquad m = 1, 2, \ldots . \end{aligned} \tag{A.1d}$$

References

[1] Hosoi A E and Mahadevan L 2004 *Phys. Rev. Lett.* **93** 137802
[2] Peng G G and Lister J R 2020 *J. Fluid Mech.* **905** A30
[3] Michaut C 2011 *J. Geophys. Res.: Solid Earth* **116** B05205
[4] Bunger A P and Cruden A R 2011 *J. Geophys. Res.: Solid Earth* **116** B02203
[5] Bunger A P and Cruden A R 2011 *J. Geophys. Res.: Solid Earth* **116** B08211
[6] Kodio O, Griffiths I M and Vella D 2017 *Phys. Rev. Fluids* **2** 014202 (*Preprint* arXiv:1609.04598)
[7] Pedersen C, Niven J F, Salez T, Dalnoki-Veress K and Carlson A 2019 *Phys. Rev. Fluids* **4** 124003 (*Preprint* arXiv:1902.10470)
[8] Huang R and Suo Z 2002 *J. Appl. Phys.* **91** 1135–1142
[9] King J R 1989 *SIAM. J. Appl. Math.* **49** 1064–1080
[10] Boyko E, Eshel R, Gommed K, Gat A D and Bercovici M 2019 *J. Fluid Mech.* **862** 732–752 (*Preprint* arXiv:1703.06820)
[11] Tulchinsky A and Gat A D 2016 *J. Fluid Mech.* **800** 517–530 (*Preprint* arXiv:1512.00730)
[12] Richards J A, Hodgson D J M, O'Neill R E, DeRosa M E and Poon W C K 2024 *Proc. Natl Acad. Sci. USA* **121** e2317832121 (*Preprint* arXiv:2311.08280)
[13] Carlson A and Mahadevan L 2016 *Phys. Fluids* **28** 011702 (*Preprint* arXiv:1507.03912)
[14] Poulain S, Carlson A, Mandre S and Mahadevan L 2022 *J. Fluid Mech.* **947** A16
[15] Papanicolaou N C and Christov I C 2023 *J. Phys.: Conf. Ser.* **2675** 012016 (*Preprint* arXiv:2308.00673)
[16] Sæter T, Pedersen C, Snoeijer J H, Salez T and Carlson A 2024 *Phys. Rev. Lett.* **132** 074001 (*Preprint* arXiv:2308.01774)
[17] Longley J E, Mahadevan L and Chaudhury M K 2013 *EPL* **104** 46002 (*Preprint* arXiv:1402.0262)
[18] Pedersen C, Salez T and Carlson A 2021 *Proc. R. Soc. A* **477** 20210354 (*Preprint* arXiv:2011.10297)
[19] Papanicolaou N C and Christov I C 2024 *submitted* (*Preprint* arXiv:2402.18740)
[20] Oron A, Davis S H and Bankoff S G 1997 *Rev. Mod. Phys.* **69** 931–980
[21] Craster R V and Matar O K 2009 *Rev. Mod. Phys.* **81** 1131–1198
[22] Leal L G 2007 *Advanced Transport Phenomena: Fluid Mechanics and Convective Transport Processes* (*Cambridge Series in Chemical Engineering* vol 7) (New York, NY: Cambridge University Press)
[23] Hewitt I J, Balmforth N J and De Bruyn J R 2015 *Eur. J. Appl. Math.* **26** 1–31
[24] Elbaz S B and Gat A D 2014 *J. Fluid Mech.* **758** 221–237
[25] Marchenko V A 2014 *Sturm-Liouville Operators and Applications* Operator Theory: Advances and Applications (Birkhäuser Basel)
[26] Greenberg L and Marletta M 1998 *SIAM J. Numer. Anal.* **35** 2070–2098
[27] Greenberg L and Marletta M 2000 *J. Comput. Appl. Math.* **125** 367–383
[28] *NIST Digital Library of Mathematical Functions*, Release 1.2.1 of 2024-06-15. F. W. J. Olver, A. B. Olde Daalhuis, D. W. Lozier, B. I. Schneider, R. F. Boisvert, C. W. Clark, B. R. Miller, B. V. Saunders, H. S. Cohl, and M. A. McClain, eds. URL `https://dlmf.nist.gov/`
[29] Wolfram Research, Inc 2024 Mathematica, Version 14.0 Champaign, IL URL `https://www.wolfram.com/mathematica`

Journal of Physics: Conference Series **2910** (2024) 012033 doi:10.1088/1742-6596/2910/1/012033

On the Lagrange and Hermite Quadrature Formula

Shuxia Li[1] and Yonghong Chen*[1]

[1]School of Mathematical and Statistical Sciences, University of Texas Rio Grande Valley, Edinburg, TX 78539, USA

Abstract

This paper builds on the error analysis method for Newton-Cotes quadrature formulas developed by D. R. Hayes and L. Rubin in 1970, which utilizes Lagrange interpolation polynomials. By adopting and extending their approach, this work derives the error estimate for Hermite interpolation quadrature. Specifically, we construct a polynomial $P(x)$ analogous to the scaling function $A(x)$ used by Hayes and Rubin, and prove its non-negativity over the interval. This allows us to establish a precise error formula for Hermite interpolation quadrature. The results provide a novel application of Hayes and Rubin's methodology, offering new insights into error analysis for high-order interpolation quadrature formulas.

Classification: Primary: 65D30, 41A55; Secondary: 65G20
Keywords: Numerical analysis, Interpolation, Quadrature formulas

1 Introduction

In 1970, D. R. Hayes and L. Rubin presented a rigorous analysis of the error associated with Newton-Cotes quadrature formulas derived from Lagrange interpolation [1]. They began by addressing the cases where the number of nodes n is either even or odd. For the equally spaced nodes $a = x_0 < \cdots < x_n = b$, we define

$$k = \begin{cases} n/2, & n \text{ is even}, \\ (n-1)/2, & n \text{ is odd}. \end{cases} \tag{1}$$

Central to their approach is the definition of the scaling function:

$$A(x) = \prod_{i=0}^{2k}(x - x_i), \tag{2}$$

*(Corresponding author)Email: yonghong.chen01@utrgv.edu
[1]Email: shuxiali@utrgv.edu

Content from this work may be used under the terms of the Creative Commons Attribution 4.0 licence. Any further distribution of this work must maintain attribution to the author(s) and the title of the work, journal citation and DOI.
Published under licence by IOP Publishing Ltd

Note that x_n does not necessarily equal x_{2k}; this depends on whether n is odd or even. For a positive even number and its subsequent odd number, the value of k is the same, and therefore, the scaling function $A(x)$ is also the same. The function $A(x)$ captures the structure of the interpolation nodes and plays a pivotal role in their error analysis.

Hayes and Rubin then introduced the function:

$$g(x) = \frac{f(x) - Q(x)}{A(x)}, \quad x \in [a, b], \quad \text{except } x = x_0, \cdots, x_n, \tag{3}$$

where $f(x)$ is the target function, and $Q(x)$ is the Lagrange interpolation polynomial. It was shown that $g(x)$ has removable discontinuity at $x = x_0, \cdots, x_n$.

To estimate the error, Hayes and Rubin performed integration by parts on the integral $\int_a^b g(x)A(x)\,dx$, leading to the expression:

$$\int_a^b g(x)A(x)\,\mathrm{d}x = g(x)A^*(x)\Big|_a^b - \int_a^b g'(x)A^*(x)\,\mathrm{d}x, \tag{4}$$

where the auxiliary function $A^*(t)$ is defined as:

$$A^*(x) = \int_a^x A(t)\,\mathrm{d}t. \tag{5}$$

This function $A^*(t)$ is non-negative over the interval, which is crucial for the subsequent error estimation.

Next, Hayes and Rubin applied the mean value theorem to the integral $\int_a^b g'(x)A^*(x)\,dx$, yielding:

$$\int_a^b g'(x)A^*(x)\,\mathrm{d}x = g'(y)\int_a^b A^*(x)\,\mathrm{d}x, \tag{6}$$

where y is some point in the interval $[a, b]$.

A key result in their analysis is the observation that the boundary terms $g(b)A^*(b)$ and $g(a)A^*(a)$ are always zero. This simplification allows the error formula to be expressed as:

$$\int_a^b f(x)\mathrm{d}x - \int_a^b Q(x)\mathrm{d}x = -\frac{f^{(2k+2)}(\xi)}{(2k+2)!}\int_a^b A^*(x)\,\mathrm{d}x. \tag{7}$$

This result highlights the dependence of the error on the derivative of $g(x)$ at some point y within the interval, as well as the integral of the scaling function $A^*(x)$. Hayes and Rubin's work provides a fundamental framework for understanding the error in quadrature formulas derived from Lagrange interpolation, and it serves as a foundation for extending these concepts to other types of interpolation, such as Hermite interpolation.

2 Error Analysis in Hermite Quadrature Formulas

Inspired by a result from Walter Rudin [2], we consider the following theorem. This result reveals that if a function f has sufficiently high order of differentiability, it is possible to derive a Hermite quadrature formula with error terms. The method for this derivation closely follows the proof technique presented by Hayes and Rubin. The theorem establishes a foundational relationship that enables the construction of such quadrature formulas.

Theorem 1. *Let $S(\lambda) = (\lambda - \alpha_1)^{m_1} \cdots (\lambda - \alpha_s)^{m_s}$ be a polynomial of degree $n = m_1 + \cdots + m_s$, and let Ω be an open set in $\mathbb{C}$ that contains the zeros $\alpha_1, \ldots, \alpha_s$ of S.*

Then, for every $f \in H(\Omega)$, there exists a polynomial R of degree less than n and a function $z \in H(\Omega)$ such that

$$f(\lambda) - R(\lambda) = S(\lambda)z(\lambda) \quad (\lambda \in \Omega).$$

This theorem implies that, given f has sufficient differentiability, we can derive a Hermite quadrature formula with error terms by employing a technique similar to that of Hayes and Rubin. Building upon the work of Hayes and Rubin, we extend their method to analyze the error in Hermite Quadrature Formulas. For a sufficiently differentiable function $f(x)$, we not only require the Hermite polynomial $H(x)$ to satisfy the interpolation conditions at the nodes x_i:

$$H(x_i) = f(x_i), \quad i = 0, 1, \cdots, n, \tag{8}$$

but also to match the function $f(x)$ and its derivatives up to order m at these nodes:

$$H^{(l)}(x_i) = f^{(l)}(x_i), \quad i = 0, 1, \cdots, n, \; l = 0, 1, \cdots, m. \tag{9}$$

It is established that the Hermite interpolation polynomial is unique and can be expressed in the form [3]

$$H(x) = \sum_{i=0}^{n} \sum_{l=1}^{m} f^{(l)}(x_i) L_{il}(x), \tag{10}$$

where

$$L_{ij}(x) = l_{ij}(x) - \sum_{l=j+1}^{m} l_{ij}^{(l)}(x_i) L_{il}(x), \tag{11}$$

and

$$l_{ij}(x) = \frac{(x - x_i)^j}{j!} \prod_{\substack{l=0 \\ l \neq i}}^{n} \left(\frac{x - x_l}{x_i - x_l} \right)^{m+1}. \tag{12}$$

For equally spaced nodes x_i, we define the polynomial

$$P(x) = \prod_{i=0}^{2k} (x - x_i)^{m+1}, \tag{13}$$

where k is defined as in (1). This polynomial $P(x)$ accounts for the polynomial factors and order $m + 1$ at each node x_i.

Similarly, we define

$$h(x) = \frac{f(x) - H(x)}{P(x)}. \tag{14}$$

We then construct the function

$$F(t) = f(t) - H(t) - P(t)[g(x) + (t - x)g'(x)]. \tag{15}$$

Applying Rolle's Theorem, we can prove the following result:

Lemma 2. *For each $x \in [a, b]$, there exists a number ξ with $a < \xi < b$ such that*

$$h'(x) = \frac{f^{(K)}(\xi)}{K!}, \tag{16}$$

where $K = (2k + 1)(m + 1) + 1$.

This theorem shows that for each point x in the interval $[a, b]$, there exists a point ξ such that the derivative $g'(x)$ can be expressed in terms of the K-th derivative of the function f. This result provides a crucial link between $g'(x)$ and the higher-order derivatives of f, supporting our error analysis for Hermite Quadrature Formulas.

Continuing from Hayes and Rubin's work, we aim to show that the polynomial $P^*(x)$ possesses similar properties to $A^*(x)$ as defined by Hayes and Rubin. Specifically, we seek to demonstrate that $P^*(x) \geq 0$ over the interval $[a, b]$.

Given that the equally spaced nodes $x_0, x_1, \ldots, x_{2k}$ of the polynomial $P(x)$ are odd in number, our analysis follows a parallel approach to that of Hayes and Rubin. However, in this case, the linear factor $x - x_i$ is replaced by $(x - x_i)^m$. Thus, we establish the following theorem:

Lemma 3. *The integral of the scaling function satisfies*

$$P^*(x) = \int_a^x P(t)\,\mathrm{d}t \geq 0 \tag{17}$$

for all $t \in [a, b]$.

With the above lemma established, we can now proceed to prove our main result:

Theorem 4. *If $f \in C^K[a, b]$, and H is the corresponding Hermite interpolation polynomial, then the quadrature error is*

$$\int_a^b f(x)\,\mathrm{d}x - \int_a^b H(x)\,\mathrm{d}x = -\frac{f^{(K)}(\xi)}{K!} \int_a^b P^*(x)\,\mathrm{d}x, \tag{18}$$

where $K = (2k + 1)(m + 1) + 1$.

Proof. With the definition of $h(x)$, the following relation holds

$$f(x) - H(x) = h(x)P(x). \tag{19}$$

To estimate the error, we perform integration by parts on the integral $\int_a^b h(x)P(x)\,dx$, leading to the expression:

$$\int_a^b h(x)P(x)\,\mathrm{d}x = h(x)P^*(x)\Big|_a^b - \int_a^b h'(x)P^*(x)\,\mathrm{d}x, \tag{20}$$

The function $P^*(t)$ is non-negative over the interval by Lemma 3.

Next, applying the mean value theorem to the integral $\int_a^b h'(x)P^*(x)\,\mathrm{d}x$ yields:

$$\int_a^b h'(x)P^*(x)\,\mathrm{d}x = h'(y)\int_a^b P^*(x)\,\mathrm{d}x, \tag{21}$$

where y is some point in the interval $[a,b]$.

By definition, either $h(b)$ or $P^*(b)$ is always equal to 0, while $P^*(a)$ is always equal to 0. Therefore, the boundary terms $h(b)P^*(b)$ and $h(a)P^*(a)$ are always zero. It follows that the error formula can be expressed as:

$$\int_a^b f(x)\mathrm{d}x - \int_a^b H(x)\mathrm{d}x = -h'(y)\int_a^b P^*(x)\,\mathrm{d}x. \tag{22}$$

Thus the theorem follows by Lemma 2. □

Acknowledgements

We would like to express our gratitude to the organizers of the Sixteenth Conference of the Euro-American Consortium for Promoting the Application of Mathematics in Technical and Natural Sciences for their kind invitation to present this work. The opportunity to participate in such a prestigious event has been invaluable, and we appreciate the effort and dedication of the conference committee in organizing this conference.

References

[1] D. Hayes and L. Rubin, "A proof of the newton-cotes quadrature formulas with error term," *The American Mathematical Monthly*, vol. 77, no. 10, pp. 1065–1072, 1970.

[2] W. Rudin, *Functional analysis.* International Series in Pure and Applied Mathematics, McGraw-Hill, Inc., New York, second ed., 1991.

[3] P. J. Davis, *Interpolation and approximation.* Courier Corporation, 1975.

[4] A. Quarteroni, R. Sacco, and F. Saleri, *Numerical mathematics*, vol. 37. Springer Science & Business Media, 2006.

Journal of Physics: Conference Series **2910** (2024) 012034 doi:10.1088/1742-6596/2910/1/012034

Effective signal denoising with adaptive polynomials

Byulent Idirizov [1,2]

[1] Department of Informational Modelling, Institute of Mathematics and Informatics, Bulgarian Academy of Sciences, 8 Acad. Georgie Bonchev Str., 1113 Sofia, Bulgaria
[2] Department of Mathematics, Angel Kanchev University of Ruse, 8 Studentska Str., 7004 Ruse, Bulgaria

*e-mail: [1] bidirizov@math.bas.bg, [2] bidirizov@uni-ruse.bg

Abstract. The study presents an informatic model for signal processing and noise reduction, harnessing the assistance of mathematics. Advanced noise reduction algorithms are applied to improve signal quality. A *MATLAB* program was developed integrating algorithms designed to reduce noise regardless of its parameter. Employment of adaptive polynomials was also utilized to enhance the reduction of noise interference. All algorithms within the model have been thoroughly automated, ensuring seamless execution. The results demonstrate significant improvements in signal clarity, with filtered signals closely matching the expected noiseless signals. One of the model's aims was to achieve a significant reduction in noise emitted by rapidly moving objects, an aim that was effectively achieved. The model is applicable in various fields requiring precise signal analysis, demonstrating robust noise reduction capabilities. As an example, it could encompass security-oriented applications such as radar systems and communication networks.

1. Introduction and model description

Removing noise from a signal has always been a significant challenge for engineers and researchers. The intricate process of isolating and eliminating unwanted noise while preserving the integrity of the original signal is crucial in various fields, including telecommunications, audio processing, biomedical engineering and security. Effective noise reduction techniques are essential for enhancing the clarity and accuracy of the transmitted information. The noise problem becomes even more complicated when the signal is transmitted from a high-speed moving source [6, 10].

To develop the model, a scenario was considered in which a signal is received from a fast-moving aircraft with a significant low target echo (Figure 1) [2]. The aim of the algorithms in the model is to significantly reduce the received noise, thereby maximizing the clarity of the received signal through mathematical techniques. In the absence of real data, the research employs computer-simulated data. In the example of the model, the receiver of the signal is regarded as a monostatic radar system.

Content from this work may be used under the terms of the Creative Commons Attribution 4.0 licence. Any further distribution of this work must maintain attribution to the author(s) and the title of the work, journal citation and DOI.
Published under licence by IOP Publishing Ltd

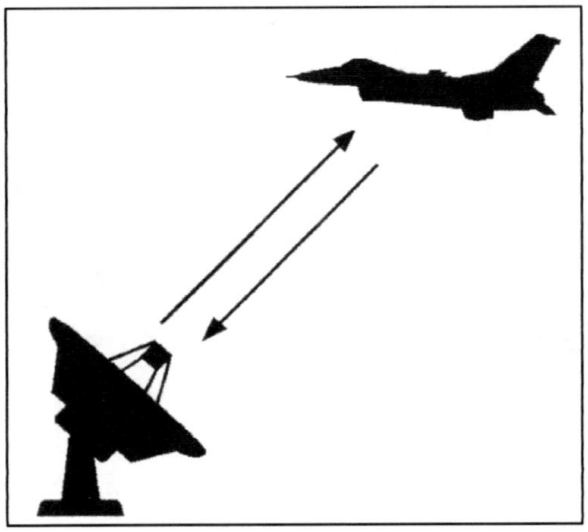

Figure 1. Radar system detecting an aircraft.

Figure 2 illustrates the clear target echo, noise, and the received signal, which is a function of the previous two. It is evident that the target echo could be effectively hidden by the noise, making the object undetectable. One of the reasons for this is the low amplitude of the target echo, which is effectively masked by the noise [1]. Considering that some aircrafts can produce noise pollution to mask the target echo and make it difficult to distinguish the target from the background noise, then the task becomes even more challenging. A comparative example between the target echo and the resulting signal is shown in Figure 3. Additional complicating factors may include the movement of the object, its shape and others [8].

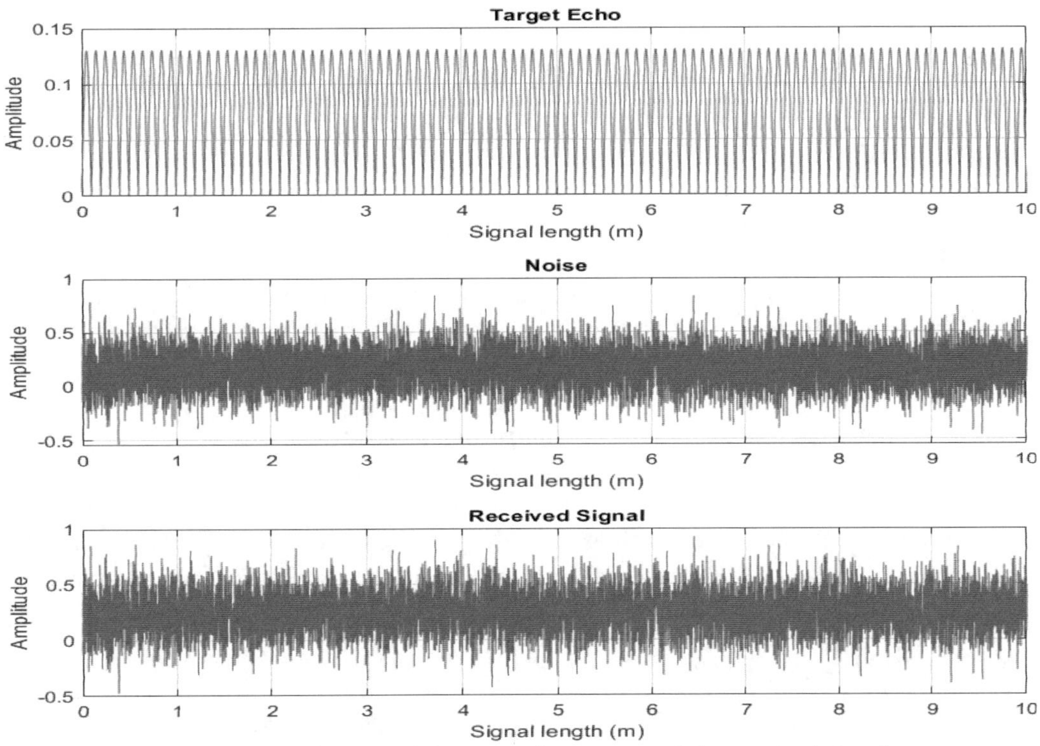

Figure 2. Illustration of the target echo, noise, and the received signal.

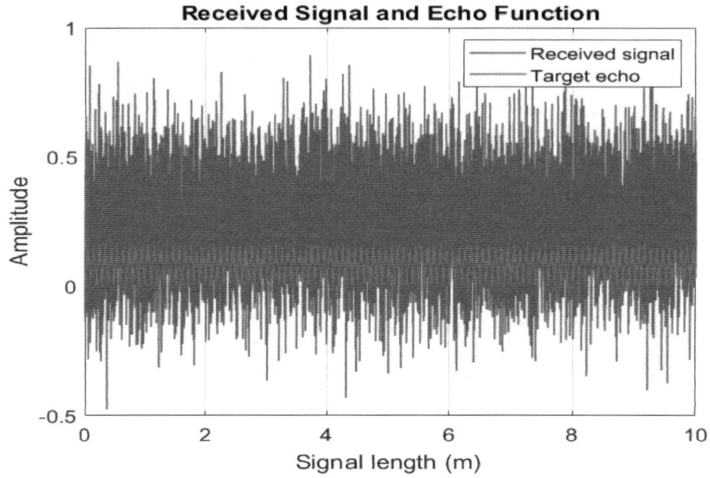

Figure 3. Comparison between target echo and received signal.

2. Solution Approach

2.1 An overview of the received signal

Every signal noise, whether of natural or artificial origin, has an average value over a given observation period [7]. The noise function is denoted by N. The algorithms will provide an approximate estimate of $E(N)$, which will be denoted as $\widehat{E(N)}$. If a transmitted signal s has a given length L_s and amplitude λ, then by generating a series of signals s_n, such that a similar or uniform reflection of the signal from the object is obtained, and knowing $\widehat{E(N)}$, the noise could be approximately eliminated from the signal. Naturally, the quality of this noise reduction depends on the number of (identical) signals in the generated series n.

The algorithm generates a non-negative sinusoidal signals and segments it into multiple parts, organizing them into a matrix. Incorporating the object's velocity v_o, results in additional modifications of the signal, particularly in its frequency. These changes must be considered during its analysis and processing.

As previously mentioned, the received signal R_s is a function of the target echo H and the noise N. In our model, this is represented as a sum of both and is simulated using the following formula:

(1) $$R_s = H + N.$$

The noise function N is defined by the following formula:

(2) $$N = N_c * (\sigma_N * U_1(x) + \mu_N),$$

where N_c is the constant scaling factor, σ_N is the volatility of noise and μ_N is the expected value of noise. The variable x is a range of values starting from x_{step}, with increments of x_{step}, up to L_s. Here x_{step} is the step of variable x and $x_{step} = \frac{L_s}{Q}$, where Q is the number of empirical data points employed for one signal measurement. The value of Q must be even and must not be less than s_n. The range x can be expressed as:

$$x = \left\{x_{step}, 2x_{step}, 3x_{step}, \dots, kx_{step} | k = \left\lfloor \frac{nL_s}{x_{step}} \right\rfloor \right\},$$

where $\lfloor . \rfloor$ is the round down function, which rounds down to the nearest integer, and nL_s is the length of the generated series of signals. The number of elements in the vector x is defined as v and can be expressed as: $v = \left[1, \left\lceil \frac{nL_s}{x_{step}} \right\rceil\right]$, where 1 is the number of rows, and $\left\lceil \frac{nL_s}{x_{step}} \right\rceil$ the number of columns, where $\lceil \cdot \rceil$ is the round up function. The number of signals n in the series of signals is obtained as $n = \frac{nL_s}{L_s}$. The values that are be used in our subsequent experiments and are as follows: $\mu_N = 1.50,\ \sigma_N = 0.35,\ N_c = 0.20,\ Q = 100, L_s = 0.10\ meters,\ nL_s = 10\ meters$ and $\lambda = 1.5$.

The function U generates a normally distributed random noise matrix with dimensions matching those of x. U generates random numbers from a standard normal distribution, which has a mean of 0 and a variance of 1. If we denote the standard normal distribution by $N(0,1)$, then U can be expressed as:

$$U \sim N(0,1).$$

When used to generate a matrix of random numbers with dimensions specified by another matrix x, it creates a matrix U where each element $u_{i,j}$ is independently derived from the standard normal distribution $N(0,1)$.

Thus, if x is an $m \times n$ matrix, then U can be written as:

$$U = \begin{pmatrix} u_{1,1} & u_{1,2} & \cdots & u_{1,n} \\ u_{2,1} & u_{2,1} & \cdots & u_{2,n} \\ \vdots & \vdots & \ddots & \vdots \\ u_{m,1} & u_{m,1} & \cdots & u_{m,n} \end{pmatrix}, \qquad where\ U \sim N(0,1)^{m \times n}.$$

Briefly, U generates an $m \times n$ matrix of independent and identically distributed random variables, each following a standard normal distribution. The received target echo is defined by the following formula:

(3) $$H = (1 + CV_t U_2)\frac{\sqrt{\sigma_{RCS}}}{2\pi} s_t, \qquad U_2 \neq U_1$$

CV_t represents the target amplitude variation due to target dynamics, σ_{RCS} represents the radar cross section of the target and s_T is the clear reflection of the transmitted signal s, which in the current example is expressed as $s = \lambda \left|sin\left(\frac{\pi x}{L_s}\right)\right|$. s_T could be described with the formula:

(4) $$s_T = \lambda \left|sin\left(\left(\frac{\pi x}{L_s}\right)^{\gamma_3}\right)\right|,$$

where

$$\gamma_3 = \begin{cases} \gamma_2 + \left(\frac{|1-\gamma_2|}{Q_n - 1}\right) * (j-1), & for\ j = 1,2,\dots,Q_n, \quad if\ \gamma_2 \le 1, \\ \gamma_2 + \left(\frac{|1-\gamma_2|}{Q_n - 1}\right) * (Q_n - (j-1)), & for\ j = 1,2,\dots,Q_n, \quad if\ \gamma_2 > 1, \end{cases}$$

and

$$\gamma_2 = \begin{cases} \gamma_1 & if the\ object\ is\ approaching\ the\ radar; \\ 1/\gamma_1 & if\ the\ object\ is\ moving\ away\ from\ the\ radar, \end{cases}$$

where

$$\gamma_1 = 1 + \left(\frac{v_o n L_s}{v_l}\right).$$

In the above formulas Q_n is the total empirical data points employed for measurement of series of signals and is obtained as $Q_n = n * Q$ and v_l is the speed of light. Some of the target values that will be used in our subsequent experiments are as follows: $\sigma_{RCS} = 0.1, CV_t = 0.09$ and $v_o = 3430\ m/s$.

There are several conditions that must be met for the current informatic model to function flawlessly, specifically:

I. $\lfloor n \rfloor = \lceil n \rceil \Leftrightarrow n \in \mathbb{Z}$; nL_s must be integer.

II. $Q \leq n, \quad Q = 2k, \; k \in \mathbb{Z}$; Q must be integer and even

III. $\frac{nL_s}{x_{step}} - Q_n = 0$

IV. $\left|\frac{d_o}{L_s}\right| \leq tol$, where d_o is the distance, which the object has moved and $d_o = \gamma_1 - 1$, tol is the tolerance, we use $tol = 0.01$.

The final goal of our task is to clean the reflected from the object signal by removing the effects of its movement and the impact of some non-Gaussian noise. This requires using signal processing methods to reduce distortions and improve the data's accuracy.

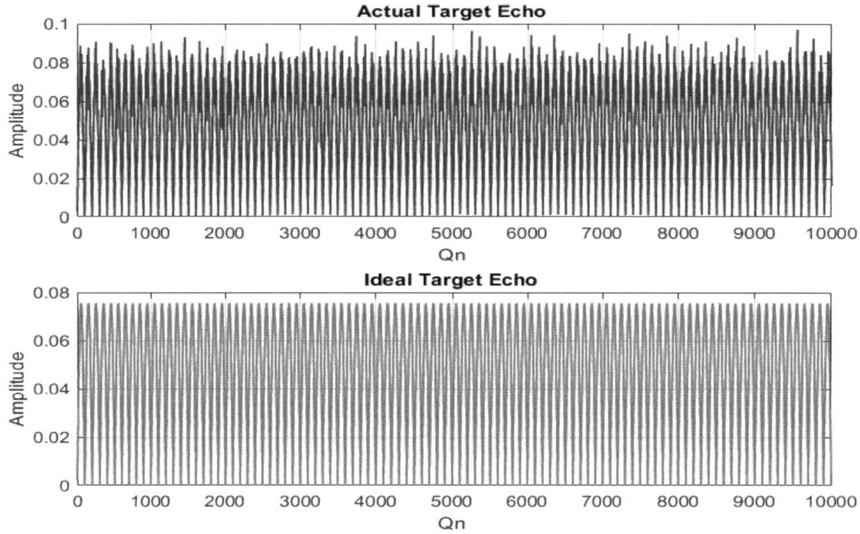

Figure 4. Illustration of the real target echo and ideal target echo.

As shown in Figure 4, there is a difference between the actual echo reflected from the object and the ideal version of the target echo reflection. The purpose of our model is to eliminate the noise and irregularities from the signal to such an extent that it approximates the ideal target echo. In some cases, the shape of the reflected signal may differ from that of the transmitted signal, but this does not pose a significant issue for the model.

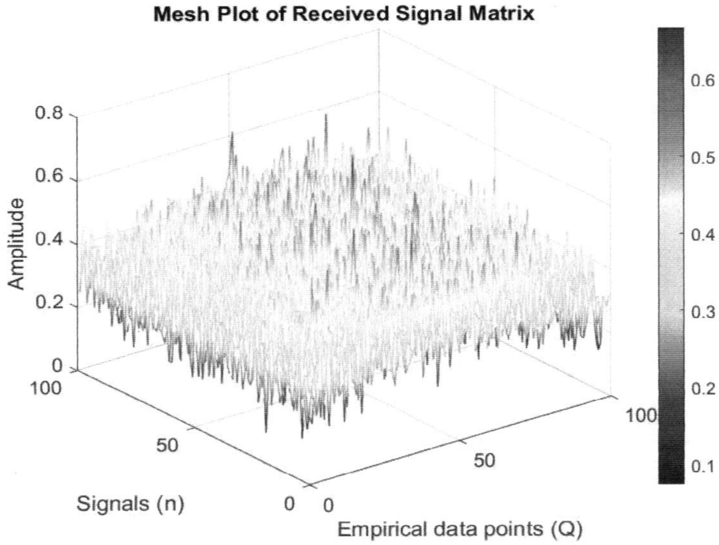

Figure 5. Mesh plot of received signal matrix.

Figure 5 show the matrix of the received signal which will be used in the algorithms. Another purpose of the model is to mitigate the effect of the object's movement, which alters the frequency of the received signal [3]. This effect is clearly demonstrated in Figure 6.

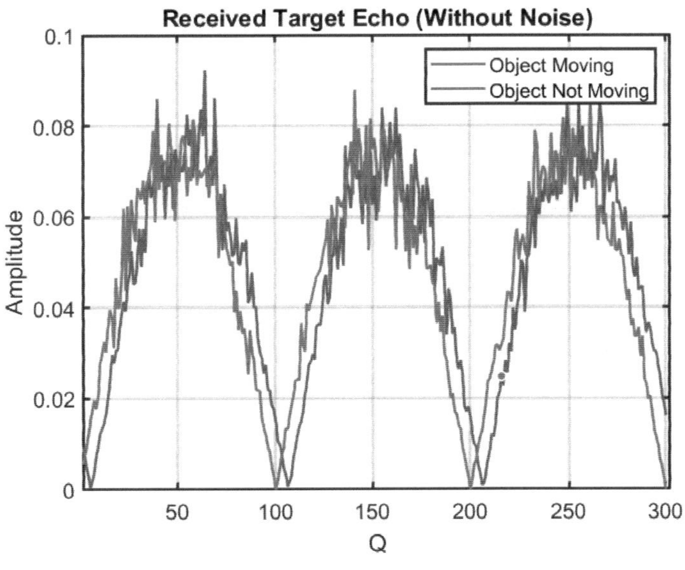

Figure 6. Comparison of Target Echo - Moving Object (blue) vs. Stationary Object (red)

2.2 Algorithms for effective signal denoising

Firstly, the received signal matrix MRs, which is shown in Figure 5 will be obtained as:

$$MRs = \begin{bmatrix} R_s[(1+(1-1)Q):(1Q)] \\ R_s[(1+(2-1)Q):(2Q)] \\ \vdots \\ R_s[(1+(n-1)Q):(Q_n)] \end{bmatrix}. \quad (5)$$

Each row of MRs is obtained by extracting a part of R_s corresponding to the indices from $(1+(i-1)*Q)$ to $(i*Q)$ for $i = 1,2,\ldots,n$. The first assessment of the target echo vector $\widehat{H}_1$ is derived from the formula:

$$\widehat{H}_1(j) = \frac{\left(\sum_{i=1}^{n} MRs_{ij}\right)}{n}, \quad j = \overline{1,Q}. \quad (6)$$

The aim of the model is to reduce noise, regardless of its expected value. For this purpose, there will be a second assessment of the target echo vector $\widehat{H}_2$ (Figure 7 a), with the formula:

$$\widehat{H}_2 = \begin{cases} \widehat{H}_1 - \min(\widehat{H}_1), & \min(\mathrm{s_n}) > 0 \\ \widehat{H}_1 - \frac{1}{2}(\max(\widehat{H}_1) + \min(\widehat{H}_1)), & \min(\mathrm{s_n}) \le 0 \end{cases} \quad (7)$$

where

$$\widehat{E(N)} = (\widehat{H}_1 - \widehat{H}_2).$$

The average simulated noise value is $E(N) \cong 0.29965$ and the estimated average noise value is $\widehat{E(N)} \cong 0.30086$, where the error is 0.00121. => $\widehat{E(N)} \approx E(N) \approx N_c * \mu_N$. If the sum of the elements in the right half of $\widehat{H}_2(j)$ for $j = \overline{1,Q}$. is greater than the sum of the elements in the left half, it follows that the object is approaching the radar. Conversely, if the sum is greater in the left half, it follows that the object is moving away, as follows:

- If $\sum_{j=\left(\frac{Q}{2}+1\right)}^{Q} \widehat{H}_2(j) > \sum_{j=1}^{Q/2} \widehat{H}_2(j)$, the object is approaching the radar.
- If $\sum_{j=1}^{Q/2} \widehat{H}_2(j) < \sum_{j=\left(\frac{Q}{2}+1\right)}^{Q} \widehat{H}_2(j)$, the object is moving away from the radar.

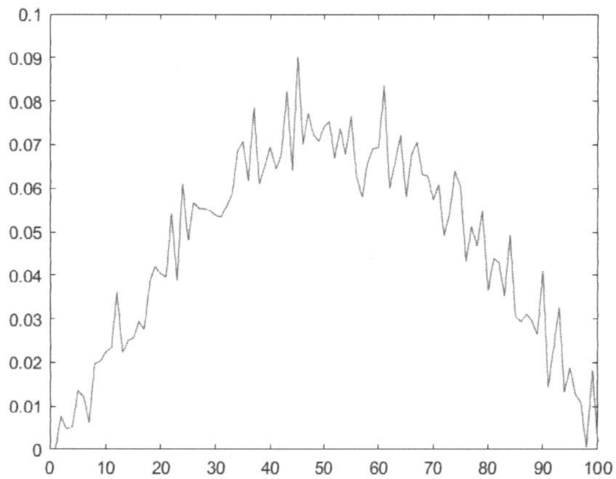

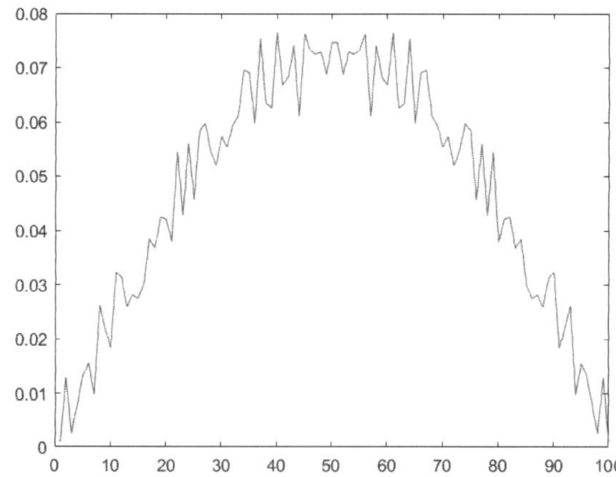

Figure 7 a. $\widehat{H}_2(j), \quad j = \overline{1,Q},$ **7 b.** $\widehat{H}_3(j), \quad j = \overline{1,Q}.$

The effect of the object's movement could sometimes pose a challenge in denoising processes[4]. When s is symmetric with respect to its central element, then formula 8 could be applied to reduce the effect of the object's movement, which helps to obtain a symmetric mirrored function $\widehat{H}_3(j),\ \ j = \overline{1,Q}$, with respect to its central element (Figure 7 b).

(8) $$\widehat{H}_3(j) = \frac{1}{2}\left(\widehat{H}_2(j) + \widehat{H}_2(Q - j + 1)\right), \quad j = \overline{1,Q}$$

Must be added that $\widehat{H}_3(j)$, where $j = \overline{1,Q_n}$ is a repeated application of $\widehat{H}_3(j)$, where $j = \overline{1,Q}$, repeated n times.

2.3 Adaptive polynomials

The next step in the algorithm involves applying of **adaptive polynomials.** They provide **local polynomial approximations** and adapt to changing signal characteristics. This technique, part of adaptive signal processing, adjusts polynomial coefficients in response to the data, enhancing the model's ability to reduce noise and handle variability in the signal, such as those caused by object movement or other conditions. **Adaptive polynomials** are particularly effective in managing dynamic environments and improving the accuracy of signal reconstruction [5, 9]. **Adaptive polynomials** are applied to $\widehat{H}_3(j)$, where $j = \overline{1,Q}$. $\widehat{H}_3(j),\ \ j = \overline{1,Q}$ is used instead of $\widehat{H}_3(j),\ \ j = \overline{1,Q_n}$, because the computation time is significantly reduced without a substantial difference in the resulting output. The model was calibrated with *MATLAB*. This is an iterative algorithm, and several iterations are made to achieve a good result. P_o is the order of the polynomials that will be applied, k is the number of abovementioned iterations. P_s is an even constant that is defined in the following rows.

After each iteration k, the values of the vector $\widehat{H}_4(j,k)$ are be obtained, where $\widehat{H}_4(j,k = 1) = \widehat{H}_3(j), j = \overline{1,Q}$. In brief, the model fits a polynomial curve P to a specified odd number $(P_s + 1)$ of empirical values, using their central element for the vector $\widehat{H}_4(j,k)$. For the outermost $\frac{P_s}{2}$ empirical values at both ends of the vector, $\overline{1,\frac{P_s}{2}}$ and $\overline{Q - \frac{P_s}{2},Q}$ values from the fitted polynomial are utilized. Other consecutive empirical values of $\widehat{H}_3(j), j = \overline{1,Q_n}$ might also be used as long as their number is equal to Q. Therefore, the vector $\widehat{H}_4(j,k+1)$ is obtained by:

(9) $$\widehat{H}_4(j,k+1) = \left[\left[\widehat{H}_4(j,k), j = \overline{1,\frac{P_s}{2}}\right], \quad [V_H], \quad \left[\widehat{H}_4(j,k), j = \overline{Q - \frac{P_s}{2},Q}\right]\right],$$

where

(10) $$[V_H] = \left[\mathrm{C}\left[P\left(\widehat{H}_4(j,k), j = \overline{1,P_s+1}\right)\right], \quad \mathrm{C}\left[P\left(\widehat{H}_4(j,k), j = \overline{2,P_s+2}\right)\right], \dots, \mathrm{C}\left[P\left(\widehat{H}_4(j,k), j = \overline{Q - P_s - 1,Q}\right)\right]\right].$$

Here $\mathrm{C}\left[P\left(\widehat{H}_4(j,k),\ j = \overline{1,P_s+1}\right)\right]$ is the central empiric element C of a polynomial vector P calculated by the elements of $\widehat{H}_4(j,k),\ j = \overline{1,P_s+1}$. The values of the parameters that were used are: $P_o = 4$ $(P_o \geq 3)$, $P_s = 8$ (P_s must be even number and $P_s > P_o$), $k \geq 1$. Calculations with different values of k are made. As k increases, $\widehat{H}_4(j,k)$ approaches the ideal target echo more closely. It should be noted, that k should not be extremely large, as this slows down computation time and leads to anomalies in the obtained result. Determining the optimal value of k is a topic for future researches.

Journal of Physics: Conference Series **2910** (2024) 012034 doi:10.1088/1742-6596/2910/1/012034

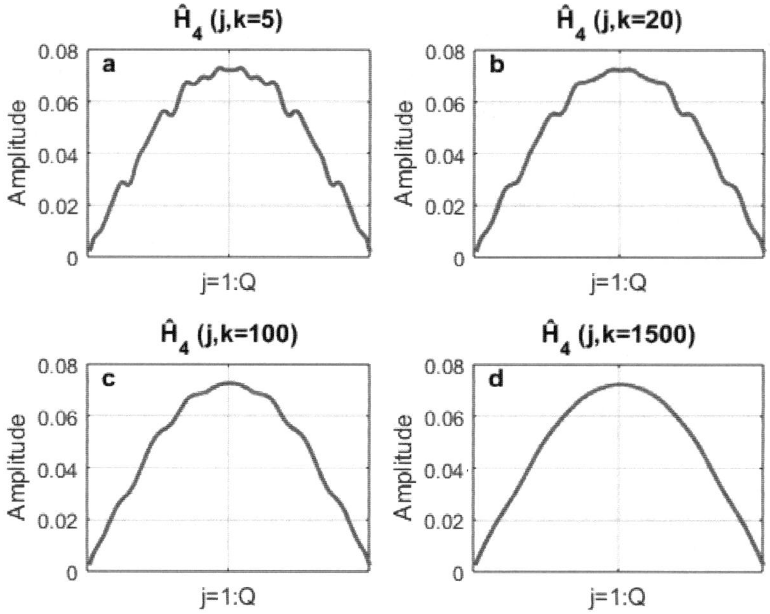

Figure 8 a. $\widehat{H}_4(j, k = 5)$, **b.** $\widehat{H}_4(j, k = 20)$ **c.** $\widehat{H}_4(j, k = 100)$ **d.** $\widehat{H}_4(j, k = 1500)$

Figure 8 visually demonstrates the effects of applying **adaptive polynomials** for further noise reduction and signal smoothing. The effect of applying **adaptive polynomials** is clearly evident in Figure 9, where a significant improvement in denoising and the similarity of the result to the Ideal Target Echo could be observed. The mean square error of $\widehat{H}_4(j, k = 1500)$ from the simulation, compared to the Ideal Target Echo is 4.744054×10^{-6}. This error metric provides an indication of the deviation of the estimated values from the ideal target, thereby assessing the accuracy and performance of the estimation process. The relatively low mean square error value suggests that the estimation method is highly accurate and closely approximates the ideal target.

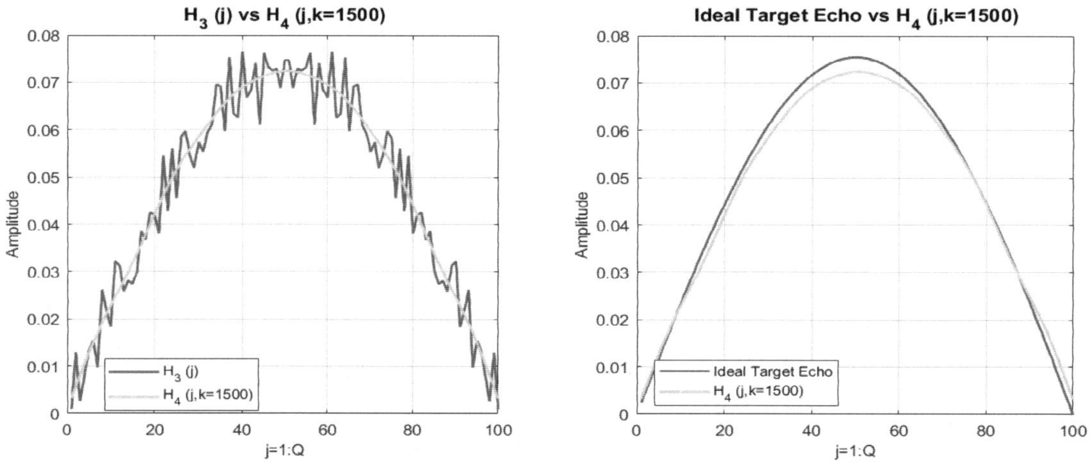

Figure 9. Comparison **a.** $\widehat{H}_3(j)$ vs $\widehat{H}_4(j, k = 1500)$, **b.** Ideal Target Echo vs $\widehat{H}_4(j, k = 1500)$

Journal of Physics: Conference Series **2910** (2024) 012034 doi:10.1088/1742-6596/2910/1/012034

3. Conclusion

In conclusion, the study successfully developed a model for signal denoising using adaptive polynomials. By integrating fully automated and advanced noise reduction algorithms within a *MATLAB* framework, the model enhances signal clarity, demonstrating its suitability for applications requiring precise signal analysis, such as radar systems and communication networks. The iterative application of polynomial fitting techniques effectively minimizes noise interference, as evidenced by the close match between the filtered and expected noiseless signals. The low mean square error indicates a high level of accuracy in the estimation process, confirming the model's ability to approximate the ideal target echo with precision. Future research should focus on optimizing the algorithm to reduce computation time and address potential anomalies. Additionally, exploring optimal parameter values could further improve performance. Overall, this research provides valuable algorithms for effective noise reduction in signal processing, contributing to enhanced accuracy in various technological applications.

4. Acknowledgements

Byulent Idirizov is supported by the project FNSE–03 "Study of mathematical and didactic models with analytical and numerical methods," funded by the "Scientific Research" Fund of the University of Ruse.

References:

[1] Chen, X.; Chen, B. Anti-jamming Approach Based on Radar Transmitted Waveform Matching. EURASIP J. Adv. Signal Process. 2021, 59 (2021). https://doi.org/10.1186/s13634-021-00776-w.

[2] Garvanov, I.; Kabakchiev, H.; Behar, V.; Vladimirova, M.; Dimitrov, K.; Vladimirov, S. Detection of Low-Flying Aircraft by GPS Signals. In Proceedings of the First Scientific Conference with International Participation "Single European Airspace - Nature and Challenges for the Air Sovereignty and Businesses", Sofia, Bulgaria, 11–12 October 2016; National Defense College "G. S. Rakovsky": Sofia, Bulgaria.

[3] Islam, M.S.; Chong, U. Noise Reduction of Continuous Wave Radar and Pulse Radar Using Matched Filter and Wavelets. J. Image Video Proc. 2014, 2014, 43. https://doi.org/10.1186/1687-5281-2014-43.

[4] Jiang, L.; Wang, Z.; Wang, C.; Li, M.; Leng, J.; Wu, X. ConsistencyDet: A Robust Object Detector with a Denoising Paradigm of Consistency Model. arXiv 2024. Available online: https://arxiv.org/abs/2404.07773 (accessed on 16 June 2024).

[5] Katkovnik, V.; Egiazarian, K.; Astola, J. Local Approximation Techniques in Signal and Image Processing; SPIE Press: Bellingham, WA, USA, 2006; Volume PM157, ISBN 0-8194-6092-3.

[6] Kirch, N.; Zhu, N. A Discourse on the Effectiveness of Digital Filters at Removing Noise from Audio. J. Acoust. Soc. Am. https://doi.org/10.1121/1.4950680.

[7] Nenov, G.D. Theory of Signals; Sofia, Bulgaria, 1990.

[8] Niu, R.; Zhi, X.; Jiang, S.; Gong, J.; Zhang, W.; Yu, L. Aircraft Target Detection in Low Signal-to-Noise Ratio Visible Remote Sensing Images. Remote Sens. 2023, 15, 1971. https://doi.org/10.3390/rs15081971.

[9] Tretter, S.A. Communication System Design Using DSP Algorithms; University of Maryland: College Park, MD, USA.

[10] Emerging Trends in EEG Signal Processing: A Systematic Review. Springer. Available online: https://link.springer.com/chapter/10.1007/978-3-030-54932-9_2.

Object-Oriented Zero-Information-Optimised Product of Monte Carlo Integrations

T Plachetka, D Bernát, P Bokes

Faculty of Mathematics, Physics and Informatics, Comenius University, Bratislava, Slovakia

E-mail: {tomas.plachetka, dusan.bernat, pavol.bokes}@fmph.uniba.sk

Abstract. We propose an object-oriented Monte Carlo framework for computation of definite integrals with a primary focus on generality and simplicity, improving in find points existing software libraries. We introduce an optimised sampling method for a product of Monte Carlo integrations. The method does not require any a-priori knowledge of the integrands or the integration domains, and can be combined with other optimisation methods. Using our framework, we implement a naive uniform sampling method and the optimised one, and empirically compare them on the Monte Carlo computation of volumes of N-dimensional balls. The experiments confirm our theoretical expectations.

1 Introduction

Consider a numerical computation of a definite integral

$$I = \int_{\Omega} f(x)\,dx \tag{1}$$

Throughout this paper, we will assume that the integrand f is a real function. Apart of this, we do not impose any limitation on the domain or other properties of f (except of that f is integrable on Ω). For example, f can be a real function of a real argument $\mathcal{R} \to \mathcal{R}$; but it can as well be e.g. a real function on vectors $\mathcal{R}^d \to \mathcal{R}$, where d is the dimension of the real vector space.

Deterministic integration techniques which use a regular sampling of the domain Ω are intractable when the dimension of Ω is large. The reason is that the number of samples which guarantees an absolute error ε is proportional to $(1/\varepsilon)^d$, where d is the dimension of Ω [1].

Monte Carlo integration uses a stochastic sampling of Ω, which turns I into a random variable $\hat{I}$ and gives only statistical guarantees on the error bound. However, the error bound is proportional to $\sqrt{\mathrm{Var}(\hat{I})/N}$, where Var is the variance of $\hat{I}$, i.e. the error does not depend on the dimension of Ω.

What also makes Monte Carlo integration very attractive is its universality, i.e. applicability in situations where the integrand f and the domain Ω are apriori unknown. Such a situation arises when the integration is implemented as a library which is subsequently used in various programs.

This work was partially supported by the grant VEGA 1/0755/22.

Content from this work may be used under the terms of the Creative Commons Attribution 4.0 licence. Any further distribution of this work must maintain attribution to the author(s) and the title of the work, journal citation and DOI.
Published under licence by IOP Publishing Ltd

In Section 2, we describe an implementation of a generic Monte Carlo integration library which requires only that the integrand returns a real value (represented as `double`). The optimisation of a product of Monte Carlo integrations is explained in Section 3. The proposed method exploits only the product structure and independence of the integrals. Nevertheless, it improves the convergence rate of the integration with only a small computational overhead. Empirical experiments with the integration library using the optimisation method are presented in Section 4. Section 5 concludes the paper.

2 Object-oriented Monte Carlo Integration

Apart from specialised integrated mathematical environments such as Mathematica [2] and *R* [3], there are several software libraries which offer Monte Carlo integration, e.g. C GSL (GNU Scientific Library) [4], C++ HEP-MC [5] (High Energy Physics Monte Carlo), Python Sci-Py [6], Python VegasFlow [7], Python ZMCIntegral [8].

The framework which we propose in this section offers a more universal interface than aforementioned libraries. It implements the Monte Carlo method to compute definite integrals from Eq. 1 for an arbitrary function f and an arbitrary domain Ω, using an arbitrary sampling probability density function (pdf) p, with an arbitrary user-controlled termination criterion, and with no internal parameters in the interface. In order to illustrate substantial differences, we compare it to the C interface of GSL (GNU Scientific Library):

```
int gsl_monte_plain_integrate(gsl_monte_function *f,
const double xl[], const double xu[], size_t dim,
size_t calls, gsl_rng *r, gsl_monte_plain_state *s, double *result,
double *abserr)
```

GSL requires the integration domain to be a `dim`-dimensional hypercubic region, i.e. a cartesian product of intervals. The lower and upper bounds of the intervals are stored in the arrays `xl` and `xu`. Our framework does not restrict the domain to a hypercubic region.

In GSL, the user must in advance specify a fixed number of Monte Carlo iterations in the argument `calls`. Our framework provides the user with a full control over the decision when to terminate the iterations.

We will use the Java language to show how the concepts used in Monte Carlo integration can be directly expressed as classes or at least fragments of code. The choice of Java is not important, any decent object-oriented language would serve equally well. The framework can also be translated to a modular (not object-oriented) imperative language such as C without much effort, as we intentionally avoid features such as generic types in its interface. We strive to minimise the set of concepts in the framework, and keep their interfaces simple and general. In a larger software project, their functionality can easily be extended in derived classes, using class inheritance.

2.1 Class `Domain`

An integration domain is represented as an object of type `Domain` which must be able to generate a random sample upon calling the abstract method `getElement()`. The returned sample is of type `Object` which is the most abstract type in Java (i.e. any other type can be recast to `Object`). Note that concrete implementations of all abstract methods must be provided before compilation into an executable code.

```
public abstract class Domain {
    public abstract Object getElement();
}
```

2.2 Class `Function`

An integrand is represented as an object of type `Function` defined (at least) on the integration domain, which upon calling the method `getValue()` returns the function value for a given argument x.

```
public abstract class Function {
    public abstract Double getValue(Object x);
}
```

2.3 Class `Stats`

An object of type `Stats` keeps statistics over a potentially infinite sequence of real numbers. The statistics include average, variance, standard deviation, maximum, minimum, count, sum etc. Upon its construction, the object is initialised with statistics of an empty sequence. A subsequent number of the sequence is passed to the object by a call to the method `update()`. Upon calling e.g. the method `getMean()` at any time, the object returns the current estimation of mean (of course, it does not make sense to call `getMean()` before the first call to `update()`). An implementation of `Stats` can use the online algorithm by Welford [9].

```
public class Stats {
    public void update(double v);
    public double getCount();
    public double getMean();
    public double getVar();
    public double getCVar2();
    ...
}
```

2.4 Class `Integrator`

The aforementioned abstract classes suffice for a concrete implementation of Monte Carlo integration which uses an unbiased estimator $\hat{I}^{(N)}$ to estimate the integral I from Eq. (1):

$$\hat{I}^{(N)} = \frac{1}{N} \sum_{i=1}^{N} \frac{f(x_i)}{p(x_i)} \tag{2}$$

where x_i are N random samples from the domain Ω, drawn from the probability density function p defined on Ω. Independently on the choice of p, it holds (because of the law of large numbers) that

$$\lim_{N \to \infty} \hat{I}^{(N)} = I \tag{3}$$

The estimator $\hat{I}^{(N)}$ is unbiased, i.e. for every N

$$\mathbb{E}\left(\hat{I}^{(N)}\right) = I \tag{4}$$

An object of type `Integrator` computes $\hat{I}^{(N)}$ for the next N upon a call to the method `update`. Here we want to emphasize an important point: it is intentionally left to the user (or the application) to decide whether to compute $\hat{I}^{(N)}$ for the next N or not. This flexibility is very desirable, as there is no single termination criterion which suits the needs of all applications (for example, we do not want to prescribe a fixed number of iterations N to compute an estimate of I).
The current estimate of the integral I can be obtained from an integrator object `mci` by calling the methods `mci.getStats().getMean()`.

```
public class Integrator {
    Domain d;
    Function f;
    Function p;
    Stats stats;

    public void update() {
        Object x;
        x = d.getElement();
        stats.update(f.getValue(x) / p.getValue(x));
    }

    public Stats getStats() {
        return stats;
    }

    public Integrator(Domain d, Function f, Function p) {
        this.d = d;
        this.f = f;
        this.p = p;
        this.stats = new Stats();
    }
}
```

3 Product of Monte Carlo Integrations and its Optimised Computation

Consider a product of definite integrals

$$P = \int_{\Omega_1} f_1(x)\,dx \cdot \int_{\Omega_2} f_2(x)\,dx \cdot \ \dots \ \cdot \int_{\Omega_K} f_K(x)\,dx = \prod_{j=1}^{K} I_j \tag{5}$$

Expressions of this form often appear in the context of definite high-dimensional integrals, where a direct Monte Carlo computation is preceded by a factorisation into lower-dimensional integrals [10], [11], [12]. Let $\hat{I}_j^{(N_j)}$ be an unbiased Monte Carlo estimator of I_j:

$$\hat{I}_j^{(N_j)} = \frac{1}{N_j} \sum_{i=1}^{N_j} \frac{f_j\left(x_j^{(i)}\right)}{p_j\left(x_j^{(i)}\right)} \tag{6}$$

This means that for every j and every N_j

$$\mathbb{E}\left(\hat{I}_j^{(N_j)}\right) = I_j \tag{7}$$

As the estimators are independent on one another,

$$\hat{P}^{(N)} = \prod_{j=1}^{K} \hat{I}_j^{(N_j)} \tag{8}$$

(where $N = \sum_{j=1}^{K} N_j$) is an unbiased Monte Carlo estimator of P, i.e. for each N

$$\mathbb{E}\left(\hat{P}^{(N)}\right) = P \tag{9}$$

In the sequel, we focus on this estimator of P. In [13], properties of various product form estimators are studied in a more general setting, where $\hat{I}_j^{(N)}$ in Eq. 8 are independent random variables (which do not necessarily estimate definite integrals).
A naive computation (uniform sampling of Monte Carlo estimators) of P updates the estimators $\hat{I}_j^{(N_j)}$ in a round-robin fashion, generating in one iteration one random sample in all the estimators of I_j. Hence, after R iterations, it estimates for each j the value of I_j as $\hat{I}_j^{(R)}$, using the same number of random samples in all the estimators. The following code fragment implements the computation of R rounds of the naive method, assuming that the estimators of type `Integrator` are stored in an array `estimators` (indexed from 0 to $K-1$):

```
int i, j;
for (i = 0; i < R; i++)
     for (j = 0; j < K - 1; j++)
          estimators[j].update();
```

An optimised computation of P profits solely from the structure of P (namely, that P is the product of I_j) and on-the-fly statistics collected during the Monte Carlo computations of integrals I_j. The idea is to improve the convergence rate of the naive computation by using different number of samples in different estimators. In other words, if the naive method uses $R \cdot K$ samples (after R iterations), so does the optimised method, but does not distribute them uniformly among the estimators. The following code fragment partially implements the optimised method with $R \cdot K$ samples:

```
int i, j;
for (i = 0; i < R * K; i++) {
     j = select_estimator();
     estimators[j].update();
}
```

What remains open is the choice of the index j, i.e. the implementation of `select_estimator()`. A single update of the Monte Carlo estimator $\hat{I}_j^{(N_j)}$ changes N_j to N_{j+1}, $\hat{I}_j^{(N_j)}$ to $\hat{I}_j^{(N_j+1)}$, and $\hat{P}^{(N)}$ to $\hat{P}^{(N+1)}$. Our goal is to choose the index j which minimises $\mathrm{Var}(\hat{P}^{(N+1)}) - \mathrm{Var}(\hat{P}^{(N)})$.
Let us compute

$$\frac{\hat{P}^{(N)}}{P} = \prod_{j=1}^{K} \frac{\hat{I}_j^{(N_j)}}{I_j} = \prod_{j=1}^{K} \left(1 + \frac{\hat{I}_j^{(N_j)} - I_j}{I_j}\right) \tag{10}$$

Because of Eq. (3), the term $(\hat{I}_j^{(N_j)} - I_j)/I_j$ is small if the sample size N_j is sufficiently large. A first-order approximation yields

$$\frac{\hat{P}^{(N)}}{P} \approx 1 + \sum_{j=1}^{K} \frac{\hat{I}_j^{(N_j)} - I_j}{I_j} \tag{11}$$

Usual variance calculations then imply (note that the following approximation of variance can also be derived using Eq. (1) from [14] and Eq. (3) from [14])

$$\frac{\mathrm{Var}\left(\hat{P}^{(N)}\right)}{P^2} \approx \sum_{j=1}^{K} \frac{\mathrm{Var}\left(\hat{I}_j^{(N_j)}\right)}{I_j^2} \tag{12}$$

i.e.

$$\mathrm{CV}\left(\hat{P}^{(N)}\right)^2 \approx \sum_{j=1}^{K} \mathrm{CV}\left(\hat{I}_j^{(N_j)}\right)^2 = \sum_{j=1}^{K} \frac{\mathrm{CV}_j^2}{N_j} \tag{13}$$

where $\mathrm{CV}(X)$ is a coefficient of variation of the random variable X, the square of which is defined as $\mathrm{CV}(X)^2 = \mathrm{Var}(X)/\mathbb{E}(X)^2$. By CV_j^2 we understand the coefficient of variation of the single-sample estimator $\mathrm{CV}_j^2 = \mathrm{CV}(f_j(X_j)/p_j(X_j))^2$, with X_j being drawn from the pdf $p_j(x_j)$.
As P^2 in Eq. (12) is a positive number, the minimisation of $\mathrm{Var}(\hat{P}^{(N+1)}) - \mathrm{Var}(\hat{P}^{(N)})$ can be replaced with the minimisation of $\mathrm{CV}(\hat{P}^{(N+1)})^2 - \mathrm{CV}(\hat{P}^{(N)})^2$. Using the approximation from Eq. (13),

$$\begin{aligned} &\mathrm{CV}\left(\hat{P}^{(N+1)}\right)^2 - \mathrm{CV}\left(\hat{P}^{(N)}\right)^2 \\ &\approx \frac{\mathrm{CV}_j^2}{N_j+1} - \frac{\mathrm{CV}_j^2}{N_j} = -\frac{\mathrm{CV}_j^2}{N_j(N_j+1)} \end{aligned} \tag{14}$$

Hence, the implementation of the method `select_estimator()` computes and returns

$$j = \mathrm{argmax}_j \frac{\mathrm{CV}_j^2}{N_j(N_j+1)} \tag{15}$$

Since the theoretical values of the coefficients of variation CV_j^2 would typically be unknown, they are replaced in the implementation by their empirical estimates. For a given j, the current estimates of CV_j^2 and N_j are available by the calls `getCVar2()` and `getCount()` on the object `estimators[j-1].getStats()`. The method `select_estimator()` returns the index j for which the expression in Eq. (15) attains a maximum value.

3.1 Offline optimisation of sampling proportions

If the coefficients of variation were apriori known, we could solve a different optimisation problem. For a given number of samples N, the task would be finding an optimal distribution of the samples in the computations of integrals I_j which minimises $\hat{P}^{(N)}$:

$$\min_{N_1 \dots N_K} \sum_{j=1}^{K} \hat{P}^{(N)}, \text{ where } \sum_{j=1}^{K} N_j = N \tag{16}$$

Using Eq. 11–13, we can solve an approximation of Eq. 16:

$$\min_{N_1 \dots N_K} \sum_{j=1}^{K} \frac{\mathrm{CV}_j^2}{N_j}, \text{ where } \sum_{j=1}^{K} N_j = N \tag{17}$$

The optimisation problem in Eq. (17) can be solved using the method of Lagrange multipliers. Its relaxed version is

$$\min_{N_1 \dots N_K} \sum_{j=1}^{K} \frac{\mathrm{CV}_j^2}{N_j} + \lambda \sum_{j=1}^{K} N_j, \text{ where } \sum_{j=1}^{K} N_j = N \tag{18}$$

Setting the partial derivates with respect to N_j to zero

$$-\frac{\mathrm{CV}_j^2}{N_j^2} + \lambda = 0 \tag{19}$$

implies that N_j are proportional to CV_j. Their optimal proportions are

$$N_j = \frac{N\,\mathrm{CV}_j}{\sum_{i=1}^{K} \mathrm{CV}_i} \tag{20}$$

A substitution of these proportions back to Eq. (13) yields

$$\mathrm{CV}_{\mathrm{opt\_offline}}\left(\hat{P}^{(N)}\right)^2 \approx \frac{\left(\sum_{j=1}^{K} \mathrm{CV}_j\right)^2}{N} \tag{21}$$

This allows for a direct comparison with the naive uniform sampling, for which

$$\mathrm{CV}_{\mathrm{uni}}\left(\hat{P}^{(N)}\right)^2 \approx \frac{K\sum_{j=1}^{K}\mathrm{CV}_j^2}{N} \tag{22}$$

It follows from Cauchy-Schwarz inequality that $\mathrm{CV}_{\mathrm{opt\_offline}}(\hat{P}^{(N)})^2 \leq \mathrm{CV}_{\mathrm{uni}}(\hat{P}^{(N)})^2$. Although this relies on approximations of the coefficients of variance, it strengthens our expectation that the optimised sampling method generally outperforms the uniform one.

4 Experiments

For an empirical comparison of the uniform and optimised sampling methods, we chose a computation of the volumes of K-dimensional balls using Monte Carlo integration. Explicit formulas for the volumes are available, which allows for a direct comparison of both the methods with the ground truth.

The volume $\|\Omega_K\|$ of a K dimensional ball Ω_K of radius R is equal to: 1 if $K = 0$, 2R if $K = 1$, $\frac{2\pi R^2}{K}\|\Omega_{K-2}\|$ otherwise. The volume can also be expressed using Euler's gamma function as

$$\|\Omega_K\| = \frac{\pi^{K/2}R^K}{\Gamma\left(K/2+1\right)} \tag{23}$$

In our experimental setup, we use in Eq. (5)

$$f_j(x_j) = \frac{2}{\|\Omega_j\|}\sqrt{R^2 - \|x_j\|^2} \tag{24}$$

and a pdf of uniform distribution in Eq. (2)

$$p_j(x_j) = \frac{1}{\|\Omega_j\|} \tag{25}$$

The Monte Carlo estimator in Eq. (6) becomes

$$\hat{I}^{(N_j)} = \frac{2}{N_j}\sum_{i=1}^{N_j}\sqrt{R^2 - \left\|x_j^{(i)}\right\|^2} \tag{26}$$

By Eq. (7), this is an unbiased estimator of

$$I_j = \frac{2}{\|\Omega_j\|}\int_{\Omega_j}\sqrt{R^2 - \|x_j\|^2}dx_j = \frac{\|\Omega_{j+1}\|}{\|\Omega_j\|} \tag{27}$$

The product estimator $P^{(N)}$ in Eq. (8) is an unbiased estimator of

$$P = \prod_{j=1}^{K} I_j = \frac{\|\Omega_2\|}{\|\Omega_1\|}\cdots\frac{\|\Omega_{K+1}\|}{\|\Omega_K\|} \tag{28}$$

Hence, $P^{(N)}$ is an unbiased estimator of $\frac{\|\Omega_{K+1}\|}{\|\Omega_1\|}$. We use $R = 1$, for which $\|\Omega_1\| = 2$. In the denominators in Eq. 24, we use exact values of $\|\Omega_j\|$ to ensure independence of the estimators I_j. A Monte Carlo computation of I_j involves uniform sampling of a $j+1$ dimensional ball, i.e. computation of $j+1$ dimensional random vectors uniformly distributed in the ball. We use the method described in [15] and [16]. An alternative method can be found in [17]. The implementation of the optimised sampling method needs initial estimates of CV_j^2 in Eq. 15. To obtain these initial estimates, we distribute 10% of the samples in the round-robin fashion in the beginning of the computation.

The results of experiments are shown in Fig. 1. It can be observed that with a growing dimension the effect of the optimisation diminishes (it also diminishes with a growing number of samples, but this is not surprising as in the limit both the methods converge to the exact value). Nevertheless, there is a noticeable increase of the convergence rate for a relatively small number of samples.

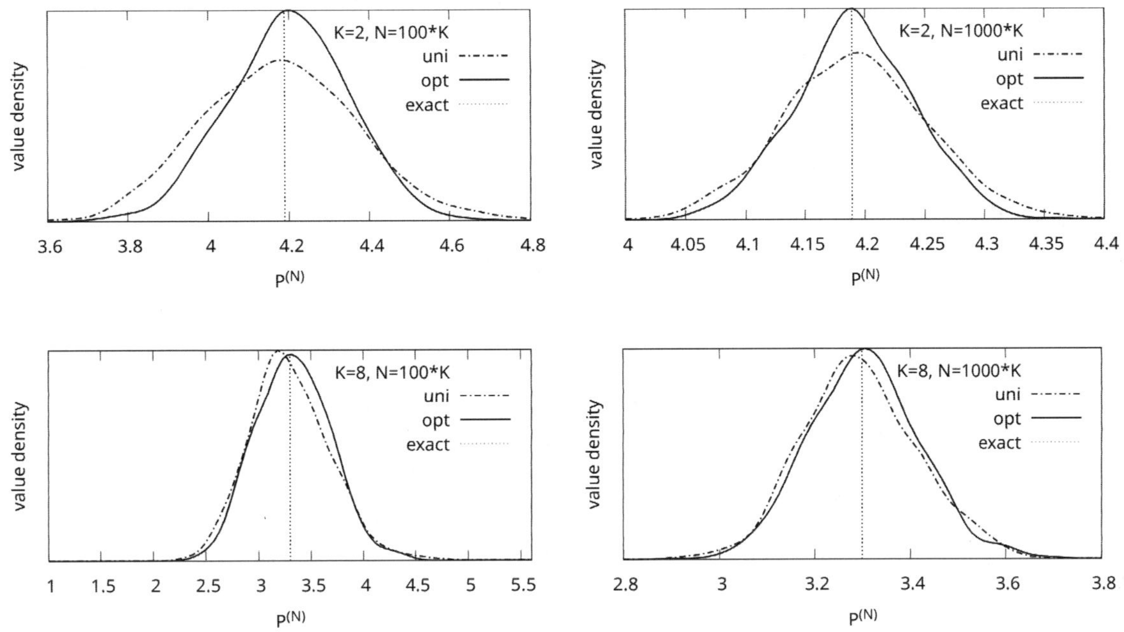

Figure 1: Monte Carlo computation of volumes of $K+1$ dimensional unit spheres using the uniform sampling method (uni) and the optimised sampling method (opt). N denotes the total number of samples. The graphs show kernel density estimations for 1000 runs. The computed volume estimates are shown on the x axis. The values on the y axis correspond to the number of runs which resulted in a value close to x. The higher a graph is and the narrower it is, the better the corresponding method performed in the experiment. The exact values of the volumes are marked with vertical lines.

The following analysis sheds some light on these results. Let us calculate CV_j^2 for our experimental setup (with $R=1$):

$$\mathrm{CV}_j^2 = \frac{\mathbb{E}\left(4\left(1-\|x_j^2\|\right)\right)}{\mathbb{E}\left(2\sqrt{1-\|x_j^2\|}\right)^2} - 1 \tag{29}$$

In Eq. 30, we use the spherical-radial transformation [18] (Section 4.1), and the formula for the surface area of a j-dimensional sphere $S_j = (j+1)\|\Omega_{j+1}\|$ [19] (Section 10.4):

$$\mathbb{E}\left(4\left(1-\|x_j^2\|\right)\right) = \frac{4S_{j-1}}{\|\Omega_j\|}\int_0^1 (1-u^2)u^{j-1}du = \frac{8S_{j-1}}{((j+1)^2-1)\|\Omega_j\|} = \frac{8j\|\Omega_j\|}{j(j+2)\|\Omega_j\|} = \frac{8}{j+2} \tag{30}$$

In Eq. 31, we use the Γ function to express the volume $\|\Omega_j\|$ [19], and subsequently its Stirling's approximation [20] (Section 6.1):

$$\begin{aligned}\mathbb{E}\left(2\sqrt{1-\|x_j^2\|}\right) &= \frac{\|\Omega_{j+1}\|}{\|\Omega_j\|} = \sqrt{\pi}\frac{\Gamma(\frac{j+1}{2}+\frac{1}{2})}{\Gamma(\frac{j+1}{2}+1)} \\ &\simeq \sqrt{\pi}\exp\left(\frac{j+1}{2}\ln\left(\frac{j+1}{2}+\frac{1}{2}\right) - \left(\frac{j+1}{2}+\frac{1}{2}\right)\ln\left(\frac{j+1}{2}+1\right)\right) \simeq \sqrt{\frac{2\pi}{j+1}}\end{aligned} \tag{31}$$

Journal of Physics: Conference Series **2910** (2024) 012035 doi:10.1088/1742-6596/2910/1/012035

where S_j is the surface area of a j-dimensional sphere of radius 1. In the latter derivation, we asymptotically approximated the gamma function using Stirling's formula. A substitution back into Eq. 29 yields

$$\mathrm{CV}_j^2 \simeq \frac{4}{\pi} - 1 \tag{32}$$

Hence, beginning with a high enough "dimension" j, the coefficients of variations CV_j^2 used in Eq. 15 are approaching the same constant. The closer their empirical estimates are to that constant, the more the optimised method behaves like the naive method, making updates almost in the round-robin fashion in those "dimensions". We emphasize, however, that this holds particularly for this particular setup, not for general products of integrations.

4.1 A-posteriori comparison with the optimal offline sample distribution

Our optimised method decides online which of the integrals in the product to update. In each step it selects an index j and updates I_j. The statistics collected during the computation can be easily extended by counting how many times each I_j was updated. Furthermore, when the computation terminates, the empirical estimates of coefficients of variation CV_j are also available. This allows for a comparison of the collected sample distribution with an optimal offline sample distribution in Eq. 20, Section 3.1 (the accuracy of the offline method is diminished in this comparison by using the a-posteriori estimates CV_j in Eq. 20, as their exact values are unknown).

Four comparisons are shown in Fig. 2 for single runs randomly chosen from the runs shown in Fig. 1. Recall that the derivation of the optimal offline sample distributions uses an a-priori knowledge of the parameters CV_j^2, whereas the optimised online method relies only on their empirical estimates. Another difference is that the offline method outputs the optimal distribution at once, whereas the optimised online method repeats greedy single-sample optimisations, contributing to the distribution with one sample in each step. The sample distributions in Fig. 1 match almost perfectly for $K = 2$, and reasonably well for $K = 8$. This suggests that in spite of the aforementioned differences, the optimised (online) method does not perform much worse than the offline method.

5 Conclusion

A generic object-oriented approach to Monte Carlo integration in Section 2 is useful in applications which use Monte Carlo integration as a library subroutine. In Section 3, we introduced an optimisation method for a product of integrals computed by Monte Carlo and implemented the method using the objected-oriented framework. We also implemented a naive uniform sampling method for Monte Carlo computation of a product of integrals and compared it to the optimised one, both analytically and empirically. The empirical comparison in Section 4 involved Monte Carlo computations of volumes of K-dimensional balls, the results match our expectations.

The optimisation of the product integral series does not use any a-priori information on the integrated functions or the integration domains. Therefore it can be seamlessly combined with optimisation techniques used in the Monte Carlo computations of individual integrals in the product series. The proposed integration framework is sufficiently versatile to incorporate also variants of the Monte Carlo method, such as Quasi-Monte Carlo [21], [22] and Markov Chain Monte Carlo [23].

Future work can involve improvements in terms of efficiency (e.g. importance sampling, quasi-Monte Carlo methods), performance (e.g. GPU support, parallelisation, quantum computing) and user experience (e.g. automatic termination criteria, a better initial estimation of coefficients of variation). While the current framework has been validated using a geometric problem, we expect that its practical utility can be further strengthened by its application to real-world domains such as financial modeling [24], [25] or physics-based simulations [26].

Journal of Physics: Conference Series **2910** (2024) 012035
doi:10.1088/1742-6596/2910/1/012035

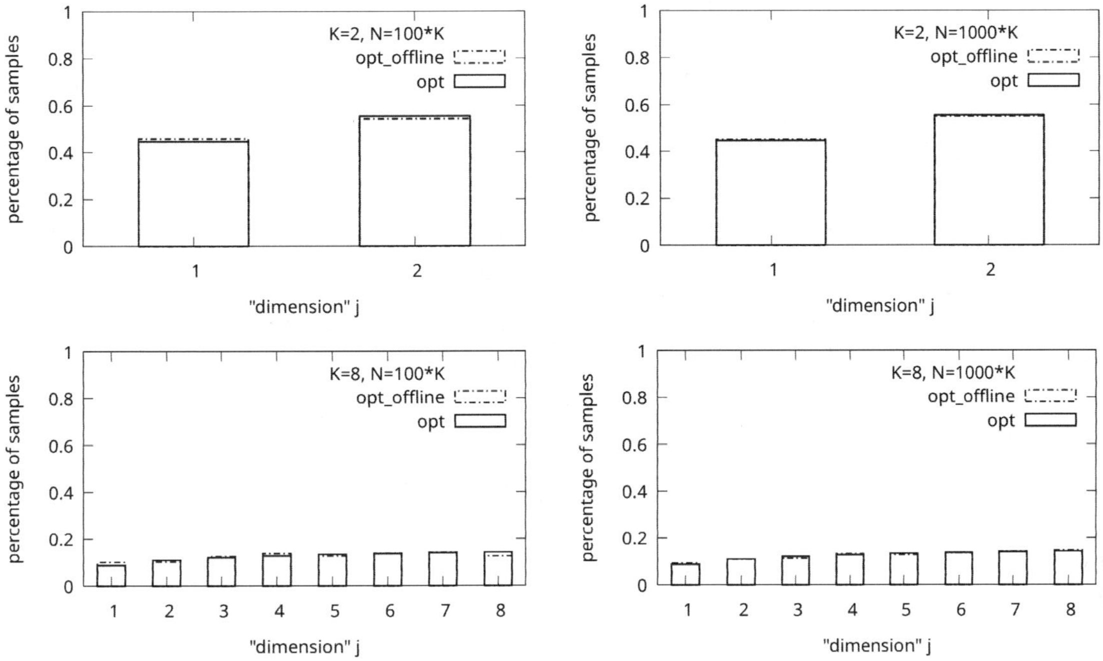

Figure 2: Optimal offline and empirical sample distributions over "dimensions" j for randomly chosen single runs of the optimised (online) method. In the top row, $K = 2$; in the bottom row, $K = 8$. In the left column, $N = 8$

References

[1] Novak E, Ritter K. The curse of dimension and a universal method for numerical integration. Multivariate Approximation and Splines. 1997 11.

[2] Wolfram Research, Inc. Mathematica. Champaign, IL, USA; 2024. Available from: `https://www.wolfram.com/mathematica`.

[3] R Core Team. R: A language and environment for statistical computing. Vienna, Austria: R Foundation for Statistical Computing; 2024. Available from: `https://www.R-project.org/`.

[4] Gough B. GNU scientific library reference manual. Network Theory Ltd.; 2009.

[5] Buckley A, Ilten P, Konstantinov D, Lönnblad L, Monk J, Pokorski W, et al. The HepMC3 event record library for Monte Carlo event generators. Computer Physics Comm. 2021;260:107310.

[6] Virtanen P, Gommers R, Oliphant TE, Haberland M, Reddy T, Cournapeau D, et al. SciPy 1.0: Fundamental algorithms for scientific computing in Python. Nature Methods. 2020;17:261-72.

[7] Carrazza S, Cruz-Martinez JM. VegasFlow: Accelerating Monte Carlo simulation across multiple hardware platforms. Comput Phys Commun. 2020;254:107376.

[8] Wu HZ, Zhang JJ, Pang LG, Wang Q. ZMCintegral: A package for multi-dimensional Monte Carlo integration on multi-GPUs. Computer Physics Comm. 2020;248:106962.

[9] Welford BP. Note on a method for calculating corrected sums of squares and products. Technometrics. 1962;4(3):419-20.

[10] Griebel M, Holtz M. Dimension-wise integration of high-dimensional functions with applications to finance. Journal of Complexity. 2010;26(5):455-89.

[11] Sun Y, Kumar M. Numerical solution of high dimensional stationary Fokker–Planck equations via tensor decomposition and Chebyshev spectral differentiation. Computers & Mathematics with Applications. 2014;67(10):1960-77.

[12] Genton MG, Keyes DE, Turkiyyah G. Hierarchical decompositions for the computation of high-dimensional multivariate normal probabilities. Journal of Computational and Graphical Statistics. 2018;27(2):268-77.

[13] Kuntz J, Crucinio FR, Johansen AM. Product-form estimators: exploiting independence to scale up Monte Carlo. Statistics and Computing. 2022;32(12).

[14] Goodman LA. The variance of the product of K random variables. Journal of the American Statistical Association. 1962;57(297):54-60.

[15] Muller ME. A note on a method for generating points uniformly on N-dimensional spheres. Comm ACM. 1959;2(4):19-20.

[16] Marsaglia G. Choosing a point from the surface of a sphere. Annals of Math Statistics. 1972;43:645-6.

[17] Harman R, Lacko V. On decompositional algorithms for uniform sampling from n-spheres and n-balls. Journal of Multivariate Analysis. 2010;101(10):2297-304.

[18] Monahan J, Genz A. Spherical-radial integration rules for Bayesian computation. Journal of the American Statistical Association. 1997;92:664-74.

[19] Styer DF. Statistical Mechanics. LibreTexts; 2024. Available from: `https://phys.libretexts.org/Bookshelves/Thermodynamics_and_Statistical_Mechanics/Statistical_Mechanics_(Styer)`.

[20] Abramowitz M, Stegun I. Handbook of mathematical functions with formulas, graphs, and mathematical tables. New York: Dover; 1964.

[21] Todorov V, Dimov I, Dimitrov Y, Ostromsky T, Georgieva R. A comparison of quasi-Monte Carlo methods based on Faure and Sobol sequences for multidimensional integrals in air pollution modeling. AIP Conference Proceedings. 2019 10;2164:030002.

[22] Todorov V, Dimov I, Fidanova S, Poryazov S. A new optimized stochastic approach for multiple integrals in option pricing. Communication Papers of the Federated Conference on Computer Science and Information Systems, ACSIS. 2020 09:21-4.

[23] Todorov V, Georgiev S. Advanced stochastic approaches for Fredholm integral equations. Journal of Physics: Conference Series. 2023;2675(1):012035.

[24] Kaneko K, Miyamoto K, Takeda N, Yoshino K. Quantum speedup of Monte Carlo integration with respect to the number of dimensions and its application to finance. Quantum Information Processing. 2021;20(5):185.

[25] Todorov V, Georgiev S. Advanced stochastic Monte Carlo optimization methods for two-dimensional European style options. In: International Conference on New Trends in the Applications of Differential Equations in Sciences. Springer; 2023. p. 371-81.

[26] Todorov V, Georgiev S, Georgiev I, Zaharieva S, Dimov I. Optimizing air pollution modeling with a highly-convergent quasi-Monte Carlo method: A case study on the UNI-DEM framework. Mathematics. 2023;11(13):2919.

Journal of Physics: Conference Series **2910** (2024) 012036 doi:10.1088/1742-6596/2910/1/012036

Refined Unbiased Stochastic Approach for Fredholm Integral Equations

Venelin Todorov[1], Ivan Georgiev[1,2], Milen Chechev[3], Yuri Dimitrov[4,1]

[1]Department of Information Modeling, Institute of Mathematics and Informatics, Bulgarian Academy of Sciences, 8 Acad. Georgi Bonchev Str., 1113 Sofia, BG
[2]Department of Applied Mathematics and Statistics, University of Ruse, 8 Studentska Str., 7004 Ruse, BG
[3]Department of Software Technologies, University of Sofia, 8 James Bourchier Blvd., 1164 Sofia, BG
[4]Department of Mathematics, Physics and Informatics, Faculty of Forest Industry, University of Forestry, 10 St. Kliment Ohridski Blvd., 1756 Sofia, BG

E-mail: vtodorov@math.bas.bg,
irgeorgiev@uni-ruse.bg,irgeorgiev@math.bas.bg,
milen.chechev@fmi.uni-sofia.bg
yuri.dimitrov@ltu.bg

Abstract. Integral equations are highly relevant across various fields such as applied mathematics, physics, engineering, geophysics, electromagnetism, the kinetic theory of gases, quantum mechanics, mathematical economics, and queuing theory. This underscores the importance of creating and examining efficient and dependable methods to solve integral equations. In the context of multidimensional issues, traditional biased stochastic algorithms, which rely on a finite set of integration points, are particularly challenged by the complexities of higher dimensionality. Therefore, it is critical to develop sophisticated unbiased algorithms to address these challenges, as discussed in this paper. We introduce and evaluate a novel unbiased stochastic approach for solving multidimensional Fredholm integral equations of the second kind.

1. Introduction

One of the main advantages of Monte Carlo methods over deterministic approaches [10] is their capability to directly compute an unknown linear functional of the solution to integral equations, requiring computational resources only to solve the equation at a single point within the domain [8, 13]. Current Monte Carlo methods for integral equations (MCM-IE) are based on probabilistic interpretations of the Liouville-Neumann series corresponding to Fredholm integral equations of the second kind [3, 8, 9, 12]. Unbiased approaches typically work with infinite series, whereas biased MC methods rely on probabilistic representations of truncated Liouville-Neumann series. A common and well-known biased approach is the Markov chain Monte Carlo method (e.g., [11]), in which the Markov chain is designed to terminate after a fixed number of steps.

In this paper, we present an unbiased stochastic algorithm that builds on a probabilistic representation of the infinite series by introducing a probability of absorption. The construction of this algorithm mirrors the technique outlined in [7] for solving linear systems.

Content from this work may be used under the terms of the Creative Commons Attribution 4.0 licence. Any further distribution of this work must maintain attribution to the author(s) and the title of the work, journal citation and DOI.
Published under licence by IOP Publishing Ltd

2. Formulation of the Problem

Consider the second kind Fredholm integral equation:

$$u(\mathrm{x}) = \int_{\Omega} k(\mathrm{x}, \mathrm{x}')u(\mathrm{x}')\mathrm{dx}' + f(\mathrm{x}) \tag{1}$$

or equivalently,

$$u = \mathcal{K}u + f,$$

where $\mathcal{K}$ is an integral operator, $k(\mathrm{x}, \mathrm{x}') \in L_2(\Omega \times \Omega)$, $f(\mathrm{x}) \in L_2(\Omega)$ are known functions, and $u(\mathrm{x}) \in L_2(\Omega)$ is the unknown function. Here, $\mathrm{x}, \mathrm{x}' \in \Omega \subset \mathbb{R}^n$, and Ω is a bounded domain.

We aim to use Monte Carlo methods to evaluate linear functionals of the solution of the form:

$$J(u) = \int \varphi(\mathrm{x})u(\mathrm{x})\mathrm{dx} = (\varphi, u), \tag{2}$$

where $\varphi(\mathrm{x}) \in L_2(\Omega)$ is a known function.

To solve the integral equation, we can use the method of successive approximations:

$$u^{(i)} = \sum_{j=0}^{i} \mathcal{K}^{(j)} f = f + \mathcal{K}f + \ldots + \mathcal{K}^{(i)} f, \quad i = 1, 2, \ldots, \tag{3}$$

with $u^{(0)}(\mathrm{x}) \equiv f(\mathrm{x})$. The condition $\|\mathcal{K}\|_{L_2} < 1$ ensures the convergence of the Liouville-Neumann series [3, 4], so that $u^{(i)} \to u$ as $i \to \infty$ in the L_2 norm.

This approach allows us to estimate a linear functional of the solution u, represented as the inner product of u with a fixed element φ in the Banach space. This is a generalization of the Picard iteration method for solving ordinary differential equations by converting them into equivalent integral equations [1].

$$J(u) = (\varphi, u) = \lim_{i\to\infty} (\varphi, u^{(i)}) = \lim_{i\to\infty} \sum_{j=0}^{i} (\varphi, \mathcal{K}^{(j)} f).$$

For further details, see [4, 8].

To approximate the unknown value (φ, u), one can use a truncated Liouville-Neumann series (3) for a sufficiently large i:

$$(\varphi, u^{(i)}) = (\varphi, f) + (\varphi, \mathcal{K}f) + \ldots + (\varphi, \mathcal{K}^{(i-1)} f) + (\varphi, \mathcal{K}^{(i)} f).$$

This transforms the problem of solving integral equations into approximating a finite number of multidimensional integrals. Let $I(j) = (\varphi, \mathcal{K}^{(j)} f)$, where $I(j)$ represents the value obtained after integrating over Ω^{j+1}. Thus, the estimate $(\varphi, u^{(i)})$ can be expressed as a sum of linear functionals of the form $(\varphi, \mathcal{K}^{(j)} f)$:

$$(\varphi, \mathcal{K}^{(j)} f) = \int_{\Omega} \varphi(\mathrm{t}_0)\mathcal{K}^{(j)} f(\mathrm{t}_0)\mathrm{dt}_0 = \int_{G} \varphi(\mathrm{t}_0)k(\mathrm{t}_0, t_1)\ldots k(\mathrm{t}_{j-1}, \mathrm{t}_j)f(\mathrm{t}_j)\mathrm{dt}_0 \ldots \mathrm{dt}_j, \tag{4}$$

where $\mathrm{t} = (\mathrm{t}_0, \ldots, \mathrm{t}_j) \in G \equiv \Omega^{j+1} \subset \mathbb{R}^{n(j+1)}$.

Defining the integrand function $F_j(\mathrm{t})$ as

$$F_j(\mathrm{t}) = \varphi(t_0)k(t_0, t_1)\ldots k(\mathrm{t}_{j-1}, \mathrm{t}_j)f(\mathrm{t}_j), \quad t \in \Omega^{j+1},$$

we can rewrite (4) as:

$$I(j) = (\varphi, \mathcal{K}^{(j)} f) = \int_G F_j(\mathrm{t})\mathrm{dt}, \quad \mathrm{t} \in G \subset \mathbb{R}^{n(j+1)}. \tag{5}$$

Thus, the problem reduces to approximating multiple integrals of the form (5). The domain G is crucial for solving the problem as a finite set of multiple integrals (FNI). Both (4) and (5) can define a biased algorithm, where the approximation $(\varphi, \mathcal{K}^{(j)} f)$ to (φ, u) is expressed as a set of $j+1$ integrals $I(j)$ with integrands $F_j(\mathrm{t})$, where t is a point in the $n(j+1)$ dimensional domain G.

There are two primary stochastic techniques. The first is a biased approach, where the expectation of a random variable approximates the solution using a truncated Liouville-Neumann series (3) for large i. The second is an unbiased approach, where the mean value of the random variable converges to the exact solution. In the biased method, two types of errors arise: a systematic truncation error $R_{sys}$and a stochastic error R_N, which is dependent on the number N of random variable evaluations or chains. In contrast, unbiased methods only involve probabilistic error, whereas biased methods must carefully balance both errors to optimize computational efficiency (for further details, see [4]).

The calculation of linear functionals (φ, u) for an integral equation reduces to computing a finite sum of linear functionals $(\varphi, \mathcal{K}^j f)$ for $j = 0, \ldots, i$ (see [13, 4]). We begin by constructing a random Markov chain T_i of length i, starting from state x_0 in the domain Ω:

$$T_i : \mathrm{x}_0 \longrightarrow \mathrm{x}_1 \longrightarrow \ldots \longrightarrow \mathrm{x}_i,$$

according to initial $\pi(\mathrm{x})$ and transition $p(\mathrm{x}, \mathrm{x}')$ probabilities, where $\pi(\mathrm{x})$ and $p(\mathrm{x}, \mathrm{x}')$ are non-negative and compatible with $\varphi(\mathrm{x})$ and the kernel $k(\mathrm{x}, \mathrm{x}')$, respectively. The transition probability $p(\mathrm{x}, \mathrm{x}')$ is chosen to be proportional to $|k(\mathrm{x}, \mathrm{x}')|$, with the constant c computed as:

$$c = \left(\int_\Omega |k(\mathrm{x}, \mathrm{x}')| \mathrm{dx}' \right)^{-1} \quad \text{for any} \quad \mathrm{x} \in \Omega.$$

This constant c acts as a scaling factor. The expected value $\mathbf{E}\theta_i[\varphi]$ is given by:

$$\mathbf{E}\theta_i[\varphi] = (\varphi, u^{(i)}), \quad \text{where} \quad \theta_i[\varphi] = \frac{\varphi(\mathrm{x}_0)}{\pi(\mathrm{x}_0)} \sum_{j=0}^{i} W_j f(\mathrm{x}_j),$$

with $W_0 = 1$ and $W_j = W_{j-1} \frac{k(\mathrm{x}_{j-1}, \mathrm{x}_j)}{p(\mathrm{x}_{j-1}, \mathrm{x}_j)}$ for $j = 1, \ldots, i$. Thus, the Monte Carlo estimate of $(\varphi, u^{(i)})$ can be expressed as:

$$(\varphi, u^{(i)}) \approx \frac{1}{N} \sum_{l=1}^{N} \theta_i[\varphi]_l.$$

The random variable $\theta_i[\varphi]$ provides a biased estimate of (φ, u) for sufficiently large i, with a statistical error of $\mathcal{O}(N^{-1/2})$, where N is the number of chains.

Importantly, the same Markov chain trajectories T_i can be reused for biased approximate evaluations of $(\varphi, u^{(i)})$ for different functions $\varphi(\mathrm{x})$ and for various integral equations with the same kernel $k(\mathrm{x}, \mathrm{x}')$ but different right-hand sides $f(\mathrm{x})$.

When using biased stochastic approaches, it is essential to balance stochastic and systematic errors [4]. Consider a linear functional of the solution of an integral equation (φ, u). Let $u^{(i)}$ denote the i-th iterative approximation of $u \, (i \geq 0)$ and $\tilde{u}$ represent the Monte Carlo approximation. Two types of errors arise: the systematic (truncation) error R_{sys} and the stochastic (probabilistic) error R_N. In unbiased algorithms, the truncation error is zero because they directly approximate the full Liouville-Neumann series.

Given a small positive parameter ε, we have:

$$|(\varphi, u) - (\varphi, \tilde{u})| \leq |(\varphi, u) - (\varphi, u^{(i)})| + |(\varphi, u^{(i)}) - (\varphi, \tilde{u})| = \varepsilon_i + \varepsilon_N < \varepsilon,$$

where ε_i is the truncation error, and ε_N is the stochastic error. A lower bound for the number of iterations i can be obtained using [3]:

$$i > \frac{\ln \frac{\varepsilon}{2|(\varphi, u^{(0)}) - (\varphi, u)|}}{\ln \|\mathcal{K}\|_{L_2}}, \tag{6}$$

where $u^{(0)}$ is the initial approximation of u. The Cauchy inequality gives:

$$|(\varphi, u^{(0)}) - (\varphi, u)| \leq \|\varphi\|_{L_2(\Omega)} \|u^{(0)} - u\|_{L_2(\Omega)}.$$

The difference $\|u^{(0)} - u\|_{L_2(\Omega)}$ can be estimated as:

$$\|u^{(0)} - u\|_{L_2(\Omega)} \leq (\|I\|_{L_2} + \|\mathcal{K}\|_{L_2} + \ldots + \|\mathcal{K}\|^i_{L_2} + \ldots)\|r^{(0)}\|_{L_2(\Omega)},$$

where $r^{(j)} = f - u^{(j)} - \mathcal{K}u^{(j)}$, for $j = 0, 1, \ldots$.

Taking the limit as $i \to \infty$, we obtain:

$$\|u^{(0)} - u\|_{L_2} \leq \frac{\|r^{(0)}\|_{L_2(\Omega)}}{1 - \|\mathcal{K}\|_{L_2}}.$$

This holds if $\|\mathcal{K}\|_{L_2} < 1$. The inequality (6) is crucial for estimating the computational complexity of biased Monte Carlo algorithms and helps define well-balanced biased algorithms. The complexity is of order $\mathcal{O}(i)$.

3. Unbiased Stochastic Approach

This section presents and evaluates an Unbiased Stochastic Approach (USA) for solving integral equations. The USA method is built upon expressing the functional as an infinite series. This approach expands on our earlier work [7], now adapted for continuous state spaces with a continuous transition density function and an absorption probability. Here, the random trajectory (Markov chain) T_i introduced in Section 2 is infinite. The USA method generalizes the approach from [7] for integral equations, replacing the discrete Markov chain with a continuous one where transition probabilities depend on the kernel $k(\mathrm{x}, \mathrm{x}')$. Subsection 3.1 covers the case where $0 \leq k(\mathrm{x}, \mathrm{x}') \leq 1$, and subsection 3.2 addresses more general kernels.

3.1. Unbiased approach for a simplified problem

Consider the simplified case where $0 \leq k(\mathrm{x}, \mathrm{x}') \leq 1$ for all $\mathrm{x}, \mathrm{x}' \in \Omega \equiv [0, 1]^n$. Let $Y \equiv U[0, 1]^n$ be a uniformly distributed random variable in $[0, 1]^n$. The function $u(\mathrm{x})$ can be represented stochastically as:

$$u(\mathrm{x}) = E\{k(\mathrm{x}, Y)u(Y) + (1 - k(\mathrm{x}, Y)) \cdot 0\} + f(\mathrm{x}),$$

This representation follows the same idea as in [7] for solving linear systems. Here, the absorption probability is $1 - k(\mathrm{x}, Y)$, and the score is the function $f(\mathrm{x})$. If the trajectory is not absorbed, it continues at point Y. The random trajectory (Markov chain) T_i starts from state $X = \mathrm{x} = \mathrm{x}_0$ in the domain Ω, with a cemetery state $\partial \in \mathbb{R}^n \setminus \Omega$ representing absorption:

$$T_i : \mathrm{x}_0 \longrightarrow \mathrm{x}_1 \longrightarrow \ldots \longrightarrow \mathrm{x}_i,$$

where

$$P(\mathrm{x}_{i+1} \in \Omega | \mathrm{x}_i \neq \partial) = k(\mathrm{x}, U_i) P(U_i \in \Omega),$$
$$P(\mathrm{x}_{i+1} = \partial | \mathrm{x}_i \neq \partial) = 1 - k(\mathrm{x}, U_i),$$
$$P(\mathrm{x}_{i+1} = \partial | \mathrm{x}_i = \partial) = 1, \quad f(\partial) = 0.$$

Thus, the solution $u(\mathrm{x})$ can be expressed as:

$$u(\mathrm{x}) = E\left(\sum_{i=0}^{\infty} f(\mathrm{x}_i) | \mathrm{x}_0 = \mathrm{x}\right) = E\left(\sum_{i=0}^{\tau} f(\mathrm{x}_i) | \mathrm{x}_0 = \mathrm{x}\right),$$

where $\tau = \inf_{i \geq 0}(\mathrm{x}_i = \partial)$. This leads to:

$$u(\mathrm{x}) = E\left(\sum_{i=0}^{\infty} f(\mathrm{x}_{i+1}) | \mathrm{x}_0 = \mathrm{x}\right) + f(\mathrm{x}) \quad = \quad \int_{\Omega} k(\mathrm{x}, \mathrm{y}) u(\mathrm{y}) d\mathrm{y} + f(\mathrm{x}) \tag{7}$$

The simplified case algorithm is as follows:

Algorithm 1 *:*

Unbiased stochastic algorithm for $0 \leq k(\mathrm{x}, y) \leq 1, \quad \mathrm{x}, \mathrm{y} \in [0,1]^n$

1. ***Initialization:*** *Input the kernel* $k(\mathrm{x}, \mathrm{y})$*, the function* $f(\mathrm{x})$*, and the number of random trajectories* N*.*
2. ***Calculations:***

   2.1. ***Initialize*** *score = 0*
   2.2. ***Set*** $\mathrm{x}_0 = \mathrm{x}$, *test = 1*
     For $j = 1$ *to* N
       While *(test* $\neq 0$*)*
         score = score + $f(\mathrm{x}_0)$
         $U = rand[0,1]^n, \quad V = rand[0,1]$
         If $k(\mathrm{x}_0, U) < V$ ***then***
           test = 0 ***else***
           $\mathrm{x}_0 = U$
         Endif
       Endwhile
     Endfor

3. ***Compute the solution:***
   $u(\mathrm{x}) = \frac{score}{N}$

3.2. Unbiased approach for the general problem

This subsection describes the method for cases where the kernel $k(\mathrm{x}, \mathrm{y})$ can be negative and $|k(\mathrm{x}, \mathrm{y})|$ may exceed 1, provided that $k(\mathrm{x}, \mathrm{y}) \in L_2(\Omega \times \Omega)$ and the Neumann series converges. When $-1 \leq k(\mathrm{x}, Y) \leq 0$, $|k(\mathrm{x}, Y)|$ is used as the probability to stop, but now $-u(Y)$ is computed instead of $u(Y)$, with a score of $-f(\mathrm{x})$. If $|k(\mathrm{x}, Y)| \geq 1$, the random walk continues, and the score is adjusted by $k(\mathrm{x}, Y) f(\mathrm{x})$. This leads to the following method, which modifies the previous approach. Consider the Markov chain T_i of length i, starting from state $X = \mathrm{x} = \mathrm{x}_0$ in the domain Ω, with a cemetery state $\partial \in \mathbb{R}^n \setminus \Omega$ representing absorption:

$$T_i : \mathrm{x}_0 \longrightarrow \mathrm{x}_1 \longrightarrow \ldots \longrightarrow \mathrm{x}_i,$$

where in this case:

$$P(\mathrm{x}_{i+1} \in \Omega | \mathrm{x}_i \neq \partial) = \min\left(|k(\mathrm{x}_i, \mathrm{x}_{i+1})|, 1\right),$$
$$P(\mathrm{x}_{i+1} = \partial | \mathrm{x}_i \neq \partial) = 1 - \min\left(|k(\mathrm{x}_i, \mathrm{x}_{i+1})|, 1\right),$$
$$P(\mathrm{x}_{i+1} = \partial | \mathrm{x}_i = \partial) = 1.$$

The solution $u(\mathrm{x})$ is then given by:

$$u(\mathrm{x}) = f(\mathrm{x}) + E\left(\sum_{k=1}^{\infty}\prod_{i=1}^{k} \hat{k}(\mathrm{x}_{i-1}, \mathrm{x}_i) f(\mathrm{x}_{i-1}) | \mathrm{x}_0 = \mathrm{x}\right),$$

where $\hat{k}(\mathrm{x}, \mathrm{x}')$ is defined as:

$$\hat{k}(\mathrm{x}, \mathrm{x}') = \begin{cases} k(\mathrm{x}, \mathrm{x}') & \text{if } |k(\mathrm{x}, \mathrm{x}')| \geq 1 \\ 1 & \text{if } 0 \leq k(\mathrm{x}, \mathrm{x}') \leq 1 \\ -1 & \text{if } -1 \leq k(\mathrm{x}, \mathrm{x}') \leq 0 \end{cases}$$

The general case algorithm is as follows:

Algorithm 2 *:*

Unbiased stochastic algorithm for the general case:

1. ***Initialization:*** *Input the kernel* $k(\mathrm{x}, \mathrm{y})$*, the function* $f(\mathrm{x})$*, and the number of random trajectories* N*.*
2. ***Calculations:***

   2.1. ***Initialize*** *score = 0*
   2.2. ***For*** $j = 1$ *to* N
   2.3. ***Set*** $\mathrm{x}_0 = \mathrm{x}$*, test = 1, prod = 1*
   While $(test \neq 0)$
   $score = score + f(\mathrm{x}_0) \times prod$
   $U = rand(0, 1)^n$
   If $|k(\mathrm{x}_0, U)| > 1$ ***then***
   $prod = prod \times k(\mathrm{x}_0, U)$
   Else
   $\mathrm{x}_0 = U$
   $V = rand[0, 1]$
   If $|k(\mathrm{x}_0, U)| < V$ ***then***
   $test = 0$
   Else
   $prod = prod \times sign(k(\mathrm{x}_0, U))$, $\mathrm{x}_0 = U$
   Endif
   Endif
   Endwhile
   Endfor

3. ***Compute the solution:***
   $u(\mathrm{x}) = \frac{score}{N}$

3.3. Some extensions of the unbiased approach

For various applications, it is often necessary to directly obtain the value of the functional of the solution to the Fredholm integral equation, particularly when it depends on a given function $\varphi(\mathrm{x})$. In situations such as sensitivity analysis [5], the interest may not be in the pointwise value of the solution, but rather in mean values like $\int_\Omega \varphi(\mathrm{x})u(\mathrm{x})d\mathrm{x}$. Here, $\varphi(\mathrm{x})$ may act as a weight function, often with support in a small subdomain of Ω. In other cases, the solution at specific points, such as lattice points, is required.

3.3.1. Weak approximation A double randomization technique can be employed to compute such quantities. If $\varphi(\mathrm{x})$ is a probability density function over Ω, the integral can be expressed as the mean value of $u(\mathrm{x})$, i.e., $\int_\Omega u(\mathrm{x})\varphi(\mathrm{x})d\mathrm{x} = E(u(Y))$, where Y is a random variable with density $\varphi(\mathrm{x})$. This requires only a minor modification of the algorithm, with the starting point x_0 as the random variable Y. If sampling from φ is difficult or if φ is not a density function, an alternative is to start with a uniformly distributed x_0 and multiply the sum of the random variable (the score) by $\varphi(\mathrm{x}_0)$.

3.3.2. Global approximation Another extension involves computing the solution of the Fredholm integral equation at all points in Ω using the same unbiased USA algorithm. During the random walk, many points in the domain are visited, and the information gathered can be used (with proper statistical treatment) to approximate the solution at all points. This concept was also applied in [7] for solving linear systems, where all visited states of a discrete random walk were reused as new starting points. The current scenario is more complex due to the continuous nature of the walk, but the following procedure can be simulated N times:

(i) Initial Step: The starting point x_0 is chosen uniformly at random.

(ii) Distribution of Visited Points: All visited points across the random trajectories are uniformly distributed. Each visited point provides a score, whose mean value is an unbiased estimate of the solution at that point.

(iii) Approximation via Taylor Expansion: The solution at any point x can be approximated using a Taylor expansion. This requires assuming higher regularity of the solution, which is not restrictive in certain cases (e.g., classical boundary integral equations in potential theory). Suppose we have values $u(\mathrm{x}_i)$ of the solution in a hypercube of size $2h$ centered at x, at points $\mathrm{x}_i, i = 1, \dots, N_h$ where $\mathrm{h}_i \in \mathbb{R}^n$ is a vector such that $\mathrm{h}_i = \mathrm{x}_i - \mathrm{x}$. We can write:

$$\begin{aligned} u(\mathrm{x}_i) = u(\mathrm{x}+\mathrm{h}_i) &= u(\mathrm{x}) + \nabla u(\mathrm{x}) \cdot \mathrm{h}_i \\ &+ \tfrac{1}{2}\mathrm{h}_i^T[D^2u(\mathrm{x})]\mathrm{h}_i + \mathcal{O}(h^3), \end{aligned}$$

where $\nabla u(\mathrm{x})$ is the gradient, $[D^2u(\mathrm{x})]$ is the Hessian matrix, and h_i^T is the transposed vector. The mean value of the solution at points x_i can be expressed as:

$$\frac{1}{N_h}\sum_{i=1}^{N_h} u(\mathrm{x}_i) = u(\mathrm{x}) + \frac{1}{N_h}\sum_{i=1}^{N_h} \mathrm{h}_i \cdot \nabla u(\mathrm{x}) + \frac{1}{N_h}\sum_{i=1}^{N_h} \frac{\mathrm{h}_i^2}{2}[D^2u(\mathrm{x})] + \mathcal{O}(h^3), \quad (8)$$

where the Hessian matrix can be estimated from above by the maximum of all second derivatives at point x.

The value of h should be small enough to ensure that the second-order term in the Taylor expansion is negligible, and the number of points in the hypercube should not be too small to avoid large variance in the first-order term. If the hypercube is within the domain, the mathematical expectation of the first-order term in (8) is zero, and its variance decreases with

 doi:10.1088/1742-6596/2910/1/012036

$O(\frac{1}{N_h})$. Finally, $u(\mathrm{x})$ can be estimated by the mean value $\frac{1}{N_h}\sum_{i=1}^{N_h} score(\mathrm{x}_i)$, where $score(\mathrm{x}_i)$ is the score of a point x_i visited during one of the N random walks and lying within the hypercube. In practice, all visited points and their associated scores are stored to compute the solution at any point x later.

4. Numerical Results and Discussion

4.1. Comparison with the Nystrom method

Among the various deterministic methods for solving Fredholm integral equations [6], the Nystrom method is one of the simplest to implement. Detailed descriptions of these methods and numerical tests on kernels with varying smoothness can be found in [2]. In this section, we compare the Nystrom method, based on the Simpson rule, with the USA (Unbiased Stochastic Approach) on one-dimensional integral equations with kernels and solutions of different smoothness. This comparison helps demonstrate the USA's capability to handle irregular kernels and assess its performance on examples typically favorable to deterministic methods.

Applying the Simpson quadrature formula to the integral term in the Fredholm equation gives:

$$u(x) \simeq \sum_{k=1}^{n} \alpha_k k(x, x_k) u(x_k) + g(x),$$

where α_k and x_k are the weights and nodes of the Simpson rule, respectively. Substituting x with the nodes x_i leads to a linear system:

$$u_i = \sum_{k=1}^{n} \alpha_k k(x_i, x_k) u_k + g(x_i),$$

where $u_i \simeq u(x_i)$. Solving this linear system iteratively provides an approximation of the solution at the node points and any other point using the continuous formula. The complexity of inverting this linear system is $\mathcal{O}(n^2)$, so we allow the same $\mathcal{O}(n^2)$ simulations for the USA to ensure a fair comparison. The Simpson rule is applied on p sub-intervals, leading to $n = 2p + 1$.

We consider the following regular: $k_1(x, y) = xy$ and discontinuous: $k_2(x, y) = 1_{x \le a} + \frac{1}{2} 1_{x > a}$ kernels and, respectively, a regular: $u_1(x) = \exp(x)$ and a discontinuous: $u_2(x) = 1_{x \le b} \exp(x)$ solution.

This results in four integral equations: RKRS, RKDS, DKRS, and DKDS, where R stands for Regular, K for Kernel, D for Discontinuous, and S for Solution. For $a = \frac{1}{3}$ and $b = \frac{\sqrt{2}}{2}$, the corresponding functions $g(x)$ are:

$$g_{RKRS}(x) = \exp(x) - x, \quad g_{RKDS}(x) = 1_{x \le b} \exp(x) - x(b \exp(x) + 1 - \exp(b)),$$

$$g_{DKRS}(x) = \exp(x) + 1 - \frac{1}{2}(\exp(b) + \exp(1)),$$

$$g_{DKDS}(x) = 1_{x \le b} \exp(x) + 1 - \frac{1}{2}(\exp(b) + \exp(a)).$$

We compute the solution approximation at $x = 0.5$ and compare it to its exact value, $\exp(0.5)$, for all four equations. Three different values of p are chosen: $p_1 = 50$, $p_2 = 400$, and $p_3 = 1000$, to compute the absolute error, as shown in Table 1.

The table presents the relative errors for two methods, Nystrom and the Unbiased Stochastic Approach (USA), at the point $x = 0.5$. The results are shown for different parameter configurations (p_1, p_2, p_3) and for various error calculation methods: RKRS, RKDS, DKRS, and DKDS.

Table 1: Relative error for Nystrom and USA at point $x = 0.5$

	Nystrom			USA		
	p_1	p_2	p_3	$(2p_1+1)^2$	$(2p_2+1)^2$	$(2p_3+1)^2$
RKRS	3E-10	8E-14	1E-15	7E-3	3E-4	5E-4
RKDS	4E-3	8E-4	4E-4	3E-3	2E-5	1E-5
DKRS	7E-3	8E-4	4E-4	1E-2	3E-4	1E-4
DKDS	5E-3	1E-3	2E-4	9E-3	1E-3	2E-4

The Nystrom method yields very small relative errors across all cases. For RKRS, the errors decrease from 3E-10 to 1E-15 as the parameters p_1, p_2, p_3 change, showing extremely high accuracy. For RKDS, the errors range from 4E-3 to 4E-4, indicating that the performance is slightly worse than for RKRS but still maintains a reasonable level of accuracy. DKRS and DKDS show errors on the order of 7E-3 to 2E-4, which are slightly larger but still relatively small, confirming the Nystrom method's high precision for these error calculation methods.

The USA method shows higher relative errors compared to the Nystrom method. For RKRS, the errors start at 7E-3 for p_1 and drop to 5E-4 for p_3, indicating that USA is less accurate than Nystrom in this case, but still performs reasonably well. For RKDS, USA shows a significant improvement as the parameters change, with errors decreasing from 3E-3 for p_1 to 1E-5 for p_3, suggesting that USA improves for certain parameter configurations. In the DKRS case, USA shows errors ranging from 1E-2 to 1E-4, again demonstrating a drop in error as p_3 increases, but the accuracy remains lower than Nystrom. For DKDS, the errors are similar to DKRS, ranging from 9E-3 to 2E-4, suggesting consistent performance across these error metrics.

The Nystrom method consistently delivers much lower errors across all error methods (RKRS, RKDS, DKRS, and DKDS) compared to the USA method. Nystrom shows exceptional accuracy, especially in RKRS, where the errors are on the order of 1E-15 for p_3 The USA method, while showing higher errors overall, improves as the parameters p_1, p_2, p_3 increase. It demonstrates better performance in RKDS compared to RKRS, but overall, it is less accurate than the Nystrom method. Both methods show decreasing error as the parameters p_1, p_2, p_3 progress, suggesting that both methods perform better with larger parameter values, particularly p_3. The USA method shows a significant reduction in error between RKDS and DKDS, indicating that it might perform better for certain types of problems or error metrics.

Nystrom Method method excels in accuracy, with consistently lower relative errors across all configurations, making it the preferred choice for applications requiring high precision. While the USA method is less accurate overall, it shows acceptable performance, particularly for RKDS and DKDS metrics, and may be more suited for cases where lower precision is acceptable or where other advantages, such as computational efficiency, outweigh the need for extreme accuracy. The overall conclusion is that Nystrom is the superior method in terms of accuracy, but USA provides a reasonable alternative, especially in cases where trade-offs between accuracy and other factors need to be considered.

4.2. Further investigation of the USA algorithm

4.2.1. General USA algorithm To validate the general version of the unbiased algorithm, we consider a *second numerical example* where the kernel can take negative values and its absolute

Journal of Physics: Conference Series **2910** (2024) 012036 doi:10.1088/1742-6596/2910/1/012036

value can exceed 1:

$$u(x) = \int_0^1 k(x, x')u(x')\mathrm{d}x' + f(x), \tag{9}$$

where

$$k(x, x') = \beta(2x' - 1)x^2 \mathrm{e}^{x'(x-1)},$$
$$f(x) = \mathrm{e}^x + \beta(2\mathrm{e}^x - 2 - x\mathrm{e}^x - x),$$
$$\varphi(x) = \delta(x - x_0),$$

where x_0 takes values 0.1, 0.5, and 0.99 in different numerical experiments. The solution to the equation is $u(x) = \mathrm{e}^x$. We introduce the parameter β to handle various challenges, such as negative kernels and kernels with absolute values greater than one. As β increases, we approach instability as the variance becomes infinite.

Following the ideas from subsection 3.3, Section 3, we compute two quantities:

$$I_1 = \int_0^1 u(x)dx,$$

$$I_2 = \int_0^1 2xu(x)dx$$

to illustrate the new unbiased approach.

For the first quantity, the initial value is picked uniformly. For the second, there are two approaches:

(i) Pick the starting point x_0 according to the density $2x$; this results in a random variable distributed as $\sqrt{U}$ (where U is uniformly distributed) using the inversion method.

(ii) Start uniformly at random from x_0 and multiply the sum of the values of the random variable (score) by $2x_0$.

Table 2: Relative error (Err.), standard deviation (σ), and mean number of absorption steps i for the USA approach with various values of β; $N = 10^6$.

	$\beta = 1$			$\beta = 2$			$\beta = 4.15$		
	Err.	σ	i	Err.	σ	i	Err.	σ	i
$x = 0.1$	7e-05	0.105	1.003	1e-05	0.17	1.009	2e-04	0.62	1.02
$x = 0.5$	2e-04	0.600	1.116	1e-04	0.97	1.27	0.003	9.35	1.66
$x = 0.99$	5e-04	1.388	1.588	3e-04	2.76	2.05	0.008	47.1	2.51
$\bar{U}$ meth.1	6e-04	0.900	1.178	8e-04	1.39	1.38	0.018	15	1.72
I_2 meth.(i)	1e-03	1.95	1.176	1e-03	2.63	1.38	0.022	26	1.72
I_2 meth.(ii)	2e-04	1.03	1.27	8e-04	1.70	1.58	0.015	21	2.03

Some results for the numerical solution of the second example (9) are presented in Table 2.

For all values of x, increasing β generally increases the relative error, except for $x = 0.1$, where $\beta = 2$ gives the lowest relative error (1e-05). For $x = 0.1$, the error remains small across different β values, with $\beta = 2$ yielding the lowest error (1e-05). However, for $x = 0.99$, the error increases significantly as β increases, with $\beta = 4.15$ resulting in an error of 0.008. Overall, the smallest error tends to occur at $\beta = 2$, especially at $x = 0.1$ and $x = 0.5$.

The standard deviation increases significantly with increasing β, particularly for larger values of x. For example, at $x = 0.99$, σ increases from 1.388 for $\beta = 1$ to 47.1 for $\beta = 4.15$. This suggests that the variability of the method increases with larger β values, particularly for points further from the origin. At $x = 0.1$, the standard deviation remains relatively low across all β values, indicating that the USA method is more stable near the origin. However, as x increases, the variability grows significantly, particularly for higher β values.

The mean number of absorption steps (i) increases as β and x increase, indicating that more steps are required to reach convergence for larger β and for points farther from the origin. For $x = 0.99$, the number of steps ranges from 1.588 for $\beta = 1$ to 2.51 for $\beta = 4.15$, showing that the method becomes slower for larger values of β and x. For smaller values of x, such as $x = 0.1$, the number of steps remains close to 1 for all β, indicating faster convergence for points near the origin.

The U method 1 has higher relative error and standard deviation compared to the USA approach, especially at $\beta = 4.15$, where the error rises to 0.018 and the standard deviation reaches 15. I_2 method (i) and (ii) show mixed performance, with (ii) showing lower relative error at smaller values of β (2e-04 for $\beta = 1$, but overall performing worse as β increases. The standard deviation is consistently higher for the I_2 methods compared to USA, particularly at higher values of β.

The USA approach performs well at small to moderate values of β, particularly at $\beta = 2$, where the relative error is minimized for most values of x. However, as β increases to 4.15, the relative error increases, especially for larger values of x, such as $x = 0.99$. Additionally, the standard deviation and number of absorption steps increase significantly with larger β, indicating greater variability and slower convergence.

The USA method is more stable (lower standard deviation and fewer absorption steps) for points closer to the origin ($x = 0.1$), but as x increases, the method requires more steps and shows higher variability, particularly at higher values of β.

The USA approach generally outperforms both the U method and the I_2 methods, offering lower relative errors and better stability, especially for smaller β. The U method shows higher errors and variability, while the I_2 methods struggle with both standard deviation and relative error at higher values of β.

4.2.2. Global approximation We apply the algorithm described in Subsection 3.3 of Section 3 to compute an approximation at all points in the domain. The goal is to observe how accuracy improves as the number of random trajectories N increases. We compute the solution at two reference points, $x_1 = 0.5$ and $x_2 = 0.95$, using two values of the parameter h: $h_1 = 0.1$ and $h_2 = 0.05$. Results for the relative error, standard deviation σ, and mean number of absorption steps i are presented in Table 3.

For all cases, increasing N significantly reduces the relative error, indicating that more random trajectories improve the accuracy of the USA method. At $x_1 = 0.5$, the relative error drops from 3e-03 at $N = 10^4$ to 1e-05 at $N = 10^7$, showing a clear improvement in accuracy as the number of random trajectories increases. Similarly, for $x_2 = 0.95$, the error decreases from 2e-03 at $N = 10^4$ to 8e-05 at $N = 10^7$. This suggests that for both values of x, the USA method becomes significantly more precise as the sample size grows. For combined variables such as (x_1, h_1) and (x_2, h_2), a similar trend is observed. The relative error consistently reduces as N increases, with cases like (x_2, h_2) showing a substantial improvement from 1e-02 at $N = 10^4$ to 2e-04 at $N = 10^6$

The standard deviation remains fairly constant across all values of N, hovering around 0.76 to 0.92. This indicates that while the accuracy (relative error) improves with increasing N, the

Table 3: Relative error, standard deviation, and mean number of absorption steps i for the USA approach with various values of random trajectories N.

	σ	i	$N = 10^4$	$N = 10^5$	$N = 10^6$	$N = 10^7$
$x_1 = 0.5$	0.76	1.27	3e-03	1e-03	7e-04	1e-05
$x_2 = 0.95$	0.92	2.37	2e-03	1e-03	8e-04	8e-05
(x_1, h_1)	0.91	1.41	1e-02	2e-03	2e-03	2e-03
(x_1, h_2)	0.91	1.41	1e-02	4e-04	4e-04	8e-05
(x_2, h_1)	0.91	1.41	1e-02	2e-02	2e-02	2e-02
(x_2, h_2)	0.91	1.41	1e-02	1e-03	2e-04	4e-04

variability in the results remains steady. For all cases, σ stays within a small range, suggesting that the stochastic nature of the USA method has consistent variability across different values of N, x and (x, h).

The number of absorption steps i remains relatively consistent for each value of x and (x, h), showing little to no change as N increases. For $x_1 = 0.5$, the absorption steps remain around 1.27, while for $x_2 = 0.95$, it stays around 2.37. This suggests that while more random trajectories improve accuracy, they do not necessarily require more absorption steps, indicating an efficient convergence behavior for the method. Similarly, for combinations like (x_1, h_1) and (x_2, h_2), the number of absorption steps remains around 1.41, showing stability in the mean absorption steps as N increases.

The USA method clearly benefits from larger numbers of random trajectories (N). With $N = 10^6$ or $N = 10^7$, the relative error becomes significantly smaller, especially for values like $x_2 = 0.95$ or (x_2, h_2). For larger sample sizes, the method demonstrates a noticeable increase in precision.

The standard deviation (σ) and the mean number of absorption steps (i) remain consistent, suggesting that the performance of the USA method is stable in terms of variability and convergence steps, regardless of how many random trajectories are used. This is a favorable aspect of the method, as it implies that increasing N directly impacts accuracy without affecting the computational complexity in terms of absorption steps.

While the relative error significantly decreases with larger values of N, the fact that the mean absorption steps remain almost unchanged suggests that the USA method maintains computational efficiency even as it becomes more accurate.

The USA method benefits greatly from increasing the number of random trajectories, with a marked reduction in relative error, particularly for higher values of x and the (x, h) combinations.The standard deviation remains largely consistent across different values of N, indicating that the randomness in the stochastic approach is controlled and stable. The number of absorption steps remains relatively constant across all cases, suggesting that the method does not require additional computational complexity in terms of convergence steps as more trajectories are added.

5. Conclusion

In this research, we carried out an in-depth evaluation of unbiased stochastic approaches for addressing Fredholm integral equations of the second kind. Through a series of thorough numerical experiments, the proposed algorithm consistently showcased both high accuracy and computational efficiency. Its performance across a range of test cases, even those involving irregular kernels and solutions, highlights the algorithm's robustness and adaptability.

In conclusion, the USA approach offers robust performance for moderate values of β, especially near the origin, but its performance degrades in terms of variability and convergence as β and x increase. The method remains competitive compared to other approaches, particularly in terms of accuracy. Improvement with larger N can be observed. Overall, the table demonstrates that the USA method scales well with larger numbers of random trajectories, leading to better accuracy without increasing the mean number of absorption steps or variability. This makes it an efficient and reliable approach for solving complex integral problems.

Acknowledgments
Venelin Todorov is supported by the Bulgarian National Science Fund (BNSF) under Project KP-06-N62/6 "Machine learning through physics-informed neural networks" and Project KP-06-N52/5 "Efficient methods for modeling, optimization and decision making".

CRediT roles: Conceptualization, V.T.; data curation, V.T. and M.C.; formal analysis, V.T.; funding acquisition, V.T.; investigation, V.T. and Y.D.; methodology, V.T. and Y.D.; project administration, V.T.; resources, V.T.; software, V.T. and I.G.; supervision, V.T.; validation, V.T.; visualization, V.T.; writing–original draft, V.T.; writing–review and editing, V.T., I.G., M.C. and Y.D.

References

[1] Arnold, I.: Ordinary differential equations, The MIT Press. ISBN 0-262-51018-9 (1978)
[2] Atkinson, K.E., Shampine, L.F.: Algorithm 876: Solving Fredholm integral equations of the second kind in Matlab. ACM Trans. Math. Software 34(4), 21 (2007)
[3] J. H. Curtiss. Monte Carlo methods for the iteration of linear operators, J. Math Phys., vol. 32, 209-232, 1954.
[4] Dimov, I.T.: Monte Carlo methods for applied scientists., *World Scientific*, New Jersey, London, Singapore, World Scientific, 2008.
[5] I. T. Dimov, R. Georgieva. Multidimensional Sensitivity Analysis of Large-scale Mathematical Models. O.P. Iliev et al. (eds.), Numerical Solution of Partial Differential Equations: Theory, Algorithms, and Their Applications, *Springer Proceedings in Mathematics and Statistics*, 45, Springer Science+Business Media, New York, 137–156, 2013. ISBN: 978-1-4614-7171-4 (book chapter).
[6] Dimov, I.T., Maire, S. A new unbiased stochastic algorithm for solving linear Fredholm equations of the second kind. Adv Comput Math 45, 1499–1519 (2019). https://doi.org/10.1007/s10444-019-09676-y
[7] I.T. Dimov, S. Maire, J.M. Sellier, A New Walk on Equations Monte Carlo Method for Solving Systems of Linear Algebraic Equations, *Applied Mathematical Modelling*, Vol. 39, Issue 15, 2015, pp. 4494-4510, http://dx.doi.org/10.1016/j.apm.2014.12.018.
[8] S.M. Ermakov, G.A. Mikhailov, Statistical Modeling, Nauka, Moscow, 1982 (in Russian).
[9] S.M.Ermakov, A.S.Sipin, Monte Carlo method and parametric separation of algorithms, Publishing house of Sankt-Petersburg University, 2014 (in Russian).
[10] S. Georgiev, L. Vulkov, Parameters identification and numerical simulation for a fractional model of honeybee population dynamics, *Fractal and Fractional*, Vol. 7, Issue 4, no. 311, (2023).
[11] Kalos, Malvin H.; Whitlock, Paula A. (2008). Monte Carlo Methods. Wiley-VCH. ISBN 978-3-527-40760-6.
[12] K. Sabelfeld, Algorithms of the Method Monte Carlo for Solving Boundary Value Problems, Nauka, Moscow, 1989 (in Russian).
[13] I. M. Sobol. *Monte Carlo Numerical Methods*, Nauka, Moscow, 1973, (in Russian).

Journal of Physics: Conference Series **2910** (2024) 012037 doi:10.1088/1742-6596/2910/1/012037

Enhanced Monte Carlo Techniques for Solving Linear Systems

Stoyan Apostolov[1], Ivan Georgiev[2,3], Nikita Nikitov[3], Velichka Traneva[4], Stoyan Tranev[4], Mihai Petrov[4], Yuri Dimitrov[5,2]

[1]Department of Mathematical Analysis, University of Sofia, 8 James Bourchier Blvd., 1164 Sofia, BG

[2]Department of Information Modeling, Institute of Mathematics and Informatics, Bulgarian Academy of Sciences, 8 Acad. Georgi Bonchev Str., 1113 Sofia, BG

[3]Department of Applied Mathematics and Statistics, University of Ruse, 8 Studentska Str., 7004 Ruse, BG

[4]Prof. Dr. Asen Zlatarov University – Bourgas, 1 Prof. Yakim Yakimov Blvd., 8010 Bourgas, BG

[5]Department of Mathematics, Physics and Informatics, Faculty of Forest Industry, University of Forestry, 10 St. Kliment Ohridski Blvd., 1756 Sofia, BG

E-mail: sapostolov@fmi.uni-sofia.bg, irgeorgiev@uni-ruse.bg, irgeorgiev@math.bas.bg, nnikitov@uni-ruse.bg, veleka13@gmail.com, tranev@abv.bg, tranev@abv.bg, mihpetrov@abv.bg, yuri.dimitrov@ltu.bg

Abstract. A novel hybrid Monte Carlo algorithm for solving systems of linear algebraic equations is introduced and evaluated in this study. This algorithm builds on the recent "Walk on Equations" Monte Carlo method developed by Ivan Dimov, Sylvain Maire, and Jean Michel Sellier. For the first time, this approach has been compared with the Gauss-Seidel method for matrices of big size. Significant improvements have been made to the algorithm by selecting optimal values for the relaxation parameters, resulting in a considerable reduction in computational time and lower relative errors across a specified number of iterations. The convergence of the algorithm has been rigorously established through a new theorem. It has been demonstrated that the original algorithm can be optimized by properly balancing the iteration matrix. Additionally, a sequential Monte Carlo method developed by John Halton, which utilizes the control variate method iteratively, has also been implemented.

1. Introduction

Linear systems play a fundamental role across numerous fields, including mathematics, science, engineering, and technology, emphasizing their critical significance. Solving and analyzing linear algebraic equations provides a powerful framework for modeling and interpreting a wide range of real-world problems. Beyond their theoretical importance, linear systems offer practical solutions to everyday challenges [1, 2, 3, 4]. Their inherent linearity makes them highly compatible with both analytical and computational methods, simplifying complex problems. In engineering, for instance, linear systems are found in control theory, signal processing, and structural analysis, where efficiently solving these systems is essential for the design and optimization of everything from electronic circuits to mechanical structures. Likewise, in economics and finance [5, 6, 7],

Content from this work may be used under the terms of the Creative Commons Attribution 4.0 licence. Any further distribution of this work must maintain attribution to the author(s) and the title of the work, journal citation and DOI.
Published under licence by IOP Publishing Ltd

linear models are frequently employed to investigate variable relationships, aiding in decision-making processes. The integration of Bayesian statistics [8] with linear systems provides a robust framework for dealing with uncertainty, incorporating prior knowledge, and making probabilistic inferences about linear models, which are critical in many scientific and engineering applications. Air pollution modeling frequently involves the use of linear systems to estimate and predict the concentrations of various pollutants over time and space [9, 10]. In scientific research, linear systems are integral to areas like data analysis, image processing, and simulations. Their broad applicability across various disciplines underscores their importance, establishing them as a cornerstone for both theoretical exploration and practical application in solving real-world problems.

The structure of the paper is as follows: the next section discusses the old, new, and hybrid algorithms in detail, along with the introduction of the dominancy number. Section 3 presents a series of tests conducted with random dense matrices, and the paper concludes with the final section.

2. Overview of the WE MC Algorithms

In this section, we present the descriptions of the new WE MC algorithm, the original WE MC algorithm, and a hybrid approach that combines both algorithms.

2.1. The Original Monte Carlo Algorithm for Estimating a Single Component of the Solution

In this subsection, we outline the first (original) WE algorithm for solving linear systems, as detailed in [11]. The input required includes the matrix B, the vector $\mathbf{f}$, the constant parameter $\gamma \in (0, 1]$, and the number of random walks N.

Algorithm 1 *:*

Estimating the Component x_{i_0} of the Solution Vector x_i, $i = 1, \ldots, n$

1. *Initialization* **Input***:* B, $\mathbf{f}$, γ, N.
2. *Preprocessing Steps:*

   2.1. **Calculate** *the matrix A using $\gamma \in (0, 1]$:*

$$\{a_{ij}\}_{i,j=1}^{n} = \begin{cases} 1-\gamma & \textit{if } i = j, \\ -\gamma \frac{b_{ij}}{b_{ii}} & \textit{ifi} \neq j. \end{cases}$$

   2.2. **Compute** *the vector asum:*

$$asum(i) = \sum_{j=1}^{n} |a_{ij}| \quad for \quad i = 1, 2, \ldots, n.$$

   Calculate *the vector of initial probabilities,*

$$\pi_i = \frac{v_i}{\sum_{j=1}^{m} v_j}, \qquad i = 1, 2, \ldots, m;$$

   2.3. **Compute** *the transition probability matrix $P = \{p_{ij}\}_{i,j=1}^{n}$,*

$$p_{ij} = \frac{|a_{ij}|}{asum(i)}, \qquad i = 1, 2, \ldots, n, \quad j = 1, 2, \ldots, n.$$

3. **Initialize** $S := 0$.
4. **For** $k = 1$ **to** N **do** *(Monte Carlo loop):*

4.1 **Set** $m := i_0$.
4.2 **Update** $S := S + b(m)$.
4.3. $test = 0$; $sign = 1$.
4.4 **While** $(test \neq 0)$ **do**:
  4.4.1. *Generate a uniformly distributed random variable* $r \in (0, 1)$.
  4.4.2. **If** $r \geq asum(m)$ **then** $test = 1$;
  4.4.3. **Else** *select the index* j *according to* p_{ij};
  4.4.4. **Update** $sign = sign * sign(a_{mj})$; $m = j$;
  4.4.5 **Modify** $S := S + sign * b_m$;
  4.4.6. **End If**.
4.5 **End While**.

5. **End For**.

6. $x_{i_0} = \frac{S}{N}$.

2.2. The Original Monte Carlo Algorithm for Estimating All Components of the Solution
Without loss of generality, the entries of the matrix are assumed to be nonnegative, i.e. $a_{ij} \geq 0$, for $i, j = 1, \ldots n$.

Algorithm 2 :

Estimating all components x_i, $i = 1, \ldots, n$ of the solution

1. *Initialization.*

2. *Preprocessing.*

3. **For** $i = 1$ **to** n **do**:

3.1. *Set* $S(i) := 0$; $V(i) := 0$.

3.2. **For** $k = 1$ **to** N **do** *(Monte Carlo loop)*:

3.2.1 **Set** $m := rand(1 : n)$.
3.2.2 **Initialize** $test := 0$; $m_1 := 0$.
3.2.3 **While** $(test \neq 1)$ **do**:
  3.2.3.1. *Update* $V(m) := V(m) + 1$; $m = m_1 + 1$; $l(m_1) = m$.
  3.2.3.2. **For** $q = 1$ *to* m_1 **do**:
   Update $S(l(q)) := S(l(q)) + b(l(q))$.
  3.2.3.3. **If** $r > asum(m)$, **then** *set* $test = 1$;
   Else *select* j *according to* $p_{i,j}$;
  3.2.3.4. **Set** $m = j$.
  3.2.3.5. **End If**.
3.2.4. **End While**.

3.3. **End For**.

3.4. **For** $j = 1$ *to* n **do**:
 Set $V(j) := \max\{1, V(j)\}$;
 Update $S(j) = \frac{S(j)}{V(j)}$.

2.3. The New WE Monte Carlo Algorithm for Computing a Single Component of the Solution
In this subsection, we introduce the novel Monte Carlo algorithm specifically developed to calculate a particular component x_{i_0} of the solution vector in a linear system with real coefficients. The algorithm begins its process at row i_0 of the system, where b_{i_0} is added to the cumulative score of the walk. The walk then either terminates or transitions to another equation

Journal of Physics: Conference Series **2910** (2024) 012037 doi:10.1088/1742-6596/2910/1/012037

based on the transition probabilities $p_{i_0 j}$ as detailed in Algorithm 3. The score contributions are adjusted according to the signs of the system's coefficients. The algorithm is described in detail below:

Algorithm 3 *:*

Calculating a Single Component x_{i_0} **of the Solution Vector** x_i, $i = 1, \ldots, n$

1. *Initialization*
   Input *initial data: the matrix* A*, the vector* $\mathbf{b}$*, and the number of random trajectories* N.
2. *Preliminary Computations (Preprocessing):*

   2.1. **Calculate** *the vector asum:*

$$asum(i) = \sum_{j=1}^{n} a_{ij}, \quad for \quad i = 1, 2, \ldots, n.$$

   2.2. **Compute** *the sum of elements in vector b:*

$$sb = \sum_{i=1}^{n} b_i, \quad for \quad i = 1, 2, \ldots, n.$$

3. **Set** $S := 0$.
4. **For** $k = 1$ **to** N **do** *(Monte Carlo loop):*

   4.1. *Initialize* $test = 0$.
   4.2. *Choose the index* j *according to the probability density vector* $(b_1, b_2, \ldots, b_n)$*; set* $pj = j$.
   4.3 **While** $test \neq 1$ **do**:
      4.3.1. *Select a new value for the index* j *according to the probability density vector* $\big(a_{pj,1}, a_{pj,2}, \ldots, a_{pj,n}, 1 - asum(pj)\big)$.
      4.3.2. **If** $j \neq n+1$ **then** *set* $pj = j$.
      4.3.3. **Else** *set* $test = 1$*; update* $S = S + b(i_0) + sb * \frac{a_{pj,i_0}}{1 - asum(pj)}$.
      4.3.4. **End If**.
   4.4 **End While**.
5. **End For**.
6. $x_{i_0} = \frac{S}{N}$.

2.4. The New WE Monte Carlo Algorithm for Computing All Components of the Solution
In this subsection, we introduce the novel algorithm designed to compute all components of the solution vector. The central concept is to accumulate scores for each state, which are treated as new starting points encountered during a trajectory. The initial equation is selected randomly from the n equations with equal probability. For each state i, the total score $S(i)$ and the total number of visits $V(i)$ are updated whenever state i is visited during a walk. The parameter $cc(i)$ represents the number of trajectories contributing to the estimation of component x_i of the solution. The algorithm's key feature is the application of the Sequential Monte Carlo method, as introduced by John Halton [12, 13, 14, 15], leveraging the control variate technique [16].

Algorithm 4 *:*

Computing All Components of the Solution x_i, $i = 1, \ldots, n$

1. *Initialization*
   Input *initial data: the matrix* A*, the vector* $\mathbf{b}$*, and the number of random trajectories* N.

2. *Preliminary Computations (Preprocessing):*

   2.1. **Calculate** *the vector asum:*

$$asum(i) = \sum_{j=1}^{n} a_{ij}, \quad for \quad i = 1, 2, \ldots, n.$$

   2.2. **Compute** *the sum of elements in vector b:*

$$sb = \sum_{i=1}^{n} b_i, \quad for \quad i = 1, 2, \ldots, n.$$

3. **Set** $S(i) = 0$ *and* $cc(i) = 0$ *for* $i = 1$ *to* n.
4. **For** $k = 1$ **to** N **do** *(Monte Carlo loop):*

   4.1 *Randomly choose the component* i_0 *to estimate from* $1, 2, \ldots, n$.
   4.2 *Increment* $cc(i_0)$ *by 1.*
   4.3. *Initialize* $test = 0$.
   4.4. *Select the index* j *according to the probability density vector* $(b_1, b_2, \ldots, b_n)$*; set* $pj = j$.
   4.5. **While** $test \neq 1$ **do**:
      4.5.1. *Choose a new value for the index* j*, based on the probability density vector* $(a_{pj,1}, a_{pj,2}, \ldots, a_{pj,n}, 1 - asum(pj))$.
      4.5.2. **If** $j \neq n + 1$ **then** *set* $pj = j$.
      4.5.3. **Else** *set* $test = 1$*; update* $S(i_0) = S(i_0) + b(i_0) + sb * \frac{a_{pj,i_0}}{1-asum(pj)}$.
      4.5.4. **End If**.
   4.6 **End While**.

5. **End For**.
6. **For** $i = 1$ *to* n **do** $x_i = \frac{S(i)}{cc(i)}$.

2.5. The Hybrid Monte Carlo Algorithm for Computing One Component of the Solution

In this subsection, we describe the hybrid WE algorithm for solving linear systems. The hybrid WE algorithm resembles the original WE method but incorporates an absorbing state mechanism, similar to the new algorithm, when the trajectory reaches a breaking point.

Algorithm 5 *:*

Computing One Component x_{i_0} **of the Solution Vector** x_i, $i = 1, \ldots, n$

1. *Initialization* **Input**: B, $\mathbf{f}$, γ, N.
2. *Preliminary Computations (Preprocessing):*

   2.1. **Compute** *the matrix* A *using* $\gamma \in (0, 1]$:

$$\{a_{ij}\}_{i,j=1}^{n} = \begin{cases} 1 - \gamma & \text{if } i = j, \\ -\gamma \frac{b_{ij}}{b_{ii}} & \text{if } i \neq j. \end{cases}$$

   2.2. **Compute** *the vector asum:*

$$asum(i) = \sum_{j=1}^{n} |a_{ij}| \quad for \quad i = 1, 2, \ldots, n.$$

Calculate *the vector of initial probabilities,*

$$\pi_i = \frac{v_i}{\sum_{j=1}^{m} v_j}, \quad i = 1, 2, \ldots, m;$$

2.3. **Compute** *the transition probability matrix* $P = \{p_{ij}\}_{i,j=1}^{n}$, *where*

$$p_{ij} = \frac{|a_{ij}|}{asum(i)}, \quad i = 1, 2, \ldots, n, \quad j = 1, 2, \ldots, n.$$

3. **Set** $S := 0$.

4. **For** $k = 1$ **to** N **do** *(Monte Carlo loop):*

4.1 **Set** $m := i_0$.
4.2 **Update** $S := S + b(m)$.
4.3. Initialize $test = 0$; $sign = 1$.
4.4 **While** $(test \neq 0)$ **do**:
4.4.1. Generate a uniformly distributed random variable $r \in (0, 1)$.
4.4.2. Select a new value for the index j*, based on the probability density vector* $(a_{pj,1}, a_{pj,2}, \ldots, a_{pj,n}, 1 - asum(pj))$.
4.4.3. **If** $j = n + 1$ **then** *set* $test = 1$.
4.4.4. **Else** *update* $sign = sign * sign(a_{mj})$; *set* $m = j$.
4.4.5 **Update** $S := S + sign * b_m$.
4.4.6. **End If**.
4.5 **End While**.

5. **End For**.

2.6. The Hybrid WE Monte Carlo Algorithm for Computing All Components of the Solution

In this subsection, we present the hybrid WE algorithm for calculating all components of the solution vector. We omit the initialization and preprocessing steps, as they are identical to those described in the previous subsection.

Algorithm 6 :

Computing All Components x_i, $i = 1, \ldots, n$ **of the Solution**

1. Initialization.

2. Preprocessing.

3. **For** $i = 1$ **to** n **do**:

3.1. Set $S(i) := 0$; $V(i) := 0$.

3.2. **For** $k = 1$ **to** N **do** *(Monte Carlo loop):*

3.2.1 **Set** $m := rand(1 : n)$.
3.2.2 **Initialize** $test := 0$; $m_1 := 0$.
3.2.3 **While** $(test \neq 1)$ **do**:
3.2.3.1. Increment $V(m) := V(m) + 1$; *set* $m = m_1 + 1$; $l(m_1) = m$.
3.2.3.2. **For** $q = 1$ *to* m_1 **do**:
Update $S(l(q)) := S(l(q)) + b(l(q))$.
3.2.3.3. Select a new value for the index j*, based on the probability density vector* $(a_{pj,1}, a_{pj,2}, \ldots, a_{pj,n}, 1 - asum(pj))$.
3.2.3.4. **If** $j = n + 1$, **then** *set* $test = 1$.
3.2.3.5. **Set** $m = j$.
3.2.3.6. **End If**.

3.2.4. **End While**.

3.3. **End For**.

3.4. **For** $j = 1$ *to* n **do**:
Set $V(j) := \max\{1, V(j)\}$;
Update $S(j) = \frac{S(j)}{V(j)}$.

2.7. Dominancy Number
In this paper, we introduce a special parameter called the dominancy number, denoted by $\mathcal{D}$:

$$\mathcal{D} = \min\left\{ \frac{|b_{ii}| - \sum_{j\neq i} |b_{ij}|}{|b_{ii}|} \;\middle|\; 1 \leq i \leq n \right\}.$$

This parameter remains positive for diagonally dominant matrices and provides a rough measure of how strongly the matrix's diagonal elements dominate the off-diagonal elements. The concept of dominancy is closely related to the absorption probabilities in the Markov chain used for the stochastic computation of solutions to linear systems. Notably, when $\mathcal{D} \geq 0.5$, the newly developed algorithm achieves significantly higher accuracy–by several orders of magnitude–compared to the older algorithm. In later sections, we will present experiments with $\mathcal{D}$ values both below and above 0.5 to analyze the performance of the three algorithms under different conditions.

We now introduce the following result:

Theorem 2.1 *If the dominancy number* $\mathcal{D}$ *is close to 1, the new algorithm significantly outperforms the previous algorithm.*

When the dominancy number is close to 1, the preconditioned matrix exhibits high absorption probabilities at each step, resulting in shorter trajectories. In the new algorithm, the variable used to estimate the solution is computed only once, at the end of each trajectory, regardless of the trajectory's length. Conversely, in the old algorithm, the solution estimate is updated at each step of the trajectory. Therefore, longer trajectories typically yield better estimates in the old algorithm. However, since the new algorithm's estimation is unaffected by trajectory length, it naturally performs better when the dominancy number is closer to 1.

3. Numerical Tests

In this section, we investigate under which conditions the new algorithm outperforms the old algorithm.

We consider randomly generated matrices of sizes 100×100 and 1000×1000 with a dominancy parameter value of $\mathcal{D} = 0.95$. While at lower values of $\mathcal{D}$ the two algorithms yield similar results, with the new algorithm showing better performance at higher sequential steps, the advantage of the new algorithm becomes clearly evident as $\mathcal{D}$ increases–see Table 1.

Our analysis is primarily focused on the 1000×1000 matrix, with the results presented in Table 1. Remarkably, within just 5 iterations, the new Monte Carlo algorithm achieved a relative error as low as 10^{-12}. In comparison, the old WE algorithm registered a relative error of 10^{-9}. Therefore, for very large-scale problems, the newly developed Monte Carlo approach demonstrates a significant improvement in accuracy over the old algorithm.

The table presents the weighted residuals for two different matrices, using two methods: a hybrid weighted residuals (WE) method and an older weighted residuals method. The results are shown across five iterations ($N = 1$ to 5), and a dominancy parameter is used to quantify the behavior of the matrix for each case.

Table 1: Weighted residuals for the matrix $B \in \mathbb{R}^{100\times100}$ (1) with a dominancy parameter $\mathcal{D} = 0.94$ and $B \in \mathbb{R}^{1000\times1000}$ (2) with a dominancy parameter $\mathcal{D} = 0.95$.

N	hybrid WE (1)	old WE (1)	hybrid WE (2)	old WE (2)
1	5.61e-3	2.48e-2	5.14e-3	2.24e-2
2	2.26e-5	5.18e-4	2.75e-5	5.36e-4
3	1.35e-7	1.05e-5	1.28e-7	1.17e-5
4	5.61e-10	3.11e-7	6.28e-10	2.69e-7
5	3.06e-12	7.05e-9	3.09e-12	6.63e-9

Initial performance at $N = 1$ shows that the hybrid WE method shows a significantly lower residual compared to the old WE method in both cases (matrices (1) and (2)). For matrix (1), the hybrid method produces a residual of 5.61e-3, while the old method yields 2.48e-2, representing an improvement of about four times in accuracy. For matrix (2), the hybrid method again shows better accuracy with a residual of 5.14e-3 compared to 2.24e-2 from the old WE method. This confirms that the hybrid method offers superior performance in the early iteration. Improvement over iterations ($N = 2$ to 5) show that as the iterations progress, both methods show reductions in residuals, but the hybrid WE method consistently achieves lower residuals than the old WE method. At N = 2, the hybrid method's residual for matrix (1) decreases to 2.26e-5, a significant improvement compared to 5.18e-4 for the old method. Similarly, for matrix (2), the hybrid method's residual is 2.75e-5, outperforming the old method's residual of 5.36e-4. By N = 5, the hybrid method reaches residuals as low as 3.06e-12 for matrix (1) and 3.09e-12 for matrix (2), showing extremely high accuracy. The old method's residuals, while improving, remain higher at 7.05e-9 and 6.63e-9, respectively. Consistency Across Different Matrix Sizes:

The hybrid method performs consistently well across different matrix sizes. The residuals for the larger matrix (2) are very similar to those for the smaller matrix (1), suggesting that the method scales effectively with matrix size. The old method, while also improving as matrix size increases, shows larger residuals compared to the hybrid method at every iteration.

The hybrid weighted residuals method consistently outperforms the old weighted residuals method across all iterations and both matrix sizes. The improvements are particularly significant in the early iterations ($N = 1$ and $N = 2$), where the hybrid method achieves much lower residuals. The hybrid method converges more quickly than the old method, reaching very low residuals by $N = 5$. This makes it more suitable for applications requiring fast and accurate convergence. The hybrid method maintains its performance across different matrix sizes, indicating that it is effective for both small and large systems. The old method, while improving, lags behind in terms of accuracy, especially in larger matrices.

The comparison between the three algorithms is illustrated in Figure 1. The corresponding dominancy numbers are $\mathcal{D} = 0.47$ and $\mathcal{D} = 0.48$. We observe similar behavior among the three algorithms, with a slight advantage for the hybrid and new WE algorithms due to the dominancy number $\mathcal{D}$ being closer to 0.5.

As we increase the dominancy number, the advantage of the new algorithm over the old algorithm becomes more pronounced–see the results in Figure 2. The corresponding dominancy numbers are $\mathcal{D} = 0.56$ and $\mathcal{D} = 0.59$.

As the dominancy number increases, the superiority of the new algorithm over the old algorithm becomes increasingly evident. The dominancy number characterizes the influence of the diagonal elements within a system, and as this influence grows, the effectiveness and efficiency of the new algorithm become more apparent compared to the older method. This trend indicates that the performance gap between the two algorithms widens as the system's key features become more dominant, highlighting the improved adaptability and efficacy of the new algorithm in scenarios with higher dominancy numbers. This observation supports the conclusion that the advantages of the new algorithm are particularly significant in cases where dominant factors play a more critical role.

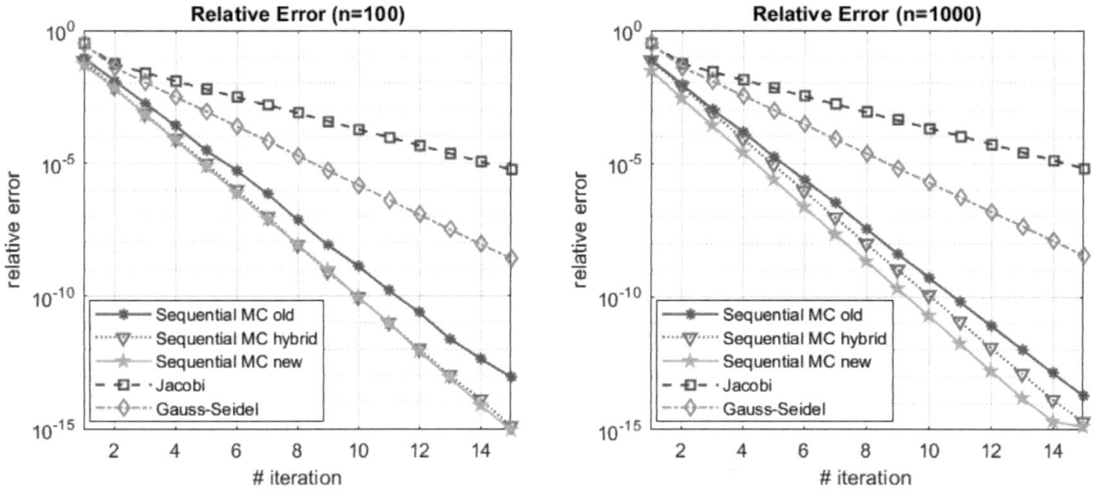

Figure 1: Relative error of random dense matrices $n \times n$ for $n = 100$ (top left), $n = 1000$ (top right), $n = 5000$ (bottom left) and $n = 10000$ (bottom right) with lower dominancy numbers

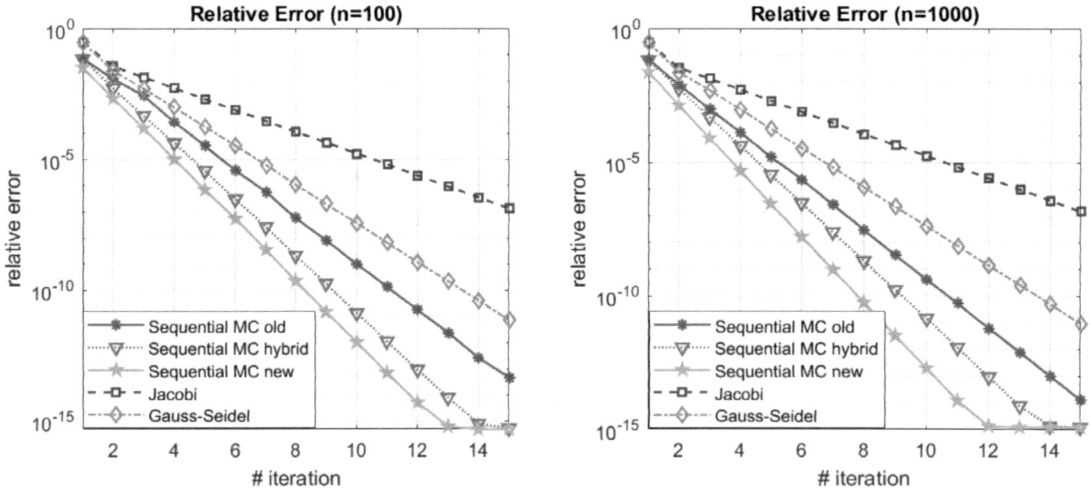

Figure 2: Relative error of random dense matrices $n \times n$ for $n = 100$ (top left), $n = 1000$ (top right), $n = 5000$ (bottom left) and $n = 10000$ (bottom right) with higher dominancy numbers

4. Conclusion

In this study, we introduced a Monte Carlo method specifically tailored for solving large-scale systems of linear algebraic equations, building upon the "Walk on Equations" Monte Carlo model, a recent advancement credited to Ivan Dimov, Sylvain Maire, and Jean Michel Sellier.

Additionally, we proposed a novel hybrid algorithm that combines features of both the original and the new methods, aiming to enhance computational efficiency and accuracy.

In conclusion, the hybrid WE method shows clear advantages in both accuracy and convergence speed, making it a superior choice for solving linear systems where precision is essential. The consistent performance across matrix sizes also suggests that it is robust and scalable for more complex problems.

Acknowledgments

This paper contains results of the work on project No 2024 - FNSE - 02, financed by "Scientific Research" Fund of Ruse University.

CRediT roles: Conceptualization, S.A.; data curation, S.A. and N.N.; formal analysis, S.A.; funding acquisition, I.G.; investigation, S.A. and Y.D.; methodology, S.A. and Y.D.; project administration, I.G.; resources, I.G., V.T. and S.T.; software, I.G, N.N. and M.P.; supervision, V.T.; validation, S.A., V.T., S.T. and M.P.; visualization, S.A., I.G. and N.N.; writing–original draft, S.A.; writing–review and editing, S.A., I.G., N.N., V.T., S.T., M.P and Y.D.

References

[1] Dimov, I., Tonev, O.: *Performance Analysis of Monte Carlo Algorithms for Some Models of Computer Architectures*, International Youth Workshop on Monte Carlo Methods and Parallel Algorithms – Primorsko (Eds. Bl. Sendov, I. Dimov), World Scientific, Singapore, 1990, 91–95.

[2] Spanier, J., Gelbard, E.:*Monte Carlo Principles and Neutron Transport Problem*, Addison–Wesley, 1969.

[3] Curtiss J.H.: *Monte Carlo methods for the iteration of linear operators*, J. Math. Phys., vol 32, No 4 (1954), pp.209–232.

[4] Curtiss J.H.: *A Theoretical Comparison of the Efficiencies of two classical methods and a Monte Carlo method for Computing one component of the solution of a set of Linear Algebraic Equations.* Proc. Symposium on Monte Carlo Methods, John Wiley and Sons, 1956, pp.191–233.

[5] Georgiev, S.G.: *Numerical and analytical computation of the implied volatility from option price measurements under regime-switching*, AIP Conf. Proc., 2172, 070007 (2019), https://doi.org/10.1063/1.5133543

[6] Todorov, V., Dimov, I.T.: *Monte Carlo methods for multidimensional integration for European option pricing*, AIP Conf. Proc., 1773, 100009 (2016), https://doi.org/10.1063/1.4965003

[7] Georgiev, S.G., Vulkov, L.G.: *Simultaneous identification of time-dependent volatility and interest rate for European options*, AIP Conf. Proc., 2333, 090006 (2021), https://doi.org/10.1063/5.0041788

[8] Todorov, V., Dimov, I. (2020). *Efficient Stochastic Approaches for Multidimensional Integrals in Bayesian Statistics.* In: Lirkov, I., Margenov, S. (eds) Large-Scale Scientific Computing. LSSC 2019. Lecture Notes in Computer Science(), vol 11958. Springer, Cham. https://doi.org/10.1007/978-3-030-41032-2_52

[9] Todorov, V., Dimov, I., Ostromsky, T., Zlatev, Z., Georgieva, R., Poryazov, S. (2022). *Optimized Quasi-Monte Carlo Methods Based on Van der Corput Sequence for Sensitivity Analysis in Air Pollution Modelling.* In: Fidanova, S. (eds) Recent Advances in Computational Optimization. WCO 2020. Studies in Computational Intelligence, vol 986. Springer, Cham. https://doi.org/10.1007/978-3-030-82397-9_20

[10] Z. Zlatev, I.T. Dimov, K. Georgiev, Three-dimensional version of the Danish Eulerian model, Zeitschrift für Angewandte Mathematik und Mechanik, 76(S4) (1996) 473-476.

[11] Dimov, I.T., Maire, S., Sellier, J.M.: *A New Walk on Equations Monte Carlo Method for Linear Algebraic Problems*, Applied Mathematical Modelling 39 (15), 4494–4510 (2015),

[12] Halton, J.: *Sequential Monte Carlo, Proceedings of the Cambridge Philosophical Society*, Vol. 58 (1962) pp. 57–78.

[13] Halton, J.: *Sequential Monte Carlo. University of Wisconsin*, Madison, Mathematics Research Center Technical Summary Report No. 816, 1967.

[14] Halton, J. and Zeidman, E. A.: *Monte Carlo integration with sequential stratification.* University of Wisconsin, Madison, Comp. Scien. Departm. Tech. Report No. 61, 1969.

[15] Halton, J.: *Sequential Monte Carlo for linear systems - a practical summary*, Monte Carlo Methods & Applications, 14, 2008, pp. 1–27.

[16] Maire, S.: *Reducing variance using iterated control variates*, Journal of Statistical Computation and Simulation, Vol. 73(1), 2003, pp. 1–29.

AUTHOR INDEX